Handbook of HydroInformatics

Handbook of HydroInformatics

Volume III: Water Data Management Best Practices

Edited by

Saeid Eslamian
Department of Water Engineering, College of Agriculture, Isfahan University of Technology, Isfahan, Iran

Faezeh Eslamian
Department of Bioresource Engineering, McGill University, Montreal, QC, Canada

ELSEVIER

Elsevier
Radarweg 29, PO Box 211, 1000 AE Amsterdam, Netherlands
The Boulevard, Langford Lane, Kidlington, Oxford OX5 1GB, United Kingdom
50 Hampshire Street, 5th Floor, Cambridge, MA 02139, United States

ISBN: 978-0-12-821962-1

For information on all Elsevier publications
visit our website at https://www.elsevier.com/books-and-journals

Publisher: Candice Janco
Acquisitions Editor: Maria Elekidou
Editorial Project Manager: Rupinder K. Heron
Production Project Manager: Bharatwaj Varatharajan
Cover Designer: Greg Harris

Typeset by STRAIVE, India

To Late Prof. Lotfi Aliasker Zadeh
(Mathematician, Computer Scientist, Electrical Engineer, Artificial Intelligence Researcher, and Professor of Computer Science at the University of California, Berkeley, USA
(Azerbaijan-Iranian-American: 1921–2017)

"As Complexity Rises, Precise Statements Lose Meaning and Meaningful Statements Lose Precision"

Contents

*Sahar Hadi Pour, Shamsuddin Shahid, and
Saad Sh. Sammen*

*Mohd Afiq Harun, Aminuddin Ab. Ghani,
Saeid Eslamian, and Chun Kiat Chang*

Contributors

Numbers in paraentheses indicate the pages on which the authors' contributions begin.

Zohra Abdelkrim (219), Institute of Management of Urban Techniques, Mohamed Boudiaf University; Laboratory of City, Environment, Society and Sustainable Development, M'sila, Algeria

Ishtiyaq Ahmad (33), Department of Civil Engineering, National Institute of Technology, Raipur, India

Shammi Akther (167), Department of Geography and Environment, Jahangirnagar University, Dhaka, Bangladesh

Md. Masood Zafar Ansari (33), Department of Civil Engineering, National Institute of Technology, Raipur, India

Ayurzana Badarch (245), School of Civil Engineering and Architecture, Mongolian University of Science and Technology, Ulan Bator, Mongolia

Sami Ullah Bhat (265), Department of Environmental Science, University of Kashmir, Srinagar, Jammu and Kashmir, India

Chun Kiat Chang (379), River Engineering and Urban Drainage Research Centre, Universiti Sains Malaysia, George Town, Malaysia

Sajad Ahmad Dar (265), Department of Environmental Science, Uttarakhand Technical University, Uttarakhand, India

Shahid Ahmad Dar (265), Department of Environmental Science, University of Kashmir, Srinagar, Jammu and Kashmir, India

Saeid Eslamian (1, 11, 33, 59, 109, 167, 207, 219, 265, 317, 339, 379), Department of Water Engineering, College of Agriculture, Isfahan University of Technology, Isfahan, Iran; Center of Excellence for Risk Management and Natural Hazards, Isfahan University of Technology, Isfahan, Iran

F. Espejo (91), High Polytechnic School of Engineering, University of Salamanca, Ávila, Spain

Henry Foust (125), Department of Mathematics, University of Houston, Houston, TX, United States

Shishir Gaur (1), Department of Civil Engineering, Indian Institute of Technology, Varanasi, India

Aminuddin Ab. Ghani (379), River Engineering and Urban Drainage Research Centre, Universiti Sains Malaysia, George Town, Malaysia

Sukanya Ghosh (81), Amity Institute of Geoinformatics and Remote Sensing (AIGIRS), Amity University Uttar Pradesh (AUUP), Noida, Uttar Pradesh, India

Albrecht Gnauck (167), Brandenburg University of Technology Cottbus, Senftenbuerg, Cottbus, Germany

Yubo Guo (291), School of Design, Shanghai Jiao Tong University, Shanghai, China

Mohd Afiq Harun (379), River Engineering and Urban Drainage Research Centre, Universiti Sains Malaysia, George Town, Malaysia

Salim Heddam (299), Faculty of Science, Agronomy Department, Hydraulics Division, Laboratory of Research in Biodiversity Interaction Ecosystem and Biotechnology, Skikda, Algeria

Vlassios Hrissanthou (67), Department of Civil Engineering, Democritus University of Thrace, Xanthi, Greece

Shafi Noor Islam (109,167), Department of Geography, Environment and Development Studies, Faculty of Arts and Social Sciences, University of Brunei Darussalam, Gadong, Brunei Darussalam

M.V.V. Prasad Kantipudi (317), Symbiosis Institute of Technology, Symbiosis International (Deemed University), Pune, India

S. Sreenath Kashyap (317), Department of Electronics and Communication Engineering, Kommuri Pratap Reddy Institute of Technology, Hyderabad, India

Jahangir Abedi Koupai (207), Department of Water Engineering, College of Agriculture, Isfahan University of Technology, Isfahan, Iran

Kamran Kouzehgar (11), Department of Civil Engineering, Varzeghan Branch, Islamic Azad University, Varzeghan, Iran

Mohibul Hassan Kowshik (167), Government Titumir College, Dhaka, Bangladesh

Deepak Kumar (81), Amity Institute of Geoinformatics and Remote Sensing (AIGIRS), Amity University Uttar Pradesh (AUUP), Noida, Uttar Pradesh, India

N.S. Pradeep Kumar (317), Department of Electronics and Communication Engineering, S.E.A.CET, Bangalore, India

Rina Kumari (81), School of Environment and Sustainable Development, Central University of Gujarat, Gandhinagar, Gujarat, India

Vladan Kuzmanović (233), Serbian Hydrological Association, International Association of Hydrological Sciences, Belgrade, Serbia

Mousa Maleki (59), Department of Civil Engineering, Illinois Institute of Technology, Chicago, IL, United States

A.M. Martín-Casado (91), Department of Statistics, University of Salamanca, Campus Miguel de Unamuno, Salamanca, Spain

Jose-Luis Molina (91), High Polytechnic School of Engineering, University of Salamanca, Ávila, Spain

Krishnamurthy Nayak (185), Department of Electronics and Communication Engineering, Manipal Institute of Technology, Manipal Academy of Higher Education (MAHE), Manipal, India

Mohammad Nazmi Newaz (167), Management Department, Bangladesh Institute of Management (BIM), Dhaka, Bangladesh

Brahim Nouibat (219), Institute of Management of Urban Techniques, Mohamed Boudiaf University; Laboratory of City, Environment, Society and Sustainable Development, M'sila, Algeria

Padam Jee Omar (1), Department of Civil Engineering, Motihari College of Engineering, Motihari, India

Dhruvesh Patel (339), Sardar Vallabhbhai National Institute of Technology, Surat; Pandit Deendayal Petroleum University, Gandhinagar, Gujarat, India

Azazkhan I. Pathan (339), Sardar Vallabhbhai National Institute of Technology, Surat; Pandit Deendayal Petroleum University, Gandhinagar, Gujarat, India

María C. Patino-Alonso (91, 277), Department of Statistics, University of Salamanca, Campus Miguel de Unamuno, Salamanca, Spain

Sahar Hadi Pour (353), School of Civil Engineering, Faculty of Engineering, Universiti Teknologi Malaysia, Johor Bahru, Malaysia

Cristina Prieto (339), Environmental Hydraulics Institute IH Cantabria- Instituto de Hidraulica Ambiental de la Universidad de Cantabria, Santander, Spain

Irfan Rashid (265), Department of Geoinformatics, University of Kashmir, Srinagar, Jammu and Kashmir, India

Sandra Reinstädtler (109,167), Independent Scientist as University of Technology Dresden—Alumna, former: Faculty of Environmental Sciences and Process Engineering and Faculty of Environment and Natural Sciences, Brandenburg University of Technology, Cottbus-Senftenberg, Germany; Department of Geography, Environment and Development Studies, University of Brunei Darussalam, Gadong, Brunei Darussalam

Dipak R. Samal (339), CEPT University, Ahmedabad, Gujarat, India

Saad Sh. Sammen (353), Department of Civil Engineering, College of Engineering, University of Diyala, Diyala Governorate, Iraq

Matthaios Saridakis (67), Department of Civil Engineering, Democritus University of Thrace, Xanthi, Greece

Shamsuddin Shahid (353), School of Civil Engineering, Faculty of Engineering, Universiti Teknologi Malaysia, Johor Bahru, Malaysia

Pushpendra Kumar Singh (33), Division of Water Resources, National Institute of Hydrology, Roorkee, India

Mike Spiliotis (67), Department of Civil Engineering, Democritus University of Thrace, Xanthi, Greece

B.S. Supreetha (185), Department of Electronics and Communication, Manipal Institute of Technology, Karnataka, India

Gokmen Tayfur (155,257,325), Department of Civil Engineering, Izmir Institute of Technology, Izmir, Turkey

Hosoyamada Tokuzo (245), Graduate School of Engineering, Nagaoka University of Technology, Nagaoka, Japan

Sailaja Vemuri (317), Department of Electronics and Communication Engineering, Pragati Engineering College, Surampalem, India

Feng Yu (291), School of International and Public Affairs; School of Emergency Management, Shanghai Jiao Tong University, Shanghai, China

Nastaran Zamani (207), Department of Water Engineering, College of Agriculture, Isfahan University of Technology, Isfahan, Iran

S. Zazo (91.277), High Polytechnic School of Engineering, University of Salamanca, Ávila, Spain

About the Editors

 **Saeid Eslamian** has been a Full Professor of Environmental Hydrology and Water Resources Engineering in the Department of Water Engineering at Isfahan University of Technology since 1995. His research focuses mainly on statistical and environmental hydrology in a changing climate. In recent years, he has worked on modeling natural hazards, including floods, severe storms, wind, drought, and pollution, and on water reuse, sustainable development and resiliency, etc. Formerly, he was a visiting professor at Princeton University, New Jersey, and the University of ETH Zurich, Switzerland. On the research side, he started a research partnership in 2014 with McGill University, Canada. He has contributed to more than 600 publications in journals, books, and technical reports. He is the founder and Chief Editor of both the *International Journal of Hydrology Science and Technology* (IJHST) and the *Journal of Flood Engineering* (JFE). Dr. Eslamian is currently Associate Editor of four important publications: *Journal of Hydrology* (Elsevier), *Eco-Hydrology and Hydrobiology* (Elsevier), *Journal of Water Reuse and Desalination* (IWA), and *Journal of the Saudi Society of Agricultural Sciences* (Elsevier). Professor Eslamian is the author of approximately 35 books and 180 book chapters.

Dr. Eslamian's professional experience includes membership on editorial boards, and he is a reviewer of approximately 100 Web of Science (ISI) journals, including the *ASCE Journal of Hydrologic Engineering*, *ASCE Journal of Water Resources Planning and Management*, *ASCE Journal of Irrigation and Drainage Engineering*, *Advances in Water Resources*, *Groundwater*, *Hydrological Processes*, *Hydrological Sciences Journal*, *Global Planetary Changes*, *Water Resources Management*, *Water Science and Technology*, *Eco-Hydrology*, *Journal of the American Water Resources Association*, *American Water Works Association Journal*, etc. Furthermore, in 2015, UNESCO nominated him for a special issue of the *Eco-Hydrology and Hydrobiology Journal*.

Professor Eslamian was selected as an outstanding reviewer for the *Journal of Hydrologic Engineering* in 2009 and received the EWRI/ASCE Visiting International Fellowship at the University of Rhode Island (2010). He was also awarded prizes for outstanding work by the Iranian Hydraulics Association in 2005 and the Iranian petroleum and oil industry in 2011. Professor Eslamian was chosen as a distinguished researcher by Isfahan University of Technology (IUT) and Isfahan Province in 2012 and 2014, respectively. In 2016, he was a candidate for National Distinguished Researcher in Iran.

Dr. Eslamian has also acted as a referee for many international organizations and universities. Some examples include the US Civilian Research and Development Foundation (USCRDF), the Swiss Network for International Studies, His Majesty's Trust Fund for Strategic Research of Sultan Qaboos University, Oman, the Royal Jordanian Geography Center College, and the Research Department of Swinburne University of Technology of Australia. He is also a member of the following associations: American Society of Civil Engineers (ASCE), International Association of Hydrologic Science (IAHS), World Conservation Union (IUCN), GC Network for Drylands Research and Development (NDRD), International Association for Urban Climate (IAUC), International Society for Agricultural Meteorology (ISAM), Association of Water and Environment Modeling (AWEM), International Hydrological Association (STAHS), and UK Drought National Center (UKDNC).

Professor Eslamian finished Hakim-Sanaei High School in Isfahan in 1979. After the Islamic Revolution, he was admitted to Isfahan University of Technology (IUT) to study a BS in water engineering, and he graduated in 1986. He was subsequently offered a scholarship for a master's degree program at Tarbiat Modares University, Tehran. He finished his studies in hydrology and water resources engineering in 1989. In 1991, he was awarded a scholarship for a PhD in civil engineering at the University of New South Wales, Australia. His supervisor was Professor David H. Pilgrim, who encouraged Professor Eslamian to work on "Regional Flood Frequency Analysis Using a New Region of Influence Approach." He earned a PhD in 1995 and returned to his home country and IUT. He was promoted in 2001 to Associate Professor and in 2014 to Full Professor. For the past 26 years, he has been nominated for different positions at IUT, including University President Consultant, Faculty Deputy of Education, and Head of Department. Dr. Eslamian is now director of the Center of Excellence in Risk Management and Natural Hazards (RiMaNaH).

Professor Eslamian has made three scientific visits, to the United States, Switzerland, and Canada in 2006, 2008, and 2015, respectively. In the first, he was offered the position of visiting professor by Princeton University and worked jointly with Professor Eric F. Wood at the School of Engineering and Applied Sciences for 1 year. The outcome was a contribution to hydrological and agricultural drought interaction knowledge through developing multivariate L-moments between soil moisture and low flows for northeastern US streams.

Recently, Professor Eslamian has written 14 handbooks published by Taylor & Francis (CRC Press): the three-volume *Handbook of Engineering Hydrology* (2014), *Urban Water Reuse Handbook* (2016), *Underground Aqueducts Handbook* (2017), the three-volume *Handbook of Drought and Water Scarcity* (2017), *Constructed Wetlands: Hydraulic Design* (2019), *Handbook of Irrigation System Selection for Semi-Arid Regions* (2020), *Urban and Industrial Water Conservation Methods* (2020), and the three-volume *Flood Handbook* (2022).

An Evaluation of Groundwater Storage Potentials in a Semiarid Climate (2019) and *Advances in Hydrogeochemistry Research* (2020) by Nova Science Publishers are also among his book publications. The two-volume *Handbook of Water Harvesting and Conservation* (2021, Wiley) and *Handbook of Disaster Risk Reduction and Resilience* (2021, New Frameworks for Building Resilience to Disasters) are further publications by Professor Eslamian, as are the *Handbook of Disaster Risk Reduction and Resilience* (2022, Disaster Risk Management Strategies) and the two-volume *Earth Systems Protection and Sustainability* (2022).

Professor Eslamian was listed among the World's Top 2% of Researchers by Stanford University, USA, in 2019 and 2021. He has also been a grant assessor, report referee, award jury member, and invited researcher for international organizations such as the United States Civilian Research and Development Foundation (2006), Intergovernmental Panel on Climate Change (2012), World Bank Policy and Human Resources Development Fund (2021), and Stockholm International Peace Research Institute (2022), respectively.

Faezeh Eslamian holds a PhD in Bioresource Engineering from McGill University, Canada. Her research focuses on the development of a novel lime-based product to mitigate phosphorus loss from agricultural fields. Dr. Elsamian completed her bachelor and master's degrees in Civil and Environmental Engineering at the Isfahan University of Technology, Iran, where she evaluated natural and low-cost absorbents for the removal of pollutants such as textile dyes and heavy metals. Furthermore, she has conducted research on worldwide water quality standards and wastewater reuse guidelines. Dr. Elsamian is an experienced multidisciplinary researcher with interests in soil and water quality, environmental remediation, water reuse, and drought management.

Preface

Water Data Management Best Practices, Volume 3 of the *Handbook of HydroInformatics*, presents in 26 chapters the latest and most thoroughly updated data processing techniques that are fundamental to the water science and engineering disciplines. These include a wide range of new methods that are used in hydro-modeling, such as advantages of the grid-free analytic element method, soft-computing techniques for determining the dam outflow and breach characteristics, the hydrologic engineering center hydrologic modeling system (HEC-HMS), soft-computing methods for turbulent stormwater modeling, bed load transport assessment by conventional and fuzzy regression methods, automated flood inundation mapping, causal reasoning modeling, data assimilation and accuracy, flood routing, water resources engineering fuzzy logic applications, geographic information systems (GIS) application in flood mapping, groundwater level forecasting using hybrid soft-computing techniques, hydroinformatics methods for groundwater simulation, hydrological-hydraulic modeling of floodplain inundation, interoceanic waterways network systems, lattice Boltzmann models for hydraulic engineering problems, mathematical developments in sediment transport, wetland ecosystems simulations, multivariate linear modeling application in hydrological engineering, case-based reasoning (CBR)-supported risk response to hydrological cascading disasters, optimally pruned extreme learning machine (OP-ELM), water quality analysis based on hyperspectral remote sensor data, real-time flood hydrograph predictions, river bathymetry acquisition techniques, gene expression programming (GEP), and sediment transport by soft computing.

This volume is a true interdisciplinary work, and the intended audience includes postgraduates and early-career researchers interested in computer science, mathematical science, applied science, Earth and geoscience, geography, civil engineering, engineering, water science, atmospheric science, social science, environment science, natural resources, and chemical engineering.

The *Handbook of HydroInformatics* corresponds to courses that could be taught at the following levels: undergraduate, postgraduate, research students, and short course programs. Typical course names of this type include: HydroInformatics, Soft Computing, Learning Machine Algorithms, Statistical Hydrology, Artificial Intelligence, Optimization, Advanced Engineering Statistics, Time Series, Stochastic Processes, Mathematical Modeling, Data Science, Data Mining, etc.

The three-volume *Handbook of HydroInformatics* is recommended not only for universities and colleges, but also for research centers, governmental departments, policy makers, engineering consultants, federal emergency management agencies, and related bodies.

Key features are as follows:

- Contains contributions from global experts in the fields of data management research, climate change and resilience, insufficient data problems, etc.
- Offers thorough applied examples and case studies in each chapter, providing the reader with real-world scenarios for comparison
- Includes a wide range of new methods employed in hydro-modeling, with step-by-step guides on how to use them

Saeid Eslamian
Department of Water Engineering, College of Agriculture, Isfahan, University of Technology, Isfahan, Iran

Faezeh Eslamian
Department of Bioresource Engineering, McGill University, Montreal, QC, Canada

Chapter 1

Advantage of grid-free analytic element method for identification of locations and pumping rates of wells

Shishir Gaur[a], Padam Jee Omar[b], and Saeid Eslamian[c,d]

[a]Department of Civil Engineering, Indian Institute of Technology, Varanasi, India, [b]Department of Civil Engineering, Motihari College of Engineering, Motihari, India, [c]Department of Water Engineering, College of Agriculture, Isfahan University of Technology, Isfahan, Iran, [d]Center of Excellence for Risk Management and Natural Hazards, Isfahan University of Technology, Isfahan, Iran

1. Introduction

Satisfying the growing water demand is the most common global problem, and groundwater plays the most important role for achieving this demand. Proper management and distribution of the groundwater resources can help for fair groundwater sharing and avoid over exploitation of this resource (Tziatzios et al., 2021). Solution of groundwater management problems often needs to find the best possible location and pumping rates of wells or both, which depends on the efficiency of the simulation model to define the precise location of wells and water budgeting. Illegal extraction of this resource through pumping wells makes this problem worsen more and sets challenges in front of water agencies (Gaur et al., 2021). The unknown/illegal wells' problem increases the unequal distribution of water share and corresponding mismanagement of the natural resources (Pu et al., 2020).

Furthermore, in order to fulfill the existing and forthcoming population's water demand, pressure has built up to conserve and sustainable management of the groundwater to ensure the fulfillment of the future water demand (Kumar et al., 2021). Usually groundwater management problems are solved by reducing the differences between computed and observed groundwater level through inverse modeling approach (He et al., 2021). In north Bihar plains, for the assessment of the groundwater, Omar et al. (2021a) conceptualized and developed a transient multilayered groundwater flow model for the Koshi River basin. This developed model is capable of solving large groundwater problems and associated complexity with it. In north Bihar plains, the Koshi River is one of the biggest tributaries of the Ganga River system. Koshi originates from the lower part of Tibet and joins the Ganga River in Katihar district, Bihar, India. After model development, calibration of the model was also done, by considering three model parameters, to represent the actual field conditions. For validation of the model, 15 observation wells have been selected in the area. With the help of observation well data, computed and observed heads were compared. Comparison results have been found to be encouraging and the computed groundwater head matched with the observed water head to a realistic level of accuracy. Developed groundwater model is used to predict the groundwater head and flow budget in the concerned area. The study revealed that groundwater modeling is an important method for knowing the behavior of aquifer systems and to detect groundwater head under different varying hydrological stresses.

Different researchers used this approach for identification of unknown pollution sources (Ayvaz, 2010; Singh et al., 2004; Sun et al., 2006; Omar et al., 2021b), preventing sea water intrusion through wells (Cheng et al., 2000) pump-and-treat optimization technique (Matott et al., 2006; Huang and Mayer, 1997), unknown/illegal wells problem (Saffi and Cheddadi, 2010; Ayvaz and Karahan, 2008) and parameters optimization. Hsiao and Chang (2002) solved groundwater problems by taking fixed well installation cost and pumping cost. Genetic algorithm (GA) was used to determine the number and locations of pumping wells. Constrained differential dynamic programming (CDDP) was used to evaluate the operating costs. The study concluded that well installation costs impact the optimal number and locations of wells significantly. Uddameri and Kuchanur (2007) developed simulation model for groundwater flow in the formations of the Gulf coast aquifer. The model results were analyzed with mathematical programming scheme to estimate maximum available groundwater in the county, including prevention of saltwater intrusion in the aquifer by limiting the amount of allowable

Handbook of HydroInformatics. https://doi.org/10.1016/B978-0-12-821962-1.00003-9

drawdown in shallow aquifers. Ameli and Craig (2018) presented a new semianalytical flow and transport model for the simulation of 3D steady-state flow and particle movement between groundwater, a surface water body and a radial collector well in geometrically complex unconfined aquifers. Their presented method was grid-free based analytic element method, which handles the irregular configurations of radial wells more efficiently than grid-based methods. This method is then used to explore how pumping well location and river shape interact and together influence (1) transit time distribution (TTD) of captured water in a radial collector well and TTD of groundwater discharged into the river and (2) the percentage of well waters captured from different sources. According to Baulon et al. (2022), estimation of groundwater level development is a major issue in the context of climate change. Groundwater is a key resource and can even account in some countries for more than half of the water supply. Groundwater trend estimates are often used for describing this evolution. However, the estimated trend obviously strongly depends on available time series length, which may be caused by the existence of long-term variability of groundwater resources (Baulon et al., 2022).

Park et al. (2021) linked a groundwater flow and heat transport simulation model with a genetic algorithm (GA) as optimization technique. This coupled model can determine optimal well locations and pumping/ injection rates together or apart. Results demonstrated that simultaneous optimization of well location and flow rate can provide a better design than optimization of only well location for given flow rate. Wang and Ahlfeld (1994) performed the study by considering wells' location as explicit decision variables for a pump-and-treat optimization problem. Barrier function technique was used along with a linear objective function. Hermite interpolation function was used to represent the well locations as a continuous function of space. Huang and Mayer (1997) developed optimization formulation for dynamic groundwater remediation management using location of wells and the corresponding pumping rates as the decision variables. They found that optimal location and pumping rate of wells obtained with the moving-well model were less expensive than solutions obtained with a comparable fixed-well model. Kayhomayoon et al. (2021) proposed a new approach for the simulation of groundwater level for an arid area. Their methodology comprises three stages as clustering, simulation, and optimization. In optimization, two advanced optimization methods, i.e., particle swarm optimization (PSO) and whale optimization algorithm (WOA) were utilized to optimize the ANN results. Mohan et al. (2007) applied Simulation-Optimization (S—O) approach for opencast mine area which had dominant groundwater features and became cause of heaving and bursting of the mine floor due to excessive uplift pressure. The S—O model was used to identify optimum depressurization strategy and find capable approach for solving large-scale groundwater management problems (Mohan et al., 2007).

Identification of location and discharge of unknown/illegal wells, for groundwater quantity management, has been addressed by limited researchers (Saffi and Cheddadi, 2010; Ayvaz and Karahan, 2008; Tung and Chou, 2004; Pu et al., 2021; Shekhar et al., 2021). Saffi and Cheddadi (2010) developed an algebraic expression to generate the transient influence coefficients matrix for a 1-D model. The governing equation was solved using a mixed compartment model. In the study, objective function was to minimize the errors between observed and simulated hydraulic heads to determine the illegal groundwater pumping at fixed well locations. Ayvaz and Karahan (2008) developed a simulation/optimization model for identification of unknown location and pumping rate of wells. Finite Difference Method (FDM) based flow model and GA model were used to determine the discharge rates whereas well locations were identified by iterative moving subdomain approach (Omar et al., 2020). The model was tested for both steady and transient flow conditions on two hypothetical aquifer models. Results showed that the true well locations were identified irrespective of starting point of the search process. Finally, the performance of the proposed model was compared with that of a GA solution and found that the proposed model had smaller Residual Error (RE) than the GA solution and required 14% less simulations.

The analytic element method (AEM) is based upon superposition of analytical expressions to simulate groundwater flow by considering different hydrogeological feature like streams, lakes and wells (Strack, 1989). The AEM is a grid free method and has certain advantages over grid-based methods, example wells are directly represented by their exact co-ordinates (Omar et al., 2019; Bandilla et al., 2007). The AEM flow model gives continuous solutions over the model domain and therefore gives more accurate water budget for the area. The two-dimensional implementation of the analytic element method (AEM) is commonly used to simulate steady-state saturated groundwater flow phenomena at regional and local scales. However, unlike alternative groundwater flow simulation methods, AEM results are not ordinarily used as the basis for simulation of reactive solute transport (Craig and Rabideau, 2006).

Above explained before, the literature review shows different quantity or quality based groundwater management problems where optimal location, discharge of wells or both were taken as decision variable. Considering this, the present study is carried out to explore the benefits of AEM based flow model to compute the optimal location and pumping rates of wells for unknown/illegal wells' problem. For this, AEM and grid based approach like FDM was adopted, and flow model was developed, along with Particle Swarm Optimization (PSO) model. After that, developed AEM-based flow model, and FDM model was coupled with the (PSO) model, individually. Accuracy assessment was done to know which model provides better results between AEM-PSO and FDM-PSO. Further, the developed AEM-PSO model was applied to the real field data to compute the location and pumping rates of unknown wells.

2. Limitations of the study

In spite of the fact that the present study is based on a comprehensive scientific analysis of various hydrological and hydrogeological features, it has some limitations that need to be addressed.

In the present study, to solve the partial differential equations of the groundwater flow, two different approaches have been adopted. One method is the Finite Difference Method (FDM), in which derivatives are approximated with finite differences. Another method is the Analytic Element Method (AEM), in which the boundary conditions of the flow model are discretized instead of discretization of the whole model. Another method such as the Finite Element Method (FEM), in which meshing is performed using finite elements, has not been taken into account for the flow modeling.

To optimize the pumping rate (discharge) of the well, Particle Swarm Optimization (PSO) model was also developed. After that, The PSO model was coupled with the developed AEM-based flow model, and grid-based FDM model separately. For the optimization, multiswarm optimization techniques also can be applied.

3. Methodology and formulation of the simulation-optimization model

Optimal location and discharge of pumping wells is often considered as a decision variables to solve unknown/illegal wells' problem. Solution of these problems depends on the accuracy to forecast the position of the wells and accuracy in calculation of water budget (Eslamian and Eslamian, 2022). In the present study, The AEM and FDM flow models and the PSO model were developed. After validation of simulation models and optimization model (Gaur et al., 2011), both simulation models were coupled with PSO model to solve the unknown/illegal wells' problem. Two cases, with and without use of moving subdomain approach (Ayvaz and Karahan, 2008), were considered to identify the optimal location and discharge of pumping wells. Comparative analysis was done for the AEM-PSO and FDM-PSO models and the efficiency of both the AEM and FDM methods to compute the location and discharge of wells for both cases were investigated. Computational efficiency of both models was also measured based on convergence of model to find optimal solution.

Elçi and Ayvaz (2014) did an intensive study to present an optimization approach to determine locations of new groundwater production wells, where groundwater is relatively less susceptible to groundwater contamination. For this, they coupled a regional-scale groundwater flow model with a hybrid optimization model that uses the Differential Evolution (DE) algorithm and the Broyden–Fletcher–Goldfarb–Shanno (BFGS) method as the global and local optimizers. In this study, several constraints such as the depth to the water table, total well length and the restriction of seawater intrusion are considered in the optimization process. The optimization problem can be formulated either as the maximization of the pumping rate or as the minimization of total costs of well installation and pumping operation from existing and new wells. After the development of simulation–optimization model, they demonstrated it on an existing groundwater flow model for the Tahtalı watershed in Izmir–Turkey. The model identifies for the demonstration study locations and pumping rates for up to four new wells and one new well in the cost minimization and maximization problem, respectively. All new well locations in the optimized solution coincide with areas of relatively low groundwater vulnerability.

3.1 AEM and FDM flow models

The analytic element method (AEM) is a numerical method used for the solution of partial differential equations. The basic principle of the analytic element method is that, for linear differential equations, elementary solutions may be superimposed to obtain more complex solutions. These analytic solutions typically correspond to a discontinuity in the dependent variable or its gradient along a geometric boundary (e.g., point, line, etc.). This discontinuity has a specific functional form and may be manipulated to satisfy Dirichlet, Neumann, or Robin (mixed) boundary conditions (Shamir et al., 1984). In AEM, groundwater flow solution is obtained by the use of potential theory where the discharge potential $\Phi(x, y)$ for an aquifer is determined by principal of superposition (Reilly et al., 1987). The linear solutions of individual elements become superimposed to find final solution and further the potential is converted into head. Each solution corresponds to particular hydraulic features (e.g., river, lakes, wells, hydraulic conductivity; Strack, 1989). AEM does not require a fixed boundary condition, which makes the development of the conceptual model less complicated (Omar et al., 2019).

In AEM, the model conceptualization was done in GIS environment (Geographic Information System) using the base map files, which was created in the DXF format. Boundary conditions are defined in the hydrological element itself as the head. The model domain was defined before inputting any model parameters. The term domain is referring to the regions within which the aquifer properties are constant.

In FDM, the governing groundwater flow equation can be defined as:

$$\frac{\partial}{\partial x}\left(K_x \frac{\partial h}{\partial x}\right) + \frac{\partial}{\partial y}\left(K_y \frac{\partial h}{\partial y}\right) + \frac{\partial}{\partial z}\left(K_z \frac{\partial h}{\partial z}\right) - W = S_s \frac{\partial h}{\partial t} \tag{1}$$

where K_x, K_y, and K_z are the hydraulic conductivity (HC) of the aquifer in all three directions (x, y, and z), W is the volumetric flux (flow) per unit volume, S_S is the specific storage of aquifer's porous material, h is the potentiometric head, and t is the time.

Numerical solution, i.e., Finite Difference Method (FDM) of the equation gives the variability of groundwater head (h) in an aquifer. In the FDM, the mathematical approximation is used in solving the groundwater flow equation while in the AEM harmonic function is used to solve the groundwater flow equation (Laplace equation) which produces more accurate results. As the AEM provides a continuous groundwater level surface while the FDM provides solutions at discrete points in the grid, the AEM approach is suitable to follow the sharp changes of the groundwater level while the FDM approach provides the groundwater surface only at discrete points/grids.

3.2 Particle swarm optimization

Optimization techniques can be classified into two types. The first is *deterministic optimization* techniques, which include linear programming (LP), nonlinear programming (NLP), and dynamic programming (DP). The second type is *stochastic optimization* including Genetic Algorithm (GA), Particle Swarm Optimization (PSO), Shuffled Complex Evolution, Simulating Annealing (SA), etc. Groundwater management problems are usually nonlinear and nonconvex mathematical programming problems (McKinney and Lin, 1994). For such problems, using deterministic optimization techniques can lead to some unforeseen situations. These techniques usually require good initial solutions to produce an optimal solution. Furthermore, they rely on the local gradient of the objective function to determine the search direction, and thus, can converge to local optimal solutions (Ayvaz, 2009). Therefore, the use of stochastic optimization techniques is generally preferred because of their ability to find solutions without the need for gradients and initial solutions (El-Ghandour and Elsaid, 2013).

Particle swarm optimization (PSO) is a stochastic population-based optimization algorithm inspired by the interactions of individuals in a social world. This algorithm is widely applied in various areas of water resource problems. Particle Swarm Optimization (PSO), which is also an evolutionary computation technique, which is an efficient method for solving large and complex optimization problems (Kennedy and Eberhart, 1995). PSO is a member of wide category of swarm intelligence based methods and efficient in global optimization problems. PSO considers two factors for achieving the goal: the particle's own best previous experience (i.e., *pbest*) and the best experience of all other members (i.e., *gbest*). The model was developed on the MATLAB (Gaur et al., 2011).

4. Model application and results

Both AEM-PSO and FDM-PSO models were employed on real field data. Both cases were analyzed and comparative study was performed. Objective function of the problem was defined to minimization the Residual Error (RE) between observed and computed values.

$$f = \min \sum_{i=1}^{N_e} \left(h^i_{comupted} - h^i_{actual}\right)^2 \tag{2}$$

where, $h^i_{comupted}$ are the values computed by AEM-PSO model, h^i_{actual} are observed values in the presence of pumping wells and N_e are the total number of control points.

The study area is located in the central part of Indo-Gangetic plain of the Indian subcontinent. Varanasi is the oldest and religious district of the Utter Pradesh, India. Varanasi has been the cultural center of North India for several thousand years and is closely linked to the Ganges. It is situated at the bank of Ganga River. The study area lies in Eastern Uttar Pradesh and bounded between two major rivers the Ganga River and Gomati River. The River Ganga lies in the southern and eastern part of the study area and Gomati River lies in northern part. Two small tributaries, namely Basuhi and Morwa lie in the western side of the area. The Varuna River divides the study area into almost two equal parts. The study area covers three districts; Varanasi, and some areas of Sant Ravi Das Nagar and Jaunpur. Sant Ravi Das Nagar is also known as Bhadohi, situated in the plains of the River Ganga. Jaunpur is situated on the bank of the Gomati River.

The study area covers about 2785 km^2 area in and around Varanasi district, of which 1535 km^2 located in Varanasi district. The study area lies between the latitude 25°05′16″ N—25°40′01″ N and longitude 82°22′05″ E—83°11′35″ E

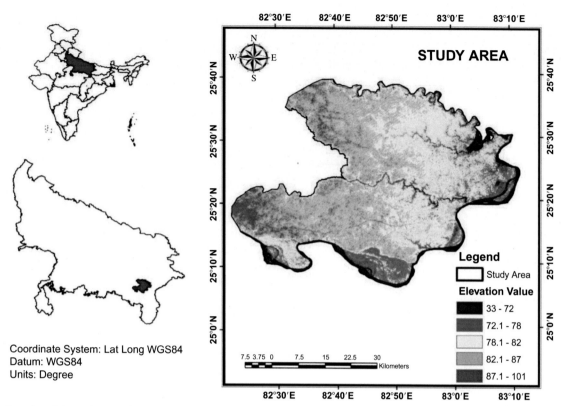

FIG. 1   Location of the study area with observation wells and pumping wells.

as shown in Fig. 1. The entire area is divided into 10 administrative blocks in which 8 blocks Pindara, Cholapur, Baragaon, Harhua, Chiraigaon, Sevapuri, Arajiline, and Kashi Vidya Peeth (KVP) exist in Varanasi District.

4.1   Physiography and topography of the area

The River Ganga is present as a trans-boundary between India and Bangladesh, the point of origin of the 2525 km (1569 mile) long River is Western Himalaya in the Indian state of Uttarakhand, and it flows to the south and east Gangetic plains of North India into Bangladesh where it drains in the Bay of Bengal (Gupta and Deshpande, 2004; Jain et al., 2007). Ganga is the longest River of India and is ranked second as the greatest River in terms of water discharge. The average annual discharge of the River Ganga is about 16,650 m^3/s (Jain et al., 2007). Ganga basin is of magnificent variation in altitude, usage of the land, the pattern of crops, and climate. One of the tributaries of River Ganga is Gomati River which originates from the Gomat taal, Pilibhit, India. The length of the river Gomati is around 900 km extending from the state of Uttar Pradesh and meets the Ganga near Saidpur, Kaithi in Ghazipur. One of the minor tributaries of the Ganga is River Varuna. Throughout the entire course, both the rivers receive a large amount of agricultural run-off consisting of pesticides and fertilizers from the catchment area.

The topography of study area is characterized by significant variation in the elevation. It varies from 33 to 101 m mean sea level (MSL). The highest elevation in the study area has been observed near around Bhadohi. The average elevation of the area is 80.71 m from mean sea level (MSL). The Ganga River has an elevation of 66.27 m MSL at the point where it enters in to the study area and 60.78, where it leaves the area in the downstream. The Gomati River elevation from top to bottom ranges from 68.32 to 60.78 m MSL, respectively.

The groundwater flow model was conceptualized on the basis of geological, climatic, and hydro-geological characteristics of the study area. Parameters have been used to develop the FDM model as well. Grid dimension has been taken 100 × 100 m for the development of FDM model. Further, the AEM-PSO and FDM-PSO models were applied on case 1 and case 2 to compute the location and discharge of unknown wells.

Initially 5 wells were placed at different locations and corresponding groundwater heads were computed by AEM and head values were recorded at predefined 12 observation points. These recorded values were considered as observed values.

Further, the AEM-PSO and FDM-PSO models were employed to identify the location and discharge of those five wells by considering different sets of 1–5 wells. Corresponding optimal values of objective functions, i.e., RE values for different set of wells have been recorded and the number of wells was finalized based on minimum value of RE. The results of AEM-PSO and FDM-PSO models were applied on two cases (1) optimal location and discharge of wells, (2) Computational efficiency of both models.

Case 1:

In this case, optimal discharge and location of pumping wells were identified by straight application of AEM-PSO and FDM-PSO model. Efficiency of model was calculated by identifying the difference in discharge values and location displacement between computed and observed values. Location displacement was computed by accumulating the difference between observed and computed location of all wells, which is defined as,

$$D = \sum_{i=1}^{N} d_i \quad (3)$$

where, d_i is the distance between two adjacent wells and N = total number of wells.

Results indicate that AEM-PSO has error of 1.9%–4.1% and FDM has error of 3.4%–7.3% in discharge rate for 5 wells. Whereas, AEM-PSO has error of 128 m and FDM has error of 437.7 m in location displacement of 5 wells. Table 1 shows the location, discharge rate of wells and groundwater head values by both AEM-PSO and FDM-PSO models. Results describe that the optimal locations differ in the AEM and FDM model, where AEM gives more accurate values in comparison of FDM. Although present model was developed by considering small size of cells, but it shows that increasing the size of cells can lead to more error in the final results.

Maximum number of iterations, i.e., 1000 was considered for the convergence of the optimization model. The model convergence showed that the AEM-PSO model converged after 905 iterations for the set of 5 wells, whereas the FDM-PSO model was found to have converged after 825 iterations for the set of 5 wells, respectively. The results show that the FDM-PSO model converged with less iteration than the AEM-PSO model. The parameters of the PSO model were considered as linearly varying inertia weight from 2.0 to 1.8, acceleration constant 2.0–2.0.

Case 2:

In this case, moving subdomain approach was applied and its efficiency in both AEM and FDM model were analyzed. Detailed description about this technique can be found in Ayvaz and Karahan, 2008. In this study, initial location of subdomain was taken same in AEM and FDM with equal size of 300 m × 300 m in AEM and 3 × 3 cells in FDM. In AEM, domain size was allowed to reduce till 20 m × 20 m from 300 m × 300 m as AEM is not bounded by minimum size of grids. Predefined well location was assumed on the center of the domain and center of its sides. Contraction of domain

TABLE 1 Actual and optimized location and discharge of the wells in case 1.

Case 1	Well ID	Exact location		Computed locations		Relative error	Pumping rates			
		X co-ordinate	Y co-ordinates	X co-ordinate	Y co-ordinates	Location displacement	Exact	Computed	Relative error	RE
AEM	W1	687,215	2,108,755	687,197	2,108,743	21.6	310	316	6	2.22
	W2	686,562	2,108,117	686,554	2,108,098	20.6	285	295	10	
	W3	687,126	2,108,085	687,111	2,108,073	19.2	320	329	9	
	W4	686,005	2,107,182	685,983	2,107,160	31.1	290	302	12	
	W5	686,836	2,107,134	686,810	2,107,109	36.1	275	281	6	
FDM	W1	687,215	2,108,755	687,143	2,108,696	93.1	310	325	15	3.48
	W2	686,562	2,108,117	686,506	2,108,051	86.6	285	298	13	
	W3	687,126	2,108,085	687,061	2,108,013	97.0	320	331	11	
	W4	686,005	2,107,182	685,966	2,107,120	73.2	290	302	12	
	W5	686,836	2,107,134	686,751	2,107,112	87.8	275	295	20	

was allowed when the location of well was not found improved in subdomain by flipping in all directions as shown in Fig. 2. In each contraction of subdomain, again it was allowed to move in all direction to find any other possible best location. The procedure remained continue till the subdomain size approaches to minimum limit, i.e., 20 × 20 m. Contraction was employed to the two sides, by the shifting rate of −10 m, having wells with more RE values than wells of remaining two sides. The RE value for addition of each subdomain was computed and used to find the optimal number of wells. Table 2 shows number of wells and the different RE values. Whereas reduction of subdomain in the FDM model was restricted and the subdomain was allowed to move until, the optimal location was not found. Optimal location was finalized when location of well, in specific cell of grid, was not improved. To finalize the location of cell, subdomain was allowed to flip in all directions (Fig. 2).

In the presence of optimal location of well, location of final cell was not found to be improved. Table 3 shows the location, discharge rate of wells and groundwater head values by both AEM-PSO and FDM-PSO models for Case 2.

Results show that optimal number of wells is five, which shows minimum RE values. Results indicate that cumulative location displacement in FDM was found 270.7 m more in comparison of the AEM model. Whereas computed pumping rates was found with the error of 1.3%–2.1% in AEM and 2.1%–3.6% in FDM. Total number of iterations in AEM-PSO model was found 22% more in comparison of FDM model. It shows that simulation model needs more iteration to get closer values to real values.

The comparison between results of Case 1 and Case 2 shows (Table 4) that moving subdomain approach is efficient to find more accurate results. It was also observed that location displacement error was more improved than pumping rates error in Case 2. It was found that location displacement error was improved by 42.7 m in AEM model and 81.7 m in FDM model from Case 1 to Case 2. Discharge values were improved by 1.1% in AEM and 1.5% in FDM. In Case 2, only 5 decision variables were taken to optimize the pumping rate of the wells, whereas total 15 decision variables were taken in Case 1 to optimize the location and pumping rates. Both AEM-PSO and FDM-PSO models were employed for single run to find the optimal solution in the Case 1. Whereas in Case 2, both models took 10 to 15 runs to find the optimal solution. Although model run was more in Case 2, but due to presence of five variables, model converges very early in comparison of Case 1.

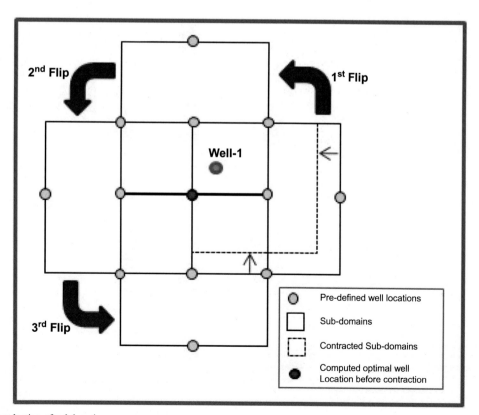

FIG. 2 Flip and reduction of subdomain.

TABLE 2 RE values at different set of wells in case 2.

Well	AEM	FDM
1	358.6	421.2
2	225.1	295
3	141.2	181
4	42.1	53
5	1.74	2.68
6	3.21	4.65

TABLE 3 Actual and optimized location and discharge of the wells in case 2.

		Exact location		Computed locations		Relative error	Pumping rates			
Case 2	Well ID	X co-ordinate	Y co-ordinate	X co-ordinate	Y co-ordinate	Location displacement	Exact	Computed	Relative error	RE
AEM	W1	687,215	2,108,755	687,203	2,108,745	15.6	310	314	4	1.74
	W2	686,562	2,108,117	686,554	2,108,102	17.0	285	291	6	
	W3	687,126	2,108,085	687,116	2,108,078	12.2	320	326	6	
	W4	686,005	2,107,182	685,993	2,107,164	21.6	290	295	5	
	W5	686,836	2,107,134	686,820	2,107,124	18.9	275	279	4	
FDM	W1	687,215	2,108,755	687,160	2,108,710	71.1	310	321	11	2.68
	W2	686,562	2,108,117	686,536	2,108,082	43.6	285	295	10	
	W3	687,126	2,108,085	687,081	2,108,030	71.1	320	329	9	
	W4	686,005	2,107,182	685,980	2,107,114	72.4	290	296	6	
	W5	686,836	2,107,134	686,751	2,107,079	101.2	275	285	10	

TABLE 4 Comparative analysis between case 1 and case 2.

	Method	Location displacement	Discharge	RE	Iterations
Case 1	AEM	128	1523	2.22	835
	FDM	437.7	1551	3.48	715
Case 2	AEM	85.3	1506	1.74	390
	FDM	356	1527	2.68	320

5. Conclusions

The present study is carried out to explore the efficiency of the Analytic Element Method (AEM) to deal with unknown/ illegal wells' problems. The AEM-based flow model and grid based FDM model was coupled with the Particle Swarm Optimization (PSO) model. Further, the developed AEM-PSO model was applied to the real field data to compute the location and pumping rates of unknown wells. The results were compared with FDM-PSO, which suggests that in some conditions The AEM model is more efficient to solve unknown wells' problems. The present study gives the motivation to apply the AEM based flow model for various groundwater resource management problems, where the exact location of wells and accurate water budgeting is the main decision criteria.

In the present study, the advantage of AEM was investigated for solving unknown/illegal wells' problem by identifying location and discharge through simulation and optimization technique. In the study, PSO model was coupled with AEM and FDM method. The AEM-PSO model was compared with the FDM-PSO model and both were applied on real field data. Two cases were considered with and without use of subdomain approach. The results show that the AEM flow model is very effective to solve unknown wells' problem with some advantage over grid based method. The AEM model computes more accurate location of wells, in both cases and provides a more accurate discharge rates as well. The results of the study show that AEM model is itself efficient to solve unknown well problems in comparison of FDM model in the both cases. It was also found that efficiency of the AEM and FDM models can be improved through subdomain approach. Although number of iterations was found more in the AEM-PSO model in comparison of the FDM-PSO model.

This study also suggested that the AEM method is very efficient to deal with groundwater management problems where location of wells and water budgeting are main decision variables. In the AEM, Computational burden depends on the number of hydrogeological features and their discretization level. Thus, the present study is more useful to perform effective study for the large geographic areas without excessive computation time. The study concluded that moving subdomain approach is effective to find the optimal well locations in less number of iterations.

References

Ameli, A.A., Craig, J.R., 2018. Semi-analytical 3D solution for assessing radial collector well pumping impacts on groundwater–surface water interaction. Hydrol. Res. 49 (1), 17–26.

Ayvaz, M.T., 2009. Application of harmony search algorithm to the solution of groundwater management models. Adv. Water Resour. 32 (6), 916–924.

Ayvaz, M.T., 2010. A linked simulation-optimization model for solving the unknown groundwater pollution source identification problems. J. Contam. Hydrol. 117 (1–4), 46–59.

Ayvaz, M.T., Karahan, H., 2008. A simulation/optimization model for the identification of unknown groundwater well locations and pumping rates. J. Hydrol. 357, 76–92.

Bandilla, K.W., Janković, I., Rabideau, A.J., 2007. A new algorithm for analytic element modeling of large-scale groundwater flow. Adv. Water Resour. 30, 446–454.

Baulon, L., Allier, D., Massei, N., Bessiere, H., Fournier, M., Bault, V., 2022. Influence of low-frequency variability on groundwater level trends. J. Hydrol. 127436. https://doi.org/10.1016/j.jhydrol.2022.127436.

Cheng, A., Halhal, D., Naji, A., Ouazar, D., 2000. Pumping optimization in saltwater-intruded coastal aquifers. Water Resour. Res. 36 (8), 2155–2165.

Craig, J.R., Rabideau, A.J., 2006. Finite difference modeling of contaminant transport using analytic element flow solutions. Adv. Water Resour. 29 (7), 1075–1087.

Elçi, A., Ayvaz, M.T., 2014. Differential-evolution algorithm based optimization for the site selection of groundwater production wells with the consideration of the vulnerability concept. J. Hydrol. 511, 736–749.

El-Ghandour, H.A., Elsaid, A., 2013. Groundwater management using a new-coupled model of flow analytical solution and particle swarm optimization. Int. J. Water Res. Environ. Eng. 5 (1), 1–11.

Eslamian, S. and F. Eslamian, 2022, Handbook of irrigation hydrology and management, Vol. vol. 2, Irrigation Methods, Taylor and Francis, CRC Group, USA.Eslamian, S., Eslamian, F., 2022. Handbook of irrigation hydrology and management. In: Irrigation Methods, Vol. 2. Taylor and Francis, CRC Group, USA.

Gaur, S., Chahar, B.R., Graillot, D., 2011. Analytic elements method and particle swarm optimization based simulation–optimization model for groundwater management. J. Hydrol. 402, 217–227.

Gaur, S., Johannet, A., Graillot, D., Omar, P.J., 2021. Modeling of Groundwater Level Using Artificial Neural Network Algorithm and WA-SVR Model. In: Groundwater Resources Development and Planning in the Semi-Arid Region. Springer, Cham, Switzerland, pp. 129–150.

Gupta, S.K., Deshpande, R.D., 2004. Water for India in 2050: first-order assessment of available options. Curr. Sci. 86 (9), 1216–1224.

He, L., Hou, M., Chen, S., Zhang, J., Chen, J., and Qi, H. (2021). Construction of spatio-temporal coupling model for groundwater level prediction: a case study of Changwu area, Yangtze River Delta region of China. Water Supply, 21, 7.He, L., Hou, M., Chen, S., Zhang, J., Chen, J., Qi, H., 2021. Construction of spatio-temporal coupling model for groundwater level prediction: a case study of Changwu area, Yangtze River Delta region of China. Water Supply 21 (7). https://doi.org/10.2166/WS.2021.140.

Hsiao, C.T., Chang, L.C., 2002. Dynamic optimal groundwater management with inclusion of fixed costs. J. Water Resour. Plan. Manag. 128 (1), 57–65.

Huang, C., Mayer, A.S., 1997. Pump-and-treat optimization using well locations and pumping rates as a decision variables. Water Resour. Res. 33 (5), 1001–1012.

Jain, S.K., Agarwal, P.K., Singh, V.P., 2007. Hydrology and water resources of India. Springer Science & Business Media 57, 56–68.

Kayhomayoon, Z., Ghordoyee Milan, S., Arya Azar, N., Kardan Moghaddam, H., 2021. A new approach for regional groundwater level simulation: clustering, simulation, and optimization. Nat. Resour. Res. 30 (6), 4165–4185.

Kennedy, J., Eberhart, R., 1995. Particle swarm optimization. Proc. IEEE Int. Conf. Neural Netw. 4, 1942–1948.

Kumar, V., Chaplot, B., Omar, P.J., Mishra, S., Azamathulla, M., H., 2021. Experimental study on infiltration pattern: opportunities for sustainable management in the northern region of India. Water Sci. Technol. 84 (10–11), 2675–2685.

Matott, L.S., Rabideau, A.J., Craig, J.R., 2006. Pump-and-treat optimization using analytic element method flow models. Adv. Water Resour. 29, 760–775.

McKinney, D.C., Lin, M.D., 1994. Genetic algorithm solution of groundwater management models. Water Resour. Res. 30 (6), 1897–1906.

Mohan, S., Sreejith, P.K., Pramada, S.K., 2007. Optimization of open-pit mine depressurization system using simulated annealing technique. J. Hydrol. Eng. 133 (7), 825–830.

Omar, P.J., Gaur, S., Dwivedi, S.B., Dikshit, P.K.S., 2019. Groundwater modelling using an analytic element method and finite difference method: an insight into lower ganga river basin. J. Earth Syst. Sci. 128 (7), 195.

Omar, P.J., Gaur, S., Dwivedi, S.B., Dikshit, P.K.S., 2020. A modular three-dimensional scenario-based numerical modelling of groundwater flow. Water Resour. Manag. 34, 1913–1932. https://doi.org/10.1007/s11269-020-02538-z.

Omar, P.J., Gaur, S., Dikshit, P.K.S., 2021a. Conceptualization and development of multi-layered groundwater model in transient condition. Appl. Water Sci. 11 (10), 1–10.

Omar, P.J., Gaur, S., Dwivedi, S.B., Dikshit, P.K.S., 2021b. Development and application of the integrated GIS-MODFLOW model. In: Fate and Transport of Subsurface Pollutants. Springer, Singapore, pp. 305–314.

Park, D., Lee, E., Kaown, D., Lee, S.S., Lee, K.K., 2021. Determination of optimal well locations and pumping/injection rates for groundwater heat pump system. Geothermics 92, 102050. https://doi.org/10.1016/j.geothermics.2021.102050.

Pu, J.H., Pandey, M., Hanmaiahgari, P.R., 2020. Analytical modelling of sidewall turbulence effect on streamwise velocity profile using 2D approach: a comparison of rectangular and trapezoidal open channel flows. J. Hydro-Environ. Res. 32, 17–25.

Pu, J.H., Wallwork, J.T., Khan, M., Pandey, M., Pourshahbaz, H., Satyanaga, A., Gough, T., 2021. Flood suspended sediment transport: combined modelling from dilute to hyper-concentrated flow. Watermark 13 (3), 379.

Reilly, T.E., Franke, O.L., Bennett, G.D., 1987. The principle of superposition and its application in ground-water hydraulics. Department of the Interior, US Geological Survey, USA, pp. 03–B6.

Saffi, M., Cheddadi, A., 2010. Identification of illegal groundwater pumping in semi-confined aquifers. Hydrol. Sci. J. 55, 8.

Shamir, U., Bear, J., Gamliel, A., 1984. Optimal annual operation of a coastal aquifer. Water Resour. Res. 20, 435–444.

Shekhar, S., Chauhan, M.S., Omar, P.J., Jha, M., 2021. River discharge study in River Ganga, Varanasi using conventional and modern techniques. In: The Ganga River Basin: A Hydrometeorological Approach. Springer, Cham, Switzerland, pp. 101–113.

Singh, R.M., Datta, B., Jain, A., 2004. Identification of unknown groundwater pollution sources using artificial neural networks. J. Water Resour. Plan. Manag. 130 (6), 506–514.

Strack, O.D.L., 1989. Groundwater Mechanics. Prentice-Hall, Englewood Cliffs, NJ, USA.

Sun, Y., Zhuang, G., Zhang, W., Zhuang, Y., 2006. Characteristics and sources of Lead pollution after phasing out leaded gasoline in Beijing. Atmos. Environ. 40 (16), 2973–2985.

Tung, C.P., Chou, C.A., 2004. Pattern classification using tabu search to identify the spatial distribution of groundwater pumping. Hydrgeol. J. 12 (5), 488–496.

Tziatzios, G., Sidiropoulos, P., Vasiliades, L., Lyra, A., Mylopoulos, N., Loukas, A., 2021. The use of pilot points method on groundwater modelling for a degraded aquifer with limited field data: the case of Lake Karla aquifer. Water Supply 21 (6), 2633–2645.

Uddameri, V., Kuchanur, M., 2007. Simulation-optimization approach to assess groundwater availability in Refugio County, TX. Environ. Geol. 51, 921–929.

Wang, W., Ahlfeld, D.P., 1994. Optimal groundwater remediation with well location as a decision variable: model development. Water Resour. Res. 30 (5), 1605–1618.

Chapter 2

Application of experimental data and soft computing techniques in determining the outflow and breach characteristics in embankments and landslide dams

Kamran Kouzehgar[a] and Saeid Eslamian[b,c]

[a]Department of Civil Engineering, Varzeghan Branch, Islamic Azad University, Varzeghan, Iran, [b]Department of Water Engineering, College of Agriculture, Isfahan University of Technology, Isfahan, Iran, [c]Center of Excellence for Risk Management and Natural Hazards, Isfahan University of Technology, Isfahan, Iran

1. Introduction

Main climatic characteristics including intensity and length of rainfall, temperature, and average annual precipitation are the most important factors that affect the water resources of a land. These factors may cause the severe floods that lead to a lot of financial, human and environmental damages, especially in dry lands (Eslamian, 2014). The probability of embankments and landslide dams resulting from the vulnerability of the downstream in the case of excessive input flow has highlighted the importance of dam breach studies.

The application of artificial intelligence (AI) has been used in many research works due to the reduction of modeling time and cost. Gene expression programming (GEP) is an evolutionary algorithm used to develop the computer programs based on search and optimization techniques (Ferreira, 2006). The recent studies suggested that GEP could be an efficient alternative to the traditional genetic programming (GP) method in soil and water engineering applications (Gholampour et al., 2017), and only a few explanations exist in the literature, related to the use of GEP in the field of breach studies (Sattar, 2013).

Physical models were constructed, overtopped, and breached in the laboratory for the present study with different features and heights in favor of a closer investigation and improving the precision of calculations. The soil combinations and grain size distribution of the particles used in this research have not been reflected in the literature. The main components of the hydrograph were also analyzed. Hereafter, the research is devoted to study the breach hydrographs by the participation of laboratory data. The failure hydrographs of some of the existing dams are simulated and compared as a case study using gene expression programming (GEP), Bayesian, regressions, and BREACH physically-based models. The use of GEP as an artificial technique has been less used in the field of embankment failure (Sattar, 2013). Explorations of the volume transformations along with two-phase flow at the downstream indicate that breach geometries are a very important factor for dam break problem.

Extensive studies on the basis of the breach formation mechanisms and peak outflow discharge (Q_p) have been carried out recently. So, several relationships have been developed related to the depth above breach (H_w) and volume stored above breach invert (V_w) based on laboratory and regression analysis (De Lorenzo and Macchione, 2014; Froehlich, 1995; Kouzehgar et al., 2021d; MacDonald and Langridge-Monopolis, 1984; Pierce et al., 2010). However, the product of the embankment length (E_l) and width (E_w) has often been able to improve the efficiency of the proposed relation (Wang et al., 2018). Regression analysis has become a suitable tool to determine the dam outflow and breach characteristics, as well as improving the efficiency of the model (Pierce et al., 2010; Thornton et al., 2011). On the other hand, several small and large-scale laboratory and field studies have been conducted to investigate the rising and falling limbs of the dam breach hydrographs (Jiang et al., 2019; Kouzehgar et al., 2021e).

However, few studies have been conducted on the use of E_l and E_w parameters on the breach outflow studies. To determine Q_p based on the parameters of V_w, H_w, E_l, and E_w, the results of sensitivity analysis indicates H_w and V_w provide

Handbook of HydroInformatics. https://doi.org/10.1016/B978-0-12-821962-1.00002-7

more accurate results than E_l and E_w, so that the contribution of these parameters significantly increases the accuracy of the calculations (Thornton et al., 2011; Wang et al., 2018). Sliding dams are another type of filled dams that often form in narrow, steep valleys in rugged and high mountainous areas (Kouzehgar et al., 2021b). Several empirical relations have been developed to determine Q_p as a function of water drop level (d) and volume released from the reservoir (V_o) using a case study of more than 18 landslide dams (Walder and O'Connor, 1997).

In the dam breach studies and analyses, small or large-scale experiments could provide more accurate results and cover many of the shortcomings identified in the technical literature (Coleman et al., 2004). Therefore, several laboratory studies have been performed to investigate the hydraulic characteristics of the breach as well as the output hydrograph during the overtopping (Dhiman and Patra, 2017; Kouzehgar et al., 2021a; Vaskinn et al., 2004) that are large-scale field experiments are also included (Ashraf et al., 2018). Physical models have also been used to investigate and compare the relations between the breach characteristics and the effect of soil properties on breach widening (Hunt et al., 2005).

In noncohesive embankments, the rate of erosion in downstream is often rapid and dramatic. The susceptibility to erosion of body materials and relationships based on flood hydrograph resulting from gradual erosion in embankments is reflected in the studies (De Lorenzo and Macchione, 2014). The variation range of the average width of breach (B_{ave}) as a function of H_d is one of the main factors that affects the overtopping and related height of flow passing through the breach invert (Johnson and Illes, 1976; Singh and Scarlatos, 1988). In physical and computational models, the main effort is to uniformize the unpredictable breach sections in the form of trapezoids, rectangles or triangles to better study the complicated erosion mechanisms. Among these models, NWS BREACH (Fread, 1988), DL BREACH (Wu, 2013), and HR BREACH (Morris, 2011) have been used more.

Accurate determination of the breach geometry including height of the breach (H_b) and the average width of the breach (B_{ave}) due to their significant effect on the resulting flood and Qp, as well as the relationship with other hydraulic parameters of the dam, were also investigated (Jiang et al., 2019; Singh and Scarlatos, 1988; Von Thun and Gillette, 1990). Determination of the eroded volume of the dam materials (V_{er}) due to failure can provide an essential information in assessing the environmental impacts assessment of the downstream and determining the gradual erosion, so, the V_{er} in the reservoir of dams with low storage capacity is closely related to the height of dam (H_d) and can affect all of the breach parameters (Hassanzadeh, 2005). Evaluation of two-phase flow transfer due to erosion along with the high volume of sediments accumulated in the reservoir resulting from the aging of most embankments and sliding dams is necessary. Hence, considering the financial and human losses, the effect of the concentration of harmful minerals and reservoir sediments on the quality of surface water and groundwater should be investigated.

Time of failure (t_f) has been studied by various input parameters. Most of the researchers were relate it to the B_{ave} (USBR, 1988), H_w or a combination of B_{ave} and H_w (Von Thun and Gillette, 1990). However, the height of breach (H_b) and V_w are considered important parameters (Froehlich, 2008).

In more than 42 cases of earth dam failures, the determination of V_{er} as the main cause of progressive erosion along with the determination of the t_f (MacDonald and Langridge-Monopolis, 1984), as well as a function of B_{ave} has been widely studied (USBR, 1988). The parameter of t_f is also considered as a function of H_w in both simple and complex erodibility modes (Von Thun and Gillette, 1990). The determination of t_f as a function of H_b and H_w (Froehlich, 1995) and Q_p has also been reflected in the technical literature (Chinnarasri et al., 2004).

2. Proposed methodology

2.1 Failure database

In this work, a comprehensive dataset of breach records including overtopping and internal erosions were collected from the various resources. The historical data, despite their limitations, provide the valuable information on how a dam erodes and fails. One of the advantages of historical data is that they are scattered around the world and do not depend on a specific geographical zone. Therefore the database of Table 1 was collected from three resources, including (Wahl, 1998), (Xu and Zhang, 2009), and (Zhang et al., 2016), providing a complete record of the available embankment, breach, and flow characteristics.

2.2 Determination of outliers

In this study, outlier data were identified and removed from the statistical population by the statistical tests. The data must follow a probability distribution. Otherwise, they are considered as an outlier and removed from the population. The probability distribution used to select the outliers from the statistical population (Table 1) was the Gaussian distribution, obtained by MATLAB codes.

TABLE 1 Historical data of 109 embankment dam failures, collected from Wahl (1998), Xu and Zhang (2009), and Zhang et al. (2016)

No.	Dam name	V_w (×10⁶ m³)	H_w (m)	H_d (m)	E_l (m)	H_b (m)	B_t (m)	B_b (m)	B_{ave} (m)	t_f (hr)	V_{er} (m³)	Q_p (m³/sec)
1	Apishapa[1,2,3,4,5]	22.2	28	34.14	200	31.1	91.5	81.5	86.5	2.5	238,000	6850
2	Baimiku[1,2]	0.2	8	8	–	8	40	–	–	–	–	–
3	Baldwin Hills[1,2,3,4,5]	0.91	12.2	71	198	21.3	31.6	18.4	25	1.3	31,700	1130
4	Banqiao[1,2,3,5]	607.5	31	24.5	2100	29.5	372	210	291	5.5	–	78,100
5	Bayi[1,2,5]	23	28	30	–	30	45	35	40	–	–	5000
6	Bearwallow Lake[1,2,4]	0.0493	5.79	10	3.1	6.4	–	–	12.2	–	1090	–
7	Belci[1]	12.7	15.5	16	400	–	–	–	–	–	–	4700
8	Big Bay[1,2,5]	17.5	13.59	17.4	200	13.56	96	70.1	83.2	–	–	4160
9	Bila Densa[1]	0.29	10.7	18	170	–	–	–	–	–	–	320
10	Bilberry[1]	0.327	23.6	20	800	–	–	–	–	–	–	725
11	Bradfield[1]	3.2	28.96	28.96	382	–	–	–	–	–	–	1150
12	Break Neck Run[3,5]	–	–	7	–	7	–	–	30.5	3	–	9.2
13	Buckhaven No.2[1,2,4]	0.0247	6.1	–	–	6.1	–	–	4.72	–	1070	–
14	Buffalo Creek[1,2,3,4,5]	0.48	14	14.02	–	14	153	125	139	0.5	319,000	1420
15	Bullock Draw[1,2,4]	0.74	3.05	5.79	–	5.79	13.6	11	12.5	–	1350	–
16	Butler[1,2,4,5]	2.38	7.16	–	850	7.16	68.6	56.4	62.5	–	4310	810
17	Castlewood II[1,2,3,4,5]	6.17	21.6	21.34	180	21.3	54.9	33.5	44.2	0.5	55,700	3570
18	Caulk Lake[1,2,4]	0.698	11.1	20	134	12.2	36.6	–	35.1	–	13,700	–
19	Centralia[1]	0.01333	5.5	–	40	–	–	–	–	–	–	71
20	Chaq Chaq[1,2,5]	2.55	14.5	14.5	–	14.5	46	29.6	38	–	–	930
21	Chenying[1,2]	5	12	12	–	12	–	–	–	–	–	1200
22	Clearwater Lake[1,2,4]	0.466	4.05	0.466	–	3.78	–	–	22.8	–	1290	–
23	Coedty[1,2]	0.311	11	10.97	262	11	67	18.2	42.7	–	–	–
24	Dalizhuang[1,2]	0.6	12	12	–	12	40	–	–	–	–	–
25	Danghe[1,2,3,5]	10.7	24.5	46	–	25	96	20	58	3	–	2500
26	Davis Reservoir[1,2,4,5]	58	11.58	11.89	–	11.9	21.3	15.4	18.3	–	6470	510
27	Delhi[1]	12.2	11.2	–	170	–	–	–	–	–	–	1950

Continued

TABLE 1 Historical data of 109 embankment dam failures, collected—cont'd

No.	Dam name	V_w (×10⁶m³)	H_w (m)	H_d (m)	E_l (m)	H_b (m)	B_t (m)	B_b (m)	B_{ave} (m)	t_f (hr)	V_{er} (m³)	Q_p (m³/sec)
28	Dells[1,2]	13	18.3	18.3	-	18.3	112.8	-	-	-	-	5440
29	Dongchuankou[1,2]	27	31	31	-	31	-	-	-	-	-	21,000
30	Dushan[1,2]	0.67	17.7	17.7	-	17.7	70	-	-	-	-	-
31	East Fork Pork[1,2,4]	1.87	9.8	13.4	-	11.4	-	-	17.2	-	7630	-
32	Elk City[1,2,3,4,5]	1.18	9.44	9.14	564	9.14	45.5	27.7	36.6	0.83	16,900	608.79
33	Emery[1,2,4]	0.425	6.55	16	130	8.23	-	1.6	10.8	-	1970	-
34	Erlangmiao[1,2]	0.196	9	12.1	-	9	36	-	18.8	-	-	-
35	Eucluides de Cunha[1,2,]	-	58.22	53.04	-	53	131	30	-	-	726,000	1020
36	Fengzhuang[1,2]	0.625	8	10	-	8	40	-	35	-	-	-
37	Fogleman[1,2,4]	0.493	11.1	-	-	12.6	-	-	7.62	-	2050	-
38	FP&L Martin Plant[1]	125	5.09	9.4	4100	-	-	-	-	-	-	2750
39	Frankfurt[1,2,3,4,5]	0.352	8.23	9.75	-	9.75	9.2	4.6	6.9	2.5	1290	79
40	Fred Burr[1,2]	0.75	10.2	10.4	100	10.4	-	-	-	-	-	654
41	French Landing[1,2,4,5]	3.87	8.53	-	350	14.2	41	13.8	27.4	-	13,800	929
42	Frenchman Dam[1,2,3,4,5]	16	10.8	12.5	600	12.5	67	54.4	54.6	3	28,400	1420
43	Frias[1,2,a]	0.25	15	15	62.2	15	62	-	-	-	-	400
44	Goose Creek[1,2,4,5]	10.6	1.37	6.1	-	4.1	30.5	22.3	26.4	-	1070	492.7
45	Gouhou[1,2,3,5] a	3.18	44	71	-	48	138	61	99.5	2.33	-	2050
46	Grand Rapids[1,2,3,4,5]	0.255	7.5	7.62	441	6.4	12.2	6	9.1	0.5	1800	7.5
47	Hastberga[1]	30	7.35	-	200	-	-	-	-	-	-	600
48	Hatchtown[1,2,3,4,5]	14.8	16.8	19.2	238	18.3	180	140	151	3	161,000	3080
49	Hatfield[1,2,3,5]	12.3	6.8	6.8	-	6.8	-	6.1	91.5	2	-	3400
50	Hell Hole[1,2,3,4,5] a	30.6	35.1	37.06	470	56.4	175	66.9	121	0.75	555,000	7360
51	Horse Creek I[1,2,3,4,5]	12.8	7.01	12.19	701	12.8	76.2	70	73.1	3	20,500	3890
52	Hougou[1,2]	0.24	8	8	-	8	-	-	20	-	-	-
53	Houshishan[1,2]	0.22	16	13	-	16	45	15	30	-	-	-
54	Huqitang[1,2,3,5]	0.424	5.1	9.9	-	9	12	3	7.5	4	-	50

#	Name											
55	Hutchinson Lake[1,2,4]	1.17	4.42	—	—	3.75	—	—	33.4	—	1750	—
56	Ireland #5 II[1,2,4,5]	0.16	3.81	—	370	5.18	15.5	11.5	13.5	—	1260	110
57	Iowa Beef Processors[1,2]	0.333	4.42	4.57	305	4.57	—	—	16.8	—	—	—
58	Jiahezi[1,2]	42	12	18	—	18	181	—	—	—	—	—
59	Johnston[1,2,4]	0.575	4.27	4.27	—	5.18	13.4	2	8.23	—	673	—
60	Johnstown[1,2,3,4,5]	18.9	22.25	38.1	284	24.4	128	61	94.5	0.75	68,800	8500
61	Kodanagar[1,2]	12.3	11.5	11.5	—	11.5	—	—	—	—	—	1280
62	Kraftsmen Lake[1,2,4]	0.177	3.66	—	—	3.2	—	—	14.5	—	376	—
63	Kelly Barnes[1,2,3,4,5]	0.777	11.3	11.58	80	12.8	35	18	27.3	0.5	9940	680
64	La Fruta[1,2,4]	78.9	7.9	—	—	14	—	—	58.8	—	32,900	—
65	Lake Avalon[1,2,3,4,5]	31.5	13.7	14.5	370	14.6	137.6	122.4	130	2	81,000	2321.9
66	Lake Barcroft[2]	—	—	21.03	—	11	23	—	—	—	—	—
67	Lake Frances[1,2,4]	0.789	14	15.24	—	17.1	30	10.4	18.9	—	12,400	—
68	Lake Genevieve[1,2,4]	0.68	6.71	7.6	—	7.92	—	—	16.8	—	2630	—
69	Lake Latonka[1,2,3,4,5]	4.09	6.25	13	—	8.69	49.5	28.9	39.2	3	9540	290
70	Lake Philema[1,2,4]	4.78	9	—	—	8.53	—	—	47.2	—	11,300	—
71	Lambert Lake[1,2,4]	0.296	12.8	16.5	—	14.3	—	—	7.62	—	5870	—
72	Laurel Run[1,2,4,5]	0.555	14.1	12.8	160	13.7	68	2.2	35.1	—	19,500	1050
73	Lawn Lake[1,2,4,5]	0.798	6.71	7.9	—	7.62	29.5	14.9	22.2	—	2400	510
74	Lijiaju[1,2]	1.14	25	25	—	25	—	—	—	—		2950
75	Liujiatai[1,2]	40.54	35.9	35.9	—	35.9	—	—	—	—	—	28,000
76	Lily Lake[1,2,5]	0.0925	3.35	—	60	3.66	11.3	10.3	10.8	—	—	71
77	Little Deer Creek[1,2,3,4,5]	1.36	22.9	26.21	400	27.1	49.9	9.3	29.6	0.33	50,600	1330
78	Longtun[1,2]	30	9.5	9.5	—	9.5	181	—	—	—	—	—
79	Long Branch Canyon[1,2,4]	0.284	3.17	—	—	3.66	—	—	9.14	—	378	—
80	Lower Latham[1,2,3,4,5]	7.08	5.79	—	—	7.01	123.4	35	79.2	1.5	14,300	340
81	Lower Otay[1,2,3,4,5]	56.9	39.6	41.15	172	39.6	172	93.8	133	0.25	107,000	15,800
82	Lower Two Medicine[1,2,5]	19.6	11.3	11.28	350	11.3	84	50	67	—	—	1800
83	Lyman[1,2,4]	35.8	16.2	19.81	—	19.8	107	87	97	—	71,900	—
84	Lynde Brook[1,2,4]	2.88	11.6	12.5	—	12.5	45.7	15.3	30.5	—	15,300	—

Continued

TABLE 1 Historical data of 109 embankment dam failures, collected—cont'd

No.	Dam name	V_w (×10⁶m³)	H_w (m)	H_d (m)	E_l (m)	H_b (m)	B_t (m)	B_b (m)	B_{ave} (m)	t_f (hr)	V_{er} (m³)	Q_p (m³/sec)
85	Machhu II[2]	–	–	60.05	4180	60	540	–	–	–	–	–
86	Mahe[1,2]	23.4	19.5	19.5	–	19.5	–	–	–	–	–	4950
87	Mammoth[5]	–	–	21.3	–	21.3	–	–	9.2	3	–	2520
88	Martin Cooling Pond[1,2,5]	136	8.53	10.4	–	12.8	–	–	186	–	–	3115
89	Melville[1,2,4]	24.7	7.92	11	243.8	9.75	40	25.6	32.8	–	10,600	–
90	Merimac[1,2,4]	0.0696	3.44	–	–	3.05	–	–	14.2	–	758	–
91	Mossy Lake[1,2,4]	4.13	4.41	2.8	–	3.44	–	–	41.5	–	2040	–
92	Niujiaoyu[1,2]	0.144	7.2	10	–	7.2	20	6	13	–	–	–
93	Oros[1,2,3,4,5,a]	660	35.8	35.36	2000	35.5	200	130	165	8.5	765,000	9630
94	Otter Lake[1,2,4]	0.109	5	6.1	–	6.1	17.1	1.5	9.3	–	1170	–
95	Potato Hill Lake[2,4]	0.105	7.77	–	–	7.77	–	–	16.5	–	3010	–
96	Prospect[1,2,3,4,5]	3.54	1.68	–	900	4.42	91.4	85.4	88.4	2.5	5120	116
97	Puddingstone Dam[1,2]	0.617	15.2	–	420	15.2	91.4	–	–	–	–	480
98	Qielinggou[1,2]	0.7	18	18	–	18	–	–	–	–	–	2000
99	Quail Creek[1,2,4]	30.8	16.7	24	550	21.3	72.1	67.9	70	–	84,400	3110
100	Rainbow Lake[1,2,4]	6.78	10	14	–	9.54	–	–	38.9	–	10,500	–
101	Renegate Resort[1,2,4]	0.0139	3.66	–	–	3.66	–	–	2.29	–	92	–
102	Rito Manzanares[1,2,4,5]	0.12	4.57	7.32	75	7.32	19	7.6	13.3	–	1290	181
103	Salles Oliviera[1,2,3,4,5]	71.5	38.4	35.05	–	35	–	–	168	2	440,000	7200
104	Scott Farm No. 2[2,4]	0.086	10.4	–	–	11.9	15	15	15	–	7020	–
105	Schaeffer[1,2,3,4,5]	4.44	30.5	30.5	335	30.5	210	64	137	0.5	227,000	4500
106	Shangliuzhuang[1,2]	0.11	14	14	–	14	30	–	–	–	–	–
107	Shanhu[1,2]	1.78	12.5	11.5	–	13	58	24	41	–	–	–
108	Sheep Creek[1,2,4]	2.91	14.02	17.07	–	17.1	30.5	13.5	22	–	18,300	–
109	Shilongshan[1,2]	2.06	14	14	–	14	50	–	–	–	–	–
110	Shimantan[1,2,3,5]	117	27.4	25	900	25.8	446	288	367	5.5	–	30,000
111	Sinker Creek[1,2,4,5]	3.33	21.34	21.34	–	21.3	92	49.2	70.6	–	84,100	926

No.	Dam											
112	South Fork[1,2,5]	0.0037	1.83	1.8	300	24.4	128.1	38.1	94.5	–	–	122
113	Spring Lake[1,2,4]	0.136	5.49	5.49	–	5.49	20	9	14.5	–	612	–
114	Statham Lake[1,2,4]	0.564	5.55	5.5	–	5.12	–	–	21	–	1350	–
115	Stevens Lake[1,2,5]	0.0789	4.27	11	–	11	12.2	2.4	7.3	–	–	5.92
116	Swift Dam No.2[1,2,4,5,a]	37	47.85	57.61	226	57.6	225	225	225	–	206,000	24,947
117	Taum Sauk[2]	5.39	31.46	–	250	–	–	–	225	–	–	7743
118	Teton[1,2,3,4,5]	310	77.4	92.96	500	86.9	237.9	64.1	151	4	3,060,000	65,120
119	Tiemusi[1,2]	0.11	12	12	–	12	60	–	–	–	–	–
120	Tongshuyuan[1,2]	0.4	10	13	–	10	30	–	–	–	–	–
121	Trial Lake[1,2,4]	1.48	5.18	9.5	–	5.18	–	–	21	–	829	–
122	Upper Pond[1,2]	0.222	5.18	5.2	–	5.18	25.4	7.6	16.5	–	–	–
123	Wanshangang[1,2]	1.5	12	13	–	12	50	30	40	–	–	–
124	Winston[1,2,4]	0.662	6.4	7.32	133	6.1	21.3	18.3	19.8	–	1480	–
125	Wetland Number[1,2,4,5]	11.6	12.2	13.6	–	13.7	46	41	35.4	–	14,600	566.34
126	Wilkinson Lake[1,2,4]	0.533	3.57	3.2	–	3.72	35.5	22.5	29	–	1420	–
127	Yuanmen[1,2]	6.4	19.2	19.2	–	19.2	–	–	–	–	–	–
128	Zhonghuaju[1,2]	0.14	16	16	–	16	–	–	–	–	–	–
129	Zhugou[1,2,3,5]	18.43	23.5	23.5	300	23.5	159	110	135	0.43	–	11,200
130	Zuocun[1,2]	40	35	35	–	35	–	–	–	–	–	23,600

[a]Rockfill dams are considered as outliers.

V_w reservoir storage volume above breach ($\times 10^6\,m^3$); H_w height of water above breach (m); E_l embankment length (m); t_f failure time (hr); V_{er} eroded volume (m^3); H_b height of breach (m); B_t top width of breach (m); B_b bottom width of breach (m); B_{ave} average breach width (m); Q_p peak outflow discharge (m^3/s).

2.3 Multivariate regression analysis

The relationship between a dependent variable and a set of independent variables in multivariate linear regression (MLR) is analyzed simultaneously, whereas the considered model expresses a linear relationship in terms of the model parameters. It is necessary to calculate the correlation coefficient in the MLR models. The relation between the independent and dependent variable can be expressed as Eq. (1):

$$y = \beta_0 + \beta_1 x_{i1} + \cdots + \beta_j x_{ij} + \varepsilon, \ i = 1,2,\ldots n, \ j = 0,1,\ldots,m \tag{1}$$

where y is the dependent variable, x_i is the independent variable, and β_j is the model parameter.

The goal in multivariate nonlinear regressions (MNLR) is to fit randomly one of the existing functions on the collected data to obtain the most efficient one. The MNLR models can support various functions including exponential, logarithmic, power, etc. but the power function is the most suitable type for the dam break problem. The relation between the variables can be expressed as Eq. (2):

$$y = \alpha x_1{}^\beta x_2{}^\gamma \ldots x_i{}^\omega, \ i = 1,2,\ldots n \tag{2}$$

where y is the dependent variable, x_i is the independent variable, and α, β, γ, ... are the constant values.

Multivariate regression analyses for the earthfill failures in the current study are performed using the IBM SPSS Statistics 22 package. Similar to GEP in the present study, 75% of the total available data were used for training the model and the rest for validation steps.

2.4 Assessment of performance indicators

The root mean square error (RMSE) is a measure of absolute error between the observed and simulated values. This statistical index varies from 0 to $+\infty$, and the smaller values have the better simulation results. The Nash-Sutcliffe efficiency (NSE) is the normal state of the least-squares error function, which indicates the ratio between residual variance and the variance of the values for assessing the efficiency of current models (Nash and Sutcliffe, 1970). The range of NSE varies from $-\infty$ to $+1$, and the optimum value is $+1$. According to Gassman et al. (2007), if this coefficient is higher than 0.5, the model has a good simulation. The coefficient of determination (R^2) is an indicator used to evaluate the accuracy of the models, which indicates the correspondence between the observed and predicted values. This index also represents a fraction of the total variance of observational values justified by the simulated values. The indices of RMSE, NSE, and R^2 can be expressed as:

$$\text{RMSE} = \sqrt{\frac{1}{N} \sum_{i=1}^{n} \left(Q_{s,i} - Q_{o,i} \right)^2} \tag{3}$$

$$\text{NSE} = 1 - \frac{\sum_{i=1}^{n} \left(Q_{s,i} - Q_{o,i} \right)^2}{\sum_{i=1}^{n} \left(Q_{o,i} - \overline{Q_o} \right)^2} \tag{4}$$

$$R^2 = \frac{\left[\sum_{i=1}^{n} \left(Q_{s,i} - \overline{Q_s} \right) \left(Q_{o,i} - \overline{Q_o} \right) \right]^2}{\sum_{i=1}^{n} \left(Q_{s,i} - \overline{Q_s} \right)^2 \sum_{i=1}^{n} \left(Q_{o,i} - \overline{Q_o} \right)^2} \tag{5}$$

In the above relations, $Q_{s,i}$ is the simulated discharge, $Q_{o,i}$ is the observed discharge, $\overline{Q}_s$ is the average of simulated discharge, $\overline{Q}_o$ is the average of the observed discharge, i is the time step, and n is the total number of time steps.

2.5 Bayesian approach

The Bayesian network is a graphical model for expressing probabilistic relationships between variables. Therefore, all of the observations and parameters of a statistical model are considered as the random quantities. This network is a set of nodes connected by the direct links and a probability function is assigned to each node. The network development is based on the use of conditional probabilities and Bayes theory (Liu et al., 2019). In the present study, the GeNIe 2.1 software package is used for the training and validation steps among laboratory and real-world failure cases, respectively. The Bayes theory states that if F and E are two assumed events such that P(A) $\neq$ 0 and P(B) $\neq$ 0, the theory can be expressed as Eq. (6):

$$P(A/B) = \frac{P(B/A)P(A)}{P(B)}$$

(6)

where $P(A/B)$ is the posterior density function; A and B are the vectors of the unknown parameters and the measured data.

2.6 Wavelet analysis

The wavelet analysis is a mathematical tool like Fourier analysis that represents a signal with functions bounded in frequency, but these functions are also bounded in time in contrary to Fourier analysis (Letelier and Weber, 2000). It is also a time-dependent spectral analysis that unravels time series in the time-frequency space to describe the processes and relationships based on the time scale. Therefore, the main advantage of wavelet transforms is the ability to obtain information on the time, location, and frequency of a signal simultaneously (Adamowski and Chan, 2011). This tool is an example of data preprocessing, which was applied for de-noising, compression, and decomposition of input data time series. Time-series preprocessing made by a wavelet-based de-noising approach was used in the present study to improve the prediction accuracy.

2.7 Gene expression programming (GEP)

GEP is a new evolutionary algorithm, first invented by Ferreira based on the genetic algorithm (GA) and genetic programming (GP). GEP incorporates both the idea of a simple linear chromosome of a fixed length used in GAs and the tree structure of different sizes and shapes used in GP. The primary difference between GEP and its predecessors, GAs and GP, stems from the nature of the individuals: in GAs, the individuals are linear strings of fixed length (chromosomes). In GP, the individuals are nonlinear entities of different sizes and shapes (parse trees). In GEP, the individuals are encoded as linear strings of fixed length (chromosomes) that are expressed as nonlinear entities of different sizes and shapes. It is able to create trees indirectly, by encoding the mass vectors of symbols and translating them into trees only in order to evaluate their fitness. This allows simple genetic operators, as found in GAs, while evolving complex and expressive trees, as GP does. It is also justified biologically in what Ferreira calls the "phenotype barrier", where the genotype must be expressed as a more complex structure in order to have an effect on the environment (Ferreira, 2002, 2006). The EC techniques are useful when the search space is large and complex, and are based upon the Darwinian evolution principle, which suggests that populations evolve through inheritance where a concept of fitness reflects the population's ability to survive. GEP was first introduced to the GP community by Ferreira. Thus it is the most recent development in the field of artificial evolutionary systems. GEP starts with allocating a fixed-length chromosome to the randomly generated initial population. Then the chromosomes are explicitly expressed, and each individual's fitness is evaluated. The individuals with high fitness are selected to improve the solution. This process is iterated for a prespecified number of generations or until an "optimal" solution has been found (Dehghani, 2018; Ferreira, 2001).

In this study, GeneXproTools 5.0 software was used to determine the breach and outflow characteristics of embankment and sliding dams using the different data resources and statistical populations. In the beginning, the data is normalized in the range of 0 and 1. Then, the operators ($\times$, /, and ˆ) was identified as suitable link functions between parameters involved in the dam break problem. Based on the current research, the best results can be obtained by selecting the multiplication ($\times$) operator as a linking function, chromosome length equal to 30, and 3 genes per chromosome. In GEP, similar to nonlinear regressions, the data were divided into training and validation steps, which is equivalent to 75% and 25% of the total data, respectively. By employment of the MATLAB 2016b software in the present study, equal amounts of large and small datasets were used in the training and validation steps.

2.8 Physical models

Experimental studies have been performed to investigate the breach mechanisms and resulting output hydrograph in mam-made embankment dams. The laboratory includes a dam hydraulic research Centre located 45 km west of Tabriz, Iran. Accordingly, the experimental setup consists of a rectangular flume with glass walls and a smooth cemented bed with the Manning roughness coefficients of 0.01 and 0.013, respectively. The flume has also a length, width, and height of 3.2, 1.5, and 0.6 m, respectively. To study the erosion process in the dam section more accurately, the flume glasses were painted in a 0.05 × 0.05 m grid. The upstream reservoir also consists of a rectangular channel with length, width and height of 10, 2, and 1.2 m, respectively (with an operating height of 0.6 m). A side (supporting) storage reservoir was also used to regulate the inflow to the upstream. Therefore, two Tavan-TMR4 submersible pumps with a maximum discharge of 6 L per second were utilized to adjust and distribute the inflow evenly upstream of the physical models. In general, a constant inflow rate is considered during the testing process. A V-notch weir with a thickness of 0.01 m was installed in the downstream basin to route the flood and corresponding outflow discharge. The steel weir was cut with CNC and the thickness was reduced to 0.0018 m along with the V-shaped edges. Additionally, the instantaneous breach outflow discharge (Q_t) was calculated using Eq. (7) (Chanson and Wang, 2013):

$$Q_t = \frac{8}{15} C_d \sqrt{2gH^5} \tag{7}$$

where C_d is the dimensionless discharge coefficient equals 0.577, g is the gravitational constant, and H is the water level above the weir.

To review the erosion mechanisms in offline mode, 4 Full-HD digital cameras were placed in fixed positions. Fig. 1 shows a side view of laboratory equipment.

In this study, 15 different physical models were constructed and breached in the laboratory at different heights and grain size distributions. So, output hydrograph components, outflow and breach characteristics were carefully extracted. Each experiment was performed twice to ensure reproducibility. A comparison of the results of the same experiments were performed in equal conditions indicated that hydrograph components, erosion mechanisms, and breach characteristics are similar. To reach the standard compaction and optimum humidity, the soil layers of physical models were compacted with a thickness of 0.05 m by hand rollers in accordance with the standard Proctor test. According to investigations, increasing soil compaction from 95% to 102% in a standard Proctor test can reduce the erosion rate by 50% (Hanson et al., 2005). Otherwise, the occurrence of progressive backward erosion during the erosion process will be inevitable. In all of the laboratory models in the current study, the width of the crest is 0.1 m, and the upstream and downstream slopes (1 vertical: 2.5 horizontal) are considered. Referring to the previous studies, slope reduction in both upstream and downstream sides will not have a significant effect on the Q_p values, as well as the output hydrographs (Walder et al., 2015).

To determine the initiation point of erosion in the dam crest, it will be necessary to consider a pilot channel in the middle of the crest. The length, width, and depth of the pilot channel are 0.1, 0.025, and 0.025 m, respectively. The physical models were made of five soil grain size distributions in three heights of 0.3, 0.4 and 0.5 m. The gradation curves of the experimental models are shown in Fig. 2A. Fig. 2B also shows the results of the Direct Shear Test (ASTM-D3080). In the present work, about 16 tons of dry grain materials were passed through laboratory sieves and weighed with a precision of 10 g and mixed again. Table 2 considers the technical properties of the materials along with the dimensions of the laboratory samples.

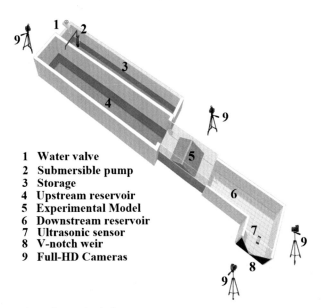

1 Water valve
2 Submersible pump
3 Storage
4 Upstream reservoir
5 Experimental Model
6 Downstream reservoir
7 Ultrasonic sensor
8 V-notch weir
9 Full-HD Cameras

FIG. 1 Schematic side view of the experimental setup (scaled).

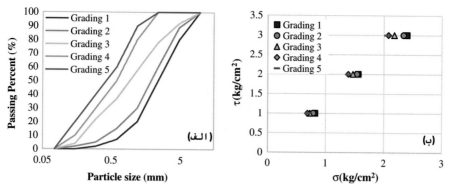

FIG. 2 (A) Gradation curves of the physical models, (B) results of the direct shear test.

TABLE 2 Technical properties of the experimental models.

Test no.	Grading no.	H_d (m)	V_w (m³)	ϕ (°)	C (kg/cm²)	D_{50} (mm)
1	1	0.3	6.3525	38.3	0.02	2.3600
2	2	0.3	6.3525	37.6	0.02	1.9666
3	3	0.3	6.3525	35.8	0.02	0.9670
4	4	0.3	6.3525	34.6	0.00	0.6000
5	5	0.3	6.3525	34.2	0.02	0.4500
6	1	0.4	8.42	38.3	0.02	2.3600
7	2	0.4	8.42	37.6	0.02	1.9666
8	3	0.4	8.42	35.8	0.02	0.9670
9	4	0.4	8.42	34.6	0.00	0.6000
10	5	0.4	8.42	34.2	0.02	0.4500
11	1	0.5	10.4625	38.3	0.02	2.3600
12	2	0.5	10.4625	37.6	0.02	1.9666
13	3	0.5	10.4625	35.8	0.02	0.9670
14	4	0.5	10.4625	34.6	0.00	0.6000
15	5	0.5	10.4625	34.2	0.02	0.4500

Where, ϕ is the internal friction angle (°), C is the cohesive strength of materials (kg/cm²), and D_{50} is the mean grain size (mm).

2.9 BREACH mathematical model

The BREACH mathematical model is one of the most common models available for evaluating outflow, hydraulic, and breach characteristics in embankments and sliding dams. The model promotes the progressive shape of the breach, the failure time, the outflow hydrographs, and the tailwater level. In overtopping, the flow into the channel is calculated by the broad-crested weir formula or by performing a mass balance in the upstream reservoir. While the flow into the pipe is calculated by the orifice flow formula and time-dependent reservoir inflow rate (Singh, 2013). To start modeling due to overtopping, the water level in the reservoir must be above the crest before the erosion begins (Hassanzadeh, 2005). The model is slightly sensitive to numerical parameters and can predict the breach initiation and subsequent deformations empirically. However, its sensitivity to the internal friction angle (ϕ), cohesion (C), and median grain size of particles (D_{50}) related to the materials of the body and the core of the dam are high. The model also assumes that the erosion rate at the breach bottom is equal to that on both sides of the breach (Hassanzadeh, 2005; Zhao et al., 2019).

Landslide dams, which are mostly located in mountainous areas and far from the monitoring facilities, are part of natural landslides with heterogeneous rock and soil materials. Due to the nature of their low density, the dams are usually less resistant than embankments. Generally, erosion and permeability in such dams are very different and often more than man-made embankments. Considering the same reservoir capacity, the length of the crest for landslides are often longer compared to the man-made embankments, as well as the mass of the body. Therefore, soil materials need more time for erosion because of the flatter and longer breach hydrograph. The failure of these dams is largely influenced by several important factors including H_d, storage volume, and dam erodibility (Shi et al., 2015; Zhang et al., 2010).

3. Landslide natural dams

Overtopping and piping are the most common causes of embankments failure while overtopping is the most important cause of landslide dam failures. According to the technical literature, the overtopping failure rates of earthen and landslide dams are 58% and 91%, respectively. On the other hand, these ratios for piping failures will be 37% and 8%, respectively (Peng et al., 2014; Shen et al., 2020), which indicates the low susceptibility of landslide dams to the internal erosion. Since the downstream and upstream faces of the dams are often gentle, the erosion rate will be slower than in the man-made embankments. On average, the application of experimental models to embankments and landslide dams leads to conservative estimates of more than 60% and 200%, to determine breach characteristics and Q_p, respectively. While the required time for the erosion of man-made embankments will be more than 2.76 times of the relationships provided to determine the t_f for the earthfill dams (Peng and Zhang, 2012; Shen et al., 2020). Fig. 3 compares the influences of S and H_d on the Q_p values from the failure of landslide dams and man-made embankment dams.

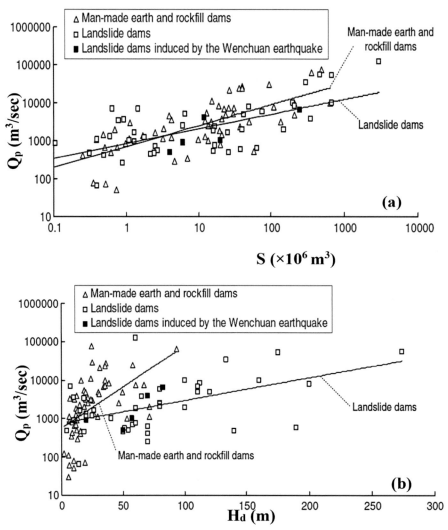

FIG. 3 Comparison of the influences of S and H_d on the Q_p values from the failure of landslide dams and man-made embankment dams.

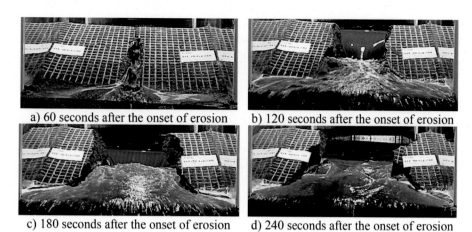

a) 60 seconds after the onset of erosion   b) 120 seconds after the onset of erosion

c) 180 seconds after the onset of erosion   d) 240 seconds after the onset of erosion

FIG. 4   Breach formation through an experimental model with a height of 0.50 m. (Experiment No. 14)

4.   Results and discussion

4.1   Experimental findings

As stated, laboratory results have been used in this research to study the evolution of erosion mechanisms. Fig. 4 shows the processes of a physical model with a height of 0.5 m and grading No. 4 (Experiment No. 14) in one-minute time steps. According to observations, the erosion process starts when the tensile forces are greater than the forces that keep the material in its primary position. Rill erosion is also a kind of erosion in which the detachment and movement of soil particles can be affected by the concentrated flowing water. This kind of erosion has an important role in the process of breach growth, as described by Mosaedi and Hosseini (2015). So, by migration of the downstream slope to upstream, the erosion process expands downward in addition to the lateral expansion, which confirms the findings of Hahn et al. (2000), Hanson et al. (2001), and Walder et al. (2015). Following the first breach in the surface, the formed cavities begin to collapse due to regressive erosion.

The process of erosion cavity expansion is determined by the tailwater conditions and the soil mechanics. Thus, the stability of the soil particles depends largely on their mechanical properties and the strength of the bottom side of the scour. Migration during headcut erosion can also lead to a series of the discrete mass failure events (Hassanzadeh, 2005). Lateral expansion of breach toward the sidewalls of the experimental flume is associated with particle separation and the collapse of large masses into the flow. Observations show that this event is related to the mechanical properties of the soil. As the particle diameter decreases, the eroded volume of the masses separated from the body of the models increases. Fig. 5 shows

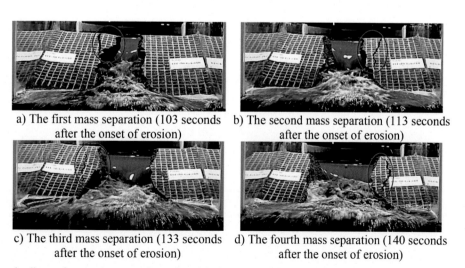

a) The first mass separation (103 seconds   b) The second mass separation (113 seconds
after the onset of erosion)   after the onset of erosion)

c) The third mass separation (133 seconds   d) The fourth mass separation (140 seconds
after the onset of erosion)   after the onset of erosion)

FIG. 5   Separation and collapse of masses in an experimental model with a height of 0.50 m (Experiment No. 14).

the mass separation in grading No. 4 (Experiment 14) on different time periods. Based on the observations, the largest mass collapse occurs in the first 2 min of erosion for all of the models, and then the erosion continues with the evolution of lateral expansion. Sensors installed downstream of all laboratory models, have recorded the insignificant variations in the flow rate passing through the breach.

4.2   Simulation of the breach characteristics

The final width of the breach and the rate of widening can significantly affect the output Q_p and downstream runoff. In the present study, to determine B_{ave} based on the available database, it was found that 48.93% of the data are in the range provided by Johnson and Illes (1976) ($0.5H_d \leq B_{ave} \leq 3H_d$). While these values are in the range recommended by Singh and Snorrason, 1984 was equal to 48.8% ($2H_d \leq B_{ave} \leq 5H_d$), on the basis of the datasets extracted from four main resources of the dam breach data available in the technical literature (Wahl, 1998, 2014; Xu and Zhang, 2009; Zhang et al., 2016). The B_{ave} values in more than 92% of the studied data are equal to 120.6% of the B_b, while this ratio is equal to 63.6% of the B_t in more than 80% of the studied data (Kouzehgar et al., 2021c). These results are also consistent with laboratory findings. Based on the field observations of the present research and review of historical data, it was determined that B_{ave} is in the following range of the H_d:

$$0.71H_d \leq B_{ave} \leq 7.86H_d \tag{8}$$

Fig.6 shows the rate of instantaneous development of the breach geometry in the constructed models. Evolution of H_b as a result of approaching B_b to the foundation level along with high flow rate through the breach, in addition to causing lateral erosion, can lead to erosion and shear surfaces on both sides of the breach in the vicinity of the bed. The direction and intensity of this eroded surface are high along the width of the crest. So, the continuation of this situation can cause breach widening, sidewall collapse, and increasing the flow concentration in the falling limbs of the hydrograph.

Based on a study on the effect of E_1 on the B_{ave}, it was found that the values of B_{ave} in rockfill dams provide the higher values compared to the earthfill dams. So that in the Swift rockfill dam, the B_{ave} value was approximately equal to 99% of E_1. In embankments, the B_{ave} is recommended in the following range of the E_1:

$$0.04E_l \leq B_{ave} \leq 0.77E_l \tag{9}$$

In a provided database (Zhang et al., 2016), the erosion of historic dams is considered. An analysis of more than 46 records of overtopping and piping failures conducted in the present work shows that the ratio of H_b to H_d in dams with high erodibility (HE) is in the range of 0.91 to 1.23, while this ratio for the dams with medium erodibility (ME) varies between 0.74 and 1.125 and for the dams with low erodibility (LE) varies between 0.54 and 1. The recent ratios show that the foundation is also affected by the rate of two-phase flow that is more severe in HE dams, as well as the base erosion. The ratio of the B_{ave} to H_d is in the range of 1.129–11.87 in HE dams. Whereas, this ratio varies between 1.24 and 5.99 in ME, and differs between 0.7 and 4.66 in LE dams, indicating that the B_{ave} is a determinant factor in the erosion process. In the embankments

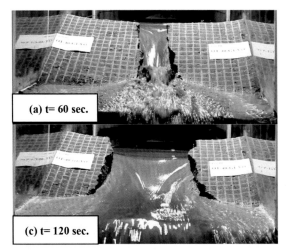
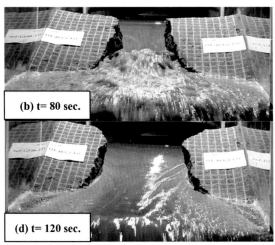

FIG. 6   Instantaneous development of the breach geometry (Experiment No. 6).

with lower H_d, the sharp slope of the walls can be attributed to the effects of cohesion. The relationship derived from laboratory observations and exploration of the technical literature indicates that B_{ave} is in the following range:

$$0.8H_w \leq B_{ave} \leq 10.4H_w \tag{10}$$

The study performed on the dimensionless ratios of historic embankment failures indicated that except for the cross-section of the Swift rockfill dam, most of the erodible sections were trapezoidal with a steep slope of sidewalls. Basically, the dimensionless ratios of B_{ave}/B_t and B_{ave}/B_b for embankments are suggested as follows:

$$0.522 \leq B_{ave}/B_t \leq 1.55 \tag{11}$$

$$0.9 \leq B_{ave}/B_b \leq 2.9 \tag{12}$$

While the mentioned ratios for landslide dams will be as follows:

$$7 \leq B_{ave}/B_t \leq 16 \tag{13}$$

$$18.75 \leq B_{ave}/B_b \leq 55 \tag{14}$$

The H_b is one of the important parameters in assessing the flow rate and erodibility of the body and subsequent sediment transformation downstream. The studies show that this depth has a direct effect on the outflow hydrograph. Since the piping erosion ultimately leads to overtopping, accurate assessment of the H_b can be used to estimate the downstream damages, as well as a number of the population at risk (PAR). Based on the studies performed on the failure database of 1443 and 1044 man-made embankments and landslide dams (Zhang et al., 2016), respectively, the following H_b/H_d ratio is proposed for the dams, respectively:

$$0.544 \leq H_b/H_d \leq 1.23 \tag{15}$$

$$0.2 \leq H_b/H_d \leq 1.755 \tag{16}$$

Findings also show that in most of the landslide dams, the eroded volume (V_{LS}) exceeds the volume of the dam body (V_D) material and covers the side separations of the dam material. Accordingly, from the total 110 records of the available data, $V_{LS} > V_D$ and $V_{LS} = V_D$ in 56.36% and 30% of cases, respectively. In landslide dams, the sidewalls seem to have a gentler slope than the man-made embankments. This can be related to the high values of E_l, t_f, and relatively high erodible volume of this type of dam. It should be noted that in the ratios presented in the above relations, most of the records in Table 1 (marked with superscript [1]) have been used.

Similar to B_{ave}, in determining H_b as a function of the flow passing through the breach, and using the GEP and various data sources that marked as superscript [2] in Table 1, the following relationship is proposed as:

$$H_b = 1.061584H_w \tag{17}$$

4.3 Simulation of the failure time

Failure time (t_f), is the time from the initiation of the first breach at the upstream of the crest to its final evolution. The t_f is considered as one of the most important characteristics of the failure parameters and plays an important role in the emergency action plan (EAP) as well as the evacuation of the downstream residents. This time was studied by the various researchers and several relations were presented that considers the time as a function of different parameters such as breach geometry, dam reservoir, etc. But there is no uniformity exists in the selection of recent parameters.

In this work, the relationship between the t_f, breach and hydraulic characteristics was investigated on all of the existing records, separately and simultaneously. According to findings the dependence of the t_f on an individual parameter cannot be assumed for the accurate evaluation. The lowest dependence of this parameter on V_{er} and the highest was related to the combined effect of H_w and V_w. Therefore, Eqs. (18) and (19) obtained from the nonlinear regression can create the most convergence among the existing data sets. Fig. 7 also shows the statistical distribution of recent relationships.

$$t_f = 0.001\, V_w^{0.967} H_w^{-0.781} B_{ave}^{-1.426} \tag{18}$$

$$t_f = 0.001\, V_w^{0.896} H_w^{-0.977} H_b^{0.332} B_{ave}^{-1.317} \tag{19}$$

The independent values of R^2, NSE, and RMSE obtained from Eq. (18) are 0.8688, 0.7779, and 0.6529, respectively, while these values for Eq. (19) are 0.8773, 0.8394, and 0.5553, respectively. The records used for the obtainment of the recent relations are marked as superscript [3] in Table 1.

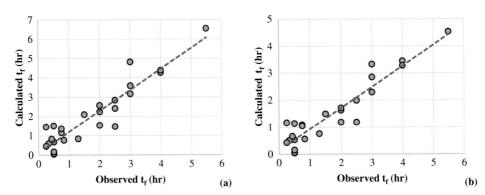

FIG. 7 Dispersion curves of observed versus calculated values of t_f (A) Eq. (18), (B) Eq. (19).

4.4 Simulations of the eroded volume of the dam

The breach dimensions increase steadily during embankment failure as the flood displaces the materials. Hence, the geometry and development of the breach are a function of reservoir and flow characteristics. On the other hand, the determination of eroded volume (V_{er}) resulting from dam failure can provide the valuable information in assessing downstream environmental impacts. Evaluation of two-phase flow transportation due to the erosion along with the high volume of sediments accumulated in the reservoir resulting from the senility of most of the earthen dams is important. The impacts of concentration of the harmful minerals as well as reservoir sediments on the quality of surface and groundwater should be taken into consideration in addition to financial and human damages. Several studies have been carried out on the mechanisms of erosion and determination of the eroded volume from the dam body. Accordingly, Eq. (20) was developed by MacDonald and Langridge-Monopolis (1984) for earthfill dams. It should be emphasized that an embankment can be eroded laterally by flow continuity during the breach. The lateral erosion rate for embankments with similar geometry for the heights (in the range) of 3–9 m can be assessed based on investigations conducted by Pugh (1985), as Eq. (21).

$$V_{er} = 0.0261 \left(V_w H_w\right)^{0.769} \tag{20}$$

$$L_{er} = 13.2 H_d + 150 \tag{21}$$

Based on 66 historical case studies, the values of R^2, NSE, and RMSE for Eq. (20) are equal to 0.65, 0.23, and 18,297.59, respectively.

In the following, 71 records of specified historical data derived from Table 1 were used to study the mechanisms and volume of transported material during erosion. Ultimately, the geometry and the ratio of V_w/H_w in the prototypes were compared with small-scaled laboratory models. The dam material is mainly eroded due to the transport capacity of flow passing from the overtopping surface. In this regard, Eqs. (22) and (23) obtained by GEP and MNLR, respectively, are proposed. Statistics show that recent equations can be used as a suitable tool for estimating downstream risks. The data used to obtain the below relations are marked as superscript [4] in Table 1.

$$V_{er} = 1.074366 \, H_w^{\,2} V_w^{\frac{1}{3}} \tag{22}$$

$$V_{er} = 0.9 \, H_w^{\,2.461} V_w^{\,0.258} \tag{23}$$

4.5 Simulation of the peak outflow discharge

In the study of maximum outflow during a dam failure that causes a great damage to the downstream residences and the existing design should be based on its criteria, it is necessary to cover a wide range of dam failures from a combination of different data sources. Using multiple datasets can cover the deficiencies of each method and provide a proper justification for the erosion process.

The first group contains historical data that exist in the literature, the actual records of embankment failures around the world, and extracted from many geographical ranges. These data give us valuable information on the dimensions and processes of erosion but they are limited and often incomplete due to the lack of proper data collection facilities in the event of dam failure. The failure data of embankment dams used in the current section are specified as superscript [3] in Table 1. The second group includes the study of hypothetical failures of real embankments that were collected from the existing prototypes located in the northwest of Iran and distributed in three Provinces of Ardabil, East, and West Azerbaijan. Table 3 provides brief specifications of 10 embankment dams that were collected as case studies. The study of hypothetical dam failures can help for the better estimation of the outflow and breach characteristics. However, these simulated data are synthetic and limited, like literature records, they almost have the similar technical features. Prediction of outflow discharge resulting from a hypothetical dam failure first requires estimating the probable maximum flood (PMF) of the upstream basin and related inflow hydrograph. This inflow hydrograph can either be estimated using a hydraulic numerical model or a software package. In the current work, this hydrograph was also obtained by the HEC-RAS package and was introduced to the BREACH as an inflow hydrograph. Hereafter, the breach characteristics and the outflow hydrograph were extracted using the BREACH mathematical model and application of the existing empirical relationships. The BREACH model has some limitations due to the simplifications it makes in simulating the erosion process and sediment transport. The breach is parallel to the dam crest and the downstream slope during the failure process (Zhao et al., 2019). The third group of data includes the physical tests described in the "Experimental setup" section. Accordingly, these small-scale structures are the simplified models of the prototypes and also follow their technical specifications. The models were constructed and over-topped in the laboratory along with the breach characteristics derived as well as flow parameters.

To use the datasets of three groups, it is necessary to use the data homogeneity test. Therefore, Levene's test was performed, which applies to three or more statistical populations. Analyzing the variances by the recent method can help significantly to understand the differences between statistical groups. To this end, the One-Way ANOVA tool was employed for the investigation of homogeneity of variances as well as analyzing the quantitative variables in the SPSS Statistics software. The results of the null hypothesis of equal variances showed a good relationship between the three groups of the statistical population ($P = 0.06 > 0.05$) and confirmed the homogeneity of the dataset. The GEP and MNLR models were employed for the development of empirical relationships as well as estimation of the main breach and flow characteristics, while data from the above three categories were divided into the training and validation processes. Table 4 provides general information on the statistical properties of data used in the training and validation steps. Although various parameters are involved in the breach and flow characteristics during an earth dam failure, the GEP model can automatically select the input variables that have the most impact on the model. Hence, Eqs. (24) and (25) obtained from GEP and MNLR, respectively, consider the Q_p as a function of breach dimensions. Based on the performance indices shown in Fig. 8 and Table 4, the proposed equations can help to reach a consequence and gain a better result in the simulation of flow and breach characteristics, volume transformations, erosion process, and sediment transport. The studied datasets for the below relations are marked as superscript [5] in Table 1.

TABLE 3 Names and specifications of the embankments that used as case studies.

No.	Dam name	River	H_d (m)	S ($\times 10^6$m^3)
(Ardabil Province)				
1	Alamdar	Aras	22	0.32
2	Khanghah	Bafrajard	30	6.9
3	Sorkhab	Aras	15	1.5
(East Azerbaijan Province)				
4	Amand	Ajichay	24	2.4
5	Ardalan	Chakichay	19	4.5
6	Param	Paramchay	34	3.6
7	Zonouz	Zonouzchay	60	6
(West Azerbaijan Province)				
8	Danalou	Sarisu-Aras	37	2.3
9	Jaldian	Jaldian	33	1.2
10	Kansepi	Nazlou	26	0.52

H_d Dam height; (m), S Reservoir capacity ($\times 10^6$m^3).

TABLE 4 Statistical distributions of the breach and flow characteristics (prototype/model).

Training data	H_b (m)	B_{ave} (m)	Q_p (m³/sec)
Max.	56.4000	186.0000	8266.3033
Min.	0.2580	0.7563	0.0159
Ave.	13.9376	45.7787	1524.0181
Std.	12.4472	46.2156	2047.9638
CV	0.8930	1.0095	1.3437
Validation data			
Max.	35	168	7200
Min.	0.2900	0.6413	0.0186
Ave.	12.9393	47.6067	1448.93
Std.	11.0742	47.8500	1959.8154
CV	0.8558	1.0051	1.3525

Max., maximum; *Min.,* minimum; *Ave.,* average; *Std.,* standard deviation; *CV,* coefficient of variation; *H_b,* height of breach (m); *B_{ave},* average breach width (m); *Q_p* = peak outflow discharge (m³/sec).

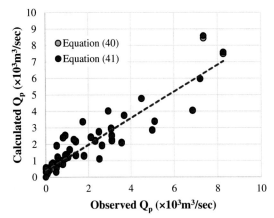

FIG. 8   The scatter plot of observed and calculated values of Q_p obtained by various Eqs. (24) and (25).

$$Q_p = 21.9522 H_b B_{ave}^{0.4} \tag{24}$$

$$Q_p = 18.461 H_b^{1.038} B_{ave}^{0.408} \tag{25}$$

4.6   Simulation of hydrograph resulting from dam failure

A hydrograph is a continuous time-based function that can be correlated and displayed statistically. In this study, using soft calculations and failure databases of 15 physical models constructed in the laboratory, the breach outflow hydrograph of the Redrock Dam has been simulated, as well as the artificial breach of Aydoghmoush embankment dam located in Mianeh, Iran, due to overtopping. Therefore, data obtained from laboratory models have been used as input in soft computing. It should be noted that the data of Redrock and Aydoghmoush embankment dams are also obtained from the BREACH mathematical model by introducing the inflow hydrographs to HEC-RAS software. On the other hand, the simulation results are compared by several methods including GEP, Bayesian, nonlinear and linear regressions with the BREACH mathematical model.

Furthermore, the effectiveness of data-driven methods depends largely on the quality of data production and accuracy in testing. In the time series, reducing or eliminating noise from the generated data is important to reduce input errors. For this

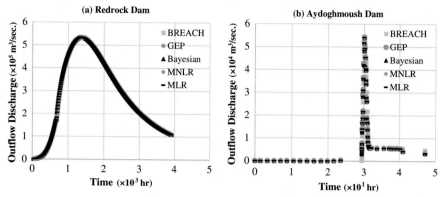

FIG. 9 Comparison of simulated embankments using different methods: (A) Redrock dam, (B) Aydoghmoush dam.

purpose, the time series of data generated in the laboratory have been de-noise by the wavelet transform before being introduced to the training stage. The wavelet transform, due to the nature of this type of time series is more appropriate in the hydrological processes. The method minimizes the problems associated with hydrograph analysis (Kouzehgar et al., 2021e). Fig. 9 shows the simulation results of recent dams.

A comparison of the results obtained from the simulated hydrographs shows a high correlation between the present models. GEP and nonlinear regression models are more conservative in estimating Q_p than other models, while Bayesian and linear regression models are more consistent in the rising and falling limbs of the hydrograph.

5.   Conclusions

Noncohesive physical modeling with different technical specifications and soil grain particles were conducted to study the breach mechanisms due to overtopping. Although the mechanical properties of the used materials are not always available in real-world failures, dam material erosion is the predominant mechanism for breaching. Therefore, the erosion rate needs lesser time to evolve in coarse particles due to lower critical shear stresses than fine-grained soils. In the following, the output hydrographs of laboratory samples were used as input to various models of artificial intelligence and regression, and the output hydrographs of the breached embankments were effectively simulated. Moreover, AI methods overestimate the hydrograph values compared to the BREACH mathematical model.

In this study, several relationships are presented using a comprehensive database of historical failures to determine breach mechanisms and output hydrograph based on the artificial intelligence and regression models. The values of Q_p due to overtopping and piping failures depends on $H_w^\alpha V_w^\beta$, H_w^γ and V_w^δ, respectively, while α, β, γ, and δ are numerical constants. Soil properties such as erodibility and strength as well as compaction of an embankment will play an essential role in the evolution rate of the erosion, breach geometry, and the resulting runoff at downstream. Soil gradations play an important role in increasing the rate of erosion and reducing breach formation and evolution time, in granular embankments. Therefore, breach formation in soils with higher D_{50} will take less time to evolve due to lower shear stresses. On the other hand, surface erosion that causes the soil layers to erode has a more significant impact than headcut erosion in noncohesive materials. The average breach width (B_{ave}) has an important effect on the output hydrograph and its expansion rate. So, the variation range will be in the range of $0.8H_w$ and $10.4H_w$. Based on observations, the B_{ave}/B_t ratio is equal to 0.816, while the B_{ave}/B_b ratio is 0.206 in the most historic failures, which implies that the final breach geometry is trapezoidal with the sharp side walls (1 Horizontal:Vertical).

Height of the breach (H_b) is an essential issue in the erodibility of the dam body and the transfer rate of sediment to the downstream. The comparison of the dimensionless ratio of H_b/H_d between the man-made embankments and natural landslide dams show higher values of the current ratio in the landslide dams which might be caused by the long breach formation time in these dams.

References

Adamowski, J., Chan, H.F., 2011. A wavelet neural network conjunction model for groundwater level forecasting. J. Hydrol. 407, 28–40. https://doi.org/10.1016/j.jhydrol.2011.06.013.

Ashraf, M., Soliman, A.H., El-Ghorab, E., Zawahry, A.E., 2018. Assessment of embankment dams breaching using large scale physical modeling and statistical methods. Water Sci. 32, 362–379. https://doi.org/10.1016/j.wsj.2018.05.002.

Chanson, H., Wang, H., 2013. Unsteady discharge calibration of a large V-notch weir. Flow Meas. Instrum. 29, 19–24. https://doi.org/10.1016/j.flowmeasinst.2012.10.010.

Chinnarasri, C., Jirakitlerd, S., Wongwises, S., 2004. Embankment dam breach and its outflow characteristics. Civ. Eng. Environ. Syst. 21, 247–264. https://doi.org/10.1080/10286600412331328622.

Coleman, S.E., Webby, M.G., Andrews, D.P., 2004. Closure to "overtopping breaching of noncohesive homogeneous embankments". J. Hydraul. Eng. 130, 374–376. https://doi.org/10.1061/(ASCE)0733-9429(2004)130:4(374).

De Lorenzo, G., Macchione, F., 2014. Formulas for the peak discharge from breached earthfill dams. J. Hydraul. Eng. 140, 56–67. https://doi.org/10.1061/(ASCE)HY.1943-7900.0000796.

Dehghani, H., 2018. Forecasting copper price using gene expression programming. J. Min. Environ. 9, 349–360. https://doi.org/10.22044/jme.2017.6195.1435.

Dhiman, S., Patra, K.C., 2017. Experimental study of embankment breach based on its construction parameters. Na.t Hazards Earth Syst. Sci. Discuss. 2017, 1–26. https://doi.org/10.5194/nhess-2017-383.

Eslamian, S., 2014. Handbook of Engineering Hydrology: Modeling, Climate Change, and Variability. CRC Press, USA.

Ferreira, C., 2001. Gene expression programming: a new adaptive algorithm for solving problems. Complex Syst. 13, 87–129.

Ferreira, C., 2002. Gene expression programming in problem solving. In: Roy, R., Köppen, M., Ovaska, S., Furuhashi, T., Hoffmann, F. (Eds.), Soft Computing and Industry: Recent Applications. Springer, London, London, UK, pp. 635–653.

Ferreira, C., 2006. Gene Expression Programming: Mathematical Modeling by An Artificial Intelligence. Studies in Computational Intelligence. Springer Berlin Heidelberg, Germany.

Fread, D.L., 1988. BREACH: An Erosion Model for Earthen Dam Failure. National Water Service (NWS) Report. Silver Spring, M.A., USA.

Froehlich, D.C., 1995. Peak outflow from breached embankment dam. J. Water Resour. Plan. Manag. 121, 90–97. https://doi.org/10.1061/(ASCE)0733-9496(1995)90:121(1).

Froehlich, D.C., 2008. Embankment dam breach parameters and their uncertainties. J. Hydraul. Eng. 134, 1708–1721. https://doi.org/10.1061/(ASCE)0733-9429(2008)134:12(1708).

Gassman, P.W., Reyes, M.R., Green, C.H., Arnold, J.G., 2007. The soil and water assessment tool: Historical development, applications, and future research directions. Trans. ASABE 50, 1211–1250. https://doi.org/10.13031/2013.23637.

Gholampour, A., Gandomi, A.H., Ozbakkaloglu, T., 2017. New formulations for mechanical properties of recycled aggregate concrete using gene expression programming. Construct. Build Mater. 130, 122–145. https://doi.org/10.1016/j.conbuildmat.2016.10.114.

Hahn, W., Hanson, G.J., Cook, K.R., 2000. Breach morphology observations of embankment overtopping tests. In: Joint Conference on Water Resource Engineering and Water Resources Planning and Management 1–10., https://doi.org/10.1061/40517(2000)411.

Hanson, G.J., Cook, K.R., Hahn, W., 2001. Evaluating headcut migration rates of earthen embankment breach tests. Paper presented at the 2001. In: ASAE Annual Meeting, St. Joseph, MI, USA.

Hanson, G.J., Cook, K.R., Hunt, S.L., 2005. Physical modeling of overtopping erosion and breach formation of cohesive embankments. Trans. ASAE 48, 1783–1794. https://doi.org/10.13031/2013.20012.

Hassanzadeh, Y., 2005. Dam-Break Hydraulics, Publication No. 63. Iranian National Committee on Large Dams, Tehran, Iran. (in Persian).

Hunt, S.L., Hanson, G.J., Cook, K.R., Kadavy, K.C., 2005. Breach widening observations from earthen embankment tests. Trans. ASAE 48, 1115–1120. https://doi.org/10.13031/2013.18521.

Jiang, X., Wei, Y., Wu, L., Hu, K., Zhu, Z., Zou, Z., Xiao, W., 2019. Laboratory experiments on failure characteristics of non-cohesive sediment natural dam in progressive failure mode. Environ. Earth Sci. 78, 538. https://doi.org/10.1007/s12665-019-8544-1.

Johnson, F.A., Illes, P., 1976. A classification of dam failures. Water Power Dam Constr. 28, 43–45.

Kouzehgar, K., Hassanzadeh, Y., Eslamian, S., 2021a. Experimental investigation and application of soft computing in the assessment of breach and outflow hydrograph in the embankment dam break. Iran. J. Marine Technol. 8, 88–103. https://doi.org/10.22034/ijmt.2021.242988.

Kouzehgar, K., Hassanzadeh, Y., Eslamian, S., Yousefzadeh Fard, M., Babaeian Amini, A., 2021b. Application of gene expression programming and nonlinear regression in determining breach geometry and peak discharge resulting from embankment failure using laboratory data (in Press). Irrig. Sci. Eng. https://doi.org/10.22055/jise.2021.35162.1931.

Kouzehgar, K., Hassanzadeh, Y., Eslamian, S., Yousefzadeh Fard, M., Babaeian Amini, A., 2021c. Correction to: experimental investigations and soft computations for predicting the erosion mechanisms and peak outflow discharge caused by embankment dam breach. Arab. J. Geosci. 14, 812. https://doi.org/10.1007/s12517-021-07101-7.

Kouzehgar, K., Hassanzadeh, Y., Eslamian, S., Yousefzadeh Fard, M., Babaeian Amini, A., 2021d. Experimental investigations and soft computations for predicting the erosion mechanisms and peak outflow discharge caused by embankment dam breach. Arab. J. Geosci. 14, 616. https://doi.org/10.1007/s12517-021-06594-6.

Kouzehgar, K., Hassanzadeh, Y., Eslamian, S., Yousefzadeh Fard, M., Babaeian Amini, A., 2021e. Physical modeling into outflow hydrographs and breach characteristics of homogeneous earthfill dams failure due to overtopping. J. Mt. Sci. 18, 462–481. https://doi.org/10.1007/s11629-020-6177-1.

Letelier, J., Weber, P., 2000. Spike sorting based on discrete wavelet transform coefficients. J. Neurosci. Methods 101, 93–106. https://doi.org/10.1016/S0165-0270(00)00250-8.

Liu, Y., Zhang, J., Zhu, C., Xiang, B., Wang, D., 2019. Fuzzy-support vector machine geotechnical risk analysis method based on Bayesian network. J. Mt. Sci. 16, 1975–1985. https://doi.org/10.1007/s11629-018-5358-7.

MacDonald, T.C., Langridge-Monopolis, J., 1984. Breaching charateristics of dam failures. J. Hydraul. Eng. 110, 567–586. https://doi.org/10.1061/(ASCE)0733-9429(1984)110:5(567).

Morris M.W. (2011) Breaching of earth embankments and dams. Open UniversityMorris, M.W., 2011. Breaching of earth embankments and dams. Open University, UK, https://doi.org/10.21954/ou.ro.0000d502.

Mosaedi, A., Hosseini, S.M., 2015. Determining the effective parameters and their optimal combination in rill erosion modeling. Arab. J. Geosci. 8, 3045–3053. https://doi.org/10.1007/s12517-014-1363-5.

Nash, J.E., Sutcliffe, J.V., 1970. River flow forecasting through conceptual models part I— a discussion of principles. J. Hydrol., 10:282-290. https://doi.org/10.1016/0022-1694(70)90255-6.

Peng, M., Zhang, L.M., 2012. Breaching parameters of landslide dams. Landslides 9, 13–31. https://doi.org/10.1007/s10346-011-0271-y.

Peng, M., Zhang, L.M., Chang, D.S., Shi, Z.M., 2014. Engineering risk mitigation measures for the landslide dams induced by the 2008 Wenchuan earthquake. Eng. Geol. 180, 68–84. https://doi.org/10.1016/j.enggeo.2014.03.016.

Pierce, M.W., Thornton, C.I., Abt, S.R., 2010. Predicting peak outflow from breached embankment dams. J. Hydrol. Eng. 15, 338–349. https://doi.org/10.1061/(ASCE)HE.1943-5584.0000197.

Pugh, C.A., 1985. Hydraulic Model Studies of Fuse Plug Embankments vol REC-ERC-B5–7. United States Department of Interior, Bureau of Reclamation, USA.

Sattar, A.M.A., 2013. Gene expression models for prediction of dam breach parameters. J. Hydroinformatics 16, 550–571. https://doi.org/10.2166/hydro.2013.084.

Shen, D., Shi, Z., Peng, M., Zhang, L., Jiang, M., 2020. Longevity analysis of landslide dams. Landslides 17, 1797–1821. https://doi.org/10.1007/s10346-020-01386-7.

Shi, Z.M., Guan, S.G., Peng, M., Zhang, L.M., Zhu, Y., Cai, Q.P., 2015. Cascading breaching of the Tangjiashan landslide dam and two smaller downstream landslide dams. Eng. Geol. 193, 445–458. https://doi.org/10.1016/j.enggeo.2015.05.021.

Singh, V.P., 2013. Dam Breach Modeling Technology. Water Science and Technology Library. Springer, Netherlands.

Singh, V.P., Scarlatos, P.D., 1988. Analysis of gradual earth-dam failure. J. Hydraul. Eng. 114, 21–42. https://doi.org/10.1061/(ASCE)0733-9429(1988)114:1(21).

Thornton, C.I., Pierce, M.W., Abt, S.R., 2011. Enhanced predictions for peak outflow from breached embankment dams. J. Hydrol. Eng. 16, 81–88. https://doi.org/10.1061/(ASCE)HE.1943-5584.0000288.

USBR, 1988. Downstream hazard classification guidelines. In: ACER, Department of the Interior, Bureau of Reclamation, Report No. 11, USA.

Vaskinn, K.A., Lövoll, A., Hoëg, K., Morris, M., Hanson, G.J., Hassan, M.A., 2004. Physical modeling of breach formation: large scale field tests, Dam safety. In: Proceedings of the Association of State Dam Safety Officials, 2004. Phoenix, Arizona, USA.

Von Thun, J.L., Gillette, D.R., 1990. Guidance on Breach Parameters, Unpublished Internal Document. U.S. Bureau of Reclamation, Denver, USA, p. 17.

Wahl, T.L., 1998. Prediction of Embankment Dam Breach Parameters: A Literature Review and Needs Assessment. DSO-98-004, Dam Safety Research Report. Denver, CO, USA.

Wahl, T.L., 2014. Evaluation of Erodibility-Based Embankment Dam Breach Equations, Hydraulic Laboratory Report HL-2014-02. States Bureau of Reclamation, Denver, Colorado, USA.

Walder, J.S., O'Connor, J.E., 1997. Methods for predicting peak discharge of floods caused by failure of natural and constructed earthen dams. Water Resour. Res. 33, 2337–2348. https://doi.org/10.1029/97WR01616.

Walder, J.S., Iverson, R.M., Godt, J.W., Logan, M., Solovitz, S.A., 2015. Controls on the breach geometry and flood hydrograph during overtopping of noncohesive earthen dams. Water Resour. Res. 51, 6701–6724. https://doi.org/10.1002/2014WR016620.

Wang, B., Chen, Y., Wu, C., Peng, Y., Song, J., Liu, W., Liu, X., 2018. Empirical and semi-analytical models for predicting peak outflows caused by embankment dam failures. J. Hydrol. 562, 692–702. https://doi.org/10.1016/j.jhydrol.2018.05.049.

Wu, W., 2013. Simplified physically based model of earthen embankment breaching. J. Hydraul. Eng. 139, 837–851. https://doi.org/10.1061/(ASCE)HY.1943-7900.0000741.

Xu, Y., Zhang, L.M., 2009. Breaching parameters for earth and rockfill dams. J. Geotech. Geoenviron. Eng. 135, 1957–1970. https://doi.org/10.1061/(ASCE)GT.1943-5606.0000162.

Zhang, L.M., Peng, M., Xu, Y., 2010. Assessing risks of breaching of earth dams and natural landslide dams. In: Indian Geotechnical Conference. GEOtrendz, India, pp. 16–18.

Zhang, L., Peng, M., Chang, D., Xu, Y., 2016. Dam Failure Mechanisms and Risk Assessment. John Wiley & Sons.

Zhao, T., Chen, S., Fu, C., Zhong, Q., 2019. Centrifugal model tests and numerical simulations for barrier dam break due to overtopping. J. Mt. Sci. 16, 630–640. https://doi.org/10.1007/s11629-018-5024-0.

Chapter 3

Hydrological modeling of Hasdeo River Basin using HEC-HMS

Md. Masood Zafar Ansari[a], Ishtiyaq Ahmad[a], Pushpendra Kumar Singh[b], and Saeid Eslamian[c,d]

[a]*Department of Civil Engineering, National Institute of Technology, Raipur, India,* [b]*Division of Water Resources, National Institute of Hydrology, Roorkee, India,* [c]*Department of Water Engineering, College of Agriculture, Isfahan University of Technology, Isfahan, Iran,* [d]*Center of Excellence for Risk Management and Natural Hazards, Isfahan University of Technology, Isfahan, Iran*

1. Introduction

India is rich in water resources where the monsoon season plays a predominant role. Having a large geographical area, uneven distribution of rainfall leads to drought and flood-like situations in some areas. Catchment systems are not stationary systems; both their characteristics and the inputs which drive the hydrology are changing over time (Milly et al., 2008). Two types of future risk need to be managed in catchment hydrology where rainfall–runoff models can play a useful role. The first is the short-term forecasting problem of whether an important flood discharge with the potential to pose a threat to life or property will occur. The second is the longer-term seasonal or decadal prediction problem of whether changes in catchment characteristics or climate might pose a threat to water resources or flood and drought frequencies. Both of these problems are dependent on the inputs from weather and climate prediction models, which are associated with significant uncertainties in their predictions. One of the main reasons to model the rainfall–runoff processes of hydrology is a result of the limitations of hydrological measurement techniques. We are not able to measure everything we would like to know about hydrological systems. We have, in fact, only a limited range of measurement techniques and a limited range of measurements in space and time. We, therefore, need a means of extrapolating from those available measurements in both space and time, particularly to ungauged catchments (where measurements are not available) and into the future (where measurements are not possible) to assess the likely impact of future hydrological change. However, the ultimate aim of prediction using models must be to improve decision-making about a hydrological problem, whether that be in water resources planning, flood protection, mitigation of contamination, licensing of abstractions, or other areas. With increasing demands on water resources throughout the world, improved decision-making within a context of fluctuating weather patterns from year to year requires improved models (Beven, 2012). Hydrological modeling is a crucial and decisive tool to estimate hydrological processes and the water resources availability. The Hydrologic Engineering Centre-Hydrological Model System (HEC-HMS) is one such model that supports both lumped parameter-based modeling and distributed parameter-based modeling (Agrawal, 2005). HEC-HMS provides a suite of hydrological modeling options, with the main components focusing on determining runoff hydrographs from subbasins and routing the hydrographs through the channels to the study outlets (Beighley and Moglen, 2003). The HEC-HMS uses the model to represent various components of the rainfall–runoff process like the Loss model for calculating infiltration loss within the subbasin, the Transform model for transforming excess precipitation into the direct surface runoff, the baseflow model for subsurface flow estimation, and the Routing model for routing the reach. The model combines a Basin model, Meteorological model, Control specification, and Time series data with the run option to obtain the model result.

Visweshwaran (2017) used the HEC-HMS model for event-based rainfall–runoff modeling for the Krishna basin using daily rainfall Runoff data. SCS-CN method was used for loss estimation and SCS unit hydrograph for transforming excess precipitation into a direct runoff hydrograph. The model was calibrated for the monsoon period of 2011 and validated for the 2007 and 2013 monsoon periods. Rathod et al. (2015) developed a lumped continuous hydrological model for estimating runoff for different rainfall events in three subbasins of the Tapi river using the Green-ampt method as a loss method and compared the SCS unit hydrograph and Snyder unit hydrograph method as a transform method and found that the SCS

Handbook of HydroInformatics. https://doi.org/10.1016/B978-0-12-821962-1.00020-9

unit hydrograph method gives better results. Halwatura (2013) made an attempt to set a rainfall–runoff model for the Attangalu Oya river basin, Sri Lanka, using the HEC-HMS model, he compared different transform and loss methods and found that the combination of Snyder unit hydrograph method as a Transform method and the deficit and constant method as a loss method gives more reliable results for Attangalu Oya river basin. The lumped models, distributed and semidistributed models have better spatial variability for the hydrologic process and watershed characteristics. Majidi and Shahedi (2012) showed that lag time is a sensitive parameter. Model validation using optimized lag time parameter showed a reasonable difference in peak flow. Knisel and Davis (2000) from their study show that runoff simulation in GLEAMS that CN is a sensitive parameter and noticed that small changes in high CNs are more sensitive than equivalent small changes in low CNs.

In this paper, the HEC-HMS model is used to develop rainfall–runoff modeling in the Hasdeo river basin. In the study, the SCS Curve Number, SCS unit hydrograph, Recession, and Muskingum method have been used as a loss model, transform model, baseflow model, and routing model respectively. The result of daily rainfall data at rain gauge stations demonstrates how rainfall change has a distinctive impact in determining discharge to peak and runoff depth with the river discharge at the outlet. The results of this study will be helpful for better watershed planning, management of basins, construction of hydraulic structures requires, understanding climate change for forecasting.

2. The rationale of the study

India is a country where the monsoon season plays a predominant role. Having a large geographical area, uneven distribution of rainfall leads to drought and flood-like situations in some areas which makes the research scope in surface water management and conservation in a sustainable way. Floods have become increasingly frequent and unpredictable in recent years as a result of human interventions in the natural environment, as well as the consequences of global climate change. India is always in news during the monsoon season because of flooding. Climate change is one of the main reasons for unexpected rainfall, which leads to the overflow of rivers and results in the inundation of banks. Climate change and land use change have raised the challenges associated with increased runoff and flood management. Hydrological modeling is one of the tools to estimate hydrological processes and the water resources availability.

Hasdeo River Basin is part of the Mahanadi middle subbasin (400–750 m above mean sea level) and is the second largest branch of Mahanadi after Sheonath River. The total catchment of this river is 10,535.96 sq. km. It flows from North to South direction and has seven watersheds located between 21°45′ N to 23°37′ N latitude and 82°00′ E to 83°04′ E longitude as shown in Fig. 1. The seven different watersheds are Upper Hasdeo, Gej Nala, Bamni Nadi, Tan Nadi, Chornai, Ahiran Nadi, and Lower Hasdeo (Prasad, 2015). The average rainfall in the Hasdeo Subbasin catchment is around 1400 mm, ranging from 900 to 1500 mm over the last 10 years 80% of the rainfall is received from June to August. Monsoon arrives slowly in the state and moves to the north, before abruptly exiting the state. So the onset of monsoon at different stations in the state is mostly delayed but the withdrawal date remains constant throughout the state. Thus there is a decrease in the length of the monsoon. The average period of monsoon is 116.87 days which is less than the normal period of 123 days (Water Year Book, 2013, Chhattisgarh). In this basin, the predominant soil type is loamy as 62% of the area consists of fine loamy to loamy soils followed by different types of clayey soils and coarse skeletal loamy varieties. Chhattisgarh plains are composed of Bhata (lateritic), Matasi (Sandy Loam), Dorsa (Clay Loam), and Kanhar (Clay) while the northern hills are composed of hilly soils, tikra, Goda Chawar, and Bahara. Among these, the Bhata, Matasi hilly soils, and tikra are the very light type of soils with very low water retentive capacity. So if ever there is a break in the monsoon for 5–7 days or just after the monsoon withdraws, a water stress-like situation occurs in the soil thus making it difficult for the plants to extract water (Hasdeo Subbasin Report, 2015).

Recently, in August 2020, Heavy rains created a flood-like situation in some areas of at least four districts and caused rivers, including the Mahanadi, to flow above the danger mark. Raipur, Janjgir-Champa, Bilaspur, Durg, and Raigarh districts witnessed "record rainfall." According to the disaster management department officials, the Jangir-Champa district received 211.9 mm rainfall (https://www.thehindu.com/news/national/other-states/flood-like-situation-in-four-chhattisgarh-districts-thousands-shifted/article32472296.ece). October 2019, due to heavy rains, the river changed its course and water entered the coal mine. The river fell into a pit that was created to extract coal (https://timesofindia.indiatimes.com/city/raipur/river-changes-course-falls-into-coalmine-in-chhattisgarhs-korba/articleshow/71403265.cms).

To take necessary steps to mitigate disaster, for designing hydraulic structures research is needed for which flood modeling is significant. So, modeling of flood events is significant for the case studies and future predictions.

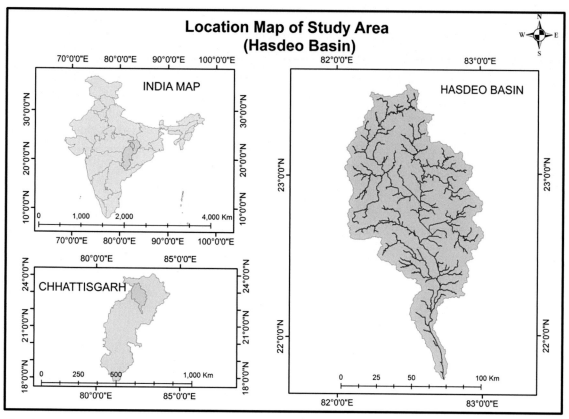

FIG. 1 Location map of study area, Hasdeo river basin.

3. Materials and methods

3.1 Data collection

The selection of an appropriate model for an application requires consideration of the suitability of the model to the catchment conditions, data requirements, and availability and assesses the quantity and quality of available data. Quite often, the available data dictate the type of model to be used more than the proposed problem itself. For analyzing and reflecting the actual hydrologic and hydraulic situation, a good grasp of the topographical, hydrological, and climatic conditions of the research area, as well as a proper set of data defining them, is important. Furthermore, because the quality of data used in modeling has a direct impact on the outcome, the data should be inspected and processed before being used (Bazrkar et al., 2017).

Table 1 shows the various data used in this study with the source of acquisition of data.

TABLE 1 List of data types used in the study.

Data type and description	Source of data
Topographical Map—Digital elevation model	https://bhuvan.nrsc.gov.in/bhuvan_links.php/
Satellite Imagery—Sentinel 2—LULC Map	https://scihub.copernicus.eu/
Soil Map	https://www.nbsslup.in/
Daily rainfall	Data Centre, WRD, Raipur
Discharge data	
Rain gauge station data	http://hydrologyproject.cg.gov.in/

3.1.1 Topographical data

Topographical information of the terrain is vital in hydrological and water management studies, as it controls rainfall–runoff responses of the basin (Horritt and Bates, 2002). DEM, a grid-based raster form, is an important source for topographical data (Wu et al., 2013). Nowadays, most of the DEMS are generated by using remote sensing techniques (Smith and Clark, 2005). The derived products are available at different spatial resolutions (e.g., 10, 30, 60, and 90 m) and data structured format. For this study, DEM data of 30 m resolution as shown in Fig. 2 was downloaded from the website of Bhuvan (https://bhuvan.nrsc.gov.in/bhuvan_links.php). It is stored in the GeoTIFF format and of spatial resolution of 1 arc s. DEM data was used in the delineation of the basin, which was used to extract various topographical and hydrological parameters. These were later used as inputs to the hydrologic model.

The detail of satellite data used as a digital elevation model is given in Table 2.

3.1.2 Multispectral satellite data

The LULC map is an integral part of any hydrological model. They are used to determine curve numbers (CN). The multispectral satellite data from Sentinel-2 (spatial resolution: 10 m) as shown in Fig. 3 were used to prepare LULC maps of the basin for the year 2018. The satellite data were collected from the website of the European Space Agency (ESA) Copernicus Open Assess Hub (https://scihub.copernicus.eu/).

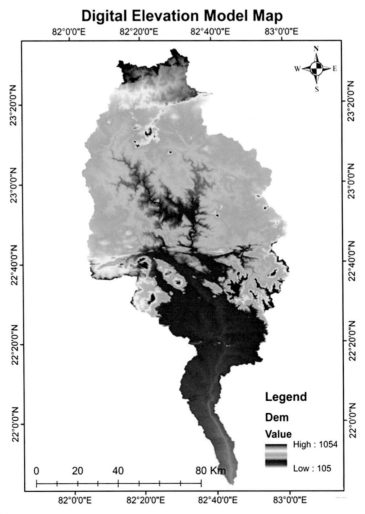

FIG. 2 Digital elevation model map.

TABLE 2 Satellite data details.

Satellite	Sensor	Tile no.	Spatial resolution	Dates of acquisition
Cartosat-1	PAN (2.5 m) Stereo data	F44E	30 m	29/04/2015
		F44K		
		F44L		
		F44Q		

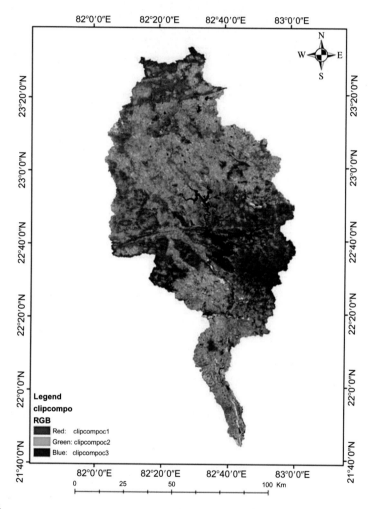

FIG. 3 Composite band map.

3.1.3 Soil data

Soil cover map was downloaded from the National Bureau of Soil Survey and Land Use planning (NBSS&LUP) (https://www.nbsslup.in/). Hydrologically Soil Group (HSG) map is prepared and grouped into four major categories, i.e., A, B, C, and D, based on the soil taxonomy as shown in Fig. 4.

3.1.4 Hydro-meteorological data

The study on flood hydrograph of different return periods requires discharge and rainfall data as input. In this study, daily rainfall data, daily discharge data, and rain gauge stations point data of Hasdeo River were collected from the Data

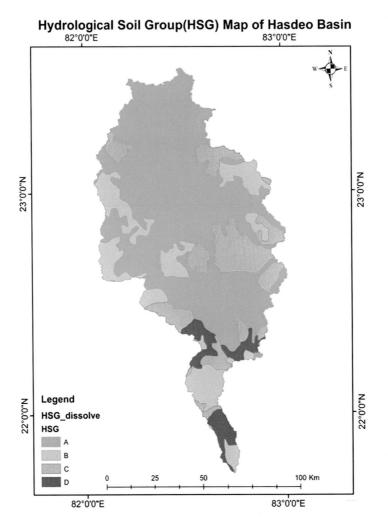

FIG. 4 HSG map.

collection center of the Water Resources Department, Raipur for Hasdeo Basin. The high-resolution daily rainfall data available on $0.25° \times 0.25°$ grids were obtained from Indian Meteorological Department (IMD). A total number of 27 grids were extracted to cover the entire area of the basin. The weighted average rainfall is calculated using the Thiessen polygon method. For which the Thiessen polygon map has been generated using ArcGIS software as shown in Figs. 5 and 6. In this method, each station's rainfall is given a weighted average based on the area closest to the station. Thus average rainfall over a catchment is given by Eq. (1).

$$\overline{P} = \sum_{i=1}^{M} P_i \frac{A_i}{A} \qquad (1)$$

where, $\overline{P}$ is the average precipitation, M is the number of stations, P_i is the rainfall at ith day, $\frac{A_i}{A}$ is the weightage factor.

3.2 Methodology

In this paper, hydrological modeling has been done using HEC-HMS 4.7 and for linking GIS environment Hec-GeoHMS extension tool of ArcGIS10.2 is used. The methodology adopted for this study is systematically explained in Fig. 7.

The following sections are used to describe Hydrological modeling: i.e., Creating Basin Model, Developing Hydrological Parameters, and finally Hydrological Modeling.

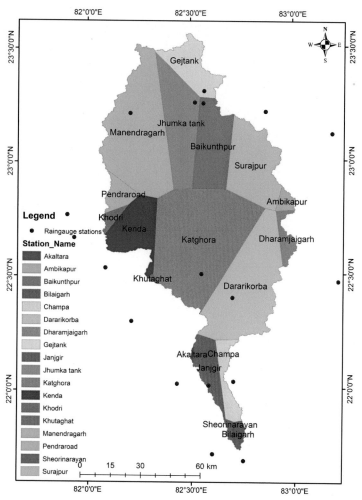

FIG. 5 Thiessen polygon for rain gauges.

3.2.1 Model development

This subheading explains how the HEC-HMS model is created. It describes how parameters are derived and then applied to the different calculation methods used in the model.

Utilization of HEC-GeoHMS in Model Development. Terrain processing and basin processing and hydrological parameters estimation are all performed with HEC-GeoHMS, a public domain extension with ArcGIS software. The project has been developed in the HEC-GeoHMS tool and then imported to HEC-HMS software for the simulation process.

The overview relationship between GIS, HEC-GeoHMS, and HEC-HMS is illustrated in Fig. 8.

HEC-GeoHMS model development process has been shown in Fig. 9.

(a) Preprocessing

The procedure is adopted to demarcate the basin boundary using HEC-GeoHMS is given in Annexure A. Finally, the catchment polygon processing generates a subbasin vector layer from the delineated grid, demarcating the study area boundary.

(b) Project setup

Generate project by adding project point at the desired outlet of basin to create project area of interest. The resulting project area for basin was $10,535.96 \, \text{km}^2$.

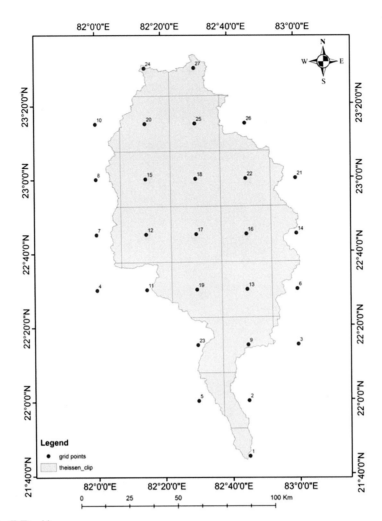

FIG. 6 Thiessen polygon for IMD grid.

(c) Basin Processing and characteristics

The delineated subbasins were merged. Physical attributes were calculated for each of the subbasins and rivers using the revised DEM.

- *River characteristics*—River length and river slope.
- *Basin characteristics*—Basin slope, longest flowpath to the basin, basin centroid, centroid elevation, and centroidal longest flowpath.

ArcHydro tool to calculate basin slope should be used.

(d) Developing hydrological parameters

The hydrological characteristics for each subbasin were calculated using land use and soil cover data. The steps involved in producing hydrological parameters are as follows:

- *Select HMS processes:*

Modeling loss, transform, base-flow, and routing procedures from HMS were chosen as shown in Table 3.

- *Generating CN grid map:*

The Curve Number is the union layer of soil cover and land use land cover map. Soil map is classified in Hydrological Soil Group (A, B, C & D) based on soil texture shown in Fig. 10, which is defined according to USDA (2009).

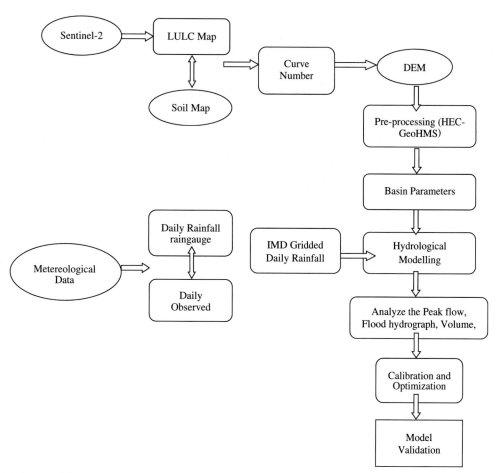

FIG. 7 Present study methodology.

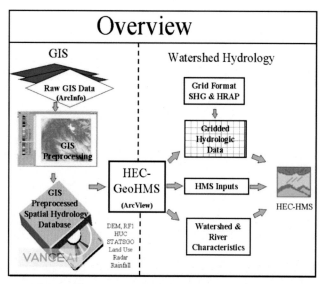

FIG. 8 Overview of the relation between HEC-GeoHMS and HEC-HMS (USACE, 2000).

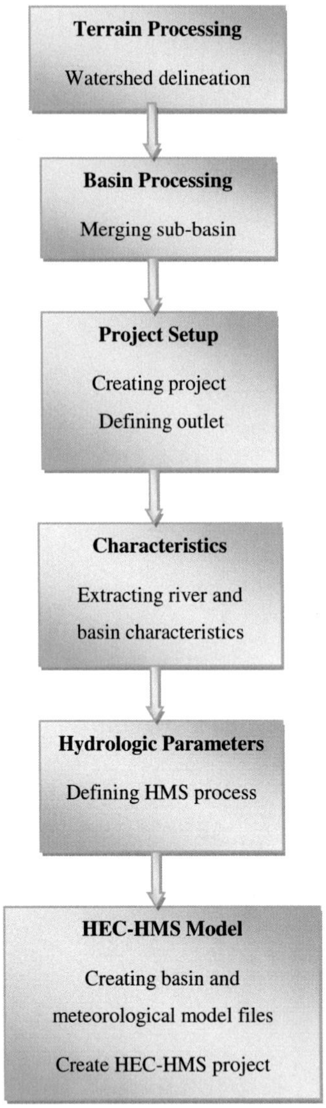

Terrain Processing

Watershed delineation

Basin Processing

Merging sub-basin

Project Setup

Creating project

Defining outlet

Characteristics

Extracting river and

basin characteristics

Hydrologic Parameters

Defining HMS process

HEC-HMS Model

Creating basin and

meteorological model files

Create HEC-HMS project

FIG. 9 Flow chart of preprocessing with HEC-GeoHMS.

TABLE 3 Model parameters.

HMS processes	Method
Loss	SCS Curve Number
Transform	SCS Unit Hydrograph
Base-flow	Recession
Routing	Muskingum

Land use land cover is created by supervised-classification of Sentinel 2 data using maximum likelihood classification method shown in Fig. 11.

Chart provided by United States Department of Agriculture (1986) to create curve numbers according to the land use and hydrologic soil groups for urban areas, cultivated agricultural lands, other agricultural lands, and arid and semiarid rangelands. Fig. 12 shows the CN grid map of the Hasdeo basin.

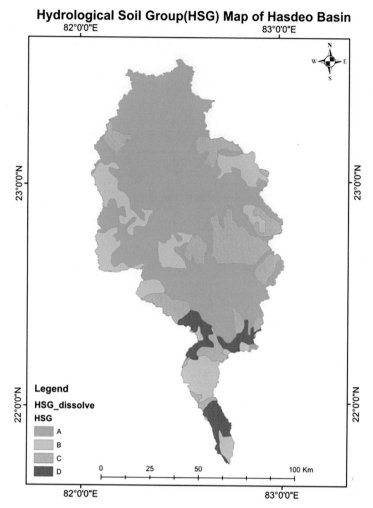

FIG. 10 Hydrological soil group map.

- *CN lag:*

The lag time for the transform method based on the CN grid was estimated using this function. Land use and soil cover layers were used to create the CN grid. Table 4 shows the CN values used for various land use types in the Hasdeo basin.

- *Developing HEC-HMS model files*:

In this step, model files for HEC-HMS were created, including the background-map file, basin model file, and meteorological model file.

3.2.2 Hydrological modeling

Modeling is done in HEC-HMS by importing files from ArcGIS after the model framework is created in HEC-GeoHMS. A basin model for study area in HEC-HMS is shown in Fig. 13.

Setting up the HEC-HMS component and estimation of model parameters:

(a) Basin model:

The primary goal of these devices is to transform atmospheric conditions into streamflow at precise places throughout the watershed.

Model parameters:
1. Loss model:

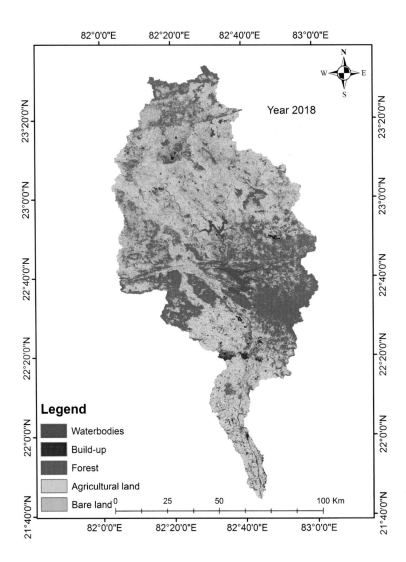

FIG. 11 LULC Map.

While a subbasin element illustrates the interaction of infiltration, surface runoff, and subsurface processes, the actual infiltration calculations are carried out by a loss method embedded within the subbasin.

The goal of the precipitation loss process is to determine what percentage of precipitation infiltrates through the ground and what percentage becomes runoff, contributing to river flow. For this study, the SCS Curve Number method was selected. The method was selected primarily because its required parameters are available. The method was developed by the Soil Conservation Service (SCS) and uses soil cover, land use, and antecedent soil moisture to determine the precipitation excess (US Army Corps of Engineers, 2010). The curve number approach for the incremental losses is implemented by the Soil Conservation Service (SCS) (NRCS, 2004).

The inputs that are required in this method are initial abstraction, curve number, and impervious (%).

Curve number: The principle parameter of the SCS Curve Number method is CN and is estimated as a function of land use and soil type. Weighted Curve number for each subbasin is assigned by the HEC-GeoHMS tool and shown in Table 5.

Initial abstraction: The maximum amount of precipitation absorbed by the ground before runoff begins to occur. The initial abstraction is calculated as a fraction of the potential maximum retention (S), which is the maximum total amount of precipitation that can be absorbed by the ground, and is a function of the *CN*. This value was estimated as the function of the curve number using the below Eqs. (2) and (3),

$$S = \frac{25400}{CN} - 254 \qquad (2)$$

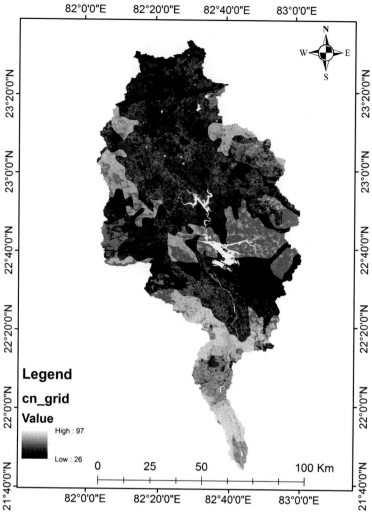

FIG. 12 CN Grid Map of Hasdeo Basin.

TABLE 4 CN lookup table.

Description	A	B	C	D
Water bodies	98	98	98	98
Buildup areas	49	69	79	89
Barren land	76	85	90	85
Agricultural land	64	75	80	87

$$I_a = 0.2*S \qquad (3)$$

where, I_a is the initial abstraction (mm), S is the maximum potential retention (mm) and CN is the curve number.

Using the curve numbers shown in Table 5, the initial abstraction for each subbasin is calculated and shown in Table 6 below.

Impervious (%): The percentage of the subbasin which is directly connected impervious area can be specified. In this study, due to the difficulty to determine precisely its value, it was related to the percent of built-up, as the built-up has the minimum infiltration. The impervious percentage is shown in Table 7.

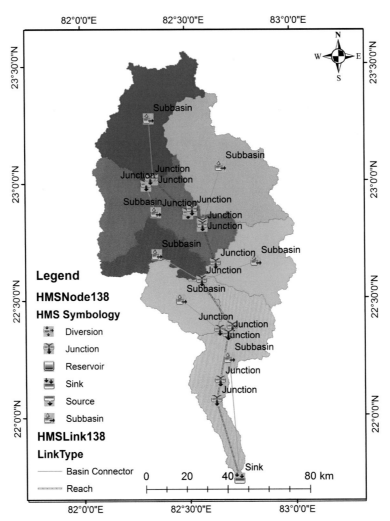

FIG. 13 Basin model for HEC-HMS.

TABLE 5 Average subbasin curve number.

	Gej Nala	Upper Hasdeo	Bamni nadi	Tan nadi	Chornai	Ahiran nadi	Lower Hasdeo
Curve number	42	52	50	46	50	54	62

TABLE 6 Initial abstraction for subbasins.

	Gej Nala (W740)	Upper Hasdeo (W760)	Bamni nadi (W870)	Tan nadi (W950)	Chornai (W980)	Ahiran nadi (W1040)	Lower Hasdeo (W1100)
I_a	70.15	46.89	50.8	59.63	50.80	43.27	31.13

TABLE 7 Impervious % for subbasins.

	Gej Nala (W740)	Upper Hasdeo (W760)	Bamni nadi (W870)	Tan nadi (W950)	Chornai (W980)	Ahiran nadi (W1040)	Lower Hasdeo (W1100)
Impervious (%)	12	10	20	13	8	15	12

2. Transform model:

A transform technique contained within the subbasin performs the actual surface runoff computations. This model depicts how much the excess precipitation is converted to runoff at a certain location. The SCS Unit Hydrograph approach is used in this study. This well-known empirical method is based on a review of the numerous researches conducted in agricultural watersheds across the United States (Bedient et al., 2008). A relationship was constructed from these studies that linked the amplitude and time of the peak hydrograph to the lag time and area of each subbasin.

The Input requires in this method is the Lag time.

T_{lag} is estimated as the Eq. (4) given below:

$$T_{lag} = T_c * 0.6 \tag{4}$$

where, T_c represents the time of concentration in hours, which is estimated using the Kirpich Eq. (5) (Eslamian and Mehrabi, 2005).

$$T_c = \frac{L^{0.8} + (S + 1)^{0.7}}{1900 + Y^{0.5}} \tag{5}$$

where, L is the hydraulic length in kilometers of the watershed and Y represents the slope of each subbasin was calculated using the HEC-GeoHMS tool in percentage.

The lag time for each subbasin has been calculated and is shown in Table 8.

The model uses these lag periods to determine how long it will take for the extra precipitation in each subbasin to travel overland into the Hasdeo River. The model directs all of the surface runoff from each subbasin to the Hasdeo River at a single site, the subbasin exit.

3. Baseflow model:

A baseflow technique embedded within the subbasin performs the real subsurface calculations. At the start of a simulation, the starting baseflow must be given.

The recession model has been used often to explain the drainage from natural storage in a watershed (Linsley et al., 1982). It defines the relationship of Q_t, the baseflow at any time t, to an initial value as is given in Eq. (6).

TABLE 8 Subbasin lag time values used in HMS model.

Subbasin	L (ft)	S (in)	Y (%)	T_c (h)	T_{lag} (h)	T_{lag} (min)
Upper Hasdeo	434,254.23	9.23	9.02	28.89	17.33	1039.93
Gej Nala	336,964.81	13.81	7.36	33.82	20.29	1217.51
Bamni nadi	457,943.65	10	7.48	34.83	20.90	1253.88
Tan nadi	252,591.82	11.74	13.71	17.71	10.63	637.68
Chornai	279,172.82	10	19.55	14.50	8.70	522.00
Ahiran nadi	255,218.79	8.52	9.34	17.64	10.59	635.15
Lower Hasdeo	418,404.69	6.13	5.97	26.77	16.06	963.83

$$Q_t = Q_0K^t \tag{6}$$

where Q_0 = initial baseflow (at time zero); and k = an exponential decay constant.

The recession constant given in Eq. (7), describes the rate at which base flow recedes between storm occurrences.

$$\text{Recession constant (k)} = \frac{\text{baseflow at the current time}}{\text{baseflow one day earlier}} \tag{7}$$

In Fig. 14, after the peak of the direct runoff, a user-specified threshold flow defines the time at which the recession model defines the total flow. That threshold may be specified as a flow rate or as a ratio to the computed peak flow.

4. Reach routing

The reach routing process converts a hydrograph at the upstream boundary of the subbasin to a resultant hydrograph at the downstream boundary of the subbasin. To route flow across the river reach, the Muskingum routing method employs a basic conservation of mass strategy.

The Muskingum method was developed by McCarthy (1938) and utilizes the continuity equation and a storage relationship that depends on both inflow and outflow. Storage in the reach is modeled as the sum of prism storage and wedge storage. As shown in Fig. 15, prism storage is the volume defined by a steady-flow water surface profile, while wedge storage is the additional volume under the profile of the flood wave. The change in the shape of a hydrograph within a reach as it moves downstream is dependent on river channel geometry and the roughness of the channel surface.

The volume of prism storage is the outflow rate, Q, multiplied by the travel time through the reach, K. The volume of wedge storage is a weighted difference between the inflow and outflow, multiplied by the travel time K. Thus, the Muskingum model defines the storage as in Eq. (9).

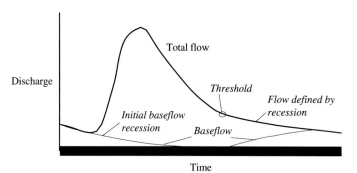

FIG. 14 Baseflow model illustration.

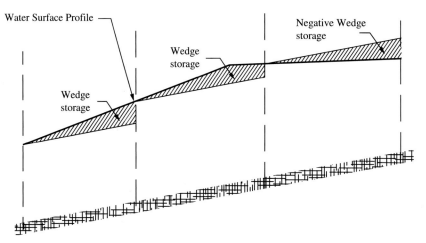

FIG. 15 Wedge storage. *(From Linsley, R.K., Kohler, M.A., Paulhus, J.L.H., 1982. Hydrology for engineers. McGraw-Hill, New York, NY.)*

$$\frac{ds}{dt} = I - Q \tag{8}$$

$$S_t = K[XI_t + (1 - X)Q_t] \tag{9}$$

where, S is the water storage at time "t," I is the inflow (m³/s), Q is the outflow (m³/s), K is the travel time of the flood wave through routing reach (h), and X is the dimensionless weighting factor ($0 \leq X \leq 0.5$).

Estimating the model parameters K and X:

The parameters K and X and the computational time step Δt also must be selected to ensure that the Muskingum model. That means that the parenthetical terms must be nonnegative; the values of K and X must be chosen so that the combination falls within the shaded region shown in Fig. 16.

According to HEC-HMS Technical reference manual (2000), the constraint for Muskingum routing parameters is tabulated in Table 9.

(b) Meteorological model:

During a simulation, the meteorological model is in responsible for preparing the boundary conditions that affect the watershed. The precipitation method utilized in this study is a specified hyetograph, which allows the subbasins to specify the exact time-series to utilize for the hyetograph. When precipitation data is processed outside of the software and then imported without modification, this method is useful.

(c) Time series data:

This study uses time-series precipitation data to estimate the basin-average rainfall and time-series of daily observed discharge, which is beneficial for calibrating a model and necessary for optimization.

(d) Control specifications:

Their primary function is to govern when simulations begin and end, as well as the simulation time period. Depending on the modeling approach, several control criteria might be utilized. Time windows and time intervals are used in the mathematical calculations (USACE, 2018).

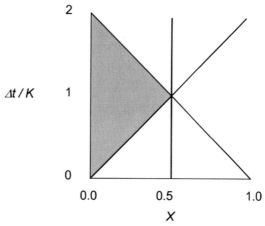

FIG. 16 Feasible region for Muskingum model parameters.

TABLE 9 Muskingum routing parameter constraint.

Model	Parameter	Minimum	Maximum
Muskingum routing	K	0.1 h	150 h
	X	0	0.5

3.3 Model calibration and validation

The process of calibrating a model involves modifying its parameters within the reasonable ranges until the simulated outcomes are close enough to the observed values (Zeckoski et al., 2015). Validation is the process of ensuring that a calibrated model can reproduce a set of observations or forecast the future conditions without requiring any additional parameter adjustments (Zheng et al., 2012).

The developed HEC-HMS model after the simulated runs was calibrated and validated against the observed discharge data at outlet. The calibration was performed in order to adjust the model parameters so that the simulated flows have had better agreement with the observed flows. After calibration, the model was validated with the observed discharge data using the same input hydrological parameters obtained from the model calibration. Fig. 17 shows the schematic diagram of calibration procedure (HEC-HMS Tech. Ref. Manual, 2000).

3.4 Model evaluation parameter

Model calibration and validation require the examination of the accuracy of results to ensure the valid representation of hydrological processes in basins. The model evaluation parameters to procure performance evaluation for a model should be used.

The Nash-Sutcliffe coefficient of efficiency (NSE) and coefficient of determination (R^2) parameters are used to assess the effectiveness of the created model. The NSE (Eq. 10) values range from $-\infty$ to 1 (Nash and Sutcliffe, 1970), and R^2 (Eq. 11) from 0 to 1.

$$NSE = 1 - \left[\frac{\sum_{i=1}^{n} (O_i - P_i)^2}{\sum_{i=1}^{n} \left(O_i - O_{avg} \right)^2} \right] \tag{10}$$

$$R^2 = \left[\frac{\sum_{i=1}^{n} \left(O_i - O_{avg} \right) \left(P_i - P_{avg} \right)}{\sum_{i=1}^{n} \left(O_i - O_{avg} \right)^2 \sum_{i=1}^{n} \left(P_i - P_{avg} \right)^2} \right]^2 \tag{11}$$

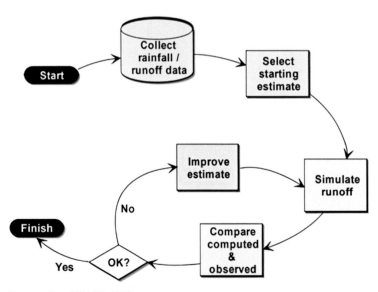

FIG. 17 Schematic of calibration procedure (USACE, 2000).

TABLE 10 Performance evaluation criteria for the basin-scale models (Moriasi et al., 2015).

Method	Very good	Good	Satisfactory	Not satisfactory
NSE	NSE>0.80	0.70<NSE≤0.80	0.50<NSE≤0.70	NSE≤0.50
R^2	R^2>0.85	0.75<R^2≤0.85	0.60<R^2≤0.75	R^2≤0.60
PBIAS (%)	PBIAS<±5	±5≤PBIAS<10	±10≤PBIAS<15	PBIAS≥±15

$$RMSR = \sqrt{\frac{1}{n_t}\sum_{t=1}^{n_t}(O_i - P_i)^2} \tag{12}$$

$$PBIAS = \frac{\sum_{t=1}^{n_t}(O_i - P_i)}{\sum_{t=1}^{n_t}(O_i)} \tag{13}$$

where, O_i is the ith observed value, O_{avg} is the average observed value for the entire study period, P_i is the ith predicted (simulated) value, and P_{avg} is the average of the predicted value over the entire study period.

The recommended list for streamflow with "Nash-Sutcliffe Efficiency," "Percent Bias," and "R squared" methods are given in Table 10 (Singh et al., 2004).

4. Result and discussions

4.1 Calibration and validation results

Using time-series data as daily rainfall stations data and IMD gridded datasets, the model has been simulated run simultaneously. To identify the best match between the simulation and observation, the HEC-HMS model was calibrated for the various flood events. Calibration has been done both manually and automatically (by optimization trial) of the simulated flow with observed daily discharge data. After the calibration from the year 2015 to 2018, the calibrated final parameters were used to validate the model for the year 2013 and 2014.

The black dotted lines indicate the observed outflow, the blue solid line indicates total simulated outflow, and the blue dashed line indicates the outflow from the junction's upstream reach.

4.1.1 For rain gauge station data

(a) Calibration results:

Fig. 18 shows the comparison between the observed and simulated hydrographs. Table 11 shows the summary of the results obtained during the model calibration for the period of 2015–18.

(b) Validated results:

Fig. 19 shows the comparison between the observed and simulated hydrographs. Table 12 shows the summary of the results obtained during the model validation for the period of 2013–14.

4.1.2 For gridded precipitation data

(a) Calibration results:

Fig. 20 shows the comparison between the observed and simulated hydrographs using the IMD gridded datasets. Table 13 shows the summary of the results obtained during the model calibration for the year 2015–2018.

(b) Validated results:

Fig. 21 shows the comparison between the observed and simulated hydrographs using the IMD gridded datasets. Table 14 shows the summary of the results obtained during the model calibration for the year 2013–14.

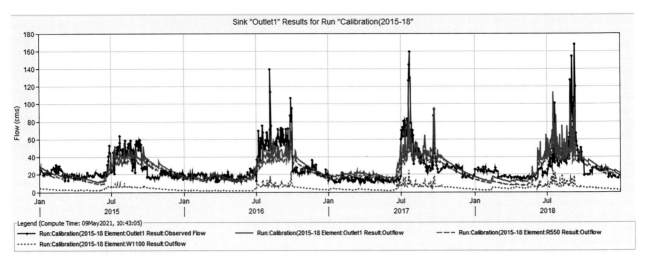

FIG. 18 Observed and Simulated flow for calibrated years (2015–18).

TABLE 11 Summary of the results for calibrated model.

Calibrated volume (mm)	352.04
Nash-Sutcliffe Efficiency(NSE)	0.61
RMSE	0.6
% Bias	1.09

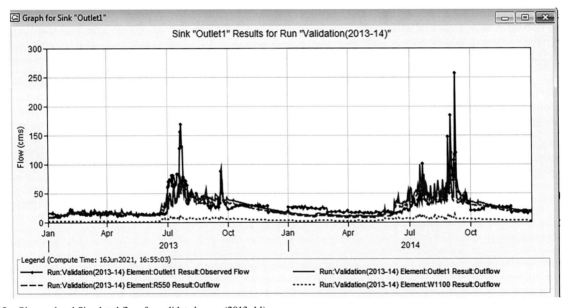

FIG. 19 Observed and Simulated flow for validated years (2013–14).

TABLE 12 Summary of results for validated model.

Calibrated volume (mm)	171.27
Nash-Sutcliffe Efficiency(NSE)	0.55
RMSE	0.7
% Bias	−0.37

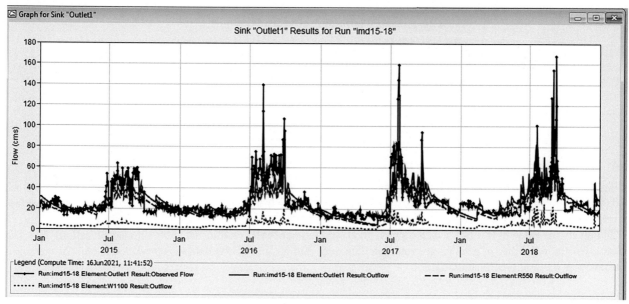

FIG. 20   Observed and Simulated flow for calibrated year 2015–18 of the IMD gridded datasets.

TABLE 13 Summary of results for calibrated model.

Calibrated volume (mm)	348.48
Nash-Sutcliffe efficiency(NSE)	0.57
RMSE	0.7
% Bias	0.08

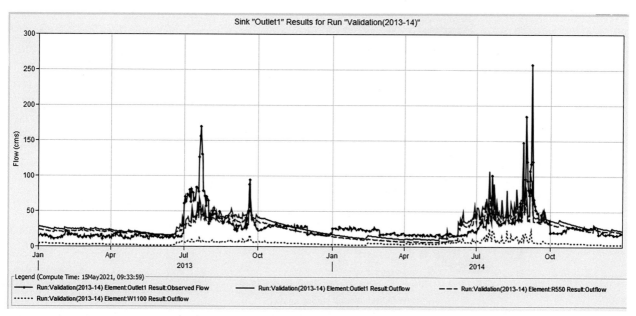

FIG. 21   Observed and simulated flow for validated year 2013–14 of IMD gridded datasets.

TABLE 14 Summary of results for validated model.	
Calibrated volume (mm)	179.03
Nash-Sutcliffe Efficiency(NSE)	0.51
RMSE	0.8
% Bias	4.16

4.2 Goodness of fit curve

In Fig. 22, the dots signify flow, while the straight dotted line denotes equality. The accuracy of the runoff forecast is close to 0.66, resulting in a satisfactory flow pattern. The R^2 number ranges from 0 to 1, with the greater R^2 values indicating less error in variance. Values greater than 0.5 are generally regarded as acceptable (Moriasi et al., 2007).

In Fig. 23, the value of R^2 is 0.8 which shows a good correlation of IMD satellite data with rain gauge stations data. The result gives the reliability of gridded precipitation data over rain gauge stations data which will be useful for ungauged catchment or missing precipitation data.

5. Limitations of the study

As the current model is a continuous-based one but due to the unavailability of input data, the SCS Curve Number is selected for the loss method which is mainly used for the event-based model is the main limitation of produced work in this paper.

● The quality and quantity of data collected and the period that they covered is another limitation. The high-quality resolution DEM for the basin would have produced a better basin parameters result.

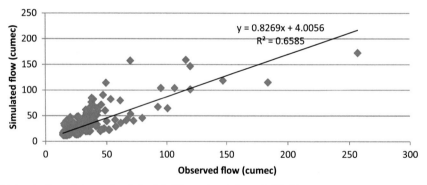

FIG. 22 Coefficient of determination of observed and simulated flow using station rainfall data.

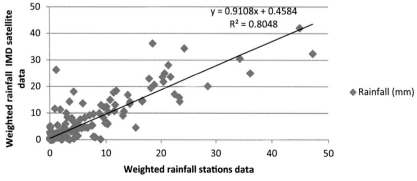

FIG. 23 Correlation coefficient between station rainfall data vs IMD gridded data.

- If the data set had been completed with evapotranspiration data, the study accuracy would have increased greatly.
- During the calibration of a model in HEC-HMS, multiple sets of parameter values can produce the comparable results. The calibrated parameter values selected may or may not have a physical impact on the field.

6. Conclusions

In this study, a hydrological model of the basin with the different climatic and geographical characteristics was developed. HEC-HMS modeling using SCS curve number method used for the actual infiltration calculations are performed by a loss model, SCS unit hydrograph method to transform the excess precipitation into the direct runoff, recession method for actual subsurface calculations are performed by a baseflow method, and routing muskingum model for routing the reach was found to be suitable methods for the Hasdeo river basin, India. The model used for rainfall–runoff simulation shows the Nash-Sutcliffe model efficiency criterion as the range of 0.43 to 0.64, the percentage error in volume as the range of 3.00–8.20 respectively. These results of the rainfall–runoff simulation model indicate the close and good correlation between the simulated and observed flow in this study area. Despite the difficulties, limitations, and uncertainties associated with the acquiring observations and measured parameters, the simulation of the rainfall–runoff process in the Hasdeo river basin produced the optimistic results. The availability and best quality of data sets at the rainfall and discharge stations in the basin were used to choose the rainfall and discharge data for the calibration and validation. The peak flow and the discharge volume obtained during validation were quite similar to the measured values, which shows the applicability of the model for utility and planning of water resource management in the Hasdeo river basin. HEC- Instead of measuring runoff in the basin, the HMS model can help saving time and money by the obtaining runoff data.

Due to higher temperatures during the summer season in Chhattisgarh, there is a higher loss of water through evapotranspiration. If the data set had been completed with evapotranspiration data, the study's accuracy would have increased greatly. Because the final calibrated set of the parameter values may not have much physical relevance, the models may be difficult to validate.

Acknowledgment

We wish our sincere thanks to the Data Collection Centre, Water Resources Department Raipur for providing the essential Meteorological data such as Daily Rainfall data and Daily discharge data for my basin. We are also grateful to the Director, National Institute of Hydrology Roorkee, and the Head Water Resources Division for providing the institutional facilities, particularly the GIS lab.

References

Agrawal, A., 2005. A Data Model with Pre and Post Processor for HEC-HMS. Report of Graduate Studies, Texas A & M Univ, College Station, Texas, USA.

Bazrkar, M.H., Adamowski, J., Eslamian, S., 2017. Water system modeling. In: Furze, J.N., Swing, K., Gupta, A.K., McClatchey, R., Reynolds, D. (Eds.), Mathematical Advances Towards Sustainable Environmental Systems. Springer International Publishing, Switzerland, pp. 61–88.

Bedient, P.B., Huber, W.C., Vieux, B.E., 2008. Hydrology and Floodplain Analysis (Vol. 816). Prentice Hall, Upper Saddle River, NJ.

Beighley, R.E, Moglen, G.E., 2003. Adjusting measured peak discharges from an urbanizing watershed to reflect a stationary land use signal. Water Resour. Res. 39 (4), 1–11.

Beven, K.J., 2012. Rainfall-Runoff Modelling: The Primer/Keith Beven.—2nd edition., ISBN: 978-0-470-71459-1.

Eslamian, S.S., Mehrabi, A., 2005. Identifying emprical equations for time of concentration in mountainous watershed. J. Agric. Nat. Resour. 12 (5), 36–45 (in Persian).

Halwatura, D., 2013. Application of the HEC-HMS model for runoff simulation in a tropical catchment. Environ. Model. Softw. 46, 155–162. https://doi.org/10.1016/j.envsoft.2013.03.006.

Horritt, M.S., Bates, P.D., 2002. Evaluation of 1D and 2D numerical models for predicting river flood inundation. J. Hydrol. 268 (1–4), 87–99. https://doi.org/10.1016/S0022-1694(02)00121-X.

Knisel, W.G., Davis, F.M., 2000. GLEAMS (Groundwater Loading Effects of Agricultural Management Systems), Version 3.0, User Manual. USDA–ARS, Tifton, GA.

Linsley, R.K., Kohler, M.A., Paulhus, J.L.H., 1982. Hydrology for engineers. McGraw-Hill, New York, NY.

Majidi, A., Shahedi, K., 2012. Simulation of rainfall-runoff process using green-ampt method and HEC-HMS model (Case Study: Abnama Watershed, Iran). J. Hydraul. Eng. 1 (1), 5–9. https://doi.org/10.5923/j.ijhe.20120101.02.

McCarthy, G., 1938. The Unit Hydrograph and Flood Routing. U.S. Army Corps of Engineers, USA, Providence.

Milly, P.C.D, Betancourt, J., Falkenmark, M., Hirsch, R.M., Kundzewicz, Z.W., Lettenmaier, D.P., Stouffer, R.J., 2008. Stationarity is dead: whither water management? Science 319, 573–574.

Moriasi, D., Arnold, J., Van Liew, M., Bingner, R., 2007. Model evaluation guidelines for systematic quantification of accuracy in watershed simulations. Trans. ASABE 50 (3), 885–900. https://doi.org/10.13031/2013.23153.

Moriasi, D., Gitau, M., Pai, N., & Daggupati, P. (2015). Hydrologic and water quality models: performance measures and evaluation criteria. Am. Soc. Agri. Biol. Eng., 58, 1763–1785.Moriasi, D., Gitau, M., Pai, N., Daggupati, P., 2015. Hydrologic and water quality models: performance measures and evaluation criteria. Am. Soc. Agri. Biol. Eng. 58, 1763–1785.

Nash, J.E., Sutcliffe, J.V., 1970. River flow forecasting through conceptual models part I—a discussion of principles. J. Hydrol. 10 (3), 282–290.

NRCS, U., 2004. Estimation of direct runoff from storm rainfall. National Engineering Handbook Part, p. 630.

Prasad, P., 2015. Report—Hasdeo sub basin, Theme: Environmental flow, Forum for Policy Dialogue on Water Conflicts in India, India.

Rathod, P., Borse, K., Manekar, V.L., 2015. Simulation of rainfall-runoff process Using HEC-HMS (case study: Tapi River, India). In: 20th International Conference on Hydraulics, Water Resources and River Engineering, (December), pp. 17–19.

Singh, J., Knapp, H., Demissie, M., 2004. Hydrologic modeling of the Iroquois River watershed using HSPF and SWAT. J. Am. Water Resour. Assoc. 41 (2), 343–360. https://doi.org/10.1111/j.1752-1688.2005.tb03740.x.

Smith, M.J., Clark, C.D., 2005. Methods for the visualization of digital elevation models for landform mapping. Earth Surf. Process. Landf. 30 (7), 885–900. https://doi.org/10.1002/esp.1210.

US Army Corps of Engineers, 2010. Hydrologic Modeling System User's Manual Version 3.5 s.l. Hydrologic Engineering Center.

USACE, 2018. Hydrologic Modeling System (HEC-HMS) User's Manual: Version 4.3.0. CA. Hydrologic Engineering Center: Davis, USA, USA.

USACE-HEC, 2000. Hydrologic modeling system HEC-HMS technical reference manual. US Army Corps of Engineers, Hydrologic Engineering Centre (HEC), Davis, USA.

USDA, 2009. Chapter 7–Hydrologic Soil Groups in: NRCS–National Engineering Handbook (NEH), Part 630–Hydrology. USDA NRCS, Washington, DC. pp. 7–1.

Visweshwaran, R., 2017. Application of the HEC-HMS model for runoff simulation in the Krishna basin. Indian Institute of Technology Bombay. https://doi.org/10.13140/RG.2.2.13326.05448.

Water Year Book, 2013. Water Resource Department. Government of Chhattisgarh, India.

Wu, P., Christidis, N., Stott, P., 2013. Anthropogenic impact on Earth's hydrological cycle. Nat. Clim. Change 3 (9), 807–810. https://doi.org/10.1038/nclimate1932.

Zeckoski, R., Smolen, M., Moriasi, D., Frankenberger, J., Feyereisen, G., 2015. Hydrologic and water quality terminology as applied to modeling. Am. Soc. Agri. Biol. Eng., 1619–1635. https://doi.org/10.13031/trans.58.10713.

Zheng, C., Hill, M., Cao, G., Ma, R., 2012. MT3DMS: model use, calibration, and validation. Am. Soc. Agri. Biol. Eng. 55, 1549–1559. https://doi.org/10.13031/2013.42263.

Further reading

Ahmad, I., Verma, M.K., (2015). Application of analytic hierarchy process in water resources planning: a GIS based approach in the identification of suitable site for water storage. Water Resour. Manag. (2018) 32:5093–5114. https://doi.org/10.1007/s11269-018-2135-x.Ahmad, I., Verma, M.K., 2015. Application of analytic hierarchy process in water resources planning: a GIS based approach in the identification of suitable site for water storage. Water Resour. Manag. 2018 (32), 5093–5114. https://doi.org/10.1007/s11269-018-2135-x.

K.M.B. Boomer, D.E. Weller, T.E. Jordan, Lewis, L.., Z.-J. Liu, J. Reilly, Gary S., and A.A. Voinov, (2012). Using multiple watershed models to predict water, nitrogen, and phosphorus discharges to the Patuxent estuary. J. Am. Water Resour. Assoc. 1-25. DOI: https://doi.org/10.1111/j.1752-1688.2012.00689.Boomer, K.M.B., Weller, D.E., Jordan, T.E., Lewis, L., Liu, Z.-J., Reilly, J., Gary, S., Voinov, A.A., 2012. Using multiple watershed models to predict water, nitrogen, and phosphorus discharges to the Patuxent estuary. J. Am. Water Resour. Assoc., 1–25. https://doi.org/10.1111/j.1752-1688.2012.00689.

Chow, V.T., Maidment, D.R., Mays, L.W., 1988. Applied Hydrology. 0-07-039732-5 McGraw-Hill, New York, USA.

Chu, X., Steinman, A., 2009. Event and continuous hydrologic modeling with HEC-HMS. J. Irrig. Drain. Eng. 135, 119–124.

Fleming, M., Brauer, T., 2016. Hydrologic Modelling System HEC-HMS Quick Start Guide. Version 4.2. US Army Corps of Engineers, Institute of Water Resources, USA.

Fleming, M., et al., 2013. HEC-GeoHMS Geospatial Hydrologic Modelling Extension. Version 10.2. US Army Corps of Engineers, Institute of Water Resources, USA.

Jaiswal, R.K., 2020. Comparative evaluation of conceptual and physical rainfall–runoff models. Appl. Water Sci. 10 (1), 48. https://doi.org/10.1007/s13201-019-1122-6.

Jarugu, A., Mahesh, R., 2020. Event-based rainfall-runoff modeling using HEC-HMS. IOSR J. Mech. Civ.Eng. https://doi.org/10.9790/1684-1704034159.

Khadka, J., Bhaukajee, J., 2018. Rainfall-Runoff Simulation and Modelling Using HEC-HMS and HEC-RAS Models: Case Studies From Nepal and Sweden. Retrieved from thesis database (TVVR-18/5009).

Lewis, D., Singer, M.J., Tate, K.W., 2000. Applicability of SCS curve number method for a California oak woodlands watershed. J. Soil Water Conserv. 55 (2), 226–230.

Lohani A.K., 2022. Lecture Notes—Rainfall-Runoff Analysis and Modeling. National Institute of Hydrology, Roorkee, IndiaLohani, A.K., 2018. Lecture Notes—Rainfall-Runoff Analysis and Modeling. National Institute of Hydrology, Roorkee, India.

Mishra, M., 2016. India River Week – Chhattisgarh, India.

NHP, 2018. Hydrological Modeling—Current Status and Future Directions. Report Volume 1.0. National Institute of Hydrology, Roorke, India.

Parhi, P.K., 2018. Flood Management in Mahanadi Basin using HEC-RAS and Gumbel's extreme value distribution. J. Inst. Eng. (India): A 99, 751–755.

Subramanya, K., 2008. Engineering Hydrology, third ed. Tata McGraw-Hill Publishing Company, New Delhi, India.

USACE-HEC, 2006. Hydrologic Modeling System HEC-HMS v3.2 User's Manual. US Army Corps of Engineers, Hydrologic Engineering Center (HEC), Davis, USA.

Chapter 4

Application of soft computing methods in turbulent storm water modeling

Saeid Eslamian[a,b] and Mousa Maleki[c]

[a]Department of Water Engineering, College of Agriculture, Isfahan University of Technology, Isfahan, Iran, [b]Center of Excellence for Risk Management and Natural Hazards, Isfahan University of Technology, Isfahan, Iran, [c]Department of Civil Engineering, Illinois Institute of Technology, Chicago, IL, United States

1. Introduction

The advent of computers and the development of hardware and software have led to the rapid emergence and growth of computing intelligence. The field of computational intelligence has revolutionized the development of new unconventional data processing and simulation techniques. Integrating intelligence by replicating human reasoning and behavior in a computing environment increases its ability to analyze information under a dynamic change environment. Soft computing is a set of computational methods inspired by human instinctual ambiguity, human intellect, intuition, and uncertainty about real life. Unlike common techniques that rely on precise solutions, soft computing uses the tolerance of inaccuracy and the trivial and uncertain nature of the problem to provide an approximate solution to a problem in a timely manner. Soft computing, a multidisciplinary field, uses a variety of statistical, probabilistic, and optimization tools to evolve distinct computational methods, including neural networks, evolutionary computing, fuzzy systems, machine learning, and probabilistic reasoning. Among the various subsets of soft computing, neural networks, genetic algorithms, and fuzzy logic are the main actors and are often used for problems related to real-life applications. Inspired by the human brain's ability to learn, artificial neural networks (ANNs) are able to create subtle relationships between independent and dependent variables whose interactions are unknown, nonlinear, or too complex to display. Genetic Algorithms (GA) is a random search and optimization computational tool that revolves around the evolutionary theories of natural genetics and natural selection. Fuzzy logic (FL) helps to solve real-life problems that are always present in one way or another and are prone to ambiguity and uncertainty (Amiri et al., 2019).

Rainfall and runoff are the most important hydrological processes that depend on physiographic, climatic and biological factors. Rainfall on a catchment is the fall of moisture from the atmosphere to the ground in the form of liquid (rainfall) or frozen (snow, hail, snow, frozen rain). The runoff of a catchment in any given period is the total amount of water discharged into a stream or reservoir in that period, which can be expressed as millimeters of water over a catchment or the total volume of water per cubic meter or hectare. Water available in a specific period and area in the river basin can be determined with the help of rainfall runoff regression model, i.e., hydrological model. The main components of a hydrological model are: precipitation, evaporation, transpiration, interception, infiltration, flow, spatial and time scale, when the intensity of rainfall exceeds the infiltration rate, runoff begins immediately. The physiography of the local area affects the amount of rainfall and runoff produced from the catchment. Earth infiltration capacity depends on soil porosity and measures the resistance of water flow to deep layers. Upper soils with a clay or loam content of more than 20% form a cap with a lower infiltration capacity and indirectly produce more runoff compared to loose sandy soils. Runoff efficiency, i.e., runoff volume per unit area increases with decreasing catchment size. Vegetation density, thick layer of leaf or grass mulch, low flow velocity in the flat area increase concentration/absorption over time,. Wavy soils have more runoff than flats due to the extra energy received from the slope and less time to penetrate. Higher drainage creates more runoff. The ratio of maximum rainfall at one point to average rainfall in a catchment called the coefficient of distribution also affects the runoff. The prevailing wind direction increases or decreases the flow rate and indirectly affects the runoff. The average characteristics of the basin, i.e., temperature, wind speed, relative humidity and annual rainfall also affect the relationship between rain runoff in the basin.

The design, operation, maintenance and optimal use of existing or proposed water resources projects in a particular river basin requires accurate knowledge of the relevant rainfall and runoff generated over a specific period/period. In addition to

Handbook of HydroInformatics. https://doi.org/10.1016/B978-0-12-821962-1.00012-X

planning and developing water resources projects, rainwater runoff models help to develop flood control, drought management, reservoir optimization, water supply, and more. In the study area, the functional relationships between rainfall and runoff are very complex due to its dependence on the factors discussed in the previous paragraph. In such cases where the interactions between the variables are complex, conventional mathematical techniques in the form of regression equations do not provide a complete picture of the rainfall-runoff phenomenon. Soft computing tools offer a simpler approach to conventional hard computing in dealing with the real-life phenomenon associated with the inaccurate, noisy, complex, and ambiguous nature of information. The review article is an attempt to introduce the three main areas of soft computing, artificial neural networks (ANN), genetic algorithms (GA) and fuzzy logic (FL) and their applications in modeling complex rainfall and runoff relationships.

Soft computing is based on techniques such as fuzzy logic, genetic algorithms, artificial neural networks, machine learning, and expert systems. Although soft computing theory and techniques were first introduced in 1980s, it has now become a major research and study area in automatic control engineering.

2. Rainfall-runoff modeling between SWMM and fuzzy logic approach

A comprehensive hydrological model, like the storm water management model (SWMM), has been widely used for rainfall-runoff simulation. In recent years, simple and effective modern modeling techniques have also brought great attention to the prediction of runoff with rainfall input (Wang and Altunkaynak, 2012).

Accurate determination of storm water is important for the analysis and planning of the catchment drainage system. It is very important to study projects related to the improvement of sewage or canal storms and the design of hydraulic structures. Conversion of rain to surface runoff of a basin is a complex hydrological process that involves nonlinear and dynamic transitions influenced by many physical and hydrological parameters. The level of complexity that characterizes hydrological models for predicting surface runoff from rainfall input is quite different due to different simulation purposes. The most rudimentary models provide only a description of the global water level of the basin. Next, experimental runoff-related formulas with effective parameters, such as rainfall, soil type, soil moisture, evapotranspiration, infiltration rate, and land use conditions, may be developed for a hydrological modeling study (Poncea and Shetty, 1995; Xu, 1999; Yokoo et al., 2001).

The simplest models of the logical equation with the "runoff coefficient" are introduced to determine the total runoff of an event by multiplying the runoff coefficient by the total rainfall (French, 2002; Şen and Altunkaynak, 2006). Through the urbanization process, the impermeability ratio plays a very important role in influencing the hydrological behavior of the catchment. Therefore, runoff coefficient can be considered as the ratio of impermeable area directly connected to the drainage network to the total area of the watershed (Desbordes, 1975) or by integration with other influential variables, such as soil characteristics, land use conditions, concentration time, and total rainfall is tabulated into a regression formula or values. Since the infiltration level depends on the previous wet and dry soil conditions, the scaling properties of duration, frequency and amount of wet and dry periods (Özger et al., 2011) also strongly influence the hydrological trend and runoff forecasts.

A black box-based conceptual modeling approach to runoff calculation has also attempted to develop linear/nonlinear formulations with calibration to illustrate the complex physical process that occurs in rainfall-runoff evolution. The unit hydrograph, which presents the response of the hypothetical runoff unit in the watershed to the rainfall input unit, is one of the famous examples of the black box linear approach. This method can make it easy to calculate the total runoff simply by performing a rotation between the rain inlet and the single hydrograph outlet. However, this type of model fails to achieve the nonlinear dynamics of the transfer process between rainfall and runoff. The inability of the single hydrograph to model the nonlinear process prompted the development of nonlinear black box systems for runoff calculation. Preliminary studies on the nonlinear runoff model using multiinput and single-output approaches were presented by Müftüoğlu (1984) and Kothyari and Singh (1999).

Next, a multiobjective runoff precipitation model with parameters uses a fuzzy optimal method with a genetic algorithm calibrated by Cheng et al. (2002). Recently, the use of artificial neural network (ANN) techniques in hydrological modeling has received more attention. The study of long-term discharge prediction using ANN and other data-driven models was presented by Cheng et al. (2008), Lin et al. (2006), Wang et al. (2009), and Wu et al. (2009). To model runoff precipitation, this knowledge-based ANN has the ability to learn the relationship between rainfall and runoff from previously measured data through a training method. This trained relationship is created in selected layers of nonlinear transfer functions to convert rainfall to runoff output. The learning process without explicit physical inputs makes the ANN modeling process very simple and effective (Rajurkar et al., 2002; Bruen and Yang, 2006).

Another black box modeling technique for rainfall–runoff simulation is the fuzzy logic (FL) approach, Zadeh (1965), in which linguistic variables rather than numerical values are often used to facilitate the expression of laws and facts, first proposed the concept of fuzzy logic modeling. Fuzzy logic modeling has been applied in the past to various engineering problems, for example, traffic junction control, water purification process, water level forecasting, flow forecasting and rainwater runoff modeling (Mamdani, 1974; Pappis and Mamdani, 1977; Takagi and Sugeno, 1985; Xiong et al., 2001; Özelkan and Duckstein, 2001; Chau et al., 2005; Şen and Altunkaynak, 2006; Alvisi et al., 2006; Altunkaynak and Şen, 2007; ÖZGER, 2009; Altunkaynak, 2010). As shown in Şen and Altunkaynak (2006), the fuzzy logic model uses linguistic expressions to establish quantitative relationships between model variables. This structure makes the modeling method transparent for calculation, interpretation and analysis. It has been shown that the fuzzy logic method is superior for modeling physical events with nonlinear, uncertain and complex systems, such as hydrological processes involving directly correlated variables. Accordingly, the fuzzy logic method can be effectively used to correlate the total rainfall input and the total runoff response with the calibrated coefficients. However, in terms of runoff time change, because the hydrological system from rainfall to runoff includes transient infiltration, ground flow, and transportation and routing in storm water and storm water canals, the conventional fuzzy logic method cannot control the details of dynamic correlations between rainfall and runoff differ from physics-based models. Therefore, with this limitation, the typical fuzzy logic approach is unable to produce the time-varying hydrograph.

Unlike black box approaches, rainwater runoff models have also been physically developed to determine the response of the runoff basin to a specific rainfall event. These comprehensive models consider the physics of hydrological processes using physical equations and influential parameters, such as physical variables (e.g., area and slope) and hydrological (e.g., impermeability percentage, Manning roughness coefficient). Other factors included in the modeling methods are hydrological losses at the watershed level (such as intrusion, depression storage, plant interception, evaporation). Physics-based models can provide a spatial and temporal shift of the hydrograph for a storm event or for a continuous simulation. Additional flow routing and transport processes are typically implemented in models. In general, they are parametric models that require initial calibration of their characteristic parameters. The following EPA storm water management model (SWMM) (Huber and Dickinson, 1988), HEC-hydrologic modeling (HEC-HMS) (US Army Engineers), MOUSE (Danish Hydraulic Institute) are among the physics-based models., Usually accepted for engineering applications.

3. Urban flood prediction using deep neural network with data augmentation

It is important to predict the likelihood of flooding in urban areas and to ensure the extent of flooding in advance. Dangerous flood events that cause damage to residential and commercial property can occur in a variety of forms and with frequent frequencies. Numerical analysis, trend analysis, and flood forecasting for different rainfall scenarios can be used as important basic data for urban planning or flood response. One-dimensional and two-dimensional flood analysis programs can be used for this purpose. However, in the case of a number-based model, setting up parameters, collecting data, and processing the output data may take some time.

For this reason, data-driven models using machine learning have recently been used to predict urban runoff (Mosavi et al., 2018). Granata et al. (2016) predicted runoff in urban areas through support vector regression (SVR) and compared it with EPA-SWMM simulation results. Although there was inaccuracy in estimating peak runoff, the possibility of improvement was confirmed through verification. If more data is used to learn SVR, the power of peak current prediction increases. Studying the input data for a machine learning method is also important to find a reliable prediction model. Talei et al. (2010) suggested the results of the evaluation of rainfall and discharge inputs with adaptive network fuzzy inference systems (ANFIS). In this study, the criteria for determining the optimal runoff rainfall analysis model were presented among the input data of successive, pruned and nonconsecutive time series. In addition to a general artificial neural network (ANN), the Deep Learning (DL) method is applied in the field of water resources to improve the predictive power of hydrological data and to incorporate more ideas into the model. With regards to Shen (2018), the use of the DL method in water resources and hydrology is becoming more common, and is likely to provide good results in water resources as well as in most scientific fields. Hu et al. (2018) used the DL model based on short-term memory (LSTM) to simulate rainfall runoff and data in 86 runoff rainfall patterns. This research was feasible because the LSTM database could be collected by hydrological observations in the target river basin.

Although many studies have been performed to predict precipitation, hydrological data, and flood events using ANN and DL techniques, research on completing input data for DL is lacking.

In fact, some studies have raised the problem of insufficient data for use in a data-driven model. Li and Willems (2018) raised the problem of input data shortages and tried to predict the likelihood of flooding in urban areas. To predict the flood, a conceptual model considering the drainage system was combined with logistic regression method. The hydraulic model was compared with the predicted results, and the confirmation results that could be developed were shown. Nikhil et al. (2019) analyzed a dam basin that had recently been affected by a catastrophe that resulted from the dam leaving. Due to lack of information, the outflow of the dam cannot be predicted. Rainfall, flow rate and water level data were used to predict downstream flooding, and by adding observation data to the ANN network, the forecasting power was reduced. It seems that if the uncertain observation data is insufficient, the function of a neural network will be destroyed.

4. Application of expert system for storm water management modeling

With the development of the economy, the migration of people from rural to industrial areas, i.e., urbanization, is accelerating around the world. As a result, due to the construction of parts, factories, sports and recreational facilities, etc., large amounts of residential areas, such as meadows, forests, agricultural lands, etc., are sealed. Thus, urbanization, on the one hand as a symbol of economic development, on the other hand has many negative effects. One of the most direct problems is overloading the municipal sewage system, which is currently the most common drainage system in the world. Overload mainly causes the following problems:

– Overflow inside the city;
– Increased flooding in nearby rivers or in cities;
– Increasing the burden of pollution of natural waters near cities;
– Reduce groundwater recharge;

To solve these problems caused by over-sewage systems, researchers and engineers are now introducing so-called storm water management measures such as green roofs, on-site detention, intrusion, etc., instead of building drainage pipes, alone in response to the rapid development of the city. In the case of on-site storms, the infiltration of rainwater from a local catchment, such as the roof of a house, parking spaces, etc., is of particular interest because it reduces not only the storm runoff volume but also the pollution. Load in runoff, as well as increase groundwater charge. However, its installation is more complicated because it involves more influential factors on the groundwater and soil conditions of the area. Diagnosing a local area for a storm water intrusion system on site usually involves assessing geological factors, hydrological factors, geographical factors, environmental factors, economic factors, and so on. This assessment is an unstructured issue. Hence, it cannot be solved with a regular algorithm or model, but is done systematically according to the term decision tree. In addition, such an assessment is usually based not only on relevant theoretical considerations but also on a large amount of practical experience as well as the availability of relevant influencing data. Therefore, the logic of a decision or decision tree varies from case to case.

To meet the needs of this complex assessment and the changeable nature of the assessment logic, a tool called the Expert System (ES) and a special expert system have been developed to assess storm seepage in situ. This tool not only facilitates the correct modification of the decision-making logic of the system, but also facilitates the automation of repeated assessments for large amounts of subcatchments in the planning of on-site storm management actions for an entire urban catchment. Because such scheduling is also related to a space issue, the tools developed, as well as the specific system, are increasingly integrated into a GIS operating system.

5. Developing a flexible expert system tool

Expert System Research, a widely used branch of artificial intelligence (AI), focuses primarily on tasks that can only be performed by experienced, trained, or in other words, by some professionals. An expert system developed is a computer program that, as an expert or a real expert, provides advice or recommendations to the user about the specified domain. In such a computer program, human expertise in the designated field is well represented and stored in the form of a knowledge base. With the implementation of a program or system, the knowledge-based reasoning process is realized through another important part of the expert system, the inference engine.

However, for the end use of a built-in system or the successful construction of an expert system, a user-friendly interface is of particular importance because the user interface provides the user with a human-computer dialog in which data entry Requested, interpretations of the argument at the time shown, hints or statements, as well as printed results. Unlike conventional programming, the development of an expert system involves the representation of knowledge and exploratory programming (Cawsey, 1998). In order to save programming costs and effort, a knowledge acquisition interface and a general inference engine are usually created using a system-specific shell or tool. A problem-specific expert system is used to communicate between professionals or users and the expert system. It should be emphasized that the user interface provided by the expert system shell is not exactly the same interface that is problematic in the expert system, part of which is actually designed by the developer of the specific expert system through the knowledge base formulator.

6. Development of ES tool "Flext"

The expert system tool presented in this article was initially based on a popular CLIP (Integrated C Language Production System) expert system. A visual C ++ application written with the integration of the CLIP inference engine. The program integrates the CLIP inference engine with a structural rule system, a question system with graphical interfaces and based on GIS data. Relevant rules are categorized into nodes in an expert system developed so that a complete system of rules can be constructed in a more flexible decision tree as a network of nodes. It makes the structural performance of a rule-based system much more comprehensible and therefore much more useful for developers in creating a large rule-based system. Along with this runtime module, a graphical knowledge formulator is written in Visual Basic that enables the visualization of the logical structure of rules in a knowledge base. In addition, this formulator provides rules that are simply written in natural mathematical language. However, although a tool based on the CLIP inference engine has the above advantages, it is not easy to achieve the internal inference process of the CLIPS embedded inference module and also to establish a relationship between the knowledge base formula and the runtime environment. An independent and structurally related system based on the above basic knowledge formulator is written in Visual Basic as a program. Thus, communication between the two modules is much easier, and hence a knowledge-based developer or end user of a developed expert system can track a reasoning process or a system implementation gradually. This additional advantage is very useful for the user in understanding the final results of the system as well as for a knowledge-based developer to control and check for errors during development. Another added benefit of this integrated tool is that it provides more flexibility for the knowledge base developer to output information during or after system execution. Knowledge base developers can use these features to write detailed interpretations for their systems. Knowledge Representation The developed tool uses knowledge-representing functions and rules. Variables are used to display information. A complex mathematical expression is used to determine the condition of a rule or one of the operations of a rule. In addition, several keywords such as Set, GoTo, Conclusion, Run, Open Database, etc. have been introduced to formulate actions. Although this tool is designed to be used in the development of rule-based systems, so-called stand-by functions can be added to a knowledge base other than rules. Standby functions are complex mathematical functions that are executed automatically whenever there is a change in the values of any of its arguments. Structure of a knowledge base as mentioned above, the most important feature in the developed Flext tool is the grouping of rules related to nodes. The whole system of rules can then be grouped into nodes, and the logical relationship or inference between nodes can be explicitly expressed through the "GoTo" keyword in the action sections of a node's rules, or implicitly through conditional The rules of a node. Thus, a system of rules can be constructed as a tree or a network of nodes.

Another feature of this tool is that it supports graphical interfaces for the question and provides an interface for designing the question system. Like rules, questions are held in nodes. With each question, a variable must be shown to calculate the answer, although it may not be processed during ES execution. Allowed answers can also be set for each year. Each node can have the required number of questions. The mechanism of inference of the general implementation cycle and the main inference philosophy in Flext is similar to that in CLIP or other stock market rule-based systems. That is, the system is driven by data or facts. After running the system, the inference engine constantly monitors the values of the knowledge base variables due to the input or operation of a rule (at the beginning of the system, all variables are empty). In the event of any change, the inference engine checks the relevant rules in the so-called active rule list and all standby functions to determine if the active rule requirements are met or if there are changes in the standby arguments. Active rules are rules of active nodes. At the beginning of the system, all the root nodes of a knowledge base (for example, a decision tree or a network of nodes) are active nodes. The status of nodes will change dynamically during the inference process due to the implementation of the rules. In order to execute the check process systematically, a special variable storage structure, rule structure and performance structure are constructed.

These rules whose conditions are met are called applicable laws and are placed in a stack. The sequence of rules applicable to the stack depends on the sequence of rules in the knowledge. Standby functions are called executable functions if they change their arguments and place them in another stack. The inference engine first executes all executable functions and examines the active rules and standby functions that are related to the variables derived from the executable functions, again repeating this process in terms of applicable rules and functions. Until there are no more usable functions. Then, the applicable rules are executed at the top of the applicable rules stack. Enforcing a rule can change the status of nodes, from active to inactive or from inactive to active, due to the GoTo function. When a node is disabled, all applicable rules belonging to that node are removed from the stack of applicable rules. Similarly, when a passive node becomes an active node, all its rules are included in the list of active rules and the inference engine examines them for application. Apart from these results, implementing a rule can also change the values of system variables due to the Set function. If the values of the system variables change, the inference engine re-examines the related standby functions and active rules for operations and applicable rules. The system performs such strict execution loops as long as no executable function and no applicable law. The questions in the active nodes are then taken turns. After each answer, the motor repeats the inference of the check loops and then asks for more input. This process continues until the result is reached. This system can also be stopped by Action Print, which prevents comments or comments made on the knowledge base by the knowledge base developer. Conclusion can be the output of the values of a series of variables or constants. The data structure of a condition usually consists of a compound logical expression that contains more mathematical functions. All operations in a situation can be separated into one operation.

7. Conclusions

Intelligent systems and therefore soft computing techniques are becoming more important as computer power Processing machines increase and their cost decreases. Intelligent systems are needed for complex decisions and use sophisticated algorithms to choose the best result from many possibilities. This requires fast processing the large power and storage space that has recently become available to many research centers in recent years.

With the power and knowledge of the Internet of Things (IoT) concept, the need to use soft computing techniques and the construction of intelligent systems have become more important than ever. Today, most soft computing applications can be managed efficiently with low-cost but very fast microcontrollers in storm water management modeling.

Rainfall and runoff interactions and their proper evaluation are an integral part of any hydrological study. Complex, ambiguous, and nonlinear factors affecting rainfall and runoff relationships make mathematical modeling difficult. Such a situation requires nature-inspired computational tools to deal with the phenomenon of real life, which is always under false, ambiguous and noisy information. Soft computing has provided a viable alternative to traditional mathematical techniques by integrating probabilistic, statistical, and optimization techniques into the computing environment. The combination of these methods has given a new dimension to computing in which human behavior similar to the ability to reason, intuition, intelligence and wisdom can be combined through software programming. Although these computational tools do not provide accurate answers, they do offer sensible decision-making solutions to problems influenced by vague and noisy information. This article provides a brief introduction to the three main methods of soft computing, including artificial neural networks, genetic algorithms and fuzzy logic, and their application in rainwater runoff modeling. Studies have shown that by applying these techniques, a somewhat complex analytical method can be avoided. By learning from empirical or historical data, the ANN network can be the backbone of complex, unknown, and difficult relationships to describe the relationship between rainfall and runoff. GA can be harnessed to calibrate these rainfall runoff models through random search capability. FL's imitation of human reasoning and decision-making ability can be used to model the problems of misinformation, ambiguity, and inaccuracy.

References

Altunkaynak, A., 2010. A predictive model for well loss using fuzzy logic approach. Hydrol. Process. 24 (17), 2400–2404.

Altunkaynak, A., Şen, Z., 2007. Fuzzy logic model of lake water level fluctuations in Lake Van, Turkey. Theor. Appl. Climatol. 90 (3–4), 227–233.

Alvisi, S., Mascellani, G., Franchini, M., Bardossy, A., 2006. Water level forecasting through fuzzy logic and artificial neural network approaches. Hydrol. Earth Syst. Sci. 10 (1), 1–17.

Amiri, M.J., Zarei, A.R., Abedi-Koupai, J., Eslamian, S., 2019. The performance of fuzzy regression method for estimating of reference evapotranspiration under controlled environment. Int. J. Hydrol. Sci. Technol. 9 (1), 28–38.

Bruen, M., Yang, J., 2006. Combined hydraulic and black-box models for flood forecasting in urban drainage systems. J. Hydrol. Eng. 11 (6), 589–596.

Cawsey, A., 1998. The Essence of Artificial Intelligence. Prentice Hall.

Chau, K.W., Wu, C.I., Li, Y.S., 2005. Comparison of several flood-forecasting models in Yangtze River. J. Hydrol. Eng. 10 (6), 485–491.

Cheng, C.T., Ou, C.P., Chau, K.W., 2002. Combining a fuzzy optimal model with a genetic algorithm to solve multi-objective rainfall–runoff model calibration. J. Hydrol. 268 (1–4), 72–86.

Cheng, C.T., Wang, W.C., Xu, D.M., Chau, K.W., 2008. Optimizing hydropower reservoir operation using hybrid genetic algorithm and chaos. Water Resour. Manag. 22 (7), 895–909.

Desbordes, M., 1975. Estimation des Coefficients de Ruissellement Urbains. Journées de l'hydraulique 13-2, 1–6.

French, R., 2002. The Rational Method: Past, Present and Future. Water Panel, The Institute of Engineers Australia (Queensland Division), Technical Session, Brisbane, Australia, pp. 1–8.

Granata, F., Gargano, R., De Marinis, G., 2016. Support vector regression for rainfall-runoff modeling in urban drainage: a comparison with the EPA's storm water management model. Water 8 (3), 69.

Hu, C., Wu, Q., Li, H., Jian, S., Li, N., Lou, Z., 2018. Deep learning with a long short-term memory networks approach for rainfall-runoff simulation. Water 10 (11), 1543.

Huber, W.C., Dickinson, R.E., 1988. Storm Water Management Model, User's Manual, Version 4.0, EPA-600/3-88-001a. U.S. Environmental Protection Agency, Athens, GA, USA.

Kothyari, U.C., Singh, V.P., 1999. A multiple-input single-output model for flow forecasting. J. Hydrol. 220 (1–2), 12–26.

Li, X., Willems, P., 2018. A data-driven hybrid urban Flood modeling approach. In: Loggia, G.L., Freni, G., Puleo, V., Marchis, M.D. (Eds.), EPiC Series in Engineering, HIC 2018, Proceedings of the 13th International Conference on Hydroinformatics, Palermo, Italy, 1–6 July 2018. vol. 3. EasyChair, Manchester, UK, pp. 1193–1200.

Lin, J.Y., Cheng, C.T., Chau, K.W., 2006. Using support vector machines for long-term discharge prediction. Hydrol. Sci. J. 51 (4), 599–612.

Mamdani, E.H., 1974. Application of fuzzy algorithms for control of simple dynamic plant. In: Proceedings of the Institution of Electrical Engineers. Vol. 121, No. 12. IET, pp. 1585–1588.

Mosavi, A., Ozturk, P., Chau, K.W., 2018. Flood prediction using machine learning models: literature review. Water 10 (11), 1536.

Müftüoğlu, R.F., 1984. New models for nonlinear catchment analysis. J. Hydrol. 73 (3–4), 335–357.

Nikhil, B.C., Arjun, N., Keerthi, C., Sreerag, S., Ashwin, H.N., 2019. Flood prediction using flow and depth measurement with artificial neural network in canals. In: Proceedings of the Third International Conference on Computing Methodologies and Communication (ICCMC 2019), Erode, India, 27–29 March, pp. 798–801.

Özelkan, E.C., Duckstein, L., 2001. Fuzzy conceptual rainfall–runoff models. J. Hydrol. 253 (1–4), 41–68.

ÖZGER, M., 2009. Comparison of fuzzy inference systems for streamflow prediction. Hydrol. Sci. J. 54 (2), 261–273.

Özger, M., Mishra, A.K., Singh, V.P., 2011. Scaling characteristics of precipitation data over Texas. J. Hydrol. Eng. 16 (12), 1009–1016.

Pappis, C.P., Mamdani, E.H., 1977. A fuzzy logic controller for a trafc junction. IEEE Trans. Syst. Man Cybern. 7 (10), 707–717.

Poncea, V.M., Shetty, A.V., 1995. A conceptual model of catchment water balance: 1. Formulation and calibration. J. Hydrol. 173 (1–4), 27–40.

Rajurkar, M.P., Kothyari, U.C., Chaube, U.C., 2002. Artificial neural networks for daily rainfall—runoff modelling. Hydrol. Sci. J. 47 (6), 865–877.

Şen, Z., Altunkaynak, A., 2006. A comparative fuzzy logic approach to runoff coefficient and runoff estimation. Hydrol.l Process. 20 (9), 1993–2009.

Shen, C., 2018. A transdisciplinary review of deep learning research and its relevance for water resources scientists. Water Resour. Res. 54 (11), 8558–8593.

Takagi, T., Sugeno, M., 1985. Fuzzy identification of systems and its applications to modeling and control. IEEE Trans. Syst. Man Cybern. 1, 116–132.

Talei, A., Chua, L.H.C., Wong, T.S., 2010. Evaluation of rainfall and discharge inputs used by adaptive network-based fuzzy inference systems (ANFIS) in rainfall–runoff modeling. J. Hydrol. 391 (3–4), 248–262.

Wang, K.H., Altunkaynak, A., 2012. Comparative case study of rainfall-runoff modeling between SWMM and fuzzy logic approach. J. Hydrol. Eng. 17 (2), 283–291.

Wang, W.C., Chau, K.W., Cheng, C.T., Qiu, L., 2009. A comparison of performance of several artificial intelligence methods for forecasting monthly discharge time series. J. Hydrol. 374 (3–4), 294–306.

Wu, C.L., Chau, K.W., Li, Y.S., 2009. Predicting monthly streamflow using data-driven models coupled with data-preprocessing techniques. Water Resour. Res. 45 (8). https://doi.org/10.1029/2007WR006737.

Xiong, L., Shamseldin, A.Y., O'connor, K.M., 2001. A non-linear combination of the forecasts of rainfall-runoff models by the first-order Takagi–Sugeno fuzzy system. J. Hydrol. 245 (1–4), 196–217.

Xu, C.Y., 1999. Estimation of parameters of a conceptual water balance model for ungauged catchments. Water Resour. Manag. 13 (5), 353–368.

Yokoo, Y., Kazama, S., Sawamoto, M., Nishimura, H., 2001. Regionalization of lumped water balance model parameters based on multiple regression. J. Hydrol. 246 (1–4), 209–222.

Zadeh, L.A., 1965. Fuzzy Sets. Inf. Control 8, 338–353.

Chapter 5

Assessment of bed load transport for steep channels on the basis of conventional and fuzzy regression

Mike Spiliotis, Vlassios Hrissanthou, and Matthaios Saridakis
Department of Civil Engineering, Democritus University of Thrace, Xanthi, Greece

1. Introduction

Sediment transport in natural streams and rivers has been studied for centuries by hydraulic engineers and geologists due to its importance in understanding river hydraulics (Ghorbani et al., 2017). The hydraulic geometry of the natural streams is altered through erosion and deposition of sediment. Especially, the excessive deposition may cause an increase of flood frequency, as well as navigation problems. Moreover, the sediment particles are the primary transporters of toxic substances that contaminate aquatic systems. The design and function of dams and reservoirs is affected by long-term sediment yield at the outlet of the corresponding basin (Kitsikoudis et al., 2015).

Depending on the motion mode, sediment transport is divided into bed load transport and suspended load transport. The bed load transport includes the rolling, sliding or sometimes jumping motion along the bed. Suspended load is the sediment that is supported by the upward components of turbulent currents and stays in suspension for an appreciable length of time. Total load is the sum of bed load and suspended load. Total load can also be defined as the sum of bed material load and wash load, based on the source of material being transported. Wash load originates from soil erosion of the corresponding basin and consists of fine materials that are finer than those found on the bed (Yang, 1996).

Bed load transport especially in steep gravel-bed streams is a complicated phenomenon that needs to be quantified for engineering design, habitat protection, and river management and restoration, among others. However, bed load transport prediction within tolerable error margins in such streams remains highly problematic (Gomez and Church, 1989) and site-specific (Barry et al., 2004; Kitsikoudis et al., 2014), and constitutes a challenge for engineers and geomorphologists. This is particularly owed to the fact that the incipience of sediment motion may occur for a relatively wide range of flow conditions and there is not a single threshold for specified flow conditions (Buffington and Montgomery, 1997; Kitsikoudis et al., 2016). This is attributed to the near-bed turbulence (e.g., Diplas et al., 2008), sediment exposure and protrusion to the flow (e.g., Kirchner et al., 1990), and relative depth (e.g., Recking, 2009).

Meyer-Peter, Favre and Einstein published in 1934 a formula related to the transport of uniform sediment on a plane bed, while Meyer-Peter and Müller published in 1948 and 1949 the definite formula related to the transport of sediment mixtures with different values of specific gravity. The main characteristic of the formula of Meyer-Peter and Müller is the distinction of bed roughness due to individual particles from bed roughness due to bed forms or the distinction of bed resistance due to skin friction from bed resistance due to bed forms. The formula of Meyer-Peter and Müller (1948, 1949) was developed on the basis of experiments with nonuniform sediments of various densities and channel slopes ranging from 0.04% to 2%. Smart and Jaeggi (1983a) and Smart (1984a) have modified the formula of Meyer-Peter and Müller, so that the modified form is valid also for steep bed slopes, from 0.04% up to 20%, and for alluvial materials with mean grain size greater than 0.4 mm.

The formula of Smart and Jaeggi (1983a) was applied to the main streams of the subbasins of the basin corresponding to the Forggensee Reservoir, in Bavaria, Germany, because of the steep bed slopes (Hrissanthou, 1986). The basin area amounts to about 1500 km^2. The largest part of the basin is located in Austria (Austrian Alps), and the main stream is the Lech River. The comparison between computed and measured annual values of sediment yield at the basin outlet, namely at the reservoir inlet, was relatively satisfactory, considering the large area and the fact that the computation was performed on a daily basis and that no runoff or sediment yield data were available for the subbasins (Hrissanthou, 1990).

Handbook of HydroInformatics. https://doi.org/10.1016/B978-0-12-821962-1.00021-0

The formula of Smart and Jaeggi (1983a) was also applied to a steep mountain stream (Rio Cordon, Italy) for the computation of the cumulated bed load volume for the exceptional flood of 14 September 1994. The agreement between computed and measured values of cumulated bed load volume was very satisfactory (D'Agostino and Lenzi, 1999).

The formula of Smart and Jaeggi (1983a) was additionally applied to the steep main streams of the mountainous sub-basins of Nestos River basin, downstream of Platanovrysi Dam (northeastern Greece). The basin area amounts to about $840\,km^2$. The comparison between computed and measured annual values of sediment yield at the basin outlet was also satisfactory (Kaffas and Hrissanthou, 2018).

Due to the scarcity of sediment transport relations for sediment transport rates on steep slopes, the traditional formulae, among others Smart and Jaeggi (1983a) formula, are frequently used in dam breach modeling. In an experimental study (Visser, 1995), the experimental results of sand-dike breach erosion were compared with the computational results according, among others, to the formulae of Smart and Jaeggi (1983a), and Bagnold-Visser (Visser, 1988). The formula of Bagnold-Visser provided the relatively best comparison results, namely this formula predicted sand transport rates within a factor two of the experimental values (Visser, 1995).

The ability of depth-averaged numerical models to simulate the complex process of breaching of a noncohesive earthen embankment was investigated in another experimental study (Van Emelen, 2014). The formula of Smart and Jaeggi (1983a), calibrated for high shear stresses and bed slopes, led to the most precise peak outflows, by a river dike overtopping.

The results of the experimental breaching of a sandy dike are also presented in Alhasan et al. (2016). The comparison showed differences between experimental and calculated sediment transport rates. In concrete terms, the comparison showed an overestimation of the breaching rate by most empirical formulae, e.g., the formula of Smart and Jaeggi (1983a). In contrast to the other empirical formulae, the formula of Bagnold-Visser (Visser, 1988) indicated smaller differences between experimental and calculated sediment transport rates (Alhasan et al., 2016).

Fluvial processes in river engineering include a lot of uncertainties (e.g., Spiliotis et al., 2017). In this chapter, the bed load transport equation of Smart and Jaeggi (1983a) for steep channels is reformulated by using fuzzy numbers as coefficients on the basis of experimental data. The reformulation of the above equation is achieved by means of the fuzzy regression approach. In concrete terms, the Tanaka (1987) fuzzy linear regression is used, which reduces to a linear programming problem. As objective function in linear programming, the total sum of the produced widths is established. Additionally, all the data must be included within the produced fuzzy band. Regarding the fuzzy nonlinear regression for crisp data, it can be reduced to a fuzzy linear regression problem by adopting auxiliary variables. An interesting point of the proposed methodology, among others, is the generated fuzzy expression for the excess shear stress, namely the difference between the exerted shear stress and the critical shear stress, on which the bed load transport rate depends. The use of a fuzzy number for the excess shear stress seems more reasonable for the calculation of the bed load transport rate, compared to the use of a crisp value.

The aim of the present chapter is to display the possibility of using the fuzzy regression instead of the conventional regression for the confrontation of uncertainty problems appeared in sediment transport formulae. As application example, the bed load transport formula of Smart and Jaeggi (1983a) is presented.

2. Bed load transport equations

2.1 Formula of Smart and Jaeggi

The formula of Smart and Jaeggi (1983a) is actually a modification of the well-known formula of Meyer-Peter and Müller (1948, 1949) for bed load transport in streams and it is valid also for steep bed slopes (from 0.04% up to 20%) and for alluvial materials with mean grain size greater than 0.4mm. The results from 77 steep channel experiments were added to the 137 experiments made by Meyer-Peter and Müller, and the complete data set was analyzed by Smart and Jaeggi (1983a).

For uniform sediment $\left(\frac{d_{90}}{d_{30}} < 1.5\right)$, the equation of Smart and Jaeggi has the following dimensionless form:

$$\Phi = 4.2S^{0.6}C\theta^{0.5}(\theta - \theta_{cr}) \tag{1}$$

while for nonuniform sediment $\left(1.5 < \frac{d_{90}}{d_{30}} < 8.5\right)$, it has the form:

$$\Phi = 4\left(\frac{d_{90}}{d_{30}}\right)^{0.2}S^{0.6}C\theta^{0.5}(\theta - \theta_{cr}) \tag{2}$$

Φ: Dimensionless sediment transport rate.

S: Channel slope.

C: Flow resistance factor (conductivity); it is the ratio of mean flow velocity to bed. shear velocity.

θ: Dimensionless shear stress at the bed.

θ_{cr}: Critical dimensionless shear stress introduced by Shields (1936).

d_{30}: Grain diameter for which 30% weight of a nonuniform sample is finer (m).

d_{90}: Grain diameter for which 90% weight of a nonuniform sample is finer (m).

The parameters Φ, θ and C are mathematically defined by the equations:

$$\Phi = q_s / \left[g(\rho\prime - 1)d^3 \right]^{1/2} \tag{3}$$

$$\theta = hS / (\rho' - 1)d \tag{4}$$

$$C = u / (ghS)^{1/2} \tag{5}$$

q_s: Volumetric sediment discharge per unit channel width (m^2/s).

g: Gravitational acceleration (m/s^2).

d: Grain diameter (m).

h: Flow depth (m).

u: Mean flow velocity (m/s)

$$\rho' = \rho_s / \rho \tag{6}$$

ρ_s: Sediment density (kg/m^3).

ρ: Water density (kg/m^3).

In Appendix I, it is shown that θ is equal to the dimensionless sediment Froude number.

For flow over a plane bed at a slope of "α" degrees to the horizontal and for incipient bed motion conditions, the Shields parameter becomes

$$\theta_{cr} = \theta_{o,cr} \cos a (1 - \tan a / \tan \beta) \tag{7}$$

β: Angle of repose of submerged bed material (°).

$\theta_{o,cr}$: Critical dimensionless shear stress for horizontal bed given by Shields diagram.

The value of $\theta_{o,cr}$ can also be given by the following equation (Schröder, 1985) describing the Shields curve for $Re^* \geq 1$:

$$1 + Ze^{-K_1 Z} + K_2 \tanh^2(K_3 Z) + \log\left(\theta_{o,cr}\right) = 0 \tag{8}$$

where

$$Z = X\left(1 + K_4 X^2\right) \tag{9}$$

$$X = \log\left(Re^*\right) \tag{10}$$

$$Re^* = u_* d / v \tag{11}$$

$$K_1 = 1.06217 \tag{12}$$

$$K_2 = 0.22185 \tag{13}$$

$$K_3 = 0.92462 \tag{14}$$

$$K_4 = 0.17000 \tag{15}$$

Re^*: Sediment Reynolds number or shear Reynolds number.
u_*: Bed shear velocity (m/s).
v: Kinematic viscosity of water (m^2/s).

The formulae of Smart and Jaeggi are valid for the values of bed slope from 0.0004 to 0.20, as mentioned above, of mean grain diameter from 0.0004 m to 0.0105 m, of flow depth from 0.01 m to 1.2 m, of specific water discharge from 0.002 m^2/s to 2.3 m^2/s and of sediment density from 2500 kg/m^3 to 3200 kg/m^3 (Smart, 1984a).

The formula of Smart and Jaeggi (1983a) for uniform sediment can be transformed into the following relationship (Appendix II):

$$q_s = 4.2 S^{0.6} q \frac{1}{\rho' - 1} (S - S_o) \tag{16}$$

$$\text{where } S_o = \frac{(\rho' - 1)d}{20h} \text{ (Appendix II)} \tag{17}$$

q_s: Volumetric sediment discharge per unit channel width (m^2/s).
S: Bed slope.
q: Stream discharge per unit channel width (m^2/s).
S_o: Bed slope for flow without sediment transport.
ρ': Fraction of the sediment density to the density of water.
d: Grain diameter of the bed load sediment (m).
h: Flow depth (m).

The formula of Smart and Jaeggi (1983a) for nonuniform sediment can be transformed into the following relationship (Appendix II):

$$q_s = 4 \left(\frac{d_{90}}{d_{30}} \right)^{0.2} S^{0.6} q \frac{1}{\rho' - 1} (S - S_o) \tag{18}$$

Eqs. (16), (17) were applied by Smart (1984b) to Stiegler's Gorge on the Rufiji River in Tanzania. The comparison results between predicted sediment transport capacity of the channel on the basis of Eqs. (16), (17) and calculated sediment transport capacity on the basis of field samples, are considered as relatively satisfactory. An interesting point is that, in Eq. (17), it is supposed that $\theta_{cr} = 0.05$, although this value is not identical with the mean value of the dimensionless critical shear stresses, determined experimentally (Smart, 1984a).

For the prediction of the mean flow velocity, Smart and Jaeggi have proposed the following conventional flow resistance equation for steep channels (in Rickenmann and Recking, 2011):

$$\sqrt{\frac{8}{f}} = 5.75 \left[1 - \exp \left(-\frac{0.05 Z_{90}}{S^{0.5}} \right) \right]^{0.5} \log (8.2 Z_{90}) \tag{19}$$

$$\text{where } Z_{90} = \frac{h}{d_{90}} \tag{20}$$

f: Darcy-Weisbach friction factor (dimensionless).

Eqs. (19), (20) used for the prediction of the mean flow velocity indicate that a feedback mechanism between flow resistance and bed load is taken into account implicitly (Recking et al., 2008).

The calculation of mean flow velocity is given in Appendix III.

2.2 Formula of Meyer-Peter and Müller

In 1934, the Laboratory for Hydraulic Research at Zurich published for the first time a formula for bed load transport based on sediment transport experiments made by Meyer-Peter et al. (1934) with bed material of uniform grain size. In 1948, the formula was extended on the basis of experiments with nonuniform sediments, with mean grain sizes ranging between

0.4 mm and 28.65 mm, of various densities and channel slopes ranging from 0.04% to 2% (Meyer-Peter and Müller, 1948). The extended formula has become known as the Meyer-Peter and Müller equation and may be written (in Smart, 1984a):

$$\Phi = 8\left[\left(\frac{k_s}{k_r}\right)^{1.5}\theta - 0.047\right]^{1.5} \tag{21}$$

k_s/k_r: Correction factor for bed form roughness; it performs the same function as the expression $CS^{0.6}$ in Eq. (1) (Smart and Jaeggi, 1983b).
k_s: Strickler coefficient related to total bed resistance, depending on both bed form roughness and grain roughness ($m^{1/3}$/s).
k_r: Strickler coefficient related to the grain resistance, depending on grain roughness ($m^{1/3}$/s).

$$k_r = \frac{26}{d_{90}} \tag{22}$$

d_{90}: Grain diameter for which 90% weight of a nonuniform sample is finer (m).

Eq. (21) is based on the concept of critical shear stress characterizing the initiation of grain motion on the bed. In this equation, 0.047 is the value of the dimensionless critical shear stress for fully developed turbulence, while, according to Shields (1936), this value amounts to about 0.06.

According to Smart (1984b), a major drawback of Eq. (21) is that it seriously underestimates the transport rates on steep slopes and is difficult to apply.

For plane bed conditions, it has been used in the form (in Smart, 1984a):

$$\Phi = 8(\theta - \theta_{cr})^{1.5} \tag{23}$$

The formulae of Meyer-Peter and Müller are valid for values of bed slope from 0.0004 to 0.02, of mean grain diameter from 0.0004 m to 0.029 m, of flow depth from 0.01 m to 1.20 m, of specific water discharge from 0.002 m^2/s to 2.3 m^2/s and of sediment density from 2500 kg/m^3 to 3200 kg/m^3 (Smart, 1984a).

Further limitations are given below (Hager and Boes, 2018):

- Fully turbulent flow, so that $Re = 4uR/v > 10^5$, Re: Reynolds number, u: mean flow velocity, R: hydraulic radius, v: kinematic viscosity.
- Particle size >1 mm, to avoid the effects of apparent cohesion.
- Flow depth >0.05 m, to assure Froude similitude.

Papalaskaris et al. (2016a) have attempted to calibrate the formula of Meyer-Peter and Müller both manually and on the basis of the least squares method, in terms of roughness coefficient, for two streams in northeastern Greece: Kosynthos River and Kimmeria Torrent. Papalaskaris et al. (2016b) have also calibrated manually the formula of Meyer-Peter and Müller, in terms of roughness coefficient, for Nestos River (northeastern Greece). In a following study, Sidiropoulos et al. (2018) have calibrated the same formula for Nestos River by means of a nonlinear optimization of two suitable parameters, while utilizing the average value of the roughness coefficient found by the manual calibration. In all three studies, the comparison between calculations and measurements of bed load transport rate was made on the basis of the following statistical criteria: root mean square error, relative error, efficiency coefficient, linear correlation coefficient, determination coefficient and discrepancy ratio. The values of the above statistical criteria for the case of manual calibration were more satisfactory, compared to the case of nonlinear optimization. However, the manual calibration was carried out on partial measurement sets, while the nonlinear optimization was carried out on the whole measurement set.

3. Fuzzy linear regression

A fuzzy set can be seen as a mapping from a general set X to the closed interval [0,1]. A fuzzy set can be expressed by a membership function, which shows to what degree an element lies in the examined fuzzy set. A membership function is confined in the interval [0, 1], with a membership degree of 0 indicating that the element does not belong to the set and a membership degree of 1 indicating that the element fully belongs to the set. Subsequently, an object with a membership degree between 0 and 1 will belong to the set to some degree.

A fuzzy number is a fuzzy set which, furthermore, satisfies the properties of convexity and normality. It is defined in the axis of real numbers and its membership function is a piecewise continuous function.

The (soft) α-cut set of the fuzzy number A, with $0 < \alpha \leq 1$, is defined as follows:

$$[A]_a = \{x | \mu_A(x) \geq a, x \in \mathbf{R}\} \tag{24}$$

where $\mu_A(x)$ the membership function of the fuzzy number A.

An interesting point is that the crisp set including all the elements with non-zero membership function is the 0-strong cut which can be defined as follows:

$$A_{0^+} = \{x | \mu_A(x) > 0, \ x \in \mathbf{R}\} \tag{25}$$

More analytically, according to Eq. (25), an open interval, that does not contain the boundaries, is above the 0-cut. For this reason, and in order to have a closed interval containing the boundaries, it is suggested the term "worst-case interval" W, which is the union of the 0-strong cut and the boundaries (Hanss, 2005; Spiliotis et al., 2020).

Linear regression analysis is used to model the linear relationship between the dependent variable and the independent variables. In the fuzzy regression model, the difference between the computational data and the actual values (measurements) is assumed to be due to the structure of the system. The proposed model carries this uncertainty back to its coefficients or, in other words, our inability to construct a precise relationship, is directly introduced into the model, on the fuzzy parameters. Based on the above reasoning, the coefficients for the independent variables are chosen to be fuzzy numbers. The problem of fuzzy linear regression is reduced to a linear programming problem according to the following steps:

1. The model is as follows

$$\tilde{Y}_j = \tilde{A}_0 + \tilde{A}_1 x_{1j} + \tilde{A}_2 x_{2j} + \ldots + \tilde{A}_n x_{nj} \tag{26}$$

where $\tilde{Y}$ is the fuzzy dependent variable; $j = 1, \ldots, m$; $i = 1, \ldots, n$; $\tilde{A}_i = (a_i, c_i)$ are fuzzy symmetric triangular numbers selected as coefficients (Fig. 1), and x is the independent variable. In addition, n is the number of independent variables, m is the number of data, a is the central value (where $\mu = 1$), and c is the semi-width.

2. Determination of the degree h at which the data $[(x_{1j}, x_{2j}, \ldots, x_{nj}), y_j]$ is aimed to be included in the estimated number Y_j:

$$\mu_{Y_j}\left(y_j\right) \geq h, \quad j = 1, \ldots, m \tag{27}$$

The constraints express the concept of inclusion in case that the output data are crisp numbers. In the examined case of the widely used model of Tanaka (1987), a more soft definition of the fuzzy subsethood is used compared to the Zadeh (1965) definition. Hence, the inclusion of a fuzzy set A into the fuzzy set B with the associated degree $0 \leq h \leq 1$ is defined as follows (Saridakis et al., 2020):

$$[A]_h \subseteq [B]_h \tag{28}$$

In our case, since the data are crisp, the set A is only a crisp value (a point which must be included within the produced fuzzy band) and the fuzzy set B is a fuzzy triangular number. Hence, Eq. (27) is equivalent to (Spiliotis and Hrissanthou, 2018):

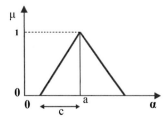

FIG. 1 Fuzzy symmetric triangular number.

$$\sum_{i=0}^{n} a_i x_{ij} - (1-h) \sum_{i=0}^{n} c_i|x_{ij}| \le y_j \le \sum_{i=0}^{n} a_i x_{ij} + (1-h) \sum_{i=0}^{n} c_i|x_{ij}|, j = 1, \dots, m \qquad (29)$$

It must be clarified that the above equations hold for a specified h-cut and not for every α-cut (e.g., Tsakiris et al., 2006; Spiliotis and Hrissanthou, 2018).

3. Determination of the minimization function (objective function) J. In the conventional fuzzy linear regression model, the objective function J is the sum of the produced fuzzy semi-widths for the data, i.e.

$$J = \left\{ mc_0 + \sum_{j=1}^{m} \sum_{i=1}^{n} c_i|x_{ij}| \right\} \qquad (30)$$

where c_0 is the semi-width of the constant term, and c_i semi-width of the other fuzzy coefficients.

Since fuzzy symmetric triangular numbers are selected as fuzzy coefficients, it can be proved that the objective function is the sum of the semi-widths of the produced fuzzy band regarding the available data:

$$J = \left\{ mc_0 + \sum_{j=1}^{m} \sum_{i=1}^{n} c_i|x_{ij}| \right\} = \frac{1}{2} \sum_{j=1}^{m} \left(Y_j^+ - Y_j^- \right) \qquad (31)$$

where Y_j^+, Y_j^- the right and the left-hand side of the worst-case interval, respectively.

4. The problem results in the following linear programming problem

$$\left. \begin{array}{c} \min \left\{ mc_0 + \displaystyle\sum_{j=1}^{m} \sum_{i=1}^{n} c_i|x_{ij}| \right\} \\[2ex] \displaystyle\sum_{i=0}^{n} a_i x_{ij} - (1-h) \sum_{i=0}^{n} c_i|x_{ij}| = y_{hj}^{L} \le y_j \\[2ex] \displaystyle\sum_{i=0}^{n} a_i x_{ij} + (1-h) \sum_{i=0}^{n} c_i|x_{ij}| = y_{hj}^{R} \ge y_j \end{array} \right\} \qquad (32)$$

where $c_i \ge 0$, for $i = 0, 1, \dots, n$.

As aforementioned, the decision variables are the central values and the semi-widths for any fuzzy coefficients (a_i, c_i).

In addition, many times, when data are classic numbers, we can easily approximate nonlinear cases with the fuzzy linear regression model with the help of auxiliary variables. In this case, the total uncertainty (cumulative width) indicates the incomplete complexity, whereas nonphysical behavior is an indicator of overtraining, due to adoption of excessive complexity in nonlinear models.

In this application, only crisp data exist (x_j, y_j). The fuzzy linear regression can be applied also in case of fuzzy data.

4. Application of the bed load transport formula of Smart and Jaeggi on the basis of conventional and fuzzy regression

The uncertainties in sediment transport prediction arise from the role of turbulence in sediment entrainment. First of all, the difference between the exerted dimensionless shear stress and the critical dimensionless shear stress is examined as main variable.

Logarithmizing both sides of Eq. (1) for uniform sediment (Smart and Jaeggi, 1983a) yields:

$$\log\left(\frac{\Phi}{S^{0.6} C \theta^{0.5}} \right) = \log(\theta - \theta_{cr}) + \log 4.2 \qquad (33)$$

By taking into account Eq. (33) and the corresponding experimental data included in Smart (1984a), the following fuzzy regression equation is obtained:

$$\log\left(\frac{\Phi}{S^{0.6}C\theta^{0.5}}\right) = (1.305, 0.0604)\log(\theta - \theta_{cr}) + (0.688, 0.0918) \tag{34}$$

Eq. (34) is of the general form $\widetilde{y}_j = \widetilde{a}_1 x_{1j} + a_o$.

The first term in the brackets expresses the central value, while the second the semi-width of the fuzzy symmetric triangular numbers. These values are produced by following a linear programming problem where the decision variables are the central values and the semi-widths of the fuzzy coefficients based on Eq. (32) for $h = 0$.

The objective function equals to:

$$\left\{40c_0 + \sum_{j=1}^{40} c_1|x_j|\right\} = \left\{40 \cdot 0.0918 + \sum_{j=1}^{40} 0.0604|\log(\theta - \theta_{cr})_j|\right\} = 4.7230 \tag{35}$$

In the brackets, the first term denotes the central value, while the second term the semi-width of the produced fuzzy coefficients, where 40 is the number of data and $\log(\theta - \theta_{cr})_j$ the considered independent variable.

Fig. 2A represents the fuzzy regression Eq. (34) together with the conventional regression Eq. (33) based on the experimental data contained in Smart (1984a). The fuzzy band contains all the data, while the conventional regression differs from the central values ($h = 1$) of the achieved fuzzy solution. The uncertainty of the fuzzy solution is functional.

Logarithmizing both sides of Eq. (2) for nonuniform sediment (Smart and Jaeggi, 1983a) results in:

$$\log\left[\frac{\Phi}{\left(\frac{d_{90}}{d_{30}}\right)^{0.2}S^{0.6}C\theta^{0.5}}\right] = \log(\theta - \theta_{cr}) + \log 4 \tag{36}$$

By taking into account Eq. (36) and the corresponding experimental data included in Smart (1984a), the following fuzzy regression equation is obtained for nonuniform sediment:

$$\log\left[\frac{\Phi}{\left(\frac{d_{90}}{d_{30}}\right)^{0.2}S^{0.6}C\theta^{0.5}}\right] = (1.1704, 0.1309)\log(\theta - \theta_{cr}) + (0.6, 0.1016) \tag{37}$$

The objective function equals:

$$\left\{37c_0 + \sum_{j=1}^{37} c_1|x_j|\right\} = \left\{37 \cdot 0.1016 + \sum_{j=1}^{37} 0.1309|\log(\theta - \theta_{cr})_j|\right\} = 4.9358 \tag{38}$$

Fig. 2B represents the fuzzy regression Eq. (37) together with the conventional regression Eq. (36) based on the experimental data included in Smart (1984a). The fuzzy band contains all the data, while the conventional regression differs from the central values ($h = 1$) of the achieved fuzzy solution. The uncertainty of the fuzzy solution is functional.

The power of the difference $(\theta - \theta_{cr})$, according to Eqs. (34), (37), lies between that of Eqs. (21), (22) (Meyer-Peter and Müller, 1948, 1949) and that of Eqs. (1), (2) (Smart and Jaeggi, 1983a), namely between 1.5 and 1.

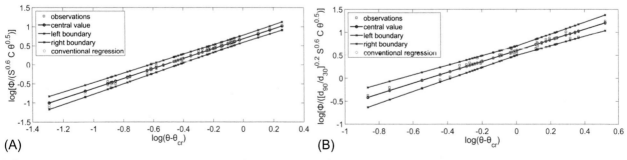

FIG. 2 Graphical representation of the fuzzy regression for (A) Eq. (34), uniform sediment, and (B) Eq. (37), nonuniform sediment.

Eq. (16) can be written also in the following form:

$$Q_s = 4.2S^{0.6}Q\frac{1}{\rho'-1}(S-S_o) \rightarrow \Phi = \frac{4.2S^{0.6}Q\frac{1}{\rho'-1}(S-S_o)}{\sqrt{b^2g(\rho'-1)d^3}} \rightarrow$$

$$\log\left(\frac{\Phi}{S^{0.6}C\theta^{0.5}}\right) = \log\left(\frac{\frac{4.2S^{0.6}Q\frac{1}{\rho'-1}(S-S_o)}{\sqrt{b^2g(\rho'-1)d^3}}}{S^{0.6}C\theta^{0.5}}\right) \tag{39}$$

In Eq. (39), b (m) is channel width, Q_s (m³/s) volumetric sediment discharge, Q (m³/s) stream discharge and Φ dimensionless sediment transport rate.

In Fig. 3, a comparison is made between the crisp Eqs. (33), (39) of Smart and Jaeggi for uniform sediment.

By comparing the crisp equations of Smart and Jaeggi (Eqs. 33, 36) with the fuzzy ones (Eqs. 34, 37) based on the experimental data included in Smart (1984a), it is obvious that:

- As aforementioned, all the data are included within the produced fuzzy band.
- The curve representing Eq. (33) is included for a significant range within the produced fuzzy band.
- The two formulae of Smart and Jaeggi (Eqs. 33, 39) are not absolutely equivalent. This is obvious since the critical dimensionless shear stress (θ_{cr}) in Eq. (33) is not equal to 0.05 (Eq. 17) regarding the experimental data included in Smart (1984a). However, the values of sediment transport rate are very similar.

Finally, a multiple fuzzy regression is implemented with uniform grain size distributions:

$$\log\left(\frac{\Phi}{\theta^{0.5}}\right) = (0.5153, 0.0719)\log S + (0.8804, 0)\log C + (1.3173, 0.0055)\log(\theta - \theta_{cr}) + (0.6834, 0.0401) \tag{40}$$

Eq. (40) is of the general form: $\widetilde{y} = \widetilde{a}_1x_1 + \widetilde{a}_2x_2 + \widetilde{a}_3x_3 + \widetilde{a}_o$.

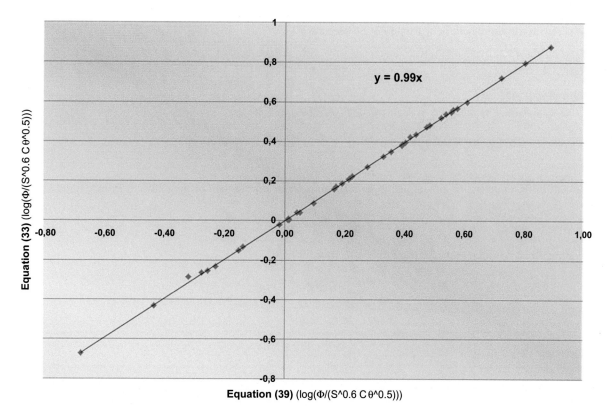

FIG. 3 Comparison between the crisp Eqs. (33), (39) of Smart and Jaeggi for uniform sediment.

These results are compatible with those of Eq. (34). In case of nonuniform grain size distributions, the results are more ambiguous.

Since more independent variables are considered, the objective function is reduced in relation to the value of Eq. (35):

$$\left\{ 40c_0 + \sum_{j=1}^{40} \left(c_1 |\log S_j| + c_2 |\log C_j| + c_3 |\log (\theta - \theta_{cr})_j| \right) \right\} =$$
$$= \left\{ 40 \cdot 0.0401 + \sum_{j=1}^{40} \left(0.0719 |\log S_j| + 0.0055 |\log (\theta - \theta_{cr})_j| \right) \right\} = 4.5246 \tag{41}$$

The zero semi-width in Eq. (40) means that the coefficient of $logC$ is a simple crisp number ($c_2 = 0$). However, the reduction of the objective function is not so significant. In this fuzzy relation, the resistance factor has no uncertainty, while the greatest uncertainty, that is the value of the semi-width, appears in the term of the bed slope.

5. Conclusions

On the basis of the experimental data contained in Smart (1984a), by taking into account the formulae of Smart and Jaeggi (1983a) for uniform and nonuniform sediment, functional fuzzy equations, that is with no high uncertainty, are achieved.

The power of the difference ($\theta - \theta_{cr}$) in both cases lies between those of formulae of Meyer-Peter and Müller (1948, 1949) and Smart and Jaeggi (1983a), namely between 1.5 and 1.

By comparing the crisp equation of Smart and Jaeggi (1983a) with the fuzzy one for uniform sediment, based on the experimental data contained in Smart (1984a), it is concluded that the curve representing the crisp equation is included for a significant range within the produced fuzzy band.

The two crisp formulae of Smart and Jaeggi (1983a) for uniform sediment (Eqs. 33, 39) are not absolutely equivalent. This is obvious since the critical dimensionless shear stress (θ_{cr}) in Eq. (33) is not equal to 0.05 (Eq. 17) regarding the experimental data contained in Smart (1984a). However, the values of sediment transport rate are very similar.

In crisp and fuzzy equations of Smart and Jaeggi (1983a), for uniform and nonuniform sediment, the difference between the exerted and the critical shear stress is applied as independent variable. Since the reduction of the objective function for uniform sediment, in case of using additional independent variables, is not appreciably significant, other more complicated fuzzy equations for uniform sediment would not be necessary.

Appendix I

Derivation of Eq. (4)

The sediment Froude number Fr^* is mathematically defined as follows:

$$Fr^* = \frac{u_*^2}{(\rho' - 1)gd} = \frac{u_*^2}{\left(\frac{\rho_s}{\rho} - 1\right)gd} = \frac{\rho u_*^2}{(\rho_s - \rho)gd} \tag{AI.1}$$

The sediment Froude number Fr^* is defined in an analogous way to the square of the classical Froude number:

$$Fr^2 = \frac{u^2}{gh} \tag{AI.2}$$

The bed shear stress u_* (m/s) is mathematically defined as follows:

$$u_* = \sqrt{\frac{\tau_o}{\rho}} = \sqrt{\frac{\rho g RS}{\rho}} = \sqrt{gRS} \approx \sqrt{ghS} \tag{AI.3}$$

ρ_s: Sediment density (kg/m^3).
ρ: Water density (kg/m^3).
g: Gravitational acceleration (m/s^2).
d: Grain diameter (m).
u: Mean flow velocity (m/s).

τ_o: Bed shear stress (N/m^2).
R: Hydraulic radius (m).
S: Energy slope.
h: Flow depth (m).

Regarding Eq. (AI.3), the energy slope can be replaced by the channel slope in the case of uniform flow. Additionally, the hydraulic radius R can be approximately replaced by the flow depth h when the channel width is larger than $30h$.

Taking into account Eq. (AI.3), Eq. (AI.1) becomes:

$$Fr^* = \frac{\rho g h S}{(\rho_s - \rho) g d} = \frac{g h S}{(\rho' - 1) g d} = \frac{h S}{(\rho' - 1) d} = \theta \tag{AI.4}$$

In other words, the sediment Froude number Fr^* is identical with the dimensionless bed shear stress θ, which is also called Shields number.

Appendix II

Derivation of Eq. (16) from Eq. (1)

By combining Eqs. (1), (3), (4), and (5) results:

$$\Phi = 4.2 S^{0.6} C \theta^{0.5} (\theta - \theta_{cr}) = 4.2 S^{0.6} \frac{u}{g^{0.5} h^{0.5} S^{0.5}} \frac{h^{0.5} S^{0.5}}{(\rho' - 1)^{0.5} d^{0.5}} \frac{h}{d(\rho' - 1)} (S - S_o) \tag{AII.1}$$

$$\text{or } \frac{q_s}{g^{0.5} (\rho' - 1)^{0.5} d^{1.5}} = 4.2 S^{0.6} \frac{uh}{g^{0.5} (\rho' - 1)^{0.5} d^{1.5}} \frac{1}{\rho' - 1} (S - S_o) \tag{AII.2}$$

$$\text{or } q_s = 4.2 S^{0.6} q \frac{1}{\rho' - 1} (S - S_o) \tag{AII.3}$$

Derivation of Eq. (17)

From Eq. (4) results:

$$\theta_{cr} = \frac{h S_o}{(\rho' - 1) d} \text{ or } \frac{S_o}{\rho' - 1} = \frac{\theta_{cr} d}{h} = \frac{0.047 d}{h} \approx \frac{0.05 d}{h} \tag{AII.4}$$

$$\text{or } S_o = \frac{(\rho' - 1) d}{20 h} \tag{AII.5}$$

It is noted that $\theta_{cr} = 0.047 \approx 0.05$ according to Meyer-Peter and Müller.

Derivation of Eq. (18) from Eq. (2)

By combining Eqs. (2), (3), (4), and (5) results:

$$\Phi = 4 \left(\frac{d_{90}}{d_{30}} \right)^{0.2} S^{0.6} C \theta^{0.5} (\theta - \theta_{cr}) = 4 \left(\frac{d_{90}}{d_{30}} \right)^{0.2} S^{0.6} \frac{u}{g^{0.5} h^{0.5} S^{0.5}} \frac{h^{0.5} S^{0.5}}{(\rho' - 1)^{0.5} d^{0.5}} \frac{h}{(\rho' - 1) d} (S - S_o) \tag{AII.6}$$

$$\text{or } \frac{q_s}{g^{0.5} (\rho' - 1)^{0.5} d^{1.5}} = 4 \left(\frac{d_{90}}{d_{30}} \right)^{0.2} S^{0.6} \frac{uh}{g^{0.5} (\rho' - 1)^{0.5} d^{1.5}} \frac{1}{\rho' - 1} (S - S_o) \tag{AII.7}$$

$$\text{or } q_s = 4 \left(\frac{d_{90}}{d_{30}} \right)^{0.2} S^{0.6} q \frac{1}{\rho' - 1} (S - S_o) \tag{AII.8}$$

Appendix III

Calculation of the mean flow velocity

The Darcy-Weisbach friction factor f can be calculated by means of Eq. (19). The mean flow velocity u (m/s) can be calculated by the following formula:

$$u = \frac{1}{\sqrt{f}} \sqrt{8gRS} \qquad\qquad \text{(AIII.1)}$$

g: Gravitational acceleration (m/s^2).
R: Hydraulic radius (m).
S: Energy slope or channel slope in case of uniform flow.

Eq. (AIII.1) results from the Darcy-Weisbach equation for closed conduits:

$$h_f = f \frac{L}{D} \frac{u^2}{2g} \text{ or } \frac{h_f}{L} = S = f \frac{1}{D} \frac{u^2}{2g} \qquad\qquad \text{(AIII.2)}$$

h_f: Head loss due to friction (m).
f: Friction factor.
L: Length of conduit (m).
D: Diameter of conduit (m).
u: Mean flow velocity (m/s).

In the case of open channels, the diameter of the conduit D (m) for a circular cross section can be replaced by $4R$, which is proved as follows:

$$R = \frac{A}{P} = \frac{\frac{\pi D^2}{4}}{\pi D} = \frac{D}{4} \qquad\qquad \text{(AIII.3)}$$

A: Cross-sectional area of flow (m^2).
P: Wetted perimeter of flow cross section (m)

Finally, Eq. (AIII.1) is obtained from Eq. (AIII.2) by solving the latter equation with respect to mean flow velocity u.

References

Alhasan, Z., Jandora, J., Riha, J., 2016. Comparison of specific sediment transport rates obtained from empirical formulae and dam breaching experiments. Environ. Fluid Mech. 16, 997–1019.

Barry, J.J., Buffington, J.M., King, J.G., 2004. A general power equation for predicting bed load transport rates in gravel bed rivers. Water Resour. Res. 40 (10), W10401.

Buffington, J.M., Montgomery, D.R., 1997. A systematic analysis of eight decades of incipient motion studies, with special reference to gravel-bedded rivers. Water Resour. Res. 33 (8), 1993–2029.

D'Agostino, V., Lenzi, M.A., 1999. Bedload transport in the instrumented catchment of the Rio Cordon – Part II: Analysis of the bedload rate. Catena 36, 191–204.

Diplas, P., Dancey, C.L., Celik, A.O., Valyrakis, M., Greer, K., Akar, T., 2008. The role of impulse on the initiation of particle movement under turbulent flow conditions. Science 322 (5902), 717–720.

Ghorbani, K., Salarijazi, M., Abdolhosseini, M., Eslamian, S., 2017. Assessment of minimum variance unbiased estimator and beta coefficient methods to improve the accuracy of sediment rating curve estimation. Int. J. Hydrol. Sci. Technol. 7 (4), 350–363.

Gomez, B., Church, M., 1989. An assessment of bed load sediment transport formulae for gravel bed rivers. Water Resour. Res. 25 (6), 1161–1186.

Hager, W.H., Boes, R.M., 2018. Eugen Meyer-Peter and the MPM sediment transport formula. J. Hydraul. Eng. ASCE 144 (6), 02518001.

Hanss M., 2005: Applied Fuzzy Arithmetic, an Introduction with Engineering Applications, Springer, Berlin, Germany, 256 p.Hanss, M., 2005. Applied Fuzzy Arithmetic, An Introduction With Engineering Applications. Springer, Berlin, Germany, p. 256.

Hrissanthou, V., 1986. Vorhersage der Feststofflieferung eines Einzugsgebietes, XIII Konferenz der Donauländer über hydrologische Vorhersagen. Jugoslawien, Belgrad, pp. 183–190.

Hrissanthou, V., 1990. Application of a sediment routing model to a Middle European watershed. Water Resour. Bull. (J. Am. Water Resour. Assoc.) 26 (5), 801–810.

Kaffas, K., Hrissanthou, V., 2018. Soil erosion, streambed deposition and streambed erosion – Assessment at the mountainous terrain. Proceedings (MDPI) 2, 626.

Kirchner, J.W., Dietrich, W.E., Iseya, F., Ikeda, H., 1990. The variability of critical shear stress, friction angle, and grain protrusion in water-worked sediments. Sedimentology 37 (4), 647–672.

Kitsikoudis, V., Sidiropoulos, E., Hrissanthou, V., 2014. Machine learning utilization for bed load transport in gravel-bed rivers. Water Resour. Manag. 28 (11), 3727–3743.

Kitsikoudis, V., Sidiropoulos, E., Hrissanthou, V., 2015. Assessment of sediment transport approaches for sand-bed rivers by means of machine learning. Hydrol. Sci. J. 60 (9), 1566–1586.

Kitsikoudis, V., Spiliotis, M., Hrissanthou, V., 2016. Fuzzy regression analysis for sediment incipient motion under turbulent flow conditions. Environ. Process. 3 (3), 663–679.

Meyer-Peter, E., Favre, H., Einstein, A., 1934. Neuere Versuchsresultate über den Geschiebetrieb. Schweizerische Bauzeitung 103 (13), 147–150.

Meyer-Peter, E., Müller, R., 1948. Formulas for bedload transport. In: Proceedings 2nd Congress IAHR, Stockholm, Sweden, pp. 39–64.

Meyer-Peter, E., Müller, R., 1949. Eine Formel zur Berechnung des Geschiebetriebs. Schweizerische Bauzeitung 67 (3), 29–32.

Papalaskaris, T., Dimitriadou, P., Hrissanthou, V., 2016b. Comparison between computations and measurements of bed load transport rate in Nestos River, Greece. Procedia Eng. 162, 172–180.

Papalaskaris, T., Hrissanthou, V., Sidiropoulos, E., 2016a. Calibration of a bed load transport rate model in streams of NE Greece. Eur. Water 55, 125–139.

Recking, A., 2009. Theoretical development on the effects of changing flow hydraulics on incipient bed load motion. Water Resour. Res. 45 (4), W04401.

Recking, A., Frey, P., Paquier, A., Belleudy, P., Champagne, J.Y., 2008. Feedback between bed load transport and flow resistance in gravel and cobble bed rivers. Water Resour. Res. 44, W05412.

Rickenmann, D., Recking, A., 2011. Evaluation of flow resistance in gravel-bed rivers through a large field data set. Water Resour. Res. 47, W07538.

Saridakis, M., Spiliotis, M., Hrissanthou, V., 2020. Assessment of bedload in sand – gravel bed rivers by using nonlinear fuzzy regression. Eur. Water 69 (70), 15–22.

Schröder R.C.M., 1985: Vergleichbarkeit von Geschiebetransportformeln, WasserWirtschaft, 75(5), 217–221.Schröder, R.C.M., 1985. Vergleichbarkeit von Geschiebetransportformeln. Wasserwirtschaft 75 (5), 217–221.

Shields, A., 1936. Anwendung der Ähnlichkeitsmechanik und der Turbulenzforschung auf die Geschiebebewegung. Mitteilungen der preußischen Versuchsanstalt für Wasser- und Schiffbau, Heft 26, Berlin, Germany.

Sidiropoulos, E., Papalaskaris, T., Hrissanthou, V., 2018. Parameter optimization of a bed load transport formula for Nestos River, Greece. Proceedings (MDPI) 2, 627.

Smart, G.M., 1984a. Sediment transport formula for steep channels. J. Hydraul. Eng. ASCE 110 (3), 267–276.

Smart, G.M., 1984b. Predicting the sediment capacity of a channel, Challenges in African Hydrology and Water Resources. In: Proceedings of the Harare Symposium, Zimbabwe, IAHS Publication No. 144, July, pp. 397–401.

Smart, G.M., Jaeggi, M.N.R., 1983a. Sedimenttransport in steilen Gerinnen, Mitteilungen der Versuchsanstalt für Wasserbau. Hydrologie and Glaziologie, Nr. 64, ETH Zürich, Switzerland.

Smart, G.M., Jaeggi, M.N.R., 1983b. Laboratory study on sediment transport on steep slopes. In: Proceedings XX Congress IAHR, Moscow, Russia, pp. 314–315.

Spiliotis, M., Angelidis, P., Papadopoulos, B., 2020. A hybrid probabilistic bi-sector fuzzy regression-based methodology for normal distributed hydrological variable. Evol. Syst. 11, 255–268.

Spiliotis, M., Hrissanthou, V., 2018. Fuzzy and crisp regression analysis between sediment transport rates and stream discharge in the case of two basins in northeastern Greece. In: Hrissanthou, V., Spiliotis, M. (Eds.), Conventional and Fuzzy Regression: Theory and Engineering Applications. Nova Science Publishers, New York, USA, pp. 2–49.

Spiliotis, M., Kitsikoudis, V., Hrissanthou, V., 2017. Assessment of bedload transport in gravel-bed rivers with a new fuzzy adaptive regression. Eur. Water 57, 237–244.

Tanaka, H., 1987. Fuzzy data analysis by possibilistic linear models. Fuzzy Sets Syst. 24, 363–375.

Tsakiris, G., Tigkas, D., Spiliotis, M., 2006. Assessment of interconnection between two adjacent watersheds using deterministic and fuzzy approaches. Eur. Water 15 (16), 15–22.

Van Emelen S., 2014: Breaching processes of river dikes: Effects on sediment transport and bed morphology, Doctoral Thesis, School of Engineering Catholic University of Louvain, Belgium, 306 p.Van Emelen, S., 2014. Breaching processes of river dikes: effects on sediment transport and bed morphology. Doctoral Thesis, Doctoral Thesis. School of Engineering, Catholic University of Louvain, Belgium. 306 p.

Visser, P., 1988. A model for breach growth in a dike-burst. In: Proceedings 21st International Conference Coastal Engineering, Malaga, Spain, pp. 1897–1910.

Visser P., 1995: Application of sediment transport formulae to sand-dike breach-erosion, Communication on Hydraulic and Geotechnical Engineering, Report no. 94-7, Delft University of Technology, The Netherlands, 78 pVisser, P., 1995. Application of sediment transport formulae to sand-dike breach-erosion. In: Communication on Hydraulic and Geotechnical Engineering. Delft University of Technology, The Netherlands. Report No. 94-7, 78 p.

Yang, C.T., 1996. Sediment Transport – Theory and Practice. McGraw-Hill.

Zadeh, L.A., 1965. Fuzzy sets. Inf. Control. 8 (3), 338–353.

Chapter 6

Automated flood inundation mapping over Ganga basin

Sukanya Ghosh[a], Deepak Kumar[a], and Rina Kumari[b]

[a]*Amity Institute of Geoinformatics and Remote Sensing (AIGIRS), Amity University Uttar Pradesh (AUUP), Noida, Uttar Pradesh, India,*
[b]*School of Environment and Sustainable Development, Central University of Gujarat, Gandhinagar, Gujarat, India*

1. Introduction

Floods mostly vary from one region to another, and their severity is determined by a variety of factors, including unpredictable weather patterns and heavy rainfall occurrences (Pham Van and Nguyen-Van, 2020; Soulard et al., 2020). Although floods are common in many places of India during monsoon seasons, the Ganga basin is particularly vulnerable (Bhatt et al., 2021; Meena et al., 2021). There are a lot of areas in the state of Bihar that get flooded due to the swelling of rivers in neighboring Nepal (Lal et al., 2020; Soulard et al., 2020; Wagle et al., 2020). This appealed to the attention of the present research. The Ganga basin spans China, Nepal, India, and Bangladesh (Agnihotri et al., 2019; Ahmad and Goparaju, 2020; Prakash et al., 2017; Sinha and Tandon, 2014).

The global emergence of COVID-19 has stopped all the activities, and it debuted as the deadliest disease with the longest nationwide lockdown. These caused enormous disruption in all aspects of people's livelihood. Besides, major obstacles got accumulated due to the effect of the flooding event during July 2020. It added misery to the people and livelihood of the people, who were trying to control the spread of COVID-19. These results in disaster-risk mitigation to other sectors. The only way to have an effective and prompt response is to have real-time information provided by space-based sensors. Using a cloud-based platform like Google earth engine (GEE), an automated technique is employed to analyze the flood inundation with Synthetic Aperture Radar (SAR) images. The study exhibits the potential of automated techniques along with algorithms applied to larger datasets on cloud-based platforms. The results present flood extent maps for the lower Ganga basin, comprising areas of the Indian subcontinent. Severe floods destroyed several parts of Bihar and West Bengal affecting a large population. This study offers a prompt and precise estimation of inundated areas to facilitate a quick response for risk assessment, particularly at times of the COVID-19.

The three states (Bihar, Jharkhand, and West Bengal), collectively known as the Lower Ganga Basin, are home to more than 30% of the population (Prakash et al., 2017). Rapid population growth and settlements resulted in changes in land use, increased soil erosion, increased siltation, and other related variables that augmented flood severity (Li et al., 2020; Pham Van and Nguyen-Van, 2020). However, floods became the most frequent disaster in recent times, what compounded the problem was the COVID-19 pandemic (Krämer et al., 2021; Lal et al., 2020). As a result, new measures were needed to manage the spread of COVID-19 as well as flood mitigation (Wang et al., 2020; Zoabi et al., 2021). Although ground data and field measurements are considered to be more accurate, they are time and money consuming. Furthermore, field surveys were impossible to conduct during this period, since social distancing has become the norm, linked with significant health concerns and trip expenditures (Jian et al., 2020; Lattari et al., 2019). Flood mitigation strategies that are ineffective may result in more human deaths, property damage, and more spread of COVID-19 (Cornara et al., 2019; Shen et al., 2019). It had disastrous impacts in 149 districts throughout Bihar, Assam, West Bengal. Since the movement was halted owing to a sudden shutdown, the only way out was to employ robust flood control techniques based on real-time information (Das et al., 2018; Dong et al., 2020; Tang et al., 2016). The dramatic increase in flood occurrence in these locations prompted specialists to implement more structured and effective flood management to address the issues, while also adhering to all COVID-19 norms and regulations (Min et al., 2020; Wang et al., 2019).

In 2020, the world is still reeling from the COVID-19 pandemic outbreak, which has caused tremendous suffering and significant economic breakdown, with disruptions in every economic aspect, including international business, supply chains, and commodity pricing (Espinosa and Rudenstine, 2020; Thomas et al., 2021). As a result of the long-term

Handbook of HydroInformatics. https://doi.org/10.1016/B978-0-12-821962-1.00006-4

economic impact, the health of a large population has drastically changed. Fighting against the spread of COVID-19 was not the only obstacle; it was also surviving monsoon floods which exacerbated the consequences on life and property was also a struggle (Ishiwatari et al., 2020; Kandekar et al., 2021; Lal et al., 2020). This highlighted the demand for geospatial technology which when combined with the advanced techniques of Artificial Intelligence (AI) algorithms, gave the automation process an extra boost (Navlakha et al., 2021; Pham Van and Nguyen-Van, 2020; Si et al., 2020). The technology-enhanced the accuracy and predictability of big spatial data processing (Xiangdong et al., 2021). In conjunction with cloud-based technology, AI has a plethora of applications in the geospatial domain (Mutanga and Kumar, 2019; Zhu et al., 2019). Meanwhile, at times of novel COVID-19 disease outbreaks, the problem may be mapped and tracked globally using spatial data delivering quick, simple, and cost-effective solutions (Dheeraj, 2020; Supriya and Chattu, 2021).

Recent developments in geospatial processing using space-based sensors like synthetic aperture radar (SAR) are mostly used for mapping and monitoring flood inundated areas. It can penetrate through any weather conditions and provide all information during the flood (Aher and Ph, 2014; Cao et al., 2019; Mullen et al., 2021).

This is why the SAR datasets are preferably used in many earlier studies for monitoring flood extent. Instead of visiting the affected areas and collecting ground samples, the generation of information in real-time using the Sentinel-1 datasets assisted in discovering inundated zones (Dong et al., 2020; Tang et al., 2016). There are different types of automated techniques for flood mapping based on threshold and machine learning methods. The machine learning-based methods need countless ground samples for training that is difficult to obtain during or immediately after a flood event. On the other hand, without the need for any training samples, the threshold-based approach aids in the separation of water pixels from non-water pixels (Duro et al., 2018; Min et al., 2020; Wedlund and Kvedar, 2021). An optimum threshold is generated to segregate the targeted pixels. Furthermore, the cloud-based infrastructure provided greater access to big datasets as well as improved management, storage, and immediate responses. The study emphasized the potential of a cloud computing framework for geoprocessing and data visualization in a dispersed surrounding. The technology facilitated in processing large Sentinel-1A datasets, greater analysis with better accuracy of the flood extent estimation, and instantaneous results at the time of emergency. This provided valuable knowledge and reports for the decision-makers for mitigation planning and sustainable management.

2. Literature review

Natural disasters occur each year, and their severity and frequency appear to have increased dramatically in the recent decades, owing primarily to the degradation of environmental quality, like deforestation, intense use of land, and increasing population (Babí Almenar et al., 2021). In terms of human and economic devastation, hydrological disasters are among the common and costly natural disasters. According to Scheuren et al., we saw a rising trend from the year 2000 to 2007, having an average annual growth rate of disasters around 8.4%. It is also added that hydrological disasters, accounted for 55% of all other disasters as reported in 2007, with a huge impact on humans (177 million victims) and significant economic losses (24.5 billion USD). Floods have become a prevalent occurrence in several regions due to natural factors such as strong monsoon rainfall, severe convective rainstorms, poor drainage, and other local variables. However, when the COVID-19 outbreak coincided with the flood and storm season, generating twin disasters, was the predominant cause of fatalities.

Natural disasters that wreaked havoc on the world during the outbreak of deadly diseases did not just happen in 2020. The 2010 Cholera pandemic breakout was followed by earthquakes in Haiti. The earthquake was recorded (7 on the Richter scale) in Haiti, the epidemic killed 7000 people, including UN humanitarian personnel. This event serves as a lesson and a warning to governments interested in providing international help to Pacific islands during typhoon season. In 2011, a similar incidence occurred. The spread of respiratory diseases was followed by earthquakes and tsunamis in Great East Japan. During this calamity, authorities reviewed methods to evaluate and manage the sickness at the evacuation point. However, the requirements in the impacted areas were constantly changing and it required relief to be delivered on time. As a result, the disaster aid was inventoried and medical teams were delivered to the affected zone in the right proportions.

This study can be referred to for the current scenario of flood calamities amid the COVID-19 pandemic. Amid the COVID-19 pandemic, several countries are struggling to cope with the effects of natural disasters, including Typhoon Hagibis in Japan, melting of snow resulting in floods in Canada, Cyclone Harold in Pacific countries, and typhoons in Bangladesh. In this context, there exist innumerable studies focusing on mitigating varied disasters among the ongoing COVID-19 pandemic situation. According to Laksmi et al., the study examined to prepare a strategy for flood catastrophe preparedness in Pekalongan city in Indonesia, so that the disaster response procedure, which includes victim evacuation, can be carried out without spreading Covid-19. Another study, it is described how the government of Kamogawa City, encouraged the public to deal with severe rain by following g the COVID-19 guideline procedure. Here, the Kamogawa administration has asked their residents to live with relatives (rather than in an evacuation center), to stay at home while conditions are safe, to avoid crowds in evacuation centers, and to follow the Covid-19 procedure, which includes wearing masks in evacuation areas.

Similarly, India is one of the world's most prone countries to flooding. Flood disasters have a bigger impact in India because of their geographic location, climate, geography, and large population. As per the national reports, floods affect 23 of the country's 35 states and union territories. Moreover, around 40 million hectares, or nearly one-eighth of India's land area, are prone to flooding. As already stated by several authors, the Indo-Gangetic-Brahmaputra basins are the world's most flood-prone places, with the Indo-Gangetic and Brahmaputra river basins regarded as the worst flood-affected region. Every year, states in the Brahmaputra basin, as well as Bihar, Uttar Pradesh, and West Bengal in the Gangetic basin, experience catastrophic flooding (Dong et al., 2020). Flooding is quite common in many parts of the Indian subcontinent, especially during the monsoon season. However, the major challenge is to mitigate the damage caused by floods as the country struggles to fight the spread of the COVID-19 pandemic. The major issue of the NDRF authorities in such a situation would be keeping our rescuer's team safe from COVID-19 infections while deploying and carrying out flood rescue operations (Ishiwatari et al., 2020).

To develop crisis and disaster maps, the earth observation (EO) proved to be a comprehensive as well as objective data source. When responding to and mitigating calamities such as floods, Earth observation (EO) data obtained from sensors enables essential information that is up to date. Satellite observations allow data collection for huge and inaccessible areas. Since it is impractical to offer this information through field surveys, we can use satellite data to determine flood areas. The flood extent is critical for hydraulic model calibration and validation (Snapir et al., 2019). Using a statistical model and GIS, the paper, presented the structure of a flood susceptible map to identify the possible flood areas around the Kelantan river basin in Malaysia. A spatial database was created from a topographical map, geological map, hydrological map, data from GPS, land use and land cover map, digital elevation model (DEM), and meteorological data to evaluate the factors related to flood susceptible analysis. Severe weather conditions, particularly the presence of clouds, obstruct flood mapping with optical imaging. SAR readings are unaffected by the time of day or weather conditions, giving vital information for flood monitoring. This is mostly because a smooth water surface in the microwave spectrum delivers no return to the antenna and appears black in SAR imaging. A wind-ruffled surface, on the other hand, can generate a stronger backscatter signal than the land around it. This makes detecting water surfaces on SAR images for flood applications more difficult.

The qualitative and quantitative monitoring of the flooded region is not possible by optical sensors as it cannot be used due to the customary overcast weather during and after the occurrence. On the other hand, flood mapping also suffers from classification issues when images from radar sensors are used, owing to the confusion caused by the similar radiometric behavior of flood pixels and hill shade pixels, resulting in misclassification of the flood extent and intensity, both qualitatively and quantitatively. Martinis et al. recently introduced a TerraSAR-X–based flood mapping service that comprises a completely automated processing chain targeted at near-real-time pixel-based flood detection based on TerraSAR-X data. Because TerraSAR-X collects data in an ad hoc manner, the flood mapping service must be triggered on-demand in the event of an emergency by programming specific acquisition of satellite imageries over flood-affected areas. Unfortunately, in many cases, the mechanisms associated with satellite-based emergency response are only engaged after a flood has become extremely severe. As a result, later tasked satellite data acquisitions may be too late to record the flood's peak, reducing the utility of obtained crisis information for managing the emergency. For successful flood monitoring, Mouratidis et al. suggested that unlike TerraSAR-X, Cosmo SkyMed, and Radarsat-2, the Sentinel-1 mission, which will be operated by the European Space Agency (ESA) as part of the European Union's Copernicus Program, with a 6-day repeat cycle. A C-Band SAR payload is carried by Sentinel-1A and B. (at 5.405 GHz). It was described and evaluated an automated Sentinel-1–based processing chain for near-real-time flood detection and monitoring (NRT). The processing chain permits developing time-critical disaster information in less than 45 min when a fresh data set is available on the European Space Agency's Sentinel Data Hub because zero user participation is needed at any stage of the flood-mapping procedure (ESA). Similarly, another paper studied the ability of SAR data in this work to analyze and map flooded zones in Sylhet, in northeast Bangladesh. The study examined how the new advances and sensor technologies have broadened the reach of earth observation for detecting the changes after the flood. Furthermore, flood maps and monitoring have been created using SAR data. SAR data from RADARSAT satellites are being used to study flood mitigation and mapping in countries like India and Bangladesh owing to monsoon rain.

There exist different methods to process SAR imageries to obtain accurate flood maps. In a research paper, Cunjian et al. used a threshold-based segmentation algorithm to recover flood extent from RADARSAT-1 imageries with the use of digital topography data. Using fuzzy connectivity principles and the coherence measure acquired from an InSAR Interferometric SAR, Dellepiane et al. introduced a new algorithm that discriminates water and land on SAR images. Likewise, Niedermeier et al. used a method to find all edges above a specific threshold in SAR pictures using an edge-detection algorithm. Based on Kohonen's SOMs application, this research proposes a neural network method for flood monitoring from SAR satellite images. SOMs have the advantage of being a powerful software tool for visualizing high-dimensional data, automatically understanding statistically significant aspects of designed vectors in a dataset, and discovering clusters in training data pattern space that may be utilized for classifying modern formats.

Several studies have found that even if the data were made widely available, there is a lack of an effective infrastructure to treat every big data volume. The ES community has created portals which are web-based services to make data access, processing, and visualization more convenient. In this context, there are two concerns required to be solved: quick access to archives of big data and on-demand near real-time processing services. An equivalent SAR processing service is presented in this work to aid in the mapping of inundated zones. The service makes use of computing technology to handle massive amounts of data and to offer the computational resources required for SAR analysis. As a result, the time-consuming process of handling satellite data can be skipped. In addition, the time between data delivery and distribution of the crisis information extracted can be greatly decreased. It is intended to give consumers an easy-to-use environment in which new technology permits them to derive ready-to-use processed items.

Hence, Google Earth Engine is one of the most advanced cloud-based geospatial processing platforms launched by Google that helps to access high-performance computational resources as referred by et al. used Sentinel datasets for flood-affected paddy rice mapping. In this study, multiple datasets are used to monitor the impact of floods and address the potential of Google Earth Engine to process the datasets and evaluate the performance of the techniques to identify the flooded areas.

3. Materials and methods

The study area (Fig. 1) selected the Lower Ganga basin by drawing an area of interest. The major focus was on determining the intense impacts caused by the floods and their spatial extent in Bihar, Jharkhand, and West Bengal, all of which were severely hit.

The period specified for the preflood (April 15–May 15, 2020) and during the postflood (June 30–July 20, 2020). Sentinel-1A images were retrieved with VH polarization in the descending direction. According to the predefined parameters, the satellite images were preprocessed and then clipped to the area of interest.

The preprocessing techniques included orbit file correction, the rectification of thermal noise, radiometric standardization, and orthorectification of the data. The Sentinel-1 images of preflood and peak-flood are preprocessed as shown in Figs. 2 and 3. All tasks were performed using Google Earth Engine (GEE) which is a powerful cloud-based platform for processing large-scale remote sensing datasets.

The structure of the method for detecting flood inundated areas is shown with the help of a flow diagram in Fig. 4. This initialized with the preprocessing of the Sentinel-1 dataset for classifying the water and nonwater pixels. To calculate the flood change per pixel, the mosaic of the postflood image was divided by the mosaic of the preflood. The light-cultured values imply large changes in the raster image, whereas the dark values indicate little changes. The raster layer is constructed using the binary image where the threshold value (1.15) was taken into account to get the inundated areas. Several other additional datasets were used to validate the results. The MODIS Land Cover Type Product having a resolution of 500 m again recategorized into six groups (Areffian et al., 2021).

To extract the affected areas under the LULC groups of urban and cropland, this was resampled and intersected with the flood extent map. In this context, to mask the areas covered by water for more than 10 months per year, the JRC Global Surface-water datasets with 30 m resolution was utilized. To estimate the number of exposed persons per pixel, the JRC Global Human Settlement Population dataset with 250 m resolution was applied in this study. Both the layers are intersected, a new raster layer is generated to calculate the population affected by the flood.

4. Results and discussion

The study demonstrated that the flood-affected a large area of 42,630 km^2 from June to July 2020. The flood wreaked havoc on the lower Ganga basin affecting sections of both India and Bangladesh. It was noted that Bihar was struggling simultaneously with two facades, which are the flood and the COVID-19 pandemic. The districts in north Bihar including the fishing and farming communities were the hardest hit. This resulted in extensive loss of livelihood to people, marooned settlements, relocated families to makeshift shelters, damaged crops, food insecurity, poor sanitation, and serious violations of rules for the spread of the COVID-19 pandemic.

The MODIS-based land use and land cover are utilized to determine the impact of flood inundation on different LULC particularly on cropland, and settlements. Forests, urban, croplands, grasslands, water bodies, and others are the six

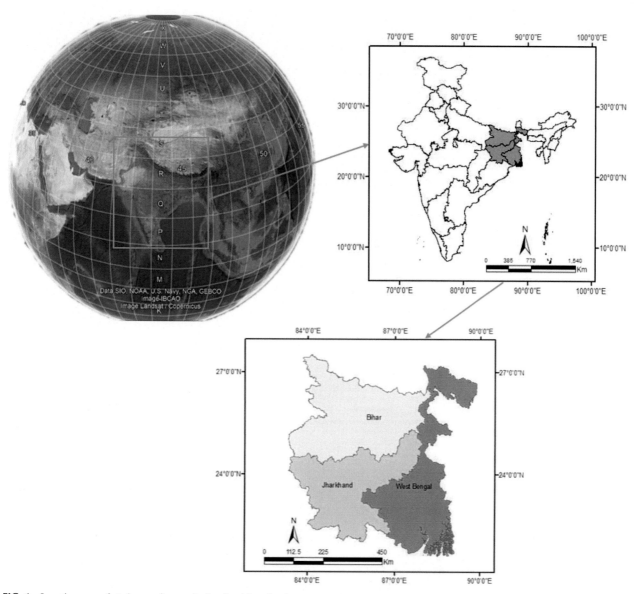

FIG. 1 Location map of study area for monitoring flood inundated areas.

categories in which the data is reclassified, as shown in Fig. 5A. The flood's impact on the LULC was investigated particularly on the cropland and settlements as shown in Fig. 5B.

Table 1 depicts the flood-affected area and population in a detailed manner. According to the maps of land use and land cover, the flood in July 2020 severely impacted the northern half of Bihar, as well as West Bengal's north and central regions. The results examined in this study are explained in the following:

1. It was noted devastating flood occurred affecting the 8763.31 km^2 area in Bihar, West Bengal covering an area of 7073.63 km^2 whereas Jharkhand with an area of 821.06 km^2 was comparatively less affected.
2. The flood inundation total area affected was 42,630.21 km^2 and among that 21% of the total area affected the state of Bihar, West Bengal affected by 17%, whereas Jharkhand was affected by 3% which is comparatively less. In the state of Bihar, most of the northern districts were affected.
3. Homes and cowsheds were entirely submerged, mostly in farming and fishing towns. The data displayed an area of 30,082.79 km^2 cropland was affected and total settlements of 6062.57 km^2 were affected due to the flood event.
4. It was also noted that affected areas under mentioned LULC groups were firstly for the state of Bihar- croplands (4.79%) and settlements (19.02%); in West Bengal–croplands (5.96%) and settlements (26%); in Jharkhand–croplands (0.95%) and settlements (0.56%).

FIG. 2 Sentinel-1A image for preflood scenario (April 15, 2020–May 15, 2020).

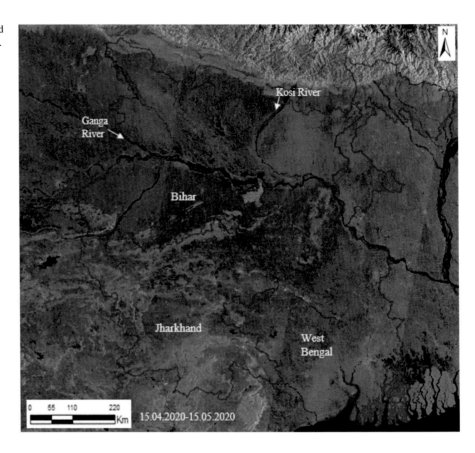

FIG. 3 Sentinel-1 A for peak-flood scenario (June 30, 2020–July 20, 2020).

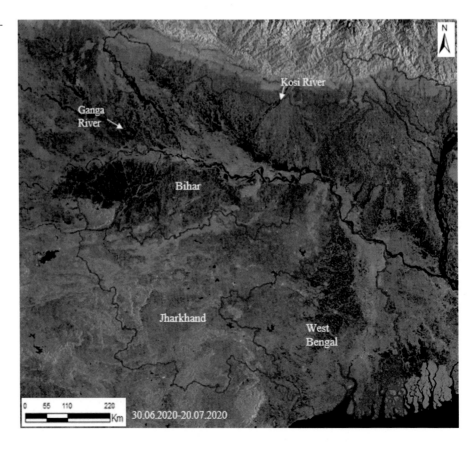

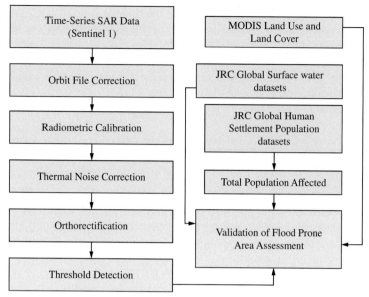

FIG. 4 Flowchart showing detailed methodology adopted for flood monitoring.

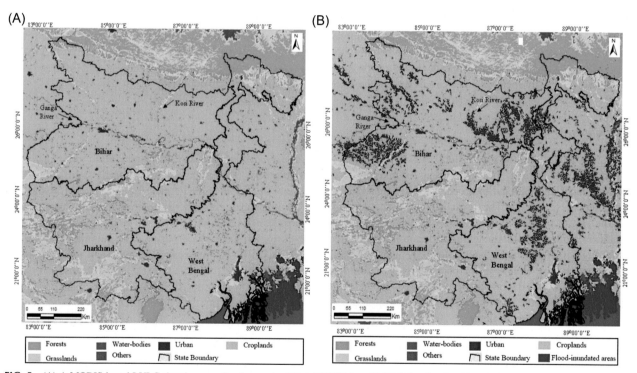

FIG. 5 (A) A MODIS-based LULC showing a preflood scenario. (B) A MODIS-based LULC showing flood inundation areas during peak flood.

TABLE 1 Impact of flood inundation in lower Ganga basin.

Categories/states	Bihar	West Bengal	Jharkhand	Sum-total
Inundated area	8763.31 km²	7072.63 km²	821.06 km²	42,630.21 km²
Affected cropland area	4.79%	5.96%	0.95%	30,081.79 km²
Affected settlement Area	19.02%	26%	0.56%	6062.57 km²
Total affected population (persons)	4,403,258	1,508,691	99,873	16, 598,223

5. According to the Global Human Settlement Population data, the flood in July 2020 affected a vast population of 16, 598,223 people over the entire region. This was followed by Bihar with 26.52%, West Bengal with 9.89%, and Jharkhand with 0.85%.
6. The information derived from this research is considered to be very significant for immediate response and rehabilitation. Furthermore, there are times when the country is on travel restrictions, and the existence in portions of affected regions has come to a halt because of the disaster calamity.
7. The study exhibited the potential of SAR imageries for allowing data retrieval in all weather conditions and mapping the water inundated areas at the time of peak flood.
8. The technique is capable of segregating water and nonwater pixels through the help of a threshold that does not require any user-dependent parameter (reference samples from the ground).
9. Lastly, the utility of the GEE platform for swift processing and accurate estimation of flood inundated areas without the need of downloading data is well highlighted in this paper.

The analysis visualized that the classified water pixels increased drastically in the peak-flood images compared to the pre-flood images. Similarly, areas submerged underwater are less in the preflood images but in the peak-flood scenario, low backscatter values are observed due to the flood inundation. It is interpreted from the flood inundated areas calculated that the flood event has affected the population majorly.

The inaccessibility of ground samples at the time of the COVID-19 pandemic is one of the limitations of this study. However, the method has shown a good performance with Sentinel-1 data, further analysis is needed to explore other high-resolution datasets. Moreover, the information shared through social media can be utilized and explored for further validation. Thus, the study can utilize different polarization and band ratio techniques that can be implemented in the different environments trying other areas for real-time flood events. As social distancing becomes the new norm in the COVID-19 pandemic, the only way to monitor flooding was to use real-time data combined with powerful AI algorithm integration. Distribution teams provided aid to people taking shelter on the roads and embankments. As mentioned by the management authorities, teams were deployed in rescue and involved in relief operations. Managing such a large population and relocating them to relief camps during the COVID-19 period was a difficult task.

5. Conclusions

This study found that the recent flood in July 2020 intensified the situation of the COVID-19 pandemic in the states, of Bihar and West Bengal. This affected an extensive area with an enormous population existing in these states, as well as causing death and damage to property. This study also showed how the flood event impacted the area of cropland and settlements respectively. The results were confirmed using multiple other imageries, also discovered that the Google Earth Engine platform aided in the rapid processing of Sentinel imageries. Since these images can be retrieved in any weather condition, with enormous potential for mapping inundated areas during the peak flood period. The automated technique highlighted its significance, well-defined by its limitations around the COVID-19 pandemic. The rapid and accurate results of this technology with new approaches can be recommended and adopted by researchers for further studies, especially response to COVID-19 or other emergencies. Because of the above, it is to be noted that flooding poses danger to life and property, thus the urgent need for actions required to assist the disaster risk mitigation communities and other sectors in achieving a resilient situation.

References

Agnihotri, A.K., Ohri, A., Gaur, S., Shivam, D.N., Mishra, S., 2019. Flood inundation mapping and monitoring using SAR data and its impact on Ramganga River in ganga basin. Environ. Monit. Assess. 191 (12). https://doi.org/10.1007/s10661-019-7903-4.

Aher, S.P., Ph, D., 2014. Synthetic aperture radar in Indian remote sensing. Int. J. Appl. Inf. Syst. 7 (2), 2012–2015.

Ahmad, F., Goparaju, L., 2020. Geospatial understanding of climate parameters within watershed boundaries of India. Spat. Inf. Res. 28 (6), 635–643. https://doi.org/10.1007/s41324-020-00323-z.

Areffian, A., Kiani Sadr, M., Eslamian, S., Khoshfetrat, A., 2021. Monitoring the effects of drought on vegetation in mountainous areas using MODIS satellite images (case study: Lorestan province). J. Environ. Sci. Stud. 5 (4), 3183–3189.

Babí Almenar, J., Elliot, T., Rugani, B., Philippe, B., Navarrete Gutierrez, T., Sonnemann, G., Geneletti, D., 2021. Nexus between nature-based solutions, ecosystem services and urban challenges. Land Use Policy 100, 104898. https://doi.org/10.1016/j.landusepol.2020.104898.

Bhatt, C.M., Gupta, A., Roy, A., Dalal, P., Chauhan, P., 2021. Geospatial analysis of September, 2019 floods in the lower gangetic plains of Bihar using multi-temporal satellites and river gauge data. Geomat. Nat. Haz. Risk 12 (1), 84–102. https://doi.org/10.1080/19475705.2020.1861113.

Cao, H., Zhang, H., Wang, C., Zhang, B., 2019. Operational flood detection using Sentinel-1 SAR data over large areas. Water (Switzerland) 11 (4). https://doi.org/10.3390/w11040786.

Cornara, S., Tonetti, S., de Vendictis, L., Papoutsis, I., Catucci, A., Kontoes, H., Bally, P., 2019. New earth observation multi-satellite mission concepts and space architectures for disaster risk reduction. In: Proceedings of the International Astronautical Congress, *IAC*, Washington D.C., USA, 21–25 October.

Das, N.N., Entekhabi, D., Kim, S., Jagdhuber, T., Dunbar, S., Yueh, S., Jackson, T.J., 2018. High-resolution enhanced product based on smap active-passive approach using sentinel 1A and 1B SAR data. Int. Arch. Photogramm. Remote Sens. Spat. Inf. Sci. - ISPRS Arch. 42 (5), 203–205. https://doi.org/10.5194/isprs-archives-XLII-5-203-2018.

Dheeraj, K., 2020. Analysing COVID-19 news impact on social media aggregation. Int. J. Adv. Trends Comput. Sci. Eng. 9 (3), 2848–2855. https://doi.org/10.30534/ijatcse/2020/56932020.

Dong, D., Wang, C., Yan, J., He, Q., Zeng, J., Wei, Z., 2020. Combing Sentinel-1 and Sentinel-2 image time series for invasive Spartina alterniflora mapping on Google earth engine: a case study in Zhangjiang estuary. J. Appl. Remote. Sens. 14 (4). https://doi.org/10.1117/1.JRS.14.044504.

Duro, J., Vicente-Guijalba, F., Centolanza, G., Iglesias, R., 2018. Innovative exploitation of long, dense and coherent InSAR sentinel-1 time series for land survey and classification. In: International Geoscience and Remote Sensing Symposium (IGARSS), 2018-July, 1569–1572. https://doi.org/10.1109/IGARSS.2018.8517966.

Espinosa, A., Rudenstine, S., 2020. The contribution of financial well-being, social support, and trait emotional intelligence on psychological distress. Br. J. Clin. Psychol. 59 (2), 224–240. https://doi.org/10.1111/bjc.12242.

Ishiwatari, M., Koike, T., Hiroki, K., Toda, T., Katsube, T., 2020. Managing disasters amid COVID-19 pandemic: approaches of response to flood disasters. Prog. Disaster Sci. 6, 100096. https://doi.org/10.1016/J.PDISAS.2020.100096.

Jian, L., Yang, X., Zhou, Z., Zhou, K., Liu, K., Eichler, M., Lee, J.S., 2020. Image segmentation based on ultimate levelings: from attribute filters to machine learning strategies. Remote Sens. Environ. 175, 163671. https://doi.org/10.1016/j.isprsjprs.2008.07.005.

Kandekar, V.U., Pande, C.B., Rajesh, J., Atre, A.A., Gorantiwar, S.D., Kadam, S.A., Gavit, B., 2021. Surface water dynamics analysis based on sentinel imagery and Google earth engine platform: a case study of Jayakwadi dam. Sustain. Water Resour. Manag. 7 (3). https://doi.org/10.1007/s40899-021-00527-7.

Krämer, A., Billaud, J.-N., Tugendreich, S., Shiffman, D., Jones, M., Green, J., 2021. The coronavirus network explorer: mining a large-scale knowledge graph for effects of SARS-CoV-2 on host cell function. BMC Bioinform. 22 (1). https://doi.org/10.1186/s12859-021-04148-x.

Lal, P., Prakash, A., Kumar, A., 2020. Google earth engine for concurrent flood monitoring in the lower basin of indo-Gangetic-Brahmaputra plains. Nat. Hazards 104 (2), 1947–1952. https://doi.org/10.1007/s11069-020-04233-z.

Lattari, F., Leon, B.G., Asaro, F., Rucci, A., Prati, C., Matteucci, M., 2019. Deep learning for SAR image despeckling. Remote Sens. (Basel) 11 (13), 1–20. https://doi.org/10.3390/rs11131532.

Li, H., Huang, C., Liu, Q., Liu, G., 2020. Accretion–erosion dynamics of the yellow river delta and the relationships with runoff and sediment from 1976 to 2018. Water (Switzerland) 12 (11). https://doi.org/10.3390/w12112992.

Meena, S.R., Chauhan, A., Bhuyan, K., Singh, R.P., 2021. Chamoli disaster: pronounced changes in water quality and flood plains using sentinel data. Environ. Earth Sci. 80 (17). https://doi.org/10.1007/s12665-021-09904-z.

Min, L., Wang, N., Wu, L., Li, N., Zhao, J., 2020. Inversion of Yellow River runoff based on multi-source radar remote sensing technology. Dianzi Yu Xinxi Xuebao/J. Electron. Inf. Technol. 42 (7), 1590–1598. https://doi.org/10.11999/JEIT190494.

Mullen, C., Penny, G., Müller, M.F., 2021. A simple cloud-filling approach for remote sensing water cover assessments. Hydrol. Earth Syst. Sci. 25 (5), 2373–2386. https://doi.org/10.5194/hess-25-2373-2021.

Mutanga, O., Kumar, L., 2019. Google earth engine applications. Remote Sens. (Basel) 11 (5), 11–14. https://doi.org/10.3390/rs11050591.

Navlakha, S., Morjaria, S., Perez-Johnston, R., Zhang, A., Taur, Y., 2021. Projecting COVID-19 disease severity in cancer patients using purposefully-designed machine learning. BMC Infect. Dis. 21 (1). https://doi.org/10.1186/s12879-021-06038-2.

Pham Van, C., Nguyen-Van, G., 2020. Assessment of the water area in the lowland region of the mekong river using MODIS EVI time series. Adv. Intell. Syst. Comput. 1121, 197–207. https://doi.org/10.1007/978-3-030-38364-0_18.

Prakash, K., Singh, S., Mohanty, T., Chaubey, K., Singh, C.K., 2017. Morphometric assessment of Gomati river basin, middle ganga plain, Uttar Pradesh, North India. Spat. Inf. Res. 25 (3), 449–458. https://doi.org/10.1007/s41324-017-0110-x.

Shen, X., Wang, D., Mao, K., Anagnostou, E., Hong, Y., 2019. Inundation extent mapping by synthetic aperture radar: a review. Remote Sens. (Basel) 11 (7). https://doi.org/10.3390/RS11070879.

Si, A., Das, S., Kar, S., 2020. Extension of topsis and vikor method for decision-making problems with picture fuzzy number. Adv. Intell. Syst. Comput. 1112, 563–577. https://doi.org/10.1007/978-981-15-2188-1_44.

Sinha, R., Tandon, S.K., 2014. Indus-Ganga-Brahmaputra Plains: the alluvial landscape. Landscapes and Landforms of India, Springer. https://doi.org/10.1007/978-94-017-8029-2_5.

Snapir, B., Momblanch, A., Jain, S.K., Waine, T.W., Holman, I.P., 2019. A method for monthly mapping of wet and dry snow using Sentinel-1 and MODIS: application to a Himalayan river basin. Int. J. Appl. Earth Obs. Geoinf. 74, 222–230. https://doi.org/10.1016/j.jag.2018.09.011.

Soulard, C.E., Walker, J.J., Petrakis, R.E., 2020. Implementation of a surfacewater extent model in Cambodia using cloud-based remote sensing. Remote Sens. (Basel) 12 (6). https://doi.org/10.3390/rs12060984.

Supriya, M., Chattu, V.K., 2021. A review of artificial intelligence, big data, and blockchain technology applications in medicine and global health. Big Data Cogn. Comput. 5 (3). https://doi.org/10.3390/bdcc5030041.

Tang, Z., Li, Y., Gu, Y., Jiang, W., Xue, Y., Hu, Q., Li, R., 2016. Assessing Nebraska playa wetland inundation status during 1985–2015 using Landsat data and Google earth engine. Environ. Monit. Assess. 188 (12). https://doi.org/10.1007/s10661-016-5664-x.

Thomas, B.K., Bhar, S., Chakravarty, S., 2021. Imagining sustainability beyond covid-19 in India. Ecol. Econ. Soc. 4 (1), 13–20. https://doi.org/10.37773/ees.v4i1.315.

Wagle, N., Acharya, T.D., Kolluru, V., Huang, H., Lee, D.H., 2020. Multi-temporal land cover change mapping using google earth engine and ensemble learning methods. Appl. Sci. 10 (22), 1–20. https://doi.org/10.3390/app10228083.

Wang, Y., Ma, J., Xiao, X., Wang, X., Dai, S., Zhao, B., 2019. Long-term dynamic of Poyang Lake surface water: a mapping work based on the Google earth engine cloud platform. Remote Sens. (Basel) 11 (3). https://doi.org/10.3390/rs11030313.

Wang, Y., Peng, D., Yu, L., Zhang, Y., Yin, J., Zhou, L., Li, C., 2020. Monitoring crop growth during the period of the rapid spread of COVID-19 in China by remote sensing. IEEE J. Sel. Top. Appl. Earth Obs. Remote Sens. 13, 6195–6205. https://doi.org/10.1109/JSTARS.2020.3029434.

Wedlund, L., Kvedar, J., 2021. New machine learning model predicts who may benefit most from COVID-19 vaccination. NPJ Digit. Med. 4 (1). https://doi.org/10.1038/s41746-021-00425-4.

Xiangdong, L., Youlong, H., Peng, Q., Wei, L., Bilalov, R.A., Smetannikov, O.Y., Miquel-Romero, M.-J., 2021. No title. J. Bus. Ethics 11 (1), 1–15. https://doi.org/10.1016/j.isprsjprs.2021.02.021.

Zhu, P., Huang, S., Yang, Y., Ma, J., Sun, Y., Gao, S., 2019. High-frequency monitoring of Inland Lakes water extent using time-series Sentinel-1 SAR data. In: ICSIDP 2019 - IEEE International Conference on Signal, Information and Data Processing 2019. https://doi.org/10.1109/ICSIDP47821.2019.9173066.

Zoabi, Y., Deri-Rozov, S., Shomron, N., 2021. Machine learning-based prediction of COVID-19 diagnosis based on symptoms. NPJ Digit. Med. 4 (1). https://doi.org/10.1038/s41746-020-00372-6.

Chapter 7

Causal reasoning modeling (CRM) for rivers' runoff behavior analysis and prediction

Jose-Luis Molina[a], S. Zazo[a], María C. Patino-Alonso[b], A.M. Martín-Casado[b], and F. Espejo[a]

[a]*High Polytechnic School of Engineering, University of Salamanca, Ávila, Spain,* [b]*Department of Statistics, University of Salamanca, Campus Miguel de Unamuno, Salamanca, Spain*

1. Introduction

Water resource management requires the proper use and management of a great range and amount of data sources, where climatic and hydrological variables are essential (Akintug and Rasmussen, 2005; Pulido-Velazquez et al., 2015; Stojkovic et al., 2015; Wang et al., 2011). It is very important to highlight that variables, such as water quality and its evolution, stream flows, rainfall, total runoff, air temperature, humidity or evapotranspiration among others, have a common feature, their random nature (Croitoru and Minea, 2015). Consequently, the challenge is to be able to design models and tools that contribute to reliably represent an unknown or poor known reality (Díaz Caballero, 2011). Models should be characterized by being robust, agile, and parsimonious, as well as by keeping observed the statistical characteristics of time series (Díaz Caballero, 2011; Myung, 2000).

More specifically, hydrological processes are undergoing an unprecedented shift toward increasing variability on a global and local scale (O'Gorman, 2015; Pfahl et al., 2017). Most of the scientific academy concludes that this is largely due to the effect of Climate Change (CC; O'Gorman, 2015; Pfahl et al., 2017). Consequently, those events, with values and trends far from the historical average behavior, are more frequent (Chang et al., 2015; Kalra et al., 2013). In order to address this new hydrological reality and aimed to anticipate and forecast it, new approaches, methodologies and techniques have emerged, incorporating this changing behavior (Chang et al., 2015; Kalra et al., 2013). Furthermore, in order to build tools for the present-future, not all the reasons that explain this increasing variability are brand new. However, there should be an improvement on the definition of climate change and projections to the future, (Molina and Zazo, 2018), but also, the historical behavior should be more deeply understood (Molina et al., 2019; Zazo et al., 2019). In this context, causal methods erect as alternative, innovative and efficient ways for analyzing and modeling the temporal, spatial and spatiotemporal behavior of hydrological records (Molina et al., 2020, 2016).

Traditionally, rainfall-runoff models are assigned to one of three broad categories: deterministic, analytic or parametric (Dawson and Wilby, 2001; Watts, 1997). Deterministic models describe the rainfall-runoff process using physical laws of mass and energy transfer. Analytic models provide simplified representations of key hydrological processes using a perceived system. On the other hand, parametric models use mathematical functions such as multiple linear regression equations to relate meteorological variables to runoff. Hydrological models are further classified as either lumped or distributed (Todini, 1988). Distributed hydrologic models feature the capability to incorporate a variety of spatially varying data from a proliferating set of databases on land, precipitation, temperature, etc., while lumped models treat the catchment as a single unit. This traditional classification shares two important features that become drawbacks on its application under this changing hydrological context. First, it is the usage of just one historical data series per gauge station and/or for the spatial interpolation procedure in the case of distributed models. This is an important drawback that it is proposed to overcome through the application of Causal Reasoning (CR) models based on the usage of huge amount of input data. Second, it is the absence of intrinsic procedures and/or methods to deal with the uncertainty.

The uncertainty quantification of hydrologic models, basically produced for the random nature of hydrological processes (Kong et al., 2017), is largely based on probabilistic method. Stochastic methods are the most powerful ones for uncertainty quantification of hydrologic models (Koutsoyiannis, 2010; Papacharalampous et al., 2020). Consequently,

Handbook of HydroInformatics. https://doi.org/10.1016/B978-0-12-821962-1.00005-2

stochastic simulation is vital tool for risk-based management of water resources systems. Parametric models have been quite useful for analyzing hydrologic time series and over the last two decades, non-parametric models have appeared in a variety of hydrological and climatological applications (Rajagopalan et al., 2010). Stochastic methods have different ways of dealing with uncertainty, for example, CR, that uses probability (Molina et al., 2016; Molina and Zazo, 2017), while other stochastic methods evaluate correlation and similarity in the datasets (Carrasco et al., 2019). In this line, methods such as Principal Component Analysis (PCA) are techniques for reducing the dimensionality of such datasets (Molina et al., 2020), demonstrating the main patterns of similarity between individuals increasing interpretability, but at the same time, minimizing information loss (Jolliffe and Cadima, 2016). Furthermore, HJ-Biplot is a graphical representation of the individuals and variables that allow the visualization of relations between them (Galindo, 1986). Other applications of different multivariate statistical techniques are cluster analysis (Forgy, 1965) and correspondence analysis which, for example, have been used to explore how the potential floodplain inundation area simulation could be used to compare physical characteristics among river reaches (Thorp et al., 2010). Other methods such as discriminant analysis have been used to predict groundwater redox status on a regional scale (Close et al., 2016) for estimation of age for a set of groundwater monitoring wells (Daughney et al., 2010).

This chapter is focused on the first category that uses probability to deal with the uncertainty. In this sense, randomness and variability of hydrological processes are incorporated and quantified by means of a type of CR modeling which is called Bayesian Causal Modeling (BCM) using the conditional probability to compute the influence among the variables.

2. Causal reasoning

Causal Reasoning (CR) should be seen as a reasoning pattern whose main goal is to predict the consequences or effects of some previous factors (Pearl, 2009). For instance, a joint distribution PB specifies the probability $PB (A = a \,|\, E = e)$ of any event a given any observations e. The probability of the event a is computed by summing the probabilities of all of the entries in the resulting posterior distribution that are consistent with a. Queries such as these, where it is about the prediction of "downstream" effects of various factors, are instances of causal reasoning or prediction.

Causality in hydrological records has not been deeply studied and it could be done by means of the joint use of different forms of reasoning patterns. These forms are Causal Reasoning (CR), Evidential Reasoning (ER) and Intercausal Reasoning (IR) (Koller and Friedman, 2009; Pearl, 2009). CR is used when the approach is done from top to bottom. In this sense, the analysis is focused on the cause and the objective comprises the prediction of the effect or consequence. Consequently, the queries in form of conditional probability, where the "downstream" effects of various factors are predicted, are instances of causal reasoning or prediction. ER comprises bottom-up reasoning, so the analysis is focused on the consequence (effect) and the cause is inferred (Bayesian Inference). IR is probably the hardest concept to understand. It comprises the interaction of different causes for the same effect. This type of reasoning is very useful in Hydrology, where a consequence can be generated or explained from several causes (Kouhestani et al., 2017). Furthermore, one of the most exciting prospects in recent years has been the possibility of using the theory of Bayesian Networks (BN) to discover causal structures in raw data (historical runoff record) (Pearl, 2014). This is performed through the usage of historical runoff data to train and populate the BN implementation. Consequently, AI techniques such as CR and ER and/or IR provide new horizons for this type of studies.

Furthermore, temporal dependence of hydrological time series has been deeply studied through classic and new approaches (Hao and Singh, 2016; Mishra and Singh, 2010, 2011; Molina et al., 2016; Molina and Zazo, 2017). Conversely, spatial and spatiotemporal dependence for hydrologic science and engineering is much poorer studied (Holmström et al., 2015) and even more through Bayesian approaches (Lasinio et al., 2007; Wikle et al., 1998). This is because of some reasons explained as follows: complexity of characterizing and differentiating water subsystems, scarcity of spatial data availability, difficulties on the application of spatial statistical methods, among others. Consequently, there is a general clear necessity of strengthening the spatiotemporal dependence studies on water systems (Holmström et al., 2015).

3. Bayesian causal modeling (BCM)

Over the last decades, remarkable progress in the ability to develop hydrologic models has been produced. Within the field of Artificial Intelligence (AI), based models have been among the most promising in simulating hydrologic processes. In recent years, Artificial Neural Networks have been widely used to solve problems including prediction, forecasting and classification (Kasiviswanathan and Sudheer, 2017). Artificial Neural Networks (ANNs) are an emerging field of research, characterized by a wide variety of techniques and a general absence of intermodel comparisons. They should be classified as parametric models that are generally lumped, and they have gained significant attention because of its ability to provide better solutions when applied to complex systems.

In Hydrology, copula applications have also been successfully applied to the hydrological prediction field. Copula started after the work of De Michele and Salvadori (2003), who tested Frank copulas for a joint study of the negatively associated storm intensity and duration. Other research, such as that of Zhang and Singh (2006), incorporated copulas for an extreme analysis of rainfall and drought events. The study of the dependence's modeling between extreme events is widely studied currently (Dutfoy et al., 2014). Dependence in extreme events needs to be evaluated through techniques focused on an extreme distribution, such as the tail-dependence coefficient. This has been commonly used in investigating hydroclimatic extremes, such as precipitation and temperature (Serinaldi et al., 2015).

Another important and recent area of research is the construction of a multivariate distribution in modeling different dependence structures (Sarabia-Alzaga and Gómez-Déniz, 2008). This is applied to Hydrology in the form of frequency analysis, downscaling, streamflow or rainfall simulation, geostatistical interpolation, bias correction, and so on. Other methods, such as multivariate parametric distribution (Balakrishnan and Lai, 2009), entropy (Singh, 2013), and copula (Nelsen, 2007) have been developed to model various dependence structures of multivariate variables through the construction of joint or conditioned distribution. On the other hand, it should be highlighted that copula method is based on the description and modeling of the dependence structure between random variables independently of the marginal laws involved. In this sense, there are high similarities with CR. However, copula is a bivariate function, while CR is based on conditioned probabilistic distributions (Molina et al., 2020).

More recently, in the fields of Water Management and Hydrology, Probabilistic Graphical Models (PGM), such as Bayesian or Markov networks, have become widespread AI techniques (Molina et al., 2013, 2016; Vogel et al., 2018). They both use Bayes' rule implementation through Bayesian inference and Bayesian Networks (BNs) or Dynamic Bayesian Networks (DBNs) for their compilation (Molina et al., 2013). In this sense, DBNs are a generalization of hidden Markov models (Molina et al., 2013; Nodelman and Horvitz, 2003). Another interesting approach in this line is the application of PGM to the identification of factors in the process of flood damage (Vogel et al., 2018). This has been later articulated through the development of BCM for the development of analytical and predictive tools, in a practical way. Main principals, general methodology and most important applicants are explained below in the next sections.

3.1 Main principles

Bayesian Network (BN) was first proposed by Pearl (1988). This is a kind of probabilistic reasoning network based on Bayesian conditional probability and graph theory. In this sense, Bayesian modeling is a powerful AI technique based on PGM (Cain, 2001; Jensen, 1996; Pearl, 1988; Sperotto et al., 2017) that has significant advantages listed as follows. There is no need of priori information of the process, being able to use raw data (Adarnowski, 2008; Zounemat-Kermani and Teshnehlab, 2008); it has high usefulness for analyzing nonlinear physical systems (Aqil et al., 2007); it is very easy to define relationships in complex systems and offers a compact representation of the joint probability distribution over sets of random variables (Molina et al., 2013).

On the other side, due to variables change over time, Dynamic Bayesian Networks (DBNs), also based on Bayes' theorem, can be seen as a BN extended with temporal dimensions to enable us to model systems dynamically. This allows defining relationships between parts in complex models through also conditional probabilities (Molina et al., 2013). Furthermore, this approach reveals the temporal and spatial evolution processes of the uncertainties. Due to the complex relationships of dependence among hydrological years, BNs arise as a very powerful tool because those are easily modeled and visualized (Zazo, 2017).

In this chapter, CRM (Causal Reasoning Modeling) is essentially addressed from a Bayesian perspective, mainly based on BNs. It should be highlighted that BNs are a kind of PGM methods that offer compact representations of the joint probability distribution over sets of random variables (Castelletti and Soncini-Sessa, 2007; Said, 2006). This easily and automatically allows defining relationships between complex systems (Molina et al., 2010) and providing more accurate results in the modeling of natural processes (Jain and Kumar, 2007; See and Openshaw, 2000). Furthermore, this approach is characterized by searching reasoning patterns focused on the cause, and where the objective comprises the prediction of the effect (Pearl, 2009). From top to bottom, this process is performed by means of queries in form of conditional probability. Here, CR is carried out over a set of random decision variables named "nodes" (water years for this research line), which are consecutively interconnected through "links," and a set of conditional probability tables between decision variables (Jensen, 1996; Molina et al., 2013; Pearl, 1988). Moreover, the quantification of the variables' relationship strength, which is propagated over time and throughout the BN, allows assessing probability distribution for each decision variable (Castelletti and Soncini-Sessa, 2007). Formally, this is expressed as:

$$P(A|B) = \frac{P(A, B)}{P(B)} = \frac{P(A \cap B)}{P(B)} \tag{1}$$

where, A represents the parent variable (independent) and B represents the child variable (dependent).

Specifically, a BN $N = (G, P)$ consists of a Direct Acyclic Graph (DAG; Lappenschaar et al., 2012) denoted by $G = (V, E)$ and a set of probability distributions P (Molina et al., 2013). The DAG is defined by a set of nodes (or variables) V and a set of links (or edges) E. The edges join the variables and are oriented, indicating the causal dependence between the connected variables ($A|B$ indicates A causes B or B is the effect of A) (Madsen et al., 2003). The set of probability distributions P includes "a priori" probabilities and a set of conditional probabilities (expressed in form of Conditional Probabilities Tables; CPTs). Bayes' theorem performs as updater of "a priori" probabilities adding the observed evidence and providing "a posteriori" probabilities. Applied to CRM, the nodes represent the hydrological variables (in this case water years; WY), and each link between two nodes is defined by a CPT containing the probabilities for the different combinations of node states (Fig. 1). In BNs, every variable is conditionally independent of its non-descendants given its parents. Therefore, a BN $N = (G, P)$ defines a joint probability distribution over all the WY present in the graph. Formally, it can be expressed:

$$P(WY_1, WY_2, \ldots, WY_n) = \prod_{i=1}^{n} P(WY_i | pa(WY_i)) \tag{2}$$

where WY_i are the variables, $pa(WY_i)$ are the parents of node WYX_i.

In this case, CRM is focused on the propagation of probability along a Bayesian implementation through a causal model. This model is defined over all decision variables (water years) who represent probability distributions for temporal runoff. One of the main advantages of this strategy is that allows computing inference omnidirectionally. Given an observation with any type of evidence on any of the networks' nodes (or a subset of nodes), BNs can compute the posterior probabilities of all other nodes in the network, regardless of arc direction, through observational inference (Cain, 2001).

The inherent ability of BNs to explicitly model uncertainty makes them suitable for temporal hydrological series analysis, like in the case of rivers' runoff behavior. Another feature of BNs is the unsupervised structural learning (Pena et al., 2004), which means probabilistic relationships between a large number of variables without having to specify input or output nodes. This can be seen as a quintessential form of knowledge discovery, as no assumptions are required to perform these algorithms on unknown datasets. Furthermore, this is strongly related to machine learning that has been applied successfully in the hydrological context (Genc and Dag, 2016; Patel and Ramachandran, 2015). Consequently, the resulting product has many similarities with a neuro-fuzzy system or Adaptive Neuro-Fuzzy Inference System (ANFIS) that has been applied in works such as (Mousavi et al., 2007).

On the other side, this methodology, presents an important triple advantage applied to the analysis of behavior of rivers that allows the identification-characterization-quantification of temporal conditioned runoff fractions (Molina and Zazo, 2018). This represents an outstanding advance over classic or alternative methods (Molina et al., 2020).

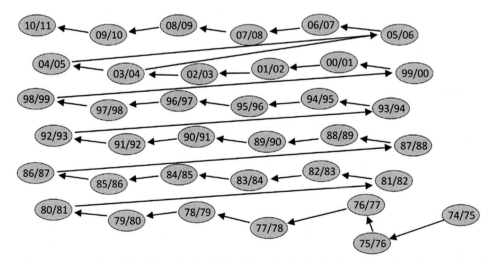

FIG. 1 Direct acyclic graph, dependent relationships. Threshold (P-value) = .10 (level of independence); dependence value to 90% (Molina and Zazo, 2018).

General methodology

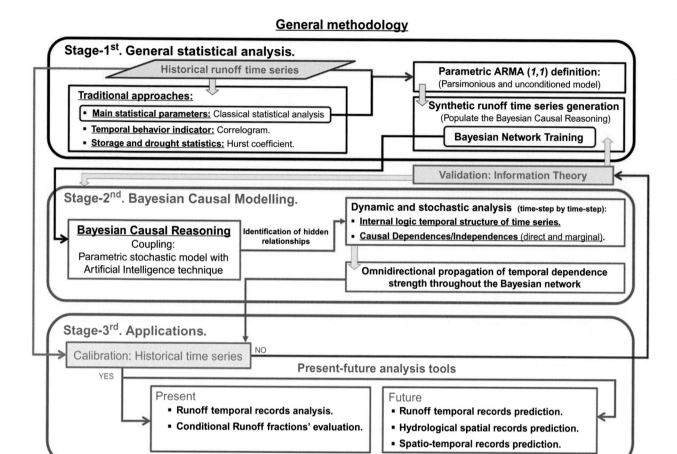

FIG. 2 General methodology.

3.2 General methodology

Causal models need to be populated by as much as information as possible to be representative (Molina et al., 2020). Consequently, after a general statistical analysis of the historical data, the process continues through a synthetic series development for the replication of the historical ones (see, Fig. 2). This new information has to be equiprobable to the historical one, and this is assured by the development of the ARMA models (Salas et al., 1980). The quality of this production is confirmed by simple quantitative calculations such as the average of the total synthetic set that is equal to the historical record (Zazo, 2017).

After this, the Bayesian Casual Modeling (BCM) begins. It is very important to remark that all variables become decision variables and represent hydrological years (water years for this case). Each variable can influence the previous and following ones, by means of their dependence strength (Molina et al., 2019).

Causal modeling design starts with a Learning Process (*LP*). Its first stage, "Acquisition Process," is based on the usage of raw data from synthetic series together with the historical ones. Then, the "Preprocessing" stage comprises a discretization process from raw data into discrete probability distributions. This is done by means of five intervals of same length. After that, the Structure Constraints process is developed. In this approach causal models are designed considering that the main relationship among variables is produced year by year. This may be seen as a preliminary design of causal model, which is based on natural relationship between years. This is also very robust as the ARMA model, previously commented, is an ARMA (1,0), a parsimonious and unconditioned model (Molina et al., 2016) and therefore, the causal design respects the inherent temporal nature of the income information from the ARMA model.

Subsequently, the potential of the CRM will reveal a non-trivial structure of relationships with time lag >1 (Zazo, 2017). By means of CRM, the logical structure of temporal interdependencies that underlies within the historical records will be discovered (Molina et al., 2019). After that, the phase "Structure Learning" is developed supported by algorithm is NPC (Necessary Path Algorithm; Madsen et al., 2003). The necessary path condition declares that, in a DAG faithful data set, in order for two variables X and Y to be independent conditional on a set S, there must exist a path between X and every $Z \in S$ (not crossing Y) and between Y and every $Z \in S$ (not crossing X). Otherwise, the inclusion of each Z in S is unexplained. Thus, in order for an independence relation to be valid, a number of edges are required to be present in the graph (Madsen et al., 2003). This is explained here by the fact that all variables are considered, a priori, dependent in this research. Then, the outcome from the compilation is useful to evaluate the degree of dependence among variables (the reader is recommended at seeing the Fig. 4). The level of significance is established in this stage at a value of 0.05. Then, "Data Dependences" phase is developed. Here, an analysis of the dependence strength among variables can be done. This is implemented through the (*1-P-value*) value (Molina et al., 2019).

The marginal dependence between A and B is defined as one minus the marginal p-value associated with $\{A,B\}$, where p-value is the measure of the strength of evidence for the independence relationship A-B. Thus, a marginal dependence of 0 means that A and B are totally independent, and 1 means that there is a lot of evidence of a total dependence. Then, the "Prior Distribution Knowledge" gives the chance of assigning different probabilities for each case. Now, the causal model is already built, and it is the time now for analyzing the dependences among variables through the model compilation.

3.3 Validation

In this methodology, a crucial stage is the validation of obtained results. Given that this is a pure quantitative BN model, its evaluation should mainly include a sensitivity analysis of results and assessments of predictive accuracy among others (Pollino et al., 2007). It should be noted that capturing and reproducing the real behavior patterns of temporal hydrological series, it is a vital issue for prediction. This is developed by means of the comparison of the results drawn from BNs versus raw runoff observed data. This internal consistency and quality of this CRM analysis maybe developed through different ways such as: (a) Information Theory by means of entropy, Mutual Information evaluation and sensitivity analysis among variables, (b) dependence strength evaluation and representation, via Dependence Mitigation Graphs (DMGs), (c) conditional runoff fractions analysis and (d) return period estimation through CRM.

Sensitivity analysis can be performed using two types of measures; entropy and Shannon's measure of mutual information (Pearl, 1988). The entropy measure assumes that the uncertainty or randomness of a variable X, characterized by probability distribution $P(x)$, can be represented by the entropy function $H(X)$:

$$H(X) = -\sum_{x \notin X} P(x) \log P(x) \tag{3}$$

Entropy of a probability distribution can be defined as a measure of the associated uncertainty to that random process that this distribution describes. Consequently, a score of uncertainty/certainty level of events can be made attending to this entropy, $H(X)$. Reducing $H(X)$ by collecting information in addition to the current knowledge about variable X is interpreted as reducing the uncertainty about the true state of X (Barton et al., 2008). The entropy measure therefore enables an assessment of the additional information required to specify an alternative. Shannon's measure of Mutual Information is used to assess the effect of collecting information about one variable (Y) in reducing the total uncertainty about variable X using:

$$I(Y.X) = H(Y) - H(Y|X) \tag{4}$$

where $I(Y.X)$ is the mutual information between variables. This measure reports the expected degree to which the joint probability of X and Y diverges from what it would be if X were independent of Y. If $I(Y.X) = 0$, X and Y are mutually independent (Pearl, 1988). $H(Y|X)$ is conditional entropy which means the uncertainty that remains about Y when X is known to be x. Indeed, two variables, that represent the annual runoff of 2 years, have a mutual information $\neq 0$ indicating that they are dependent (Pearl, 1988). On the contrary, in case the mutual information is 0 means that they are independent. This analysis represents another way for characterizing and quantifying the temporal dependence and behavior of hydrological series.

As for dependence strength evaluation over time, this can be seen as an analysis of the internal memory of the historical records. This is dynamically performed through interactions over the target year according to the time-horizon (time-lag) defined by the propagation of temporal dependence. The strength of each yearly interaction was evaluated by Dependence Rate (*DR*; from 0 to 1), expressed as (Molina and Zazo, 2017).

$$DR = 1 - (P - value) \tag{5}$$

where *P-value* is the measure of the strength of evidence against the independence relationship (Molina et al., 2016; Zazo, 2017). Therefore, a *DR* value equal to 1 expresses a lot of evidence of a total dependence, in contrast, 0 represents little evidence against the independence.

For its part, propagation of temporal dependence is assessing versus to classical correlogram by means of an innovative qualitative methodological approach, named Dependence Mitigation Graph (DMG), for assessing time dependency from a geometrical perspective, in a dynamic and continuous manner, against the classical, static and punctual analysis of a correlogram. This is done through a novel whose analysis leads to a dependent temporal behavior in the case of an asymmetric graphic, and where an independent behavior is highlight by means of symmetry graph (Zazo, 2017; Zazo et al., 2019).

The runoff of a river can be analyzed through a temporal analysis using, for instance, the historical runoff time series. Using those time series, a causal analysis (Bayes' theorem) that computes and shows how the different time steps (years) interact each other may be developed. This analysis provides the knowledge of the level of dependence relationship among time steps against traditional approaches such as Pearson correlation coefficients, Spearman rank correlation coefficient that measure the dependence in the main data body and/or a classical correlogram as well. Each relationship between time steps has two components or fractions, the nonconditioned and the conditioned components. Of course, those relationships and hence the fractions vary from almost zero to almost 100% dependence. When the whole time series and all relationships are analyzed, the temporal behavior of a river runoff, expressed by both fractions, is obtained. The sum of both fractions composes the total amount of runoff. Depending on the temporal level of dependence of a river, its runoff can be severely affected or conditioned by the past behavior. Consequently, in a temporally conditioned river, the runoff fractions can be identified, characterized, and quantified. In other words, the current flow may be composed of a conditioned and a nonconditioned fraction. The validation is carried out by means of suitability identification of time order through the ARIMA modeling, supported by the main criteria for appropriate model selection, which are the Schwarz criterion, Akaike information criterion, and Durbin-Watson criterion (Molina and Zazo, 2018).

Finally, as for Return Period (RP) estimation through CRM, this is applied on two opposite runoff fractions, TCR (Temporally Conditioned Runoff) and TNCR (Temporally Non-Conditioned Runoff), into which is divided runoff of a river and over case studies with a strong temporally dependent nature. This provides a novel stochastic approach for calculating the return period considering (a) the internal and inherent causality of runoff time series and (b) two new concepts for the RP computation for each runoff fraction, TCRP (Temporally Conditioned Return Period) and TNCRP (Temporally Non-Conditioned Return Period), respectively. This computation is attributable exclusively to the part of the time series that has some dependence from another variable. Naturally, the validation is done in a general way and exclusively computed during the duration of the dependence (Molina and Zazo, 2018).

4. Applications

The usefulness of CR is shown through different applications explained below. Those applications are classified as follows: (1) Runoff temporal records analysis, where, largely, the evaluation of runoff fractions takes place. (2) Runoff temporal records prediction, where the developed and implemented tools aimed to the prediction are shown. (3) Hydrological spatial records prediction that is described through an analysis of the potentiality of this implementation. Finally, (4) the dual spatiotemporal records prediction is described, aimed to show the potential of this double dimension hybridization and the CR potential for developing robust evaluations under this approach. Subsequently, these applications are addressed in depth.

4.1 Runoff temporal records analysis (runoff fractions' evaluation)

According to Molina and Zazo (2018), this application largely comprises the quantification of Temporally Conditioned Runoff Fractions. The application of this research to the dimensioning of regulatory storage hydraulic infrastructures, such as reservoirs, seems quite straightforward. In this sense, the assessment of all the runoff fractions derived from a temporal behavior perspective, such as TCR and TNCR fractions, will lead to reconsidering the capacity of the storage reservoirs, especially for high temporally dependent river basins and for nonstationary series. In addition, the characterization of those fractions may help to achieve an optimal design and dimensioning of hydraulic infrastructures such as reservoirs or channels.

Under this approach, on the one hand, Temporally Conditioned Runoff (TCR) is generated and computed; on the other hand, an independent fraction of runoff called Temporally Non-Conditioned Runoff (TNCR) is also quantified. Both fractions are computed through causality (temporal dependence/independence). As previously mentioned, the summation of these fractions represents the total runoff that a river carries.

In this context, dependence of river runoff largely depends on three factors: (a) river-aquifer interaction (existence and amount of baseflow), (b) geomorphology (type of materials), and (c) rainfall/evapotranspiration behavior. A quantitative assessment of temporally conditioned/non-conditioned runoff fractions is developed in this research line. This evaluation has two components; the first is focused on the average fraction of conditioned flow (TACR) of annual runoff. This comprises a weighted average in which the weights represent the dependences belonging to the year interactions as follows:

$$X_{I=0}^{n} = \frac{\sum_{j=0}^{i=1} Y_j * \theta(0, j)}{\sum_{j=0}^{i=1} \theta(0, j)} \tag{6}$$

where X is weighted average conditioned runoff for each year (i) of the series; $\theta(0,j)$ is dependence function between the studied year (j) and the beginning of the series ($j=0$); Y_j is conditioned runoff for the studied year (j); and n is the number of years of the time series.

The Total Cumulative Conditioned Runoff (TCCR) was studied for each year of the series. This is a much larger amount of runoff than the average, and its evaluation is crucial mainly for defining the dimensioning and operation rules of downstream reservoirs

$$X_{I=0}^{n} = \sum_{j=0}^{i=1} Y_j * \theta(0, j) \tag{7}$$

where X is cumulative conditioned runoff for each year (i) of the series; $\theta(0,j)$ is dependence function between the studied year (j) and the beginning of the series ($j=0$); Y_j is conditioned runoff for the studied year (j); and n is the number of years of the time series.

Obviously, the rest of the runoff value represents the independent or non-conditioned runoff. Furthermore, the summation of both fractions comprises the total runoff that occurs per year.

Molina and Zazo (2017) and Molina et al. (2016) presented exhaustive analyses of the temporal behavior of both subbasins. For their part, Molina and Zazo (2018) identified and quantify these previously shown runoff fractions depending on that temporal behavior.

Fig. 3 shows the calculation of runoff fractions as well as the computation of the total cumulative conditioned runoff. The region representing the weighted average conditioned runoff for 1-year lag is very large and needs to be considered for the river basin planning and management. This means that the component or fraction of temporally dependent runoff is very large. The variable total cumulative conditioned runoff is a key variable to take into account for the dimensioning of regulatory storage hydraulic infrastructures. Large reservoir capacity is traditionally based on return period, which is computed on a stationary and independent (non-conditioned) basis. Consequently, TCCR together with new approaches for estimating return period in multivariate and conditioned situations, are essential for achieving an optimum dimensioning of reservoirs.

Fig. 4 shows the general temporal behavior of the Porma-Esla (Fig. 4A) and Adaja (Fig. 4B) subbasins in terms of dependence and the sixth-order function that describes this wraparound function of maximum and minimum for the dependence values with approximately 0.99 coefficient of determination. Furthermore, it is also shown dependence mitigation over time, which is stabilizing at Time Lag 13. Both graphs are very different because they show two opposite subbasins' temporal behavior.

According to quantitative approach of DMGs (see the works Zazo (2017) and Zazo et al. (2019)), the Porma-Esla subbasin has a dependent temporal behavior (Fig. 4A). Results from BNs reliably confirm the time order for dependence on the short-term drawn from traditional techniques (Table 1; Molina and Zazo, 2018). The dependence persistence and mitigation can clearly be observed by analyzing the dependence behavior over time. In this sense, all series converge to 0 (y-axis), providing a very detailed method for the analysis of the dependence persistence-mitigation in the medium term (13 years). Moreover, the area within the maximum and minimum wraparounds remains in the positive area of the plot, meaning that there is dependence.

In contrast, the Adaja subbasin (Fig. 4B) has a strong independent temporal behavior. Despite this general temporal independence, the analysis presents some dependence exclusively for Time Order 1 (Fig. 4). This demonstrates that even independent rivers could have some small amount of dependence that should be assessed. The area within the maximum and minimum wraparound are balanced between the positive and negative areas of the plot except for Time Lag 1; in other

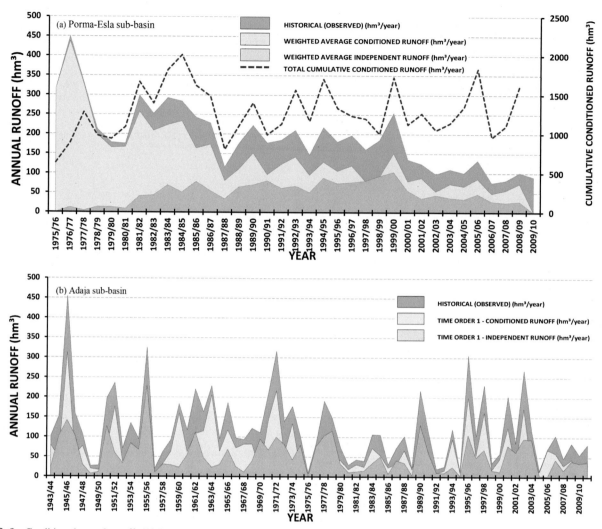

FIG. 3 Conditioned annual runoff: (A) Porma subbasin; (B) Adaja subbasin (Molina and Zazo, 2018).

words, there is an asymmetry between the maximum and minimum wraparound functions for Time Lag 1. This means that there is some dependence for Time Lag 1. Results strongly agrees with the correlogram and entropy analyses (Table 1; Molina and Zazo, 2018).

4.2 Runoff temporal records prediction

This represents the ultimate stage of this research framework. It comprises, on one side, a preliminary design of the predictive model from a probabilistic perspective, and, on the other, the assessment of the prediction reliability. The predictive model is based on the hidden temporal conditionality that inherently underlies of the historical series and that has been discovered through causality. Additionally, some studies have demonstrated the validity of stationary assumptions and stochastic parametric models applied to the generation of hydrological predictive models, such as references (Papacharalampous et al., 2019; Tyralis and Koutsoyiannis, 2014). Recently, Papacharalampous et al. (2019) presented an exhaustive study in which large-scale results versus traditional approaches focused on cases studies are provided. Its main conclusion is that stochastic and machine learning methods provide similar forecasts. Earlier, Tyralis and Koutsoyiannis (2014) justified the validity of the stationarity hypothesis applied to stochastic methods in a context of prediction of hydroclimatic variables.

The predictive model was performed through stochastic hydrological parametric ARMA models. This was generated by parsimonious ARMA models, one for each temporal fraction (TCR and TNCR) according to temporal horizons of dependence mitigation.

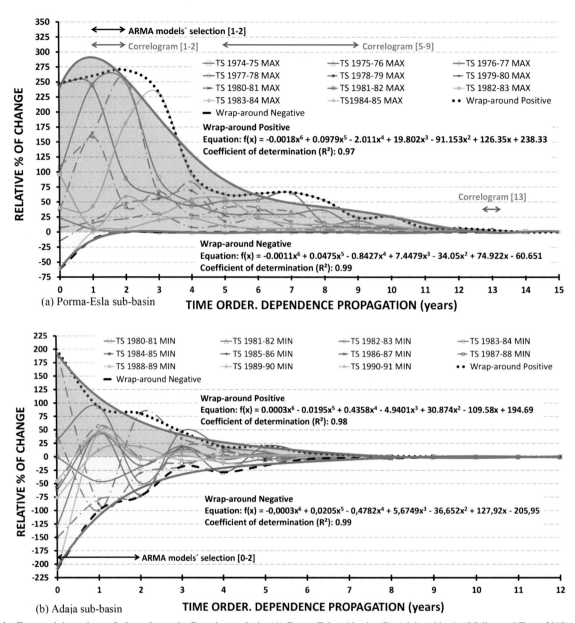

FIG. 4 Temporal dependences/independences by Bayesian analysis: (A) Porma-Esla subbasin; (B) Adaja subbasin (Molina and Zazo, 2018).

Given that the availability of water resources in Spain is characterized by a high seasonal and annual variability (CEDEX, 2017; Garrote et al., 2007; Iglesias et al., 2005), the predictive model was generated by a time-horizon (time-lag) of one (1) year (Dep-1). For that reason, the prediction was done over a water year (in this case water year 2014/15, please see Molina et al., 2019). The obtained value of this final year was used to assess the predictive reliability based on the resulting probability.

Finally, the reliability of the predictive model was assessed from a probabilistic perspective, by suggesting two alternatives. The first one general and the second one detailed. The second approach is based on the fact that the TNCR fraction does not depend on time, and, consequently, adds uncertainty to the prediction. For that, two Monte Carlo simulations from the ARMA models with 5000 data were done. The probabilistic approach was based on Gringorten probability expressed as (Jiménez Álvarez, 2016):

$$F_{(y_i)} = \frac{i - 0.44}{n + 0.12}$$

(8)

TABLE 1 Prediction analysis for a dependence propagation one year (time-lag 1).

Probability	TCR	TNCR[a]	Overall	Runoff prediction (hm³)	
				Detailed	
0.50	14.68	2.06	16.74	14.68 ± 2.06	[12.62, 16.74]
0.60	16.82	2.62	19.44	16.82 ± 2.62	[14.20, 19.44]
0.70	19.38	3.37	22.75	19.38 ± 3.37	[16.01, 22.75]
0.80	22.67	4.30	26.97	22.67 ± 4.30	[18.37, 26.97]
0.85	25.17	5.06	30.23	25.17 ± 5.06	[20.11, 30.23]
0.90	28.38	6.12	34.50	28.38 ± 6.12	[22.26, 34.50]
0.95	34.46	8.03	26.43	34.46 ± 8.03	[26.43, 42.49]

[a]TNCR, fraction uninfluenced by time.
Result from Molina, J.L., Zazo, S., Martín-Casado, A.M., 2019. Causal reasoning: towards dynamic predictive models for runoff temporal behavior of high dependence rivers. Water 11, 877.

where $F_{(yi)}$ is the value of the data distribution function y_i, i the occupied position by the data in the series ordered from lowest to highest, and n the total number of data in the series.

Fig. 5 shows the design of the dynamic predictive model over the time-horizon (time-lag) of one (1) year (target water year 2014/15, please see Molina et al., 2019). This was performed by means of the conditioned data (Molina et al., 2019), TCR and TNCR fractions time series for a one-year time dependency (Dep-1; Fig. 5) and two parsimonious ARMA (1,0) models, defined each of them over those runoff temporal fractions. It should be noted that model selection was carried out by Akaike Information Criterion, due to AIC of the ARMA models from (1.0) to (2.2) reported a slight difference in each one of the fractions (TCR: [−5.8267, −5.8265]; TNCR [−5.8469, −5.8466] (see Molina et al., 2019).

On the other hand, the reliability of the predictive model is shown in Table 1 from Molina et al. (2019), both a general approach and a detailed one developed through a definition of TNCR intervals which is an uninfluenced fraction of the time. From this approach, this adds an uncertainty component to the prediction because this fraction does not depend on time. The recorded value of target water year to assess the predictive reliability, 28.4 hm³, was supplied by the Jucar River Basin Authority (please, see MITECO, 2018).

This methodology allows knowing, in detail and jointly, the behavior of water resources in the short and medium term (Fig. 5; Molina et al., 2019; time-lag intervals [0, 4] and [4, 7]), as well as their time-lag (7 years versus 10 of the correlogram; Figs. 3B and 5; Molina et al., 2019). In this sense, it is worth highlighting the coherence between the Hurst coefficient (0.84; Molina et al., 2019) and the results obtained (asymmetry in Fig. 5; Molina et al., 2019). It is also remarkable

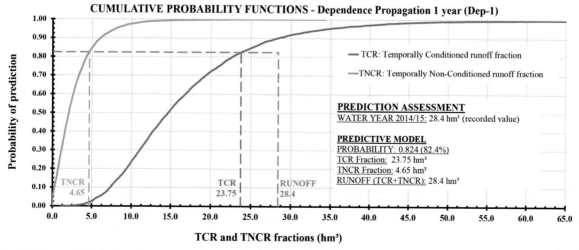

FIG. 5 Predictive model for a dependence propagation one year (Molina et al., 2019).

the existing agreement between the dual pattern of behavior presented in Fig. 5 (Molina et al., 2019) (intervals [0, 4] and [4, 7]) and the analysis of the TCR and TNCR fractions´ temporal behavior evolution showing a time-lags 1 to 4; varying from 84.8% to 70.1%.

Furthermore, the potential of this methodological framework for the basin water resources management largely became clear when the temporal behavior nuances of each fraction series (Molina et al., 2019) were discovered. It is important to mention that these nuances cannot be detected by the classical approaches. This was revealed by the variability of the ARMA model coefficients, especially noteworthy in the case of TNCR fraction (Molina et al., 2019).

Regarding the predictive model, it shows a high level of reliability, 82.4% in general perspective, and within 85% to 90% through a detailed approach ([20.11, 30.23] and [22.26, 34.50] for a time-dependency of one-year; Molina et al., 2019).

Moreover, detailed analysis over the temporal fractions presents crucial and convenient implications over dimensioning, and control operations of reservoirs and dams. This is clearly shown on the uncertainty that TNCR fraction adds to the predictive model. In a sense, for a runoff 28.4 hm^3 in the target water year, there is an uncertainty among 5.06 and 6.12 hm^2 due to the fraction not influenced by time, approximately 17.8 or 21.5% of the total runoff, respectively (please see, Molina et al., 2019).

Therefore, in a management context, that volume should be jointly provided by the reservoir-dam and its control operations in the previous year. In this way, the failure capacity of the infrastructure is reduced or minimized.

Although reservoir dimensioning involves knowing multiple key aspects such as capacity-area elevation curves, downstream impacts and sediments inflow, among others, this methodological framework might effectively contribute to achieving a better dimensioning of reservoirs and dams. This will necessarily improve service guarantees through detailed knowledge of the runoff temporal fractions (TCR and TNCR), especially TNCR fraction, which it is not temporally dependent (Molina and Zazo, 2018; Molina et al., 2019).

Furthermore, this methodology opens new perspectives for building dynamic and stochastic hydrological predictive models. Also, more river basins sustainable management approaches are expected to be designed. This may be produced through more reliable control operation of reservoirs within the current context of global warming, and for high temporal dependent river basins. Future research will incorporate the analysis of spatial dimension behavior in the design of causal models.

This research supposes a breakthrough in the temporal characterization of runoff series. Here, causality has discovered a hidden and logical structure of temporal interdependence that inherently underlie the hydrological series. This represents a latent behavior pattern that was waiting to be discovered and that the classical approaches had not been able to reveal. This was performed by means of the coupling of a stochastic hydrological parametric, parsimonious and unconditioned ARMA model, and causal reasoning. The latest was supported by Bayesian modeling, which is a powerful AI technique based on probabilistic graphical model.

The methodological framework makes possible to generate dynamic management scenarios according to the propagation (mitigation/evolution) of the dependence (time-lags), considering the runoff dependence strength over time. In that sense, two opposite temporal fractions, the Temporally Conditioned Runoff (TCR) fraction, and Temporally Non-Conditioned Runoff (TNCR) fraction were determined and quantified within the runoff series. Furthermore, the observed behavior trends of TCR/TNCR fractions made it possible to build a predictive runoff model with a high level of reliability.

The application of this new analytical strategy over annual runoff series applied to dimensioning of reservoirs, dams, and control operation of reservoirs seems quite straightforward. In this sense, a design more adjusted to the real needs of reservoirs-dams through a better knowledge and prediction on temporal behavior of TCR and TNCR fractions is expected to be achieved. This will lead to a reconsideration of the capacity of the storage reservoirs, especially their convenience in the current context of global warming and for high temporal dependent river basins. This would imply substantial economic and environmental savings in the future.

This has demonstrated that past information provides prior knowledge of the future with a high degree of reliability. Furthermore, this research reinforces the suitability of causality (causal reasoning) in modeling the temporal behavior of the water resources of a highly dependent basin from a dynamic and stochastic perspective.

4.3 Hydrological spatial records prediction

The rivers' runoff dependence is not only influenced by the temporal dimension, but also, depends on the spatial dimension. The fact that different basins interact each other is crucial to get to accurately know a general behavior of rivers. The quantification of those interactions is crucial and, consequently, this framework provides a very reliable and accurate approach for this dependence evaluation (Molina and Zazo, 2017). This general evaluation largely depends on climatic, geomorphological and hydrological parameters, among others (Molina and Zazo, 2018). Of course, to get to know the interaction and causality of basins and their hydrological/hydraulic components (infrastructures) is a very complex matter. This is proved by the formulation of questions such as interaction strength, direction and/or temporal evolution of those relationships, which may be answered through CR. This is especially important in those situations where relationships cannot be known in advance or cannot be adequately modeled with traditional statistics, or where no probability function is found adequate or the chosen one would be very complex for fitting a probability distribution to the variables, among others.

4.4 Spatiotemporal records prediction

The ultimate and most complete application may comprise a joint evaluation and prediction for time and space (Hipel and McLeod, 1994). This is technically possible through fully connected CR developments (Molina and Zazo, 2018; Molina et al., 2013). Being able to analyze, at the same time, the influence of time and space and their interaction for the runoff prediction is not a trivial issue and it can be done through CR and the large amount of data availability. These models should be built as simple as possible, based on principle of parsimony, because the inherent complexity of the logical structure is far enough.

5. Results and discussion

Hydrological processes and their associated data series are characterized by their variability, randomness, and uncertainty. These concepts are similar, but they each have their own peculiarities. Therefore, stochastic approaches are needed in order to improve the knowledge and modeling of past behavior. Rigorous and accurate knowledge of the past is necessary for building robust and reliable predictive models. This research line proved that causality through conditional probability is very useful in the analysis of temporal behavior of runoff series and identification and quantification of runoff temporal fractions.

Bayesian inference articulated by the use of BN and computed through conditional probability was used as a pure quantitative tool for analyzing and quantifying the temporal behavior of runoff series. This was demonstrated to be a suitable approach for identifying causal relationships of dependences among time steps, quantifying the strength of dependence, and quantifying those dependences in order to provide a new characterization of runoff temporal fractions.

The coupling of traditional techniques such as the ARMA models with new stochastic methods such as Bayesian approaches (approaches based on artificial intelligence) provided sufficient reliable data to populate the causal reasoning process. In addition, temporal dependence transmission from ARMA to BN was demonstrated to be insufficient for invalidating the learning through causal reasoning on temporal dependence in hydrological series. Furthermore, the fact that BN application does not require defining mathematically the relationships among model variables before establishing the analysis, makes this approach very powerful and flexible. Bayesian networks, through Bayesian observational inference via Bayesian rules, calculate prior and posterior probability distributions and/or functions. Furthermore, this approach allows analyzing temporal dependences time step by time step in a dynamic way, in contrast to traditional approaches that tend to be static. In addition, the fact that the analysis can be focused on a particular time step makes this analysis more accurate.

Causality through conditional probability assesses the influence of the time on the rest of the time series. This allows assessing the dependence persistence and the dependence mitigation/attenuation of a series by means of analyzing the intensity of temporal dependence propagation via a novel DMG. Here, dependences were mathematically modeled, with a very high coefficient of determination, R^2, which coincided with the analysis of temporal dependence and independence.

Reasonable coherence in the short-term was shown between time orders obtained from Pearson's correlation coefficient. Furthermore, validation suitability analysis is done via ARIMA models, developed for checking the stationarity of the data and for validating the results obtained from Bayesian analysis. Another method of validation and a sensitivity analysis were developed through concepts of information theory. Total entropy, conditional entropy, and mutual information were the chosen variables for this analysis. Results highly agreed with the previous analysis.

Regarding applications, the runoff fractions evaluation comprises advances over previous research. In this sense, an innovative physical identification, quantification, and interpretation of runoff fractions (TCR and TNCR) is presented. In practical terms, infrastructures placed downstream of a very high dependence river basin might be designed carefully incorporating this novel analysis. In this sense, this could be a useful advance for defining more accurately the operation rules of reservoirs. Moreover, another innovation is the ability of the approach to assess the temporal dependence in the medium term. Traditional techniques were designed exclusively for the short term through Pearson's correlation coefficient or the ARMA models and for the long term through the Hurst coefficient. Finally, once runoff fractions are interpreted, it is logical to calculate the return period of each fraction. For this, two new concepts for the RP computation for each runoff fraction were proposed, TCRP and TNCRP. This provides engineers and scientists with a novel stochastic approach for calculating the return period considering the internal and inherent causality of runoff time series. Consequently, more-optimized hydraulic fundamental human-built infrastructures dimensioning would be developed. The TCR and TNCR fractions and their associated stochastic return periods (TCRP and TNCRP, respectively) can be an efficient adaptive tool against climate change, not only for the dimensioning of new reservoirs but also for their operation and management. This research line improves the existing knowledge of hydrological series temporal behavior and better characterizes and quantifies runoff temporal fractions. This was done by means of the introduction and application of new stochastic approaches and concepts that were used successfully in two unregulated river basin stretches (Porma-Esla and Adaja).

This research line introduces a characterization of runoff fractions. In this sense, runoff can be divided into two fractions depending of its temporal behavior. The color of water is another way of doing correctly. Temporally conditioned runoff is the part of the total runoff that is conditioned by the time. Temporally non-conditioned runoff or independent runoff represents the runoff fraction that is produced without influence of the time. Consequently, total runoff is the summation of both fractions and represents the amount of water that is counted in the gauge stations. This leads to a more optimum dimensioning of water infrastructures, such as reservoirs. Because this research focused on the average temporal behavior, the more direct and interesting application would be for the design of large storage regulatory reservoirs.

Furthermore, a new preliminary approach for assessing the return period from a stochastic point of view was addressed. Temporally conditioned return period and temporally non-conditioned return period were proposed for evaluating a more accurate return period considering the temporal behavior of runoff series. Because this study did not focus on extreme events and data did not follow any extreme value distribution, the runoff for 10-, 20-, and 30-year TCRP and TNCRP were evaluated. The results showed coherent and representative differences among them that should be properly interpreted. Given the low differences in assessed return periods, differences in results were also small, but they differed in a coherent way. Thus, the largest conditioned runoff corresponded to the largest return period (30 years) and the lowest conditioned runoff corresponded to the shortest return period (10 years). Non-conditioned runoff behaved in the same logical way.

On the other hand, regarding runoff temporal records prediction, the research developed so far reveals the following main features. This methodology allows knowing, in detail and jointly, the behavior of water resources in the short and medium terms (Fig. 5; [0, 4] and [4, 7]; see Molina and Zazo, 2018), as well as their time-lag (7 years versus 10 of the correlogram). In this sense, it is worth highlighting the coherence between the Hurst coefficient (0.84) and the results obtained (asymmetry). It is also remarkable the existing agreement between the dual pattern of behavior presented in Fig. 5 (intervals [0, 4] and [4, 7]; Molina and Zazo, 2018) and the analysis of the TCR and TNCR fractions' temporal behavior evolution varying from 84.8% to 70.1% for time-lags 1 to 4 (Molina and Zazo, 2018).

Furthermore, the potential of this methodological framework for the basin water resources management largely became clear when the temporal behavior nuances of each fraction series (Molina and Zazo, 2018) were discovered. It is important to mention that these nuances cannot be detected by the classical approaches. This was revealed by the variability of the ARMA model coefficients, especially noteworthy in the case of TNCR fraction.

Regarding predictive modeling development (see, Molina et al., 2019), it shows a high level of reliability, 82.4% in general perspective, and within 85% to 90% through a detailed approach ([20.11, 30.23] and [22.26, 34.50] for a time-dependency of one-year. Moreover, detailed analysis over the temporal fractions presents crucial and convenient implications over dimensioning, and control operations of reservoirs and dams. This is clearly shown on the uncertainty that TNCR fraction adds to the predictive model. In a sense, for a runoff of 28.4 hm^3 in the target water year, there is an uncertainty among 5.06 and 6.12 Hm^3 due to the fraction not influenced by time, approximately 17.8 or 21.5% of the total runoff, respectively. Therefore, in a management context, that volume should be jointly provided by the reservoir-dam and its control operations in the previous year. In this way, the failure capacity of the infrastructure is reduced or minimized.

Although reservoir dimensioning involves knowing multiple key aspects such as capacity-area elevation curves, downstream impacts and sediments inflow, among others, this methodological framework might effectively contribute to achieving a better dimensioning of reservoirs and dams. This will necessarily improve service guarantees through detailed knowledge of the runoff temporal fractions (TCR and TNCR), especially TNCR fraction, which it is not temporally dependent.

Furthermore, this methodology opens new perspectives for building dynamic and stochastic hydrological predictive models. Also, more river basins sustainable management approaches are expected to be designed. This may be produced through more reliable control operation of reservoirs within the current context of global warming, and for high temporal dependent river basins. Upcoming research will incorporate the analysis of spatial dimension behavior in the design of causal models.

These scientific applications become a breakthrough in the temporal characterization of runoff series. Here, causality has discovered a hidden and logical structure of temporal interdependence that inherently underlies the hydrological series.

This represents a latent behavior pattern that was waiting to be discovered and that the classical approaches had not been able to reveal. This was performed by means of the coupling of a stochastic hydrological parametric, parsimonious and unconditioned ARMA model, and Causal Reasoning. The latest was supported by Bayesian modeling, which is a powerful AI technique based on a probabilistic graphical model.

The application of this new analytical strategy over annual runoff series applied to dimensioning of reservoirs, dams, and control operation of reservoirs seems quite straightforward. In this sense, a design more adjusted to the real needs of reservoirs-dams through a better knowledge and prediction on temporal behavior of TCR and TNCR fractions is expected to be achieved. This will lead to a reconsideration of the capacity of the storage reservoirs, especially their convenience in the current context of global warming and for high temporal dependent river basins. This would imply substantial economic and environmental savings in the future. This research has demonstrated that past information provides prior knowledge of the future with a high degree of reliability. Furthermore, this research reinforces the suitability of causality (causal reasoning) in modeling the temporal behavior of the water resources of a highly dependent basin from a dynamic and stochastic perspective.

The progress in regard to hydrological spatial records prediction is limited so far. We are nowadays analyzing the influence of spatial features on the river runoff behavior and extracting conclusions from that. This analysis is articulated through the three main components that drive the rivers' runoff behavior and evolution (Section 4.3). It is firmly believe that, by analyzing the interactions of those components, a much more deep and complete knowledge of runoff behavior is to be reached. This progress will be automatically derived to the development of predictive modeling that will be useful for many applications.

Spatiotemporal runoff analysis comprises the development of very complex CRM structures and models. This specific design for each case study is found to be a very qualified job that requires the involvement of expert teams. However, the evaluation of double influence of space and time on the runoff behavior is possible through the design of CRM models heavily populated from very large datasets with thousands of multidirectional observations (Section 4.4).

6. Conclusions

This chapter provides the insights of a long-term research activity that started back in 2005. Over this time, this general methodology has been evolved, improved and expanded, aimed to provide a useful modeling approach and tools largely for the advanced description and prediction of river's runoff behavior. CRM developments in Hydrology can be used from several different kinds of data and for many purposes and usages. Scientists, developers and the rest of stakeholders should put the problem in hand together, and choose the optimal and specific approach for getting as best results as possible. That is the reason why the general methodology described in this chapter concludes with the real and potential applications that this approach may be addressed. In this regard, main applications can be classified into two main groups. On the one hand, first group may comprise the applications of CRM for describing rivers' runoff in deep. In this regard, the quantification of runoff fractions becomes a milestone in Hydrology, not only on the progress of hydrological knowledge, but also on the development of reliable and detailed predictive rivers' runoff models for the description of behavior and for the dimensioning of hydraulic infrastructures. On the other hand, second group contains the usage of CRM for developing predictive hydrological models, either temporal, spatial, or spatiotemporal.

There are several challenges to be addressed in the near future that may comprise the generation of monthly CRM for description and prediction. In addition, it is very necessary to implement a specific CRM methodology for describing and predicting the extreme hydrological processes such as drought and flood events. It is also planned to develop a general, digital and interactive platform for integrating all these developments in a unique place where stakeholders can manipulate the models for their own usage. Finally, but not least, improving and refining the definition of Return Period is a basic task that is still pending in Hydrology. Consequently, the continuation of the research activity that drove to a preliminary definition of Stochastic Return Period (TCRP and TNCRP) is a crucial and relatively straightforward task. Its application is rapid to the optimal design and dimensioning of hydraulic infrastructures.

Acknowledgments

The authors especially appreciate the contribution in the phase of algebraic development of this research of Professor Dr. Marta Molina (Salamanca University, Spain). Authors acknowledge the great support given by the High Polytechnic School of Avila (Avila, Spain) and University of Salamanca to elaborate this book chapter.

References

Adarnowski, J.F., 2008. Development of a short-term river flood forecasting method for snowmelt driven floods based on wavelet and cross-wavelet analysis. J. Hydrol. 353, 247–266.

Akintug, B., Rasmussen, P.F., 2005. A Markov switching model for annual hydrologic time series. Water Resour. Res. 41, W09424.

Aqil, M., Kita, I., Yano, A., Nishiyama, S., 2007. A comparative study of artificial neural networks and neuro-fuzzy in continuous modeling of the daily and hourly behaviour of runoff. J. Hydrol. 337, 22–34.

Balakrishnan, N., Lai, C., 2009. Continuous Bivariate Distributions. Springer Science & Business Media, Berlin, Germany.

Barton, D.N., Saloranta, T., Moe, S.J., Eggestad, H.O., Kuikka, S., 2008. Bayesian belief networks as a meta-modelling tool in integrated river basin management – pros and cons in evaluating nutrient abatement decisions under uncertainty in a Norwegian river basin. Ecol. Econ. 66, 91–104.

Cain, J., 2001. Planning Improvements in Natural Resources Management. 124 Centre for Ecology and Hydrology, Wallingford, UK, pp. 1–123.

Carrasco, G., Molina, J.L., Patino-Alons, M., Castillo, M.D.C., Vicente-Galindo, M., Galindo-Villardon, M., 2019. Water quality evaluation through a multivariate statistical HJ-Biplot approach. J. Hydrol. 577. UNSP 123993.

Castelletti, A., Soncini-Sessa, R., 2007. Bayesian Networks and participatory modelling in water resource management. Environ. Model Softw. 22, 1075–1088.

CEDEX, 2017. Centro de Estudios y Experimentación de Obras Públicas. Ministerio de Fomento. Ministerio de Agricultura y Pesca, Alimentación y Medio Ambiente. Gobierno de España. Informe Técnico: Evaluación del Impacto del Cambio Climático en los recursos hídricos y sequías en España. pp. 15–19.

Chang, B., Guan, J., Aral, M.M., 2015. Scientific discourse: climate change and sea-level rise. J. Hydrol. Eng. 20, A4014003.

Close, M.E., Abraham, P., Humphries, B., Lilburne, L., Cuthill, T., Wilson, S., 2016. Predicting groundwater redox status on a regional scale using linear discriminant analysis. J. Contam. Hydrol. 191, 19–32.

Croitoru, A., Minea, I., 2015. The impact of climate changes on rivers discharge in Eastern Romania. Theor. Appl. Climatol. 120, 563–573.

Daughney, C.J., Morgenstern, U., van der Raaij, R., Reeves, R.R., 2010. Discriminant analysis for estimation of groundwater age from hydrochemistry and well construction: application to New Zealand aquifers. Hydrogeol. J. 18, 417–428.

Dawson, C.W., Wilby, R.L., 2001. Hydrological modelling using artificial neural networks. Prog. Phys. Geogr. 25, 80–108.

De Michele, C., Salvadori, G., 2003. A generalized Pareto intensity-duration model of storm rainfall exploiting 2-copulas. J. Geophys. Res.-Atmos. 108, 4067.

Díaz Caballero, F.F., 2011. Selección de modelos mediante criterios de información en análisis factorial: aspectos teóricos y computacionales. Editorial de la Universidad de Granada, Spain.

Dutfoy, A., Parey, S., Roche, N., 2014. Multivariate extreme value theory – a tutorial with applications to hydrology and meteorology. Dependence Modeling. 2, 30–48.

Forgy, E.W., 1965. Cluster analysis of multivariate data – efficiency vs interpretability of classifications. Biometrics 21, 768.

Galindo, M., 1986. An alternative for simultaneous representations. HJ-Biplot Qüestiió. 10, 13–23.

Garrote, L., De Lama, B., Martín-Carrasco, F., 2007. Previsiones para España según los últimos estudios de cambio climático. El cambio climático en España y sus consecuencias en el sector agua. Universidad Rey Juan Carlos, Spain, pp. 3–15.

Genc, O., Dag, A., 2016. A machine learning-based approach to predict the velocity profiles in small streams. Water Resour. Manag. 30, 43–61.

Hao, Z., Singh, V.P., 2016. Review of dependence modeling in hydrology and water resources. Prog. Phys. Geogr. 40, 549–578.

Hipel, K.W., McLeod, A.I., 1994. Time series modelling of water resources and environmental systems. Elsevier.

Holmström, L., Ilvonen, L., Seppa, H., Veski, S., 2015. A Bayesian spatiotemporal model for reconstructing climate from multiple pollen records. Ann. Appl. Stat. 9, 1194–1225.

Iglesias, A., Estrela, T., Gallart, F., 2005. Impactos sobre los recursos hídricos. Evaluación Preliminar de los Impactos en España for Efecto del Cambio Climático. pp. 303–353.

Jain, A., Kumar, A.M., 2007. Hybrid neural network models for hydrologic time series forecasting. Appl. Soft Comput. 7, 585–592.

Jensen, F.V., 1996. An Introduction to Bayesian Networks. UCL Press London, UK.

Jiménez Álvarez, A., 2016. Desarrollo de metodologías para mejorar la estimación de Los hidrogramas de diseño para el cálculo de los Órganos de Desagüe de las Presas. Universidad Politécnica de Madrid, España, Spain, Tesis Doctoral. 128 p.

Jolliffe, I.T., Cadima, J., 2016. Principal component analysis: a review and recent developments. Philos. Trans. Roy. Soc. A: Math. Phys. Eng. Sci. 374, 20150202.

Kalra, A., Li, L., Li, X., Ahmad, S., 2013. Improving streamflow forecast lead time using oceanic-atmospheric oscillations for Kaidu River Basin, Xinjiang. China. J. Hydrol. Eng. 18, 1031–1040.

Kasiviswanathan, K.S., Sudheer, K.P., 2017. Methods used for quantifying the prediction uncertainty of artificial neural network based hydrologic models. Stoch. Env. Res. Risk A. 31, 1659–1670.

Koller, D., Friedman, N., 2009. Probabilistic graphical models: Principles and techniques. MIT Press, Cambridge, Massachusetts, USA.

Kong, X., Huang, G., Fan, Y., Li, Y., Zeng, X., Zhu, Y., 2017. Risk analysis for water resources management under dual uncertainties through factorial analysis and fuzzy random value-at-risk. Stoch. Env. Res. Risk A. https://doi.org/10.1007/s00477-017-1382-31-16.

Kouhestani, S., Eslamian, S., Besalatpour, A., 2017. The effect of climate change on the Zayandeh-Rud River Basin's temperature using a Bayesian machine learning. Soft Computing Technique 21 (1), 203–216.

Koutsoyiannis, D., 2010. HESS opinions 'A random walk on water'. Hydrol. Earth Syst. Sci. 14, 585–601.

Lappenschaar, M., Hommersom, A., Lucas, P.J.F., 2012. Qualitative chain graphs and their use in medicine. In: Proceedings of the Sixth European Workshop on Probabilistic Graphical Models. Proceedings of the Sixth European Workshop on Probabilistic Graphical Models, Granada, Spain, pp. 179–185.

Lasinio, G.J., Sahu, S.K., Mardia, K.V., 2007. Modeling Rainfall Data Using a Bayesian Kriged-Kalman Model. Bayesian Statistics and Its Applications. Anshan, Tunbridge Wells, UK, pp. 61–86.

Madsen, A.L., Lang, M., Kjærulff, U.B., Jensen, F., 2003. The Hugin tool for learning Bayesian networks. In: European Conference on Symbolic and Quantitative Approaches to Reasoning and Uncertainty; Prague, Czechia, pp. 594–605.

Mishra, A.K., Singh, V.P., 2010. A review of drought concepts. J. Hydrol. 391, 204–216.

Mishra, A.K., Singh, V.P., 2011. Drought modeling – a review. J. Hydrol. 403, 157–175.

MITECO, 2018. https://sig.mapama.gob.es/redes-seguimiento/. (Accessed 30 October 2018).

Molina, J.L., Bromley, J., Garcia-Arostegui, J.L., Sullivan, C., Benavente, J., 2010. Integrated water resources management of overexploited hydrogeological systems using Object-Oriented Bayesian Networks. Environ. Model Softw. 25, 383–397.

Molina, J.L., Pulido-Velazquez, D., Garcia-Arostegui, J., Pulido-Velazquez, M., 2013. Dynamic Bayesian Networks as a Decision Support tool for assessing Climate Change impacts on highly stressed groundwater systems. J. Hydrol. 479, 113–129.

Molina, J.L., Zazo, S., 2017. Causal reasoning for the analysis of rivers runoff temporal behavior. Water Resour. Manag. 31 (14). https://doi.org/10.1007/s11269-017-1772-9.

Molina, J.L., Zazo, S., 2018. Assessment of temporally conditioned runoff fractions in unregulated rivers. J. Hydrol. Eng. 23, 04018015.

Molina, J.L., Zazo, S., Martín-Casado, A., Patino-Alonso, M., 2020. Rivers' temporal sustainability through the evaluation of predictive runoff methods. Sustain. For. 12, 1720.

Molina, J.L., Zazo, S., Martín-Casado, A.M., 2019. Causal reasoning: towards dynamic predictive models for runoff temporal behavior of high dependence rivers. Water 11, 877.

Molina, J.L., Zazo, S., Rodriguez-Gonzalvez, P., Gonzalez-Aguilera, D., 2016. Innovative analysis of runoff temporal behavior through Bayesian networks. Water 8, 484.

Mousavi, S.J., Ponnambalam, K., Karray, F., 2007. Inferring operating rules for reservoir operations using fuzzy regression and ANFIS. Fuzzy Sets Syst. 158, 1064–1082.

Myung, I.J., 2000. The importance of complexity in model selection. J. Math. Psychol. 44, 190–204.

Nelsen, R.B., 2007. An Introduction to Copulas. Springer Science & Business Media, Berlin, Germany.

Nodelman, U., Horvitz, E., 2003. Continuous Time Bayesian Networks for Inferring Users' Presence and Activities with Extensions for Modeling and Evaluation. Microsoft Research. July–August.

O'Gorman, P.A., 2015. Precipitation extremes under climate change. Curr. Clim. Change Rep. 1, 49–59.

Papacharalampous, G., Tyralis, H., Koutsoyiannis, D., 2019. Comparison of stochastic and machine learning methods for multi-step ahead forecasting of hydrological processes. Stoch. Environ. Res. Risk A. 33, 481–514.

Papacharalampous, G., Tyralis, H., Koutsoyiannis, D., Montanari, A., 2020. Quantification of predictive uncertainty in hydrological modelling by harnessing the wisdom of the crowd: a large-sample experiment at monthly timescale. Adv. Water Resour. 136, 103470.

Patel, S.S., Ramachandran, P., 2015. A comparison of machine learning techniques for modeling river flow time series: the case of upper Cauvery river basin. Water Resour. Manag. 29, 589–602.

Pearl, J., 1988. Probabilistic Reasoning in Intelligent Systems: Networks of Plausible Inference. Morgan Kaufmann Inc., Los Altos, CA, USA.

Pearl, J., 2009. Causality: Models, Reasoning, and Inference, second ed. Cambridge University Press, New York, UK.

Pearl, J., 2014. Graphical models for probabilistic and causal reasoning. In: Tucker, A., Gonzalez, T., Topi, H., Diaz-Herrera, J. (Eds.), Computing Handbook Computer Science and Software Engineering. CRC Press, Boca Raton, FL, USA.

Pena, J.M., Lozano, J.A., Larranaga, P., 2004. Unsupervised learning of Bayesian networks via estimation of distribution algorithms: an application to gene expression data clustering. Int. J. Uncertain. Fuzz. Knowl.-Based Syst. 12, 63–82.

Pfahl, S., O'Gorman, P.A., Fischer, E.M., 2017. Understanding the regional pattern of projected future changes in extreme precipitation. Nat. Clim. Chang. 7, 423.

Pollino, C.A., Woodberry, O., Nicholson, A., Korb, K., Hart, B.T., 2007. Parameterisation and evaluation of a Bayesian network for use in an ecological risk assessment. Environ. Model Softw. 22, 1140–1152.

Pulido-Velazquez, D., Luis Garcia-Arostegui, J., Molina, J.L., Pulido-Velazquez, M., 2015. Assessment of future groundwater recharge in semi-arid regions under climate change scenarios (Serral-Salinas aquifer, SE Spain). Could increased rainfall variability increase the recharge rate? Hydrol. Process. 29, 828–844.

Rajagopalan, B., Salas, J.D., Lall, U., 2010. Stochastic methods for modeling precipitation and streamflow. In: Advances in Data-Based Approaches for Hydrologic Modeling and Forecasting. World Scientific, pp. 17–52.

Said, A., 2006. The implementation of a Bayesian network for watershed management decisions. Water Resour. Manag. 20, 591–605.

Salas, J., Delleur, J., Yevjevich, V., 1980. Applied Modeling of Hydrologic Time Series. Water Resources Publications Edition, USA.

Sarabia-Alzaga, J., Gómez-Déniz, E., 2008. Construction of multivariate distributions: a review of some recent results. Stat. Oper. Res. Trans. 32, 3–36.

See, L., Openshaw, S., 2000. A hybrid multi-model approach to river level forecasting. Hydrol. Sci. J. Journal Des Sciences Hydrologiques 45, 523–536.

Serinaldi, F., Bardossy, A., Kilsby, C.G., 2015. Upper tail dependence in rainfall extremes: would we know it if we saw it? Stoch. Env. Res. Risk A. 29, 1211–1233.

Singh, V.P., 2013. Entropy Theory and Its Application in Environmental and Water Engineering. John Wiley & Sons, New York, USA.

Sperotto, A., Molina, J.L., Torresan, S., Critto, A., Marcomini, A., 2017. Reviewing Bayesian Networks potentials for climate change impacts assessment and management: a multi-risk perspective. J. Environ. Manag. 202, 320–331.

Stojkovic, M., Prohaska, S., Plavsic, J., 2015. Stochastic structure of annual discharges of large European rivers. J. Hydrol. Hydromech. 63, 63–70.

Thorp, J.H., Thoms, M.C., Delong, M.D., 2010. The Riverine Ecosystem Synthesis: Toward Conceptual Cohesiveness in River Science. Elsevier.

Todini, E., 1988. Rainfall-runoff modeling – past, present and future. J. Hydrol. 100, 341–352.

Tyralis, H., Koutsoyiannis, D., 2014. A Bayesian statistical model for deriving the predictive distribution of hydroclimatic variables. Clim. Dyn. 42, 2867–2883.

Vogel, K., Weise, L., Schroeter, K., Thieken, A.H., 2018. Identifying driving factors in flood-damaging processes using graphical models. Water Resour. Res. 54, 8864–8889.

Wang, Y., Traore, S., Kerh, T., Leu, J., 2011. Modelling reference evapotranspiration using feed forward backpropagation algorithm in arid regions of Africa. Irrig. Drain. 60, 404–417.

Watts, G., 1997. Hydrological modelling in practice. In: Wilby, R.L. (Ed.), Contemporary Hydrology: Towards Holistic Environmental Science. Wiley, pp. 151–193.

Wikle, C.K., Berliner, L.M., Cressie, N., 1998. Hierarchical Bayesian space-time models. Environ. Ecol. Stat. 5, 117–154.

Zazo, S., 2017. Analysis of the Hydrodynamic Fluvial Behaviour Through Causal Reasoning and Artificial Vision. Doctoral Thesis, Salamanca University, Spain. 159 p.

Zazo, S., Macian-Sorribes, H., Sena-Fael, C.M., Martín-Casado, A.M., Molina, J.L., Pulido-Velazquez, M., 2019. Qualitative approach for assessing runoff temporal dependence through geometrical symmetry. In: Covilhã, Portugal, 27–29 November.

Zhang, L., Singh, V.P., 2006. Bivariate flood frequency analysis using the copula method. J. Hydrol. Eng. 11, 150–164.

Zounemat-Kermani, M., Teshnehlab, M., 2008. Using adaptive neuro-fuzzy inference system for hydrological time series prediction. Appl. Soft Comput. 8, 928–936.

Chapter 8

Data assimilation in hydrological and hazardous forecasting

Sandra Reinstädtler[a], Shafi Noor Islam[b], and Saeid Eslamian[c,d]

[a]*Independent Scientist as University of Technology Dresden—Alumna, former: Faculty of Environmental Sciences and Process Engineering and Faculty of Environment and Natural Sciences, Brandenburg University of Technology, Cottbus-Senftenberg, Germany,* [b]*Department of Geography, Environment and Development Studies, Faculty of Arts and Social Sciences, University of Brunei Darussalam, Gadong, Brunei Darussalam,* [c]*Department of Water Engineering, College of Agriculture, Isfahan University of Technology, Isfahan, Iran,* [d]*Center of Excellence for Risk Management and Natural Hazards, Isfahan University of Technology, Isfahan, Iran*

1. Introduction

Data assimilation (DA) methods were first used in numerical weather prediction and gained popularity in the geosciences field (comp. Kalnay et al., 1996). These methods also support further environmental forecasting problems (Eslamian, 2012), such as hydrological forecasting; or Bayesian networks in data assimilating approaches for assessing natural hazards and specific landslides (Cardenas, 2019). Also, DAs appear for re-analyses of the atmospheric state of extraterrestrial planetary solar systems with so far singular applications for planet Mars (NASA, 2020).

A possible definition of DA is that "data assimilation is the process of combining observations from a wide variety of sources and forecast output from a weather prediction model. The resulting analysis is considered the best estimate of the state of the atmosphere at a particular instant in time. The process of combining the observational and model information is accomplished within a Bayesian statistical framework where probability distributions associated with observations and forecasts are combined with dynamical constraints" (NCARS, 2018).

Moreover, according to Nearing et al. (2018), DA is the application of Bayes' theorem to condition the states of a dynamical systems model on observations. However, "any real-world application of Bayes' theorem is approximate, and therefore we cannot expect that data assimilation will preserve all of the information available from models and observations" (Nearing et al., 2018). Due to proper data quality management, accurate data is one of the most important aspects of any scientific endeavor, research project, or business development (compare (comp.) Barchard and Verenikina, 2013) in order to minimize deficiencies in accurate data as well as in the predictability or general reliability grade. Adverse effects in critical research pathway decisions and reductions in the overall result strength within research or business might arise. For instance, environmental (comp. Huang and Chang, 2003), land change science and land systems modeling (Goodchild, 1992; Rindfuss et al., 2004; Singh et al., 2012), long-term ecological research (Gnauck and Fongwa, 2012), water-related long-term trend estimation (Gray et al., 2018) or general systems analysis might possibly face challenges of data inaccuracy.[a] Therefore, this challenge of data inaccuracy has to be overcome by following, not closing up, but initially, points and having to be highlighted after sources named above (comp. Fig. 1):

(1) Accuracy and consistency are two of the existing central criteria for data quality (Cong et al., 2007), especially in hydrological or hazardous forecasting database preparation;

(2) Consistency through existing data (base) quality goals: Errors and Inconsistencies in a database often approached as violations of integrity constraints (Cong et al., 2007). So setting up realistic goals for data accuracy toward improving overall data quality of databases and fundamental problems for a quality database must be understood by the data entry specialists and, or by the help of automatically-generated repair methods for databases with quality data; data entry, efficient data capturing, and effective coding are vital in that processing. Conditional functional dependencies (CFDs) proposed by Bohannon et al. (2007) help further to set goals against inconsistencies (Cong et al., 2007). Those can capture inconsistencies and errors (Cong et al., 2007) (comp. Fig. 1);

---

a. Named authors are exemplary specialists for each sector of modeling or research analysis. These authors are not named in case of data insecurities.

Handbook of HydroInformatics. https://doi.org/10.1016/B978-0-12-821962-1.00018-0

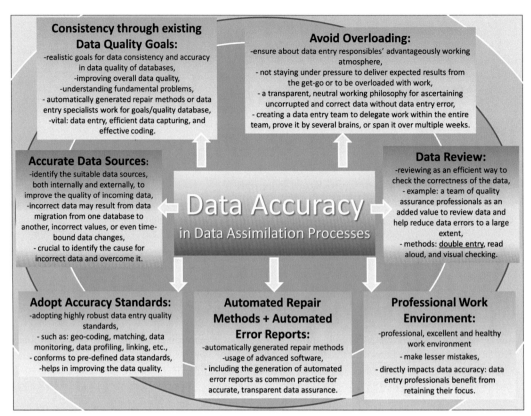

FIG. 1 Correct data accuracy (and consistency) steps in general data assimilation processes. *(After Cong, G., Fan, W., Geerts, F., Jia, X., Ma, S., 2007. Improving data quality: consistency and accuracy. Proc. Int. Conf. VLDB. 315–326; Barchard, K.A., Verenikina, Y., 2013. Improving data accuracy: Selecting the best data checking technique. Computers in Human Behavior 29(5):1917–1922. https://doi.org/10.1016/j.chb.2013.02.021; Goodchild, M. F., 1992. Geographical data modeling. Comput. Geosci. 18 (4), 401–408. https://doi.org/10.1016/0098-3004(92)90069-4; Huang et al. (2003); Rindfuss, R.R., Walsh, S.J., Turner II, B.L., Fox, J., Mishra, V., 2004. Developing a science of land change: Challenges and methodological issues. PNAS 101 (39), 13976–13981. https://doi.org/10.1073/pnas.0401545101; Gnauck and Fongwa (2012); Singh, K.K., Vogler, J.B., Shoemaker, D.A., Meente-meyer, R.K., 2012. LiDAR-Landsat data fusion for large-area assessment of urban land cover: Balancing spatial resolution, data volume and mapping Accuracy. ISPRS J. Photogramm. Remote Sens. 74, 110–121. https://doi.org/10.1016/j.isprsjprs.2012.09.009.; Nearing, G., Yatheendradas, S., Crow, W., Zhan, X., Liu, J., Chen, F., 2018. The Efficiency of Data Assimilation. Water Resour. Res. 54(9), 6374–6392. https://doi.org/10.1029/2017WR020991; Gray, D.K., Hampton, S.E., O'Reilly, C.M., Sharma, S., Cohen, R.S., 2018. How do data collection and processing methods impact the Accuracy of long-term trend estimation in lake surface-water temperatures? Limnol. Oceanogr. Methods 16, 504–515. https://doi.org/10.1002/lom3.10262.)*

(3) Avoid overloading: An endeavor-, project-, data- or enterprise manager must ensure that the data entry responsible has an advantageously working atmosphere. Next to others, that means not staying under pressure to deliver expected results from the get-go or to be overloaded with work: a transparent, neutral working philosophy in any research or business sphere helps ascertain uncorrupted and correct data without data entry error. It might be advantageous to create a data entry team to delegate work within the entire team, prove it by several brains, or span it over multiple weeks (comp. Fig. 1);

(4) Accurate data sources: Endeavor-, project-, data-, or enterprise-holder should identify the suitable data sources, both internally and externally, for improving the quality of incoming data. For example, incorrect data may result from data migration from one database to another, incorrect values, or even time-bound data changes. Therefore, it is crucial to identify the cause for incorrect data and overcome it (comp. Fig. 1);

(5) Data review: As data entry errors can have malfunction effects on the results of statistical analyses (Barchard and Verenikina, 2013), reviewing is an efficient way to check the correctness of the data. It gives endeavor-, project-, data- or enterprise-holder the chance to incorporate an efficient way to review and double entry-check the data. For example, a team of quality assurance professionals might be an added value to review data and help reduce data errors to a large extent. Methods are, next to others, double entry, read aloud, and visual checking with using types of data and data entry personnel (Barchard and Verenikina, 2013). Especially considerable effort is spent in researchers checking their data (Barchard and Verenikina, 2013). After Barchard and Verenikina (2013) participatory study with students, it can be said: "Among people with no previous experience, read aloud, and visual checking had more than

20 times as many errors as a double entry. In addition, double entry was preferred over visual checking. Thus, although double entry takes slightly longer, it is clearly worth the extra effort." (comp. Fig. 1);

(6) Adopt accuracy standards: Adopting highly robust data entry quality standards such as geo-coding, matching, data monitoring, data profiling, linking, etc., ensures that the data entered conforms to pre-defined data standards. It, in turn, helps in improving the data quality (comp. Fig. 1);

(7) Automated Repair methods and Automated Error Reports: The usage of advanced software, including the generation of automated error reports, is a common practice for accurate, transparent data assurance. Only it has to be secured that also the automatically-generated repair is the same accurate (Cong et al., 2007). Next to the automatically-generated repair methods, algorithms might help to improve the consistency of the data (Cong et al., 2007) (comp. Fig. 1);

(8) Professional work environment: Having a professional, excellent and healthy work environment helps the employees make lesser mistakes and, therefore, directly impacts data accuracy. Data entry professionals benefit from retaining their focus (comp. Fig. 1) (comp. Cong et al., 2007; Barchard and Verenikina, 2013; Nearing et al., 2018; Huang et al., 2003; Goodchild, 1992; Rindfuss et al., 2004; Gnauck and Fongwa, 2012; Singh et al., 2012; Gray et al., 2018).

So this chapter considers the main principles and technologies used in developing an operational sound DA system, which is the basis for the most needed monitoring and modeling (comp. Andreadis et al., 2017) systems. The main focus is the hydrological and hazardous (drought, flood) forecasting and, up to that spatially, through Geographic Information System (GIS) or Remote Sensing (RS), supporting DA tools based on data accuracy. Finally, what data accuracy means in the field of DA, will be declared.

2. Data assimilation for hydrological forecasting

In a world being threatened, next to other crises, by the climate crises and its climate-changing, hydrological and hazardous risks creating uncertain planning, policy, and decision-making processes (Reinstädtler, 2011, 2013, 2017a,b; comp. Reinstädtler, 2021a, b; Reinstädtler et al., 2017), it is more than ever most crucial to serve and produce data accurate, transparent, consistent, highly qualitative and having real-world simulating DA. Further, freshwater resources are imbalanced, with having only an Earth planetary amount of 2.5% or 35 Million km^3 (Reinstädtler and Schmidt, 2015; UNESCO and WWAP, 2003; Reinstädtler et al., 2021a, b). Up to that, water resources are endangered through more frequent ongoing water scarcity and drought (Islam et al., 2017a; Reinstädtler et al., 2017). Therefore, water amount, quality, and place of available water and water pollution are some of the essential substantiality and analytically as well as in instrumental way guiding factors to sustainable processes (Reinstädtler and Schmidt, 2015). Therefore, monitoring and modeling systems, nowcast, and forecast simulations in the most diverse forms are one sort of prevention and risk reduction methodology against these potential threats. Moreover, DA plays a crucial role in overcoming data insecurities in hydrological forecasting.

Overcoming these insecurities means to follow structured, vital steps in hydrological practices. Within those different steps, an advantageous research background can be built up for hydrological assessments and to apply data assimilated and hydrological forecasting information for water management purposes (comp. WMO, 1994). The following exemplary steps make up the holistic surrounding of acquiring sound hydrological data for hydrological forecasting with the highest grade of reliability (WMO, 1994): (1) General constellations around hydrological data acquiring and successful transfer to water management measures in practice (water-related activities, existing hydrological services, or international up to regional organizations involved with hydrology and water resources, hydrological standards, and regulations, also comp. Andreadis et al., 2017); (2) Hydrological instruments and methods of observation and estimation (overview on hydrological instruments and methods of observation, precipitation measurement, measurement of snow cover, evaporation and evapotranspiration, water levels of rivers, lakes, and reservoirs or discharge, sediment, ice on rivers, lakes, and reservoirs, soil moisture, groundwater, water quality or safety considerations and stream gauging stations); (3) Collection, processing, and dissemination of hydrological data (the role of hydrological data in an information system, design and evaluations of hydrological networks, collection of data, data review and coding, primary data processing, data storage, and retrieval, data dissemination); (4) Hydrological analysis (methods of analyses in hydrology, frequency analyses, rainfall frequency and intensity, storm rainfall analysis, interpretation of precipitation data, snow-melt runoff analysis, evaluation of streamflow data, rainfall-runoff relationships, streamflow routing, low-flow and drought analysis, flood-flow frequency, estimating lake and reservoir evaporation, estimating basin evapotranspiration, modeling of hydrological systems (comp. Andreadis et al., 2017), measurement of physiographic characteristics); (5) Hydrological forecasting (characteristics, effectiveness, and accuracy of forecasts, data requirements of hydrological forecasting, forecasting techniques, flood and water supply forecast, snow-melt forecast, forecasts of ice formation and break-up); (6) Applications for water management (purposes of water management projects, water resources systems, sustainable water development, water quality and resource protection, water resources assessment, estimating water demand, estimating reservoir capacity, estimating design floods, flood mitigation, irrigation and drainage, hydropower and energy-related projects, navigation and river training, urban water resources management, sediment transport and river-bed deformation) (WMO, 1994).

In this subchapter, mainly steps three to five will be focused on, although especially the last step into the application should not be underestimated in the efficiency of data having assimilated.

So, the role of hydrological data in an information system and its collection, processing, and dissemination of hydrological data in an information system, computer technology, staff, and training are vital key areas to be checked for sound DA and accuracy. Furthermore, for information systems, it has to be stated that general information about hydrologic conditions is of critical importance to real-world applications (Houser et al., 2012), information systems therefore in the same way.

Next to computer technology, staff, and training, a vital role in data collection steps plays in designing and evaluating hydrological networks (WMO, 1994). The demand for environmental, including hydrological and atmospheric information, was developing rapidly and still processing while requiring the design and development of increasingly exotic and comprehensive observing systems (comp. Daley, 1991). An example is that many regions worldwide, especially in Africa and the Middle East, are vulnerable to hydrological aspects such as drought and water and, up to that, food insecurity (comp. Arsenault, 2020). The Famine Early Warning Systems Network (FEWS NET) is the hydrological network's response to provide early warning of drought events in the region, led by the efforts of motivating organizations such as the U.S. Agency for International Development (USAID) (Arsenault, 2020): the FEWS NET reaches millions of people each year. Also, it is a new NASA multimodel, remote sensing–based hydrological forecasting and analysis system that guides as warnings guide life-saving assistance. Part of it is the NASA Hydrological Forecast and Analysis System (NHyFAS), "which has been developed to support such efforts by improving the FEWS NET's current early warning capabilities" (Arsenault et al., 2020).

From the general collection of data and processing steps, "typically, when using data assimilation to improve hydrologic forecasting, observations are assimilated up to the start of the forecast" (Leach and Coulibaly, 2019). "This is done to provide more accurate state and parameter estimates which, in turn, allows for a better forecast" (Leach and Coulibaly, 2019). Nevertheless, which processes are part of hydrological data collection? (comp. Fig. 2) (WMO, 1994): (1) Site selection; (2) Station identification: Identification of data collection sites; Descriptive information: Station description, Detailed sketch of station location, Map, Coordinates, Narrative description; (3) Frequency and timing of station visits: Manual stations, Recording stations; (4) Maintenance of collection sites; (5) Observations: Manual stations, Recording stations, Real-time reporting, Additional instructions for observers; (6) Transmission systems: General requirements, Transmission links, Factors affecting the choice of transmission systems; (7) Water quality monitoring: Station identification, Field sheets for water quality monitoring, Transportation of water quality samples, Field quality assurance in water quality monitoring; (8) Special data collection: Requirement, Bucket surveys of storm rainfall, Weather-radar and satellite data, Extreme river stages and discharges (WMO, 1994).

Furthermore, in addition to the steps involved in hydrological data collection (WMO, 1994), general DA of four basic strategies, as a function of time (NCARS, 2018; after Courtier et al., 1998) must also be considered (comp. Fig. 2): "The way the time distribution of observations ("obs") is processed to produce a time sequence of assimilated states (the lower curve in each panel) can be sequential and/or continuous" (NCARS 2018; after Courtier et al., 1998).

Data collection products were and still are, for instance, created using three-dimensional or four-dimensional (3DVAR, 4DVAR) variational DA schemes (Courtier et al., 1994; NCARS, 2018): For observations within some time interval about the target analysis time as occurring at the time of the analysis is being treated by 3DVAR; Observations distributed about the target analysis time to estimate the value are served by 4DVAR (Courtier et al., 1994; NCARS, 2018). In hydrological and Water Framework Research (WFR), 3DVAR and 4DVAR are used (Mazzarella et al., 2017).

Before part four of hydrological analysis of the DA steps is treated, data review and coding, primary data processing, data storage, retrieval, and dissemination (WMO, 1994) have to be fulfilled.

Further on, some statements about methods of analyses in hydrology will be shortly touched. Well-established principles of hydrodynamics, thermodynamics, and statistics are the basis for hydrological analysis (WMO, 2001; comp. Langat et al., 2019). However, applying these principles is difficult as a natural environment might be sparsely sampled, non-homogeneous, and only partially understood. The arouse samples are usually unplanned and uncontrolled, and analyses are performed to obtain spatial and temporal information about regional generalizations, certain variables, and relationships between the variables (WMO, 1994). Pertinent components are not accessed in their necessary measurements directly.

Moreover, analyses can be derived through the following different approaches: deterministic, probabilistic, parametric, and stochastic. An analysis based on the deterministic approach follows the laws that describe physical and chemical

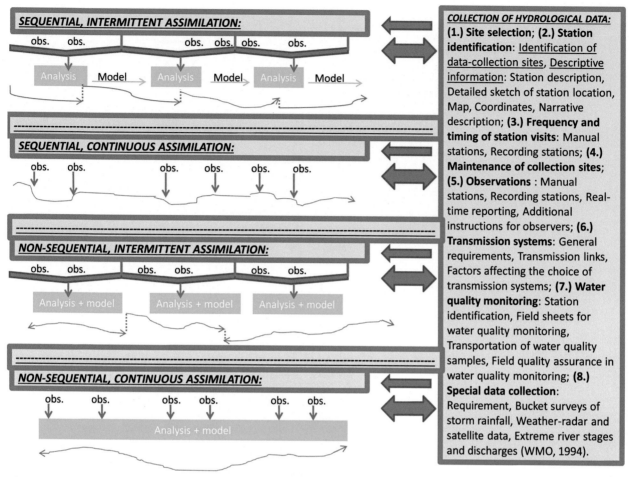

FIG. 2 Four basic strategies for general data assimilation, as a function of time (NCARS, 2018; after Courtier et al., 1998) coupling with hydrological data collection key points (WMO, 1994). The way the time distribution of observations (obs) is processed to produce a time sequence of assimilated states (the lower curve in each panel) can be sequential and/or continuous. *(Source: slightly changed by authors after WMO—World Meteorological Organization, 1994. Guide to Hydrological Practices. Data Acquisition and Processing, Analysis, Forecasting And Other Applications. WMO-No. 168, WMO, Geneva, Switzerland; NCARS—National Center for Atmospheric Research Staff (Eds.), 2018. The Climate Data Guide: Simplistic Overview of Reanalysis Data Assimilation Methods. Online-Article, last modified 11.10.2018. Accessed 23.08.2021 and retrieved from https://climatedataguide.ucar.edu/climate-data/simplistic-overview-reanalysis-data-assimilation-methods; After Courtier, P., Andersson, E., Heckley, W.A., Pailleux, J., Vasiljevic, D., Hamrud, M., Hollingsworth, A., Rabier, F., Fisher, M., 1998. The ECMWF implementation of three-dimensional variational assimilation (3D-Var). Part 1: formulation. Quart. J. Roy. Meteor. Soc. 124, 1783–1807. doi.org/10.21957/unhecz1kq.)*

processes (WMO, 2001; Yevjevich, 1974). The probabilistic approach analyses the frequency of occurrence of different magnitudes of hydrological variables (Yiping and Momcilo, 2011; WMO, 1994). In the parametric approach, an intercomparison of hydrological data is recorded at different locations and times (Sathish and Khadar Babu, 2020). Finally, the stochastic approach assesses the sequential order and frequency of occurrence of different magnitudes (Filipova et al., 2019).

Existing variables are measured directly for stage and velocity. Alternatively, they are computed directly from measurements, such as discharge. Other variables might be computed from a sample of direct measurements, for example, rainfall depth over a catchment (WMO, 2009, 2001, 1994). Other variables, such as lake evaporation, can only be evaluated indirectly (WMO, 2001, 1994).

Next to the already named forms of hydrological analysis, the water quality monitoring (point 7) of hydrological data collection, comp. Fig. 2) needs water quality assessment. The long-term research by Islam and Gnauck (2010) on salinity and its development in the Ganges-Brahmaputra-Megna (GBM) delta and Bangladesh (comp. Fig. 3) (Islam and Gnauck, 2010; Islam et al., 2017b) is one example: The map (comp. Fig. 3) shows the location of coastal zones in Bangladesh that are affected by climate change, SLR impacts and salinity intrusion. Also, it shows areas of future environmental impact that have been projected by IPCC (2007). For example, Fig. 3 shows the three broad regions of Bangladesh with the coastal

FIG. 3 Salinity grade due to climate change impacts and projected affected coastal zone and its development in the GBM Delta in Bangladesh (Islam and Gnauck, 2010; Islam et al., 2017b).

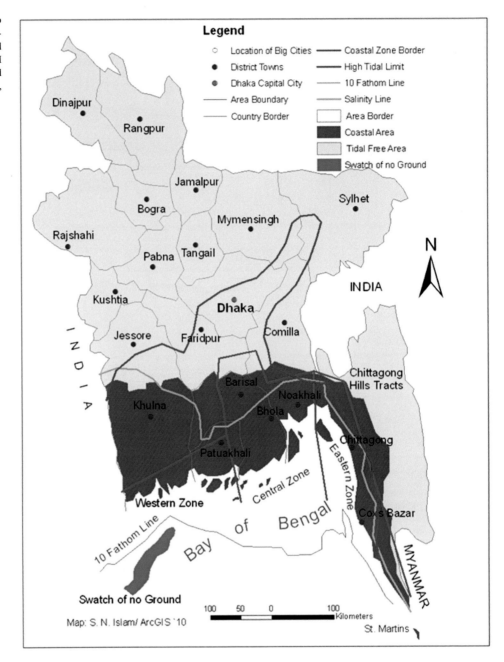

areas (brown color), the swatch of no ground (blue color), and the tidal free area (light green color), with different potential geographical boundaries such as the ten-fathom offshore limit line (light blue color), the high tidal limit line (red line), salinity line (yellow line), coastal zone border (blue line), high land area, and location of case areas (Islam and Gnauck, 2010; Islam et al., 2017b): about 20% of the net cultivable area of Bangladesh is located in the coastal zones and offshore island (Ali, 1999; Rahman and Ahsan, 2001; Islam and Gnauck, 2010; Islam et al., 2017b). In these areas, soils are affected by different degrees of salinity. For instance, about 203,000 ha very slightly, 492,000 ha slightly, 461,000 ha moderately, and 490,200 ha strongly salt-affected soils are assessed in the southwestern part of the coastal area (Rahman et al., 2000; after Islam and Gnauck, 2010; Islam et al., 2017b) (Fig. 3). It has been predicted that with about 1.5 m SLR, about 17 million (15%) of the population will be affected. It means that the urban and rural coastal people will have to be displaced or homeless, whereas 22,000 km^2 (16%) land will permanently be inundated (IPCC, 2007; Islam and Gnauck, 2007a, b; Islam, 2010; after Islam and Gnauck, 2010; Islam et al., 2017b). In addition, other problems will arise in the coastal belt, mainly environmental problems (such as water pollution and scarcity, soil degradation, deforestation, urban solid and hazardous waste discharge, and loss of urban biodiversity) and low coastal crop production. These problems could be detrimental to human livelihood and a severe health risk (Ali, 1999; Munasinghe and Swart, 2005; after Islam and Gnauck, 2010; Islam et al., 2017b).

The fifth step within the to be described processes in hydrological DA is hydrological forecasting. Many purposes are given for processing hydrological forecasts and warnings. It varies from short-term events like flash floods up to seasonal outlooks of power production, the potential water supply for irrigation, or inland navigation (WMO, 1994). Techniques include the correlations to the use of complex mathematical models representing all phases of the water balance of a river basin. Also, the use of simple empirical formulas is common (WMO, 1994). The main distinction of hydrological forecasts from statistical calculations is the calculation of the magnitudes of specific elements of the hydrological regime for a specified time in the future. Therefore the hydrologist evaluates only the expected probability of the elements (WMO, 1994).

Effectiveness and accuracy of forecasts (comp. Fig. 1) and data requirements of hydrological forecasting, forecasting techniques constantly has to be paid special attention. Flood and water supply forecast, snow-melt forecast, forecasts of ice formation and break-up should be processed as necessary for the specific needed DA project.

As had already had the example regarding an Early Warning System (EWS) against drought and to water and food insecurity, the Famine Early Warning Systems Network (FEWS NET) used the new NASA multimodel, remote sensing–based hydrological forecasting and analysis system, NHyFAS (Arsenault, 2020). This NHyFAS shows that it depends on two sources: (i) accurate initial conditions; an offline land modeling system generates these by applying and/or assimilating various satellite data (precipitation, soil moisture, and terrestrial water storage), and (ii) meteorological forcing data during the forecast period; they were generated by a state-of-the-art ocean-land-atmosphere forecasting system (Arsenault, 2020). The land modeling framework used is the Land Information System (LIS). LIS employs a suite of land surface models, allowing multimodel ensembles and multiple DA strategies better to estimate land surface conditions (Arsenault, 2020).

Lastly, the three ways of time spheres for DA in general and hydrological ones can be differentiated: the short-termed (comp. For instance, Bugaets et al., 2018; Leach and Coulibaly, 2019; Dawson and Wilby, 2021) and long-termed (Gnauck and Fongwa, 2012) or medium-ranged research and its DA. Short-termed hydrological or environmental assessments and data structures behind are more focusing on hazardous forecast, such as in river basins (flash-) flood forecasts (comp. Bugaets et al., 2018; Leach and Coulibaly, 2019; Dawson and Wilby, 2021), or rainfall-runoff modeling (comp. Dawson and Wilby, 2021). Also, long-term hydrological, environmental, land, climate systems changes and research about those need frequent data acquisition and support data security, availability for long- and short-termed DA.

Summarizing, hydrological forecasts are valuable for the rational regulation of runoff, irrigation (especially in arid regions), water supplies, the utilization of river energy, inland navigation, and water quality management (WMO, 1994). Moreover, further coping with dangerous phenomena on rivers or other water bodies strongly needs highly reliable forecasts. For example, advanced flood warnings enable taking steps to prevent loss of life and damage to property. So disruption of activities and destruction brought up by these natural calamities can be kept to a minimum (WMO, 1994). The great importance of these hydrological and other forecasts, such as hazardous forecasts, arises as to their reliability, for which data accuracy is primary.

3. Data assimilation for hazardous forecasting

After the Third UN World Conference on Disaster Risk Reduction (WSCRR), the United Nations International Strategy for Disaster Reduction (UNISDR) had adopted the Sendai Framework Program for Disaster Risk Reduction in 2015. The Sendai Framework Program made clear that monitoring systems are getting vital in sorts of enhancing the forecasting and early warning systems and developing a better understanding of various types of threats (UNISDR, 2015) and risks.

Especially rainfall-runoff modeling and flood forecasting can be named in hazardous hydrological forecasting to avoid threats and minimize risks. Within the already named example of the hydrological water quality DA regarding saltwater intrusion and salinity grade due to climate change, SLR impacts and projected affected coastal zones and its development in the GBM delta and Bangladesh (comp. Fig. 3; Islam and Gnauck, 2010; comp. Islam et al., 2017b), the great life securing necessity for accurate DA, hazardous forecasting, and early warning systems gets clear when summarizing some of the historical natural disasters having threatened coastal zones and coastal towns in Bangladesh (Islam and Gnauck, 2010; Islam et al., 2017b):

- During one storm surge in 1970, over 300,000 people died in the coastal towns and rural areas (Ali, 1999; after Islam and Gnauck, 2010; Islam et al., 2017b).

- There have already been several disastrous floods with massive losses of life in recent years, associated with storm surges caused by tropical cyclones; in 1985, 1987, 1988, and 1991 (Whyte, 1995; after Islam and Gnauck, 2010; Islam et al., 2015, 2016, 2017b).
- The 1991 cyclone killed at least 125,000 people (Whyte, 1995).
- The 2007 Cyclone Sidr destroyed mangrove resources, and almost 20,000 people were killed in the coastal zone of Bangladesh (Islam and Gnauck, 2010; Islam et al., 2015, 2017b).

Therefore climate change and SLR are expected to change the flooding and drought patterns in the Bangladesh coastal zone. This increased saltwater intrusion can probably substantially affect the availability of usable urban drinking water and soil for all kinds of human activities (Farouq, 1996; Islam, 2010; Islam and Gnauck, 2010). Furthermore, a predicted SLR, accelerated by global warming, will cause a further "squeezing" of the natural tidal land (IPCC, 2007). In Bangladesh case, it has been projected by IPCC (2007) and MoEF that 3 mm/year sea-level rise will occur before the year 2030 and 2500km^2 of land (2%) will be inundated (comp. Fig. 3; all after Islam and Gnauck, 2010; Islam et al., 2017b). With this highly exposed case of Bangladesh's highest climate change impacts on hand, it becomes clear that hazardous forecasting and early warning systems are only one small part of the needs and risks against climate change impacts while at least with these tools not saving land but being essential tools for saving lives.

So, being an essential, emerging field of research, together with artificial neural networks (ANNs), hazardous forecasting is characterized by a diversity of geographical contexts and a wide variety of techniques (Dawson and Wilby, 2021). Also, the Famine Early Warning Systems Network (FEWS NET) is an important step forward in modern (hydrological) monitoring systems (Arsenault, 2020) for supporting chances of taking the appropriate action for hazard mitigation. In soil-hazardous drought forecasting, soil moisture DA is an important example in the hydrologic sciences (De Lannoy et al., 2016). As part of the baseline mission, the NASA Soil Moisture Active/Passive (SMAP) mission (Entekhabi et al., 2010) offers a DA product (Reichle et al., 2016; Nearing et al., 2018).

For forecasting necessary monitoring or modeling systems, exemplary performance in the short-term forecast of, e.g., rainfall floods in a river basin must be demonstrated and reproduced within a complex spatiotemporal structure for the runoff formation during extreme rainfall (comp. Bugaets et al., 2018). Therefore, DA's fundamental principles, sources, and technologies must be considered while combining and developing operational modeling systems, e.g., river basins or other environmental and hydrological systems. Furthermore, the combination and processing way next to division and preprocessing of data for model calibration/validation or data standardization techniques (comp. Dawson and Wilby, 2021) is essential for receiving a better grade of forecast reliability, which might further be crucial for civil, environmental, energy, water and food security and safety purposes. Using a database of forcing data, model states, predicted streamflow, and streamflow observations might be of help (Leach and Coulibaly, 2019). The technologies can include automated systems of hydrological or environmental monitoring and data management, ANN modeling or more conventional statistical modeling, physical–mathematical modeling with distributed parameters of specific models, and possibly directly or indirectly integration of the numerical mesoscale atmosphere model WRF (Weather Research and Forecasting Model), designed for both atmospheric research and operational forecasting applications (comp. Bugaets et al., 2018; Dawson and Wilby, 2021). Alternatively, also a lookup function can be used to provide an observation during the forecast, which can be assimilated (Leach and Coulibaly, 2019). It allows for an indirect way to assimilate the numerical weather prediction forcing data.

Furthermore, it is essential to have freely combined tools, intermodel, and cross-platform interoperability in a modeling or monitoring data assimilating system implemented to allow flexible forecasting and informational components (comp. Bugaets et al., 2018). All these approaches, sort of information systems or network design, can help address prediction uncertainty: an ensemble of previous observations can be pulled from the database, corresponding to the forecast probability density function and, therefore, to given previous information (Leach and Coulibaly, 2019). Prediction uncertainties might have minimized and forecast reliability bettered with this sort of processing. For instance, Leach and Coulibaly (2019) further propose an extension to the traditional DA approach: it would allow assimilation to continue into the forecast, further improving its performance and reliability (Leach and Coulibaly, 2019). Extending data assimilation into the forecast can improve forecast performance. It showed that the forecast reliability could be improved by up to 78% throughout this extension (Leach and Coulibaly, 2019).

All these partially possible to be coupled technologies and different data systems involved might use the Simple Object Access Protocol (SOAP) web services and the Open Geospatial Consortium Open Modeling Interface (OGC OpenMI) standard to safe over this messaging protocol specification the exchange of structured information (comp. Bugaets et al., 2018).

To summarize, after Bugaets et al. (2018), the main essential features in hazardous (flood) and especially state-of-the-art hazardous hydrological forecasting and its DA are as the following. First, with sufficient data accuracy, they have to provide the maximum lead time for the users to take the appropriate action to optimize water use strategy or hazard mitigation (Bugaets et al., 2018). Second, such a systems' standard structure includes data management blocks, modeling resources, GIS tools, and more. Third, analysts must use the most accurate data (Bugaets et al., 2018). Fourth, data-rich and data-poor national hydrometeorological services struggle with quality controlling, retrieving, infilling, formatting, archiving, or redistributing data (Pagano et al., 2014). Therefore, robust algorithms are necessary to overcome data processing and preparation issues (Bugaets et al., 2018).

4. Importance of spatial precision systems in error reduction

In hydrological data assessment, applications for water management need to be observed as the last and most vital step in applying the data collection, data analyses, and results to, for instance, purposes of water management projects, water resources systems, sustainable water development (comp. WMO, 2001, 1994). Also, water quality and resource protection, water resources assessment or estimating water demands, estimating reservoir capacity, estimating design floods, flood mitigation, irrigation, and drainage (comp. WMO, 2001, 1994) were partially highlighted within the case studies of Islam and Gnauck (2010) or Islam et al. (2017b). In addition, other essential infrastructures and energy-related aspects have to be secured, such as hydropower and energy-related projects, navigation and river training, urban water resources management, sediment transport, and river-bed deformation (WMO, 1994).

Water-related and hazardous-related transferences for data accuracy crucial steps of collection, processing, and dissemination of data, data analysis, and forecasting have to be strongly supported to maximize accurate data assessments' outcomes and placement for applying data.

Spatial precision systems are possible and necessary tools to foster transference stability into practice and in the same time, force error reduction while spatially determining the results: extracting information from Earth-observing RS observations or retrievals over satellite data is one of the examples of one of the most common methods: data assimilation (De Lannoy et al., 2016; Nearing et al., 2018; comp. Dorigo et al., 2007). RS and numerical modeling of water quality and interactions or other analyses with, for instance, an ecosystem, their ecosystems responses to climate and water or further cycle variations are some of the fields spatial precision systems are a more excellent added value in securing data accuracy and transference into practice. With the example of modeling the behavior of agroecosystems in case of the severely deteriorated ecological status of many agroecosystems, RS can support such modeling by offering information on the spatial and temporal variation of different state variables (Dorigo et al., 2007).

A further vastly emerging field is the GIS technology in spatially precision, for which in before Fig. 3 and especially following Fig. 4 can stand for. On the one hand, the exemplary in the Region of Lusatia and inner part Spree Forest by Reinstädtler, 2017a, 2017b, 2019, 2021b, 2021a; Reinstädtler, 2014 derived and within the water network of Brandenburg placed parts of the Landscape Unit 1 (LU1) of "Alluvial Fan Influenced Landscape with Wetland and Fließe Structures" of Spree Forest Inland Delta and exemplary determined SDGs are showing up these more significant needs of transference activating tools (comp. Fig. 4 as part of LU1). Spatial precision systems can serve especially on informal planning level for any stakeholder, communal, regional, or international (scientific) institution, university, administration, or policymakers while in the same time possibly implanting spatially determined flagship motivations such as the SDGs as a precautionary tool (Reinstädtler, 2017a, b, 2019, 2021a). On the other hand, communicating results is vital in transferring DA into practice (Reinstädtler, 2017a, b, 2019, 2021a, b). Therefore, "straightforward" evaluations such as on the map of the forestry land-use change can show up tremendously crucial developmental outcomes while being spatially as well as temporally precise (comp. Fig. 4) not only for information systems but also for any practical stakeholder: the comparison of the years of 1751, 1846 and 1939 in Upper Spree Forest as part of LU1 and as an example for GIS-applied data assessment in the processing roadmap for sustainable development (derived by the author with the basis-map of plane-table sheets, map of Dr. P. Rupp in TVS, 2000, p. 94–95; Bönisch et al., 1994, p. 27; Krausch, 1960; Krausch, 1955; Reinstädtler, 2014). Having also highlighted by Reinstädtler, (2014, 2017a,b, 2019, 2021a,b), the spatial scale next to planning level- and instrumental flexibility is further essential in avoiding data inaccuracies. Also, Paciorek (2010) found the importance of scale for spatial-confounding bias and the precision of spatial regression estimators. Understanding the key issues and deriving informative analytic results (Paciorek, 2010) means: "one can reduce bias by fitting a spatial model only when there is variation in the covariate at a scale smaller than the scale of the unmeasured confounding" (Paciorek, 2010).

Also affected by data aggregation is next to scale the efficiency and uncertainty estimation (Paciorek, 2010). However, correct scale selection and usage should not be estimated as a disadvantage for spatial precision systems but a chance to overcome those insecurities in data accuracy on the spatial extent.

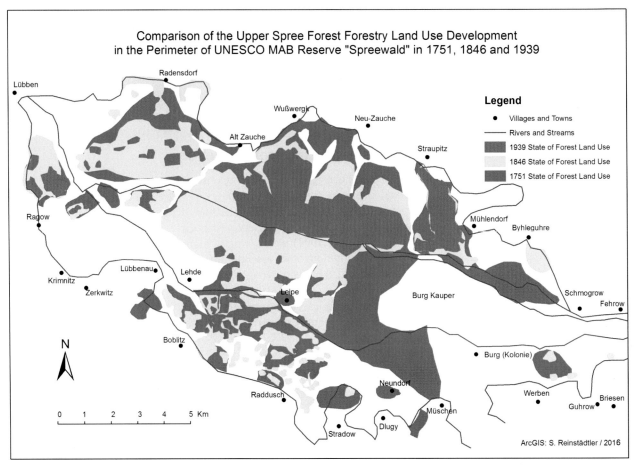

FIG. 4 Map of the forestry land-use change in comparison of the years of 1751, 1846, and 1939 in Upper Spree Forest as an example for GIS-applied assessment in the processing roadmap for sustainable development (derived by the author with the basis-map of plane-table sheets, map of Dr. P. Rupp in TVS, 2000, p. 94–95; Bönisch et al., 1994, p. 27; Krausch, 1960; Krausch, 1955 and Reinstädtler, 2014).

5. Discussion and future perspective

The inferior overview on DA and data accuracy in hydrological and hazardous forecasting shows a strong need for technologies, methods, models, and databases, which are flexible enough to deal with today's multivarious demands. Therefore, especially data accuracy and communication and networking for transparent (open sourced) data exchange stay at the forefront of supporting hydrological and hazardous forecasting. Furthermore, as hydrological and hazardous risks create uncertain planning, policy, and decision-making processes, the pathway to describe modeling or monitoring data ascertaining results on the way to decision making is further critical. "Bridging the gap between science and decision making" (von Winterfeldt, 2013) has to get smartly overcome through communication, networking, innovation, and informal planning tools (Reinstädtler, 2017a, b, 2021a). Moreover, overbridging these gaps means emphasizing the development of standard performance measures that hinder unnecessary model complexity (comp. Dawson and Wilby, 2021). As a result, general understanding and openness and non-specialists of data assessment such as decision-makers, stakeholders, etc., will be increased.

In this row of discursive improvement, possibilities stand the rainfall-runoff modeling, flood forecasting, and ANNs, which are often characterized by a general absence of intermodel comparisons and inconsistent reporting of model skills (Dawson and Wilby, 2021). Again, these deficiencies could be improved quickly, with the same possible applicability across administrative boundaries and diverse topical specifications.

Concerning possible future orientation, the modular architecture and design of any DA architecture in the form of databases, frameworks, or models could allow for potential modifications in the future that could enhance its applicability. Furthermore, coupling other models is enriching and should be investigated, especially in topical specifications of climate change-induced data assessment and modeling.

Future-oriented and sustainable topics to be supported are about the ecological status of many agroecosystems, which has been severely deteriorated during the last 50 years (comp. Dorigo et al., 2007; comp. MEA, 2005). Because, to meet the growing demand for food, timber, fiber, and fuel while having created an intensified use, the management of agroecosystems has been undergoing significant changes (Dorigo et al., 2007; comp. MEA, 2005). So data assessing and modeling the behavior of agroecosystems would be a great asset and future project for multiple worldwide existing agroecosystems, if those would allow the definition of management strategies that maximize (crop) production while minimizing the environmental impacts (comp. Dorigo et al., 2007). Similar structural requests and solutions are needed for several intense and concurring land-use forms and worldwide crisis-related fields such as climate change, biodiversity loss, or social disruption.

Up to that, Dorigo et al. (2007) states that for future agroecosystem modeling and RS data assimilation to be commonly employed on a global operational basis, future-demand is and, therefore, emphasis will have to be laid in bridging mismatches between model and user requirements on the one hand, and data availability and accuracy on the other. These challenges are similar intense in diverse other fields and have to get supported transparently.

6. Conclusions

Facilitation of the deployment of structured and transparent frameworks or models, drought, natural or water resources simulations, accurate DA design and processing assimilation of a variety of datasets (including satellite observations and GIS), and diverse environmental or further data analyses is the answer on proper DA coupled forecasting endeavors. Even scaling-, planning level challenges and instrumental and communicatory flexibility into practice are critical objectives in using spatial precision systems, the added value of correct usage is higher for securing data accuracy. A multicriteria patch of coupled databases such as RS/GIS databases at the core with various satellite and model-based datasets could be an advantage.

However, coupled technologies and different data systems should be supported through messaging protocol specification web services or modeling interfaces and their standards to safe the non-biased exchange of structured, accurate information. Moreover, today's databases need to allow different modeling components to interface against it in that constellation, simplifying the coupling of different topical necessities. So it should be encouraged with a designed framework-, modeling-database-interfaces even with extensive validation, to generate a suite of data products in relatively diverse contexts. The increasing complexity of DA methods and models describing multivarious functioning, such as agroecosystem functioning, has significantly increased computational demands. However, in the same way, the flexibility of diverse systems, DA methods, models, or frameworks have to get more flexible:

- In the case of data flexibility and tradition of open source: the RS/GIS features and other data of databases facilitate the dissemination of the data products to diverse platforms in the form of web, mobile, and desktop.
- Flexibility in multimodel combinations: other hydrologic models can be added within databases and models to supplement one database and model, creating a multimodel ensemble, improving predictability.
- Further flexibility in scale calibration: the spatial scales that databases are applicable are governed by a model at its core, which has to be implemented at different resolutions: a minimum spatial scale could be on the order of x-5 km, depending on the observation objective and spatial extent. Nonetheless, specific regional applications may require further calibration and should be achievable with external software tools of the specific model parameters and validation against either in-situ or satellite measurements.
- Flexibility in time: such as a hydrodynamic model makes the system suitable for flood forecasting applications at even sub-hourly time steps.
- Flexibility in usage style: compared to similar modeling systems that either require extensive configuration or scripting a solution custom to a specific end-user, a database and/or model should allow the implementation of a nowcast and/or forecasting system with minimal inputs from user automating and abstracting many of the details away.

Nevertheless, flexibility also means creating uncertainty in some parts, which is already apparent in other topical specifications such as within climate change. So recommendations in this instance might be:

- Overcoming insecurities means following structured, vital steps in hydrological, hazardous-mitigating practices, as shown in chapter one with the different existing steps of DA. These named exemplary steps make up a holistic surrounding of acquiring valuable data for forecasting with the highest grade of reliability:
  ○ This should increase the awareness of responding agencies, the end-users of results, stakeholders, decision-makers, and the general population.

- ○ The output data products should include an exhaustive set of different (for instance, hydrological) variables, each having an uncertainty estimate associated with it for describing the grade of accuracy of available data.
  - ○ In the context of insecurities for decision making, examples could include the use of a product to provide an outlook on various potentials: this could mean exemplary to give the information of a soil moisture change product to provide an outlook on the crop growth potential, or the drought onset/duration product to plan for food storage.
- In sorts of data collection within the data assessment process, the role of data in an information system, design, and evaluations of hydrological networks, collection of data or primary data processing is vital to anchor points for thinking about or actively implementing spatial precision systems with tools such as RS, GIS. The same anchor points can be found within DAs' analyses processes in good form on hydrological aspects with proving methods of analyses and further in chapter two named vital steps of hydrological practices and its step four of hydrological analysis such as evaluation of streamflow routing or low-flow and drought analysis. In the forecasting part, spatial precision systems should be considered in most of the data requirements of hydrological, hazardous, or further forecasting processes, also for over bridging general insecurities.
- Bridging uncertainty creating situations through mismatches between data availability and accuracy could be minimized by integrating imagery with different spatial, temporal, spectral, and angular resolutions and fusing optical data with data of different origins, such as LIDAR and radar/microwave.

There are many options for still optimizing DA and data accuracy in general and for hydrological and hazardous forecasting. However, at the same time, scientists worldwide already contribute to perfect DA through multivarious innovative ideas and traditional processing. Well-balancing minimization of uncertainty and maximizing flexibility while profoundly stabilizing the architecture of data assimilating frameworks might be access points for optimized data accuracy. Also, sustainability science endeavors and requests on sustainable data assessment structures could be gaining ones.

References

Ali, A., 1999. Climate change impacts and adaptation assessment in Bangladesh. Climate Res. 12, 109–116.

Andreadis, K.M., Das, N., Stampoulis, D., Ines, A., Fisher, J.B., Granger, S., et al., 2017. The Regional Hydrologic Extremes Assessment System: A software framework for hydrologic modeling and data assimilation. PLoS ONE 12 (5), e0176506. https://doi.org/10.1371/journal.pone.0176506.

Arsenault, K.R., Shukla, S., Hazra, A., Getirana, A., McNally, A., Kumar, S.V., Koster, R.D., Peters-Lidard, C.D., Zaitchik, B.F., Badr, H., Jung, H.C., Narapusetty, B., Navari, M., Wang, S., Mocko, D.M., Funk, C., Harrison, L., Husak, G.J., Adoum, A., Galu, G., Magadzire, T., Roningen, J., Shaw, M., Eylander, J., Bergaoui, K., McDonnell, R.A., Verdin, J.P., 2020. The NASA hydrological forecast system for food and water security applications. Bull. Am. Meteorol. Soc. 101 (7), E1007–E1025. https://doi.org/10.1175/BAMS-D-18-0264.1.

Barchard, K.A., Verenikina, Y., 2013. Improving data accuracy: Selecting the best data checking technique. Comput. Hum. Behav. 29 (5), 1917–1922. https://doi.org/10.1016/j.chb.2013.02.021.

Bohannon, P., Fan, W., Geerts, F., Jia, X., Kementsietsidis, A., 2007. Conditional functional dependencies for data cleaning. In: ICDE 2007, IEEE 23rd International Conference on Data Engineering., https://doi.org/10.1109/ICDE.2007.367920.

Bönisch, F. Breddin, R. and Krausch, H.-D., 1994. Geschichte. In: Grundmann, L. (Ed.). Burger und Lübbenauer Spreewald. Werte der deutschen Heimat 55, second ed., Institut für Länderkunde Leipzig, Weimar, p. 27.Bönisch, F., Breddin, R., Krausch, H.-D., 1994. Geschichte. In: Grundmann, L. (Ed.), Burger und Lübbenauer Spreewald. Werte der deutschen Heimat 55, 2nd. Institut für Länderkunde Leipzig, Weimar, Germany, p. 27.

Bugaets, A., Gartsman, B., Gelfan, A., Motovilov, Y., Sokolov, O., Gonchukov, L., Kalugin, A., Moreido, V., Suchilina, Z., Fingert, E., 2018. The integrated system of hydrological forecasting in the ussuri river basin based on the ECOMAG model. Geosciences 8 (1), 5. https://doi.org/10.3390/geosciences8010005.

Cardenas, I.C., 2019. On the use of Bayesian networks as a meta-modelling approach to analyse uncertainties in slope stability analysis. Georisk 13 (1), 53–65. https://doi.org/10.1080/17499518.2018.1498524.

Cong, G., Fan, W., Geerts, F., Jia, X., Ma, S., 2007. Improving data quality: consistency and accuracy. Proc. Int. Conf. VLDB, 315–326.

Courtier, P., Thépaut, J.-N., Hollingsworth, A., 1994. A strategy for operational implementation of 4D-VAR, using an incremental approach. Quart. J. Roy. Meteor. Soc. 120, 1367–1387.

Courtier, P., Andersson, E., Heckley, W.A., Pailleux, J., Vasiljevic, D., Hamrud, M., Hollingsworth, A., Rabier, F., Fisher, M., 1998. The ECMWF implementation of three-dimensional variational assimilation (3D-Var). Part 1: formulation. Quart. J. Roy. Meteor. Soc. 124, 1783–1807. https://doi.org/10.21957/unhecz1kq.

Daley, R., 1991. Atmospheric Data Analysis. Cambridge Atmospheric and Space Science Series. Cambridge University Press, ISBN: 0-521-38215-7. 457 pages.

Dawson, C.W., Wilby, R.L., 2021. Hydrological modelling using artificial neural networks. Prog Phys Geogr. Earth Environ. 25 (1), 80–108. https://doi.org/10.1177/030913330102500104.

De Lannoy, G.J.M., de Rosnay, P., Reichle, R.H., 2016. Soil moisture data assimilation. In: Handbook of Hydrometeorological Ensemble Forecasting. Springer, pp. 1–43.

Dorigo, W.A., Zurita-Milla, R., de Wit, A.J.W., Brazile, J., Singh, R., Schaepman, M.E., 2007. A review on reflective remote sensing and data assimilation techniques for enhanced agroecosystem modeling. Int. J. Appl. Earth Obs. Geoinf. 9, 165–193.

Entekhabi, D., Njoku, E.G., O'Neill, P.E., Kellogg, K.H., Crow, W.T., Edelstein, W.N., et al., 2010. The soil moisture active passive (SMAP) mission. Proceedings of the IEEE 98 (5), 704–716. https://doi.org/10.1109/JPROC.2010.2043918.

Eslamian, S., 2012. Forecasting, Encyclopedia of Energy. Salem Press, USA, pp. 461–464.

Farouq, A.A., 1996. Adaptation to climate change in the coastal resources sectors of Bangladesh: Some issues and problems. In: Smith, B.J., Bhatti, N. (Eds.), Adaptation to Climate Change Assessments and Issues. Springer, pp. 335–342.

Filipova, V., Lawrence, D., Skaugen, T., 2019. A stochastic event-based approach for flood estimation in catchments with mixed rainfall and snowmelt flood regimes. Nat. Hazards Earth Syst. Sci. 19, 1–18. https://doi.org/10.5194/nhess-19-1-2019.

Gnauck, A., Fongwa, E.A., 2012. Long-term ecological research for sustainable environmental information. In: Arndt, H.-K., Knetsch, G., Pillmann, W. (Eds.), EnviroInfo Dessau 2012, Part 1: Core Application Areas. Shaker Verlag, Aachen.

Goodchild, M.F., 1992. Geographical data modeling. Comput. Geosci. 18 (4), 401–408. https://doi.org/10.1016/0098-3004(92)90069-4.

Gray, D.K., Hampton, S.E., O'Reilly, C.M., Sharma, S., Cohen, R.S., 2018. How do data collection and processing methods impact the Accuracy of long-term trend estimation in lake surface-water temperatures? Limnol. Oceanogr. Methods 16, 504–515. https://doi.org/10.1002/lom3.10262.

Houser, P.R., De Lannoy, G.J.M., Walker, J.P., Tiefenbacher, J.P., 2012. Hydrologic data assimilation. In: Approaches to Managing Disaster—Assessing Hazards, Emergencies and Disaster Impacts. IntachOpen, https://doi.org/10.5772/31246.

Huang, G.H., Chang, N.B., 2003. Perspectives of environmental informatics and systems analysis. J. Environ. Inf. 1 (1), 1–6. https://doi.org/10.3808/jei.

IPCC—Intergovernmental Panel on Climate Change, 2007. Summary for policymakers, Climate Change 2007: Impact Adaptation and Vulnerability. Contribution of Working Group II to the Fourth Assessment Report of the IPCC, Parry, M.L., Canziani, O.F. and Palutikot, P.J. et al. (Eds.), Cambridge University Press, Cambridge. 1000pp.IPCC—Intergovernmental Panel on Climate Change, 2007. Summary for policymakers. In: Parry, M.L., Canziani, O.F., Palutikot, P.J., et al. (Eds.), Climate Change 2007: Impact Adaptation and Vulnerability. Contribution of Working Group II to the Fourth Assessment Report of the IPCC. Cambridge University Press, Cambridge, UK. 1000pp.

Islam, S.N., 2010. Threatened wetlands and ecologically sensitive ecosystems management in Bangladesh. Front. Earth Sci. China 4 (4), 438–448.

Islam, S.N., Gnauck, A., 2007a. Effects of salinity intrusion in mangrove wetlands ecosystems in the Sundarbans: An alternative approach for sustainable management. In: Okruszko, T., Jerecka, M., Kosinski, K. (Eds.), Wetlands: Monitoring Modelling and Management. Taylor & Francis/ Balkema, Leiden, The Netherlands, pp. 315–322.

Islam, S.N., Gnauck, A., 2007b. Salinity intrusion due to fresh water scarcity in the Ganges catchment: a challenge for urban drinking water and mangrove wetland ecosystems in the Sundarbans region, Bangladesh. In: Proceedings of the Worldwide Workshop WWW YES 2007 for Young Environmental Scientists. University of Paris, Paris, France, pp. 20–30.

Islam, S.N., Gnauck, A., 2010. Climate change versus urban drinking water supply and management: a case analysis on the coastal towns of Bangladesh. In: World Wide Workshop for Young Environmental Scientists: 2010, May 2010, Arcueil, France. hal-00521467.

Islam, S.N., Reinstädtler, S., Eslamian, S., 2015. Water reuse sustainability in cold climate regions (chapter 68). In: Eslamian, S., Eslamian, F. (Eds.), Urban Water Reuse Handbook, Taylor & Francis. CRC Press, New York, pp. 883–894.

Islam, S.N., Reinstädtler, S., Gnauck, A., 2016. Coastal environmental degradation and ecosystem management in the Ganges deltaic region in Bangladesh. Inter. J. Ecol. Econ. Stat. 37 (4), 59–81.

Islam, S.N., Reinstädtler, S., Gnauck, A., 2017a. Capacity building and drought management. Chapter 3. In: Eslamian, S., Eslamian, F.A. (Eds.), Handbook of Drought and Water Scarcity. Vol. 2: Management of Drought and Water Scarcity. CRC Press, Taylor & Francis Group, Boca Raton, USA.

Islam, S.N., Reinstädtler, S., Ferdaush, J., 2017b. Challenges of climate change impacts on urban water management and planning in the coastal towns of Bangladesh. In: International Journal of Environment and Sustainable Development (IJESD). Inderscience Publishers Limited, Geneva, Switzerland.

Kalnay, E., Kanamitsu, M., Kistler, R., Collins, W., Deaven, D., Gandin, L., Iredell, M., Saha, S., White, G., Woollen, J., Zhu, Y., Chelliah, M., Ebisuzaki, W., Higgins, W., Janowiak, J., Mo, K.C., Ropelewski, C., Wang, J., Leetmaa, A., Reynolds, R., Jenne, R., Joseph, D., 1996. The NCEP/NCAR 40-Year Reanalysis Project. Bull. Am. Meteorol. Soc. 77 (3), 437–472. https://doi.org/10.1175/1520-0477(1996)077<0437:TNYRP>2.0.CO;2.

Krausch, H.-D., 1955. Wälder und Wiesen im Spreewald in geschichtlicher Entwicklung. In: Wiss. Zeitschrift der Päd. Hochschule Potsdam, Mathematisch Naturwissenschaftliche Reihe 1 (2).

Krausch, H.-D., 1960. Die Pflanzenwelt des Spreewaldes. Wittenberg, Germany.

Langat, P.K., Kumar, L., Koech, R., Ghosh, M.K., 2019. Hydro-morphological characteristics using flow duration curve, historical data and remote sensing: effects of land use and climate. Water 11 (2), 309. https://doi.org/10.3390/w11020309.

Leach, J.M., Coulibaly, P., 2019. An extension of data assimilation into the short-term hydrologic forecast for improved prediction reliability. Adv. Water Resour. 134, 103443.

Mazzarella, V., et al., 2017. Comparison between 3D-Var and 4D-Var data assimilation methods for the simulation of a heavy rainfall case in central Italy. Adv. Sci. Res. 14, 271–278.

MEA—Millennium Ecosystem Assessment (Ed.), 2005. Ecosystems and Human Well-Being: Synthesis. Island Press, Washington, DC, USA.

Munasinghe, M., Swart, R., 2005. Primer on Climate Change and Sustainable Development—Facts, Policy Analysis and Applications. Cambridge University Press, Washington DC, USA.

NASA—National Aeronautics and Space Administration. Mars Climate Modeling Center. 2020. https://www.nasa.gov/mars-climate-modeling-center-ames (Accessed 17.06.2021).

NCARS—National Center for Atmospheric Research Staff (Eds.), 2018. The Climate Data Guide: Simplistic Overview of Reanalysis Data Assimilation Methods. Online-Article, last modified 11.10.2018. Accessed 23.08.2021 and retrieved from https://climatedataguide.ucar.edu/climate-data/simplistic-overview-reanalysis-data-assimilation-methods.

Nearing, G., Yatheendradas, S., Crow, W., Zhan, X., Liu, J., Chen, F., 2018. The efficiency of data assimilation. Water Resour. Res. 54 (9), 6374–6392. https://doi.org/10.1029/2017WR020991.

Paciorek, C.J., 2010. The importance of scale for spatial-confounding bias and precision of spatial regression estimators. Stat. Sci. 25 (1), 107–125. https://doi.org/10.1214/10-STS326.

Pagano, T.C., Wood, A.W., Ramos, M.-H., Cloke, H.L., Pappenberger, F., Clark, M.P., Cranston, M., Kavetski, D., Mathevet, T., Sorooshian, S., et al., 2014. Challenges of operational river forecasting. J. Hydrometeorol. 15, 1692–1707.

Rahman, M., Ahsan, A., 2001. Salinity constraints and agricultural productivity in coastal saline area of Bangladesh. In: Rahman, M. (Ed.), Soil Resources in Bangladesh: Assessment and Utilization. Soil Resource Development Institute, Prokash Mudrayan, Dhaka, Bangladesh, pp. 2–14.

Rahman, M.M., Hassan, M.Q., Islam, M.S., Shamsad, S.Z.K.M., 2000. Environmental impactassessment on water quality deterioration caused by the decreased Ganges outflow and salinewater intrusion in south-western Bangladesh. Environ. Geol. 40 (1–2), 31–40.

Reichle, R., Crow, W., Koster, R., Kimball, J., De Lannoy, G., 2016. SMAP Level 4 Surface and Root Zone Soil Moisture (L4_SM), Data Product. Soil Moisture Active Passive, NASA, USA.

Reinstädtler, S., 2011. Sustaining Landscapes – Landscape Units for Climate Adaptive Regional Planning. Conference Presentation and Proceeding of Abstracts of the International World Heritage Studies (WHS) – Alumni – Conference "World Heritage and Sustainable Development" from 16.-19.06.2011, IAWHP (International Association for World Heritage Professionals), Cottbus, Germany, Presentation-slides, p.1-107 + Online publication: http://www.iawhp.com/wp-content/uploads/2012/01/IAWHP2011_Book-Abtracts.pdf, p. 18.

Reinstädtler, S., 2013. Sustaining Landscapes – Landscape Units for Climate Adaptive Regional Planning. Conference Proceeding of the International World Heritage Studies (WHS) – Alumni – Conference "World Heritage and Sustainable Development" from 16 to 19 June 2011. IAWHP (International Association for World Heritage Professionals), Cottbus, Germany, pp. 45–65. ISBN: 978-3-00-041891-4.

Reinstädtler, S., 2014. Sustaining UNESCO MAB Reserve Spree Forest – the right for preserving landscape values in the German Lusatia region. Conference Proceeding of the International World Heritage Studies (WHS) – Alumni – Conference "The Right to [World] Heritage" from 23 to 25 October 2014, IAWHP (International Association for World Heritage Professionals), Cottbus, Germany, ISBN 978-3-00-047536-8, pp. 294–333.

Reinstädtler, S., 2017a. Spatial Determination of the SDGs for Activating Regional Management in Spree Forest and Lusatia Region and its Lusatian Traditional and Modern 21st Century Landscapes. Conference Proceeding Of Abstracts of the "SustEcon Conference—The Contribution of a Sustainable Economy to Achieving the sdgs " From 25.-26.09.2017, Organized by NaWiKo (www.nachhaltigeswirtschaften-soef.de). Berlin, Germany.

Reinstädtler, S., 2017b. Why spatially determining of SDGs matters for landscapes in the German Lusatia region. In: Oral Presentation and Forum Contribution on 20.12.17 (block 2, 11:00-12:30) in Form of a TED-talk at the "Global Landscape Forum 2017, Bonn "from 19.-20.12.2017, Organized by CIFOR, Bonn, Germany. Video Online Available (Beginning With Minute 24:49). https://events.globallandscapesforum.org/landscape-talk/bonn-2017/day-2/block-2/.

Reinstädtler, S., 2019. Bedeutung Klima-Smarter Planung für Nachhaltende Landschaften und Landnutzungen in den Transformationsprozessen der Region Lausitz—(written in German, english translation:) "Significance of Climate Smart Planning for Sustaining Landscapes and Land Uses in the Transformation Processes of Lusatia Region" at 30.09.19. In: Conference proceeding of abstracts and presentation of the IALE-D-Conference "Landschaft im Klimaschutz "– "Landscape in Climate Protection "within the section of "Change Management im Klimaschutz "– "Change Management in Climate Protection "from 30.09.-02.10.2019, organized by International Association for Landscape Ecology, Section Germany (IALE-D) and University of Potsdam, Potsdam-Golm, Germany.

Reinstädtler, S., 2021b. An innovative land approach for spatially determining SDGS —sustainably governing German spree forest and Lusatia region. J. Civ. Eng. Archit. 15 (11), 577–585. https://doi.org/10.17265/1934-7359/2021.11.004.

Reinstädtler, S., 2021a. Innovations in theoretical approaches for spatially determining SDGS on landscape scale and regional planning level—application for sustainably governing spree forest and Lusatia region in Germany. In: Johansson, C., Mauerhofer, V. (Eds.), Accelerating the Progress Towards The 2030 SDGs in Times of Crisis. Conference Proceeding of The 27th Annual ISDRS-Conference 2021 Within The Section of "I. Sustainability and Science"—"Ia. Theoretical Approaches" at 13.07.2021 (PhD-Presentation) and 14.07.2021 (Conference-Presentation), From 13.07.-15.07.2021, Organized by the International Sustainable Development Research Society (ISDRS) and Mid Sweden University, Department of Media and Communication Science and Department of Ecotechnology and Sustainable Building, Östersund, Sweden. ISBN 978–91–89341-17-3 (Article in conf. proceed.). Online available: http://miun.diva-portal.org/smash/get/diva2:1608600/FULLTEXT01.pdf.

Reinstädtler, S., Schmidt, M., 2015. Environmental impact assessment for urban water reuse – a tool for sustainable development. In: Eslamian, S. (Ed.), Handbook of Urban Water Reuse. CRC Press, Taylor & Francis Group, Geneva, New York, pp. 243–257. Chapter 21.

Reinstädtler, S., Islam, S.N., Eslamian, S., 2017. Drought management for landscape and rural security. Chapter 8. In: Eslamian, S., Eslamian, F.A. (Eds.), Handbook of Drought and Water Scarcity. Management of Drought and Water Scarcity, 2. CRC Press, Taylor & Francis Group, Boca Raton, USA.

Reinstädtler, S., Islam, S.N., Levy, J., Eslamian, S., 2021a. Social, economic and political dimensions of water harvesting Systems in Germany: Possibility, prospects and potential. Chapter 29. In: Eslamian, S., Eslamian, F. (Eds.), Handbook of Water Harvesting and Conservation: Case Studies and Application Examples, second ed. New York Academy of Sciences Series, Wiley-Blackwell, USA, pp. 429–441.

Reinstädtler, S., Islam, S.N., Levy, J., Eslamian, S., 2021b. Water harvesting case studies in Germany—urban and regional investigations. Chapter 30. In: Eslamian, S., Eslamian, F. (Eds.), Handbook of Water Harvesting and Conservation: Case Studies and Application Examples, second ed. New York Academy of Sciences Series, Wiley-Blackwell, USA, pp. 443–453.

Rindfuss, R.R., Walsh, S.J., Turner II, B.L., Fox, J., Mishra, V., 2004. Developing a science of land change: Challenges and methodological issues. PNAS 101 (39), 13976–13981. https://doi.org/10.1073/pnas.0401545101.

Sathish, S., Khadar Babu, S.K., 2020. Forecasting discharge level time series on statistical parametric approach using hydrological data. Int. J. Sci. Res. Multidiscip. Stud. 6 (11), 24–28. P-ISSN: 2454-6143.

Singh, K.K., Vogler, J.B., Shoemaker, D.A., Meentemeyer, R.K., 2012. LiDAR-Landsat data fusion for large-area assessment of urban land cover: Balancing spatial resolution, data volume and mapping Accuracy. ISPRS J. Photogramm. Remote Sens. 74, 110–121. https://doi.org/10.1016/j.isprsjprs.2012.09.009.

TVS—Tourismusverband Spreewald e.V, 2000. Reiseführer Spreewald—Spreewälder stellen ihre Heimat vor. Stadtinfoverlag Berlin, Berlin, Germany. p. 94–95; 218 pp.

UNESCO and WWAP, 2003. Water for People, Water for Life – The United Nations World Water Development Report (WWDR). UNESCO Paris and Berghahn Books, New York.

UNISDR—United Nations International Strategy for Disaster Reduction, 2015. Sendai Framework for Disaster Risk Reduction 2015–2030. Accessed 30.04.2015 under: http://www.wcdrr.org/uploads/Sendai_Framework_for_Disaster_Risk_Reduction_2015-2030.pdf.

von Winterfeldt, D., 2013. Bridging the gap between science and decision making. Proc. Natl. Acad. Sci. U. S. A. 110 (S3), 14055–14061.

Whyte, I.D., 1995. Climatic Change and Human Society. Arnold, London, UK, pp. 141–174.

WMO—World Meteorological Organization, 1994. Guide to Hydrological Practices. Data Acquisition and Processing, Analysis, Forecasting And Other Applications. WMO-No. 168. WMO, Geneva, Switzerland.

WMO—World Meteorological Organization, 2001. Guide to Hydrological Practices: Data Acquisition and Processing, Analysis, Forecasting and Other Applications. WMO-No. 168. WMO, Hydrology and Water Resources Department, Geneva, Switzerland.

WMO—World Meteorological Organization, 2009. Guide to Hydrological Practices. Volume II—Management of Water Ressources and Application of Hydrological Practices. WMO-No. 168, sixth ed. WMO, Geneva, Switzerland.

Yevjevich, V., 1974. Determinism and stochasticity in hydrology. J. Hydrol. 22 (3–4), 225–238.

Yiping, G., Momcilo, M., 2011. Analytical probabilistic approach for estimating design flood peaks of small watersheds. J. Hydrol. Eng., 847–857. https://doi.org/10.1061/(ASCE)HE.1943-5584.0000380.

Chapter 9

Flood routing computations

Henry Foust

Department of Mathematics, University of Houston, Houston, TX, United States

1. Introduction

A hydrograph (see Fig. 1) shows how the volumetric discharge changes with time at some particular location and represents in our discussion either the inflow or the outflow to a river reach (section of river). Please note Fig. 1 is the discharge for the Mississippi River at *St.* Louis, MO, for certain years.

As the flood wave travels through the reach it can experience two effects—translation and storage effects. Two extreme cases are:

- Pure translational effects (see Fig. 2)
- Pure storage effects (see Fig. 3). This chapter has the tasks of determining for a given inflow what is the corresponding outflow. In order to answer this question, a second relationship is needed between the inflow (outflow) and storage, which is the current volume of the water of the system.

Two classes of routing models will be discussed:

- Hydrological Routing Models
- Hydraulic Routing Models

Each is discussed below.

1.1 Hydrological routing models

We shall see that for hydrological routing models, two sets of relationships are established
Conservation of Mass

$$\frac{dS_i}{dt} = I_i - O_i \tag{1}$$

where S_i $[L^3]$[a] is the storage at time i, I_i $\left[\frac{L^3}{T}\right]$ is the inflow at time i, and O_i $\left[\frac{L^3}{T}\right]$ is the outflow at time i.
Storage versus *Discharge*

$$S_i = f(I_i, O_i) \tag{2}$$

1.2 Hydraulic routing models

We shall see that the modeling effort for hydraulic routing models is more sophisticated and involves some simplification of the St. Venant's Equation, which is discussed below and based on conservation of mass and momentum.
This chapter will cover the following topics:

- Reservoir Storage
- Hydrological Routing Models

---

a. L is length and T is time.

Handbook of HydroInformatics. https://doi.org/10.1016/B978-0-12-821962-1.00013-1

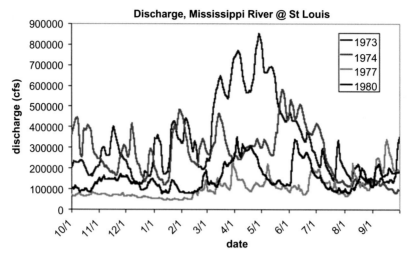

FIG. 1 Hydrograph (https://www.hec.usace.army.mil/software/, n.d.).

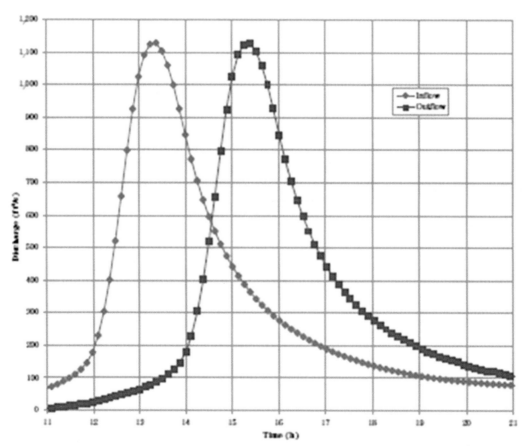

FIG. 2 Flood wave at two points along a channel exhibiting pure translation effects (USDA, 2014).

- Hydraulic Routing Models
- Derivation and simplifications of St. Venant's Equation
- Uniform Flow
- Specific Energy and the Froude Number
- Gradually Varied Flow

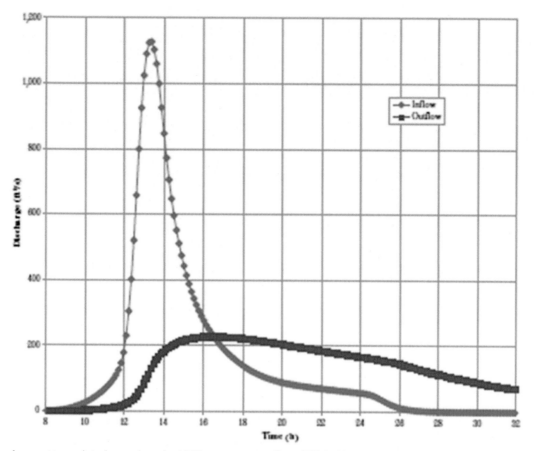

FIG. 3 Flood wave at two points along a channel exhibiting pure storage effects (USDA, 2014).

Each topic is briefly discussed here and further discussed in their respective sections.

A reservoir is an impoundment of water created by a dam or other civil works that restricts the natural flow of a stream. A stream is a flow of water in the natural world that is driven by gravity and adheres to a distinct boundary (channel).

Hydrological models will be shown to be developed around conservation of mass and utilized to determine the relationship between $Q(in)$ and $Q(out)$ for a particular reach of a particular stream. Two models will be developed: Muskingum method and Modified Puls method. These models are based on a conservation of mass equation of the form (Zareeian et al., 2017):

$$\dot{m}_{sys} = \dot{m}_{in} - \dot{m}_{out} \tag{3}$$

where $\dot{m}_{sys}$ is the change in mass with time for the system, $\dot{m}_{in}$ is the change in mass entering with time, and $\dot{m}_{out}$ is the change in mass leaving with time.

Hydraulic routing models will come out of conservation of mass and conservation of momentum, which in its most general form is known as the *St.* Venant's Equation (SVE). The conservation of momentum equation derived will be of the form

$$S_f = S_0 - \frac{\partial y}{\partial x} - \frac{v}{g}\frac{\partial v}{\partial x} - \frac{1}{g}\frac{\partial v}{\partial t} \tag{4}$$

where S_f is the slope of the water surface, S_0 is the slope of the streambed, y is depth of flow, x is distance down the reach, v is the average flow velocity, g is the gravity constant and t is time.

The SVE is a set of partial differential equations (PDEs) with no analytic solution and as such simplifications can be made for particular conditions. These simplifications are listed in Table 1.

The discussion ends with uniform flow and nonuniform flow (gradually, varied flow). Uniform flow is derived from the Darcy-Weisbach equation where it is shown for the range of conditions seen in a natural stream the friction factor is

TABLE 1 Simplifications of Saint Venant's equations (Bedient et al., 2019).

Simplified form	Name	Comments
$S_f = S_0$	Uniform Flow (kinematic wave)	Solve using Manning's Equation
$S_f = S_0 - \frac{\partial y}{\partial x}$	Steady, noninertial flow (diffusion wave)	Various Methods of Solution
$S_f = S_0 - \frac{\partial y}{\partial x} - \frac{v}{g}\frac{\partial v}{\partial x}$	Steady, Nonuniform Flow	Solve using Conservation of Mass and friction considerations
$S_f = S_0 - \frac{\partial y}{\partial x} - \frac{v}{g}\frac{\partial v}{\partial x} - \frac{1}{g}\frac{\partial v}{\partial t}$	Unsteady, Nonuniform Flow (SVE)	Numerical Solution of PDE's

TABLE 2 Reservoir storage.

Time	Input	Output	ds/dt	S
[Days]	[cfs]	[cfs]	[cfs]	[acre-ft]
0	0	0	0	0
0.5	500	250	250	124
1.5	3500	1000	2500	2851
2.5	9000	3000	6000	11,281
3.5	9750	4500	5250	22,438
4.5	8000	5750	2250	29,876
5.5	4500	6000	−1500	30,620
6.5	2250	5250	−3000	26,157
7.5	1250	4250	−3000	20,207
8.5	250	3250	−3000	14,256
9.5	0	2500	−2500	8802
10.5	0	1500	−1500	4835
11.5	0	1000	−1000	2355
12.5	0	750	−750	620
13.5	0	0	0	0

constant in terms of Reynold's number and only dependent on a relative roughness—this analysis results in the Manning Equation.

For gradually, varied flow, the slope of the water surface no longer matches the slope of the streambed. A form of the conservation of energy equation with friction considerations will be developed and an ordinary differential equation will result that is then solved numerically utilizing Euler's method (Table 2).

2. Hydrological routing

2.1 Reservoir routing

In this section, a model for reservoir inflows and outflows will be established based on conservation of mass. This simple model will be the basis to the more elaborate models for hydrological routing in the next two sections.

From conservation of mass,

$$\dot{m}_{sys} = \dot{m}_{in} - \dot{m}_{out} \tag{5}$$

and for most incompressible systems the density throughout the fluid field is constant, such that

$$\left(\frac{dV}{dt}\right)_{sys} = Q_{in} - Q_{out} \tag{6}$$

or in terms of the flood routing nomenclature

$$\left(\frac{dS}{dt}\right)_i \approx I_i - O_i \tag{7}$$

where I_i is inflow at time "i," O_i is inflow at time "i," and S is the storage
   Another form of Eq. (7) is

$$\frac{S_2 - S_1}{\Delta t} \approx \frac{I_1 + I_2}{2} - \frac{O_1 + O_2}{2} \tag{8}$$

Example 1. Reservoir Storage

An example of these concepts if the following: given the following inflows and outflows with time for a given reservoir determine the change in storage with time and the storage with time. Also show that where the inflow equals the outflow ($I=O$) is where the storage is maximized.

   Where $\left(\frac{dS}{dt}\right)_{.5} \approx 500 \ cfs - 250 \ cfs = 250 \ cfs$ and

$S_{.5} = S_0 + \frac{1}{2}[(I_{.5} - O_{.5}) - (I_0 - O_0)]\Delta t \left(\frac{24 \ hours}{1 \ Day}\right) \left(\frac{3600 \ s}{1 \ hr.}\right) \left(\frac{1 \ acre}{43,560 \ f^2}\right)$ and so on.

   We see from Fig. 4 that inflows and outflows intersect at 5 days and we also see from Fig. 5 that the storage is maximized at

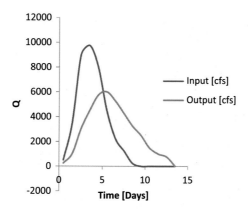

FIG. 4 Q versus time.

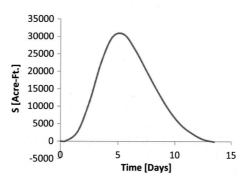

FIG. 5   Storage versus time.

   5 days.

2.2 Muskingum method

The reservoir routing method pursued in the last section assumed a level surface and the routing model in this section will allow for a sloping surface (Bedient et al., 2019; Chaudhry, 2008; Chow, 1959; USDA, 2014).

Given the central difference conservation of mass from the previous section

$$\frac{S_2 - S_1}{\Delta t} = \frac{1}{2}(I_1 + I_2) - \frac{1}{2}(O_1 + O_2) \tag{9}$$

or

$$S_2 - S_1 = \frac{I_1 + I_2}{2}\Delta t - \frac{O_1 + O_2}{2}\Delta t \tag{10}$$

and the following relationship between flows and storage

$$S = K[xI + (1-x)O] \tag{11}$$

where K is known as a storage constant and x is a weighting factor; additionally,

$$K \approx \frac{L}{v_w} \tag{12}$$

where L is the length of the reach and v_w is the average velocity through the reach.and

$$S_i = K[xI_i + (1-x)O_i] \tag{13}$$

Substitution of Eq. (13) into Eq. (10) for S_1 and S_2 results in

$$xI_2 + (1-x)O_2 - xI_1 - (1-x)O_1 = \frac{\Delta t}{2K}(I_1 + I_2) - \frac{\Delta t}{2K}(O_1 + O_2) \tag{14}$$

and solving for O_2 results in

$$(1-x)O_2 + \frac{\Delta t}{2K}O_2 = \frac{\Delta t}{2K}(I_1 + I_2) - \frac{\Delta t}{2K}O_1 + xI_1 + (1-x)O_1 - xI_2 \tag{15}$$

and

$$2(1-x)O_2 + \frac{\Delta t}{K}O_2 = \frac{\Delta t}{K}(I_1 + I_2) - \frac{\Delta t}{K}O_1 + 2xI_1 + 2(1-x)O_1 - 2xI_2 \tag{16}$$

Further,

$$\left[2(1-x) + \frac{\Delta t}{K}\right]O_2 = \left[\frac{\Delta t}{K} + 2x\right]I_1 + \left[\frac{\Delta t}{K} - 2x\right]I_2 + \left[2(1-x) - \frac{\Delta t}{K}\right]O_1 \tag{17}$$

or

$$O_{i+1} = C_1 I_i + C_2 I_{i+1} + C_3 O_i \tag{18}$$

where

$$C_1 = \frac{\left[\frac{\Delta t}{K} + 2X\right]}{C_0} \tag{19}$$

$$C_2 = \frac{\left[\frac{\Delta t}{K} - 2X\right]}{C_0} \tag{20}$$

$$C_3 = \frac{\left[2(1-X) - \frac{\Delta t}{K}\right]}{C_0} \tag{21}$$

and

$$C_0 = \frac{\Delta t}{K} + 2(1-X) \tag{22}$$

where a check is available through the fact that (Table 3)

TABLE 3 Time versus discharge.

Time (h)	Inflow (m³/s)	Outflow (m³/s)
1	93	85
2	137	91
3	208	114
4	320	159
5	442	233
6	546	324
7	630	420
8	678	509
9	691	578
10	675	623
11	634	642
12	571	635
13	477	603
14	390	546
15	329	479
16	247	413
17	184	341
18	134	274
19	108	215
20	90	170

Check 1

$$C_1 + C_2 + C_3 = 1 \tag{23}$$

Solution

Using Eq. (18),

$$O_2 = C_1 I_1 + C_2 I_2 + C_3 O_1$$
$$O_3 = C_1 I_2 + C_2 I_3 + C_3 O_2 \tag{24}$$
$$O_4 = C_1 I_3 + C_2 I_4 + C_3 O_3$$

or in matrix form

$$\begin{bmatrix} O_2 \\ O_3 \\ O_4 \\ \ddot{O}_N \end{bmatrix} = \begin{bmatrix} I_1 & I_2 & O_1 \\ I_2 & I_3 & O_2 \\ I_3 & I_4 & O_3 \\ \cdots & \cdots & \cdots \\ I_{N-1} & i_N & O_{N-1} \end{bmatrix} \begin{bmatrix} C_1 \\ C_2 \\ C_3 \end{bmatrix} \tag{25}$$

where

$$Y = X\beta \tag{26}$$

and using least squares, we can solve for β

$$\beta = \frac{1}{X'X} X'Y \tag{27}$$

From Eqs. (19), (20), (21), and (22) with the substitution of Eq. (31) for $\left\{\frac{\Delta t}{K}, 2X\right\}$ results in

$$C_1 = \frac{A + B}{A + 2 - B} \tag{28}$$

$$C_2 = \frac{A - B}{A + 2 - B} \tag{29}$$

$$C_3 = \frac{2 - A - B}{A + 2 - B} \tag{30}$$

where

$$A = \frac{\Delta t}{K}, B = 2X \tag{31}$$

Further

$$C_1[A + 2 - B] = A + B, 2C_1 = A + B - AC_1 + BC_1 \tag{32}$$

$$C_2[A + 2 - B] = A - B, 2C_2 = A - B - AC_2 + BC_2 \tag{33}$$

and

$$C_3[A + 2 - B] = 2 - A - B, 2C_3 = 2 - A - B - AC_3 + BC_3 \tag{34}$$

or

$$\begin{bmatrix} 1-C_1 & 1+C_1 \\ 1-C_2 & C_2-1 \\ -1-C_3 & -1+C_3 \end{bmatrix} \begin{bmatrix} A \\ B \end{bmatrix} = \begin{bmatrix} 2C_1 \\ 2C_2 \\ 2C_3-2 \end{bmatrix} \tag{35}$$

Once $\{A, B\}$ are known, using Eq. (31) solve for $\{K, X\}$.

There are two additional checks to ensure the method is appropriate

Check 2

$$0 \le X \le .5 \tag{36}$$

and

Check 3

$$2KX \le \Delta t \le K \tag{37}$$

Example 2. Muskingum method (Bedient et al., 2019).

An example of this method is given below.

For the given inflow and outflow records determine $\{C_1, C_2, C_3\}$ and $\{K, X\}$ associated with the Muskingum Method. Determine the matrix X and Y associated with the linear system $Y = X\beta$.

X =	93	137	85	Y =	91
	137	208	91		114
	208	320	114		159
	320	442	159		233
	442	546	233		324
	546	630	324		420
	630	678	420		509
	678	691	509		578
	691	675	578		623
	675	634	623		642
	634	571	642		635

X =	93	137	85	Y =	91
	571	477	635		603
	477	390	603		546
	390	329	546		479
	329	247	479		413
	247	184	413		341
	184	134	341		274
	134	108	274		215
	108	90	215		170

and

X'*X =	3796604	3748842	3456945
	3748842	3796055	3263502
	3456945	3263502	3485448
inv(X'X) =	7.23765E-05	−5E-05	−2.5E-05
	−5.0055E-05	3.6E-05	1.6E-05
	−2.4917E-05	1.6E-05	1E-05
X'*Y =	3592483		
	3464340		
	3461797		
Beta =	0.35	**C(1)**	
	0.06	**C(2)**	
	0.59	**C(3)**	

where $C_1 + C_2 + C_3 = 1$!
Next {A, B} are estimated and {X, K} determined.

X =	0.65	1.35	Y =	0.6923993	X' =	0.65	0.94	−1.59
	0.94	−0.94		0.1235039		1.35	−0.94	−0.41
	−1.59	−0.41		−0.8159478				
					X'*X =	3.8423114	0.6493415	
						0.6493415	2.8590056	
					inv(X'X) =	0.2706483	−0.06147	
						−0.06147	0.3637331	
					X'*Y =	1.8675784		
						1.1491159		

X =	0.65	1.35	Y =	0.6923993	X′ =	0.65	0.94	−1.59
					Beta =	0.4348207	A	
						0.3031714	B	
					X =	0.15		
					K =	2.30		

There are two final checks

$$0 \leq X \leq .5$$
$$2KX \leq \Delta t \leq K$$

where

$$0 < .15 < .5!$$

and

$$.69 < 1 < 2.3!$$

These results are consistent with an example found in (Bedient et al., 2019).

2.3 Modified Puls method

Another hydrological routing model is known as the Modified Puls Method (Bedient et al., 2019; USDA, 2014) and again utilizes the finite difference form of the conservation of mass

$$\frac{S_2 - S_1}{\Delta t} = \frac{1}{2}(I_1 + I_2) - \frac{1}{2}(O_1 + O_2) \tag{38}$$

or

$$(I_n + I_{n+1}) + \left[\frac{2S_n}{\Delta t} - Q_n\right] = \left[\frac{2S_{n+1}}{\Delta t} + Q_{n+1}\right] \tag{39}$$

Additionally, a relationship is needed between $Q(n+1)$ and $\left[\frac{2S_{n+1}}{\Delta t} + Q_{n+1}\right]$.

The method is best explained through an example where the intent is always to provide a relationship between inflows and outflows (Figs. 6–8; Table 4).

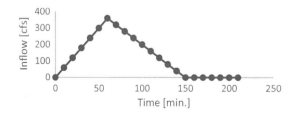

FIG. 6 Inflow versus time.

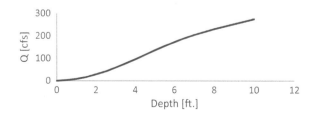

FIG. 7 Q versus depth.

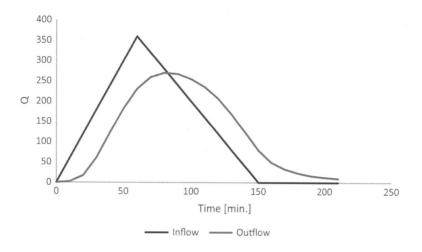

FIG. 8 *Q* versus time.

TABLE 4 Depth versus *Q*, storage.

Depth	Q(n+1)	S(n+1)	{2*S(n+1)/dt + Q(n+1)}
[feet]	[cfs]	[ft³]	[cfs]
0	0	0	0
0.5	3	21,780	75.6
1	8	43,560	153.2
1.5	17	65,340	234.8
2	30	87,120	320.4
2.5	43	108,900	406
3	60	130,680	495.6
3.5	78	152,460	586.2
4	97	174,240	677.8
4.5	117	196,020	770.4
5	137	217,800	863
5.5	156	239,580	954.6
6	173	261,360	1044.2
6.5	190	283,140	1133.8
7	205	304,920	1221.4
7.5	218	326,700	1307
8	231	348,480	1392.6
8.5	242	370,260	1476.2
9	253	392,040	1559.8
9.5	264	413,820	1643.4
10	275	435,600	1727

Example 3. Modified Puls Method.

Given the following Inflow versus Time and Depth versus Outflow where the footprint is 1 acre that does not change cross-section with depth.

From the relationship between Depth and Outflow, the following two quantities are determined:

$$S_{n+1} = 43,560 ft^2 * Depth$$

$$\left\{ \frac{2S_{n+1}}{\Delta t} + Q_{n+1} \right\}$$

Which are provided in the table below

Knowing the left-hand side of the following equation

$$(I_n + I_{n+1}) + \left[\frac{2S_n}{\Delta t} - Q_n \right] = \left[\frac{2S_{n+1}}{\Delta t} + Q_{n+1} \right]$$

Allows one to determine the value of the right-hand side. In example,

$$(I_1 + I_2) + \left[\frac{2S_1}{600} - Q_1 \right] = 60 + 0 - 0 = 60 = \left[\frac{2S_2}{\Delta t} + Q_2 \right]$$

From a linear interpolation, $Q_2 = 2.38$

X(1)	0	Y(1)	0
X(2)	76	Y(2)	3
X	60	Y	2.38

and

$$\left[\frac{2S_n}{\Delta t} - Q_n \right] = 60 - 2(2.38) = 55.24$$

And it proceeds from there. The solution is given below.

And the relationship between the inflow and outflow is

3. Hydraulic routing

3.1 Derivation of St. Venant's equation

As opposed to hydrological routing, where the developed models are based on conservation of mass, the hydraulic models come out of the principles of conservation of mass and momentum. In this section, the Saint Venant Equation (SVE) will be developed and forms the basis to all hydraulic routing models. SVE is a set of partial differential equations with no analytic solutions. And as such, simplifying assumptions are made to get to tractable solutions. Two of these simplifying solutions—uniform flow and steady, nonuniform flow will be discussed further in subsequent sections (Table 5).

From our knowledge of conservation of mass,

$$\dot{m}_{sys} = \dot{m}_{in} - \dot{m}_{out} \tag{40}$$

Eq. (40) takes on the form

$$\frac{\partial A}{\partial t} \Delta x \Delta t = \left[Q - \frac{\partial Q}{\partial x} \frac{\Delta x}{2} \right] \Delta t + q \Delta x \Delta t - \left[Q + \frac{\partial Q}{\partial x} \frac{\Delta x}{2} \right] \Delta t \tag{41}$$

Simplifying

$$\frac{\partial A}{\partial t} \Delta x \Delta t = \left[-2 \frac{\partial Q}{\partial x} \frac{\Delta x}{2} \right] \Delta t + q \Delta x \Delta t \tag{42}$$

and dividing both sides by $\Delta x \Delta t$ results in

TABLE 5 Modified Puls method.

Index	Time [min]	$I(n+1)$ [cfs]	$I(n) + I(n+1)$ [cfs]	$[2S/dt{-}Q](n)$ [cfs]	$[2S/dt + Q](n+1)$ [cfs]	$Q(n+1)$ [cfs]
1	0	0	0	0	0	0
2	10	60	60	0	60	2.38
3	20	120	180	55.24	235.24	17.07
4	30	180	300	201.1	501.1	61.09
5	40	240	420	378.92	798.92	123.16
6	50	300	540	552.6	1092.6	182.18
7	60	360	660	728.24	1388.24	230.34
8	70	320	680	927.56	1607.56	259.28
9	80	280	600	1089	1689	270
10	90	240	520	1149	1669	267.37
11	100	200	440	1134.26	1574.26	254.9
12	110	160	360	1064.46	1424.46	235.19
13	120	120	280	954.08	1234.08	206.93
14	130	80	200	820.22	1020.22	168.45
15	140	40	120	683.32	803.32	124.11
16	150	0	40	555.1	595.1	78.94
17	160	0	0	437.22	437.22	48.92
18	170	0	0	339.38	339.38	32.88
19	180	0	0	273.62	273.62	22.9
20	190	0	0	227.82	227.82	16.23
21	200	0	0	195.36	195.36	12.65
22	210	0	0	170.06	170.06	9.86

$$\frac{\partial A}{\partial t} = -\frac{\partial Q}{\partial x} + q, \frac{\partial A}{\partial t} + \frac{\partial Q}{\partial x} = q \tag{43}$$

From Newton's second law

$$\sum F_i = \frac{d}{dt}(mv) \tag{44}$$

where

$$\frac{d(mv)}{dt} = m\frac{dv}{dt} + v\frac{dm}{dt} \tag{45}$$

and for our fluid system

$$\frac{d(mv)}{dt} = \rho A \Delta x \frac{dv}{dt} + \rho v q \Delta x \tag{46}$$

and the total derivative for $\frac{dv}{dt}$ is

$$\frac{dv}{dt} = \frac{\partial v}{\partial t} + v\frac{dv}{dx} \tag{47}$$

Substitution of Eqs. (46), (47) into Eq. (44) results in

$$\rho A \Delta x \left\{ \frac{\partial v}{\partial t} + v \frac{dv}{dx} \right\} + \rho v q \Delta x = F_H + F_G + F_f \tag{48}$$

where

Hydrostatics forces are represented by

$$F_H = -\gamma \frac{\partial(\bar{y}A)}{\partial x} \Delta x \tag{49}$$

Gravity forces are represented by

$$F_G = \gamma A S_0 \Delta x \tag{50}$$

And frictional forces are represented by

$$F_f = -\gamma A S_f \Delta x \tag{51}$$

Dividing both sides by $\rho \Delta x A$ results in

$$\frac{\partial v}{\partial t} + v \frac{dv}{dx} + v \frac{q}{A} = -\frac{g}{A} \frac{\partial}{\partial x} (\bar{y}A) + g \left(S_o - S_f \right) \tag{52}$$

Assuming no lateral flows and a wide, shallow river results in

$$\frac{1}{g} \frac{\partial v}{\partial t} + \frac{v}{g} \frac{dv}{dx} = -\frac{\partial y}{\partial x} + \left(S_o - S_f \right) \tag{53}$$

or

$$S_f = S_0 - \frac{\partial y}{\partial x} - \frac{v}{g} \frac{\partial v}{\partial x} - \frac{1}{g} \frac{\partial v}{\partial t} \tag{54}$$

where Eqs. (43), (54) are known as the *St. Venant Equations* (SVE) and SVE is a set of partial difference equations with no analytic solution. Several simplifying assumptions can be made to provide tractable problems, which are given below as Table 6.

In the next section, we'll see when some of these simplifying assumptions are valid and in the subsequent sections we'll discuss uniform flow and a form of steady, nonuniform flow known as gradually, varied flow.

Numerical solutions for the full SVE system are given in (Akbari and Firoozi, 2010; Chaudhry, 2008; Chow, 1959; Moussa and Bocquillon, 2000; Keskin, 1998; Ying et al., 2004) and for kinematic, diffusion, and gravity wave (Bedient et al., 2019; Chaudhry, 2008; Chow, 1959; White, 1999).

3.2 Regimes of flow

The question asked in this section is for a given set of conditions on a particular reach of a stream (river) what form of the hydraulic routing equations is appropriate? In order to answer this question, a discussion will be made on appropriate

TABLE 6 Various hydraulic routing models (Bedient et al., 2019).

Simplified form	Name	Comments
$S_f = S_0$	Uniform Flow (kinematic wave)	Solve using Manning's Equation
$S_f = S_0 - \frac{\partial y}{\partial x}$	Steady, noninertial flow (diffusion wave)	Various Methods of Solution
$S_f = S_0 - \frac{\partial y}{\partial x} - \frac{v}{g} \frac{\partial v}{\partial x}$	Steady, Nonuniform Flow	Solve using Conservation of Energy and friction considerations
$S_f = S_0 - \frac{\partial y}{\partial x} - \frac{v}{g} \frac{\partial v}{\partial x} - \frac{1}{g} \frac{\partial v}{\partial t}$	Unsteady, Nonuniform Flow (SVE)	Numerical Solution of PDE's

nondimensional numbers which reveals the dominate forces acting in a particular situation and allow us to glen the appropriate simplifications.

In Moussa and Bocquillon (Moussa and Bocquillon, 1996), three nondimensional numbers were developed for understanding what form of the hydraulic routing model was appropriate to a given set of condition. The nondimensional numbers are

- $\eta = W_2/W_1$
- Froude Number (Fr)
- $T^* = \frac{T v_0 s_{f0}}{y_0}$

where W_2 is the top width to include overbanks, W_1 is the top width of the channel, Froude number is $Fr = \frac{v_0}{\sqrt{g y_0}}$, v_0 is the initial velocity, g is the gravity constant, y_0 is the initial depth of flow, T is the wave period, and s_{f0} is the initial slope of the water surface.

Utilizing a form of sensitivity analysis that compared the results of the full SVE equation with three approximations

- Gravity Wave
- Diffusive Wave
- Kinematics Wave

The following results were determined

And from Fig. 9, the following relationships were established (see Table 6)

In (Moussa and Bocquillon, 2000), the River Loire between Grangent and Feurs in France (see Table 7) was discussed. The following conditions prevailed

For the six flood events measured,

$$.03 \leq Fr_0^2 \leq .12 \tag{55a}$$

$$10 \leq T^* \leq 45 \tag{55b}$$

and

$$1 \leq \eta \leq 8 \tag{55c}$$

And the likely model to utilize (from Fig. 9 and Table 8) would be diffusive wave. A final point to be made is that one really needs to understand their system and carefully determine the correct model based on the best available information about the system; in example, as $\eta \gg 1$, it is likely the flow is no longer one dimensional (Bedient et al., 2019).

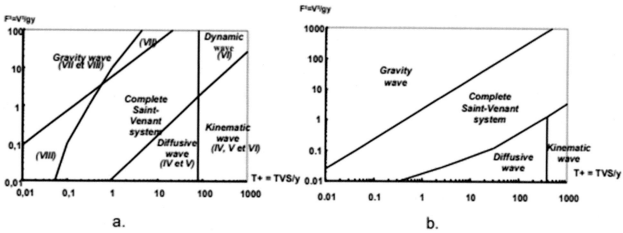

FIG. 9 Regions of flow [(a) $\eta = 1$, (b) $\eta = 8$].

TABLE 7 Hydrology & hydraulics parameters for Loire river.

Inflow Recorder	Grangent (catchment area 4113 km²)
Outflow Recorder	Feurs (catchment area 4978 km²)
Mean Annual Discharge at Feurs [cms]	41.4
River Elevation High [m]	373
River Elevation Low [m]	325
Mean Width of the river [m]	130
Reach Length [km]	43
Mean Width of the flooded river [m]	200 to 1000
A depth of 3.4 m corresponds to a flow of 500 cms	
A depth of 5.8 m corresponds to a flow of 1200 cms	
Mean Velocity [m/s]	1.1 to 1.4
Mean Slope	0.1%

TABLE 8 Relationships between nondimensional numbers for various regions of Fig. 9.

η	Gravity wave	Diffusive wave	Kinematic wave
1	$\frac{Fr_0^2}{T^*} \geq 12.41$	$\frac{Fr_0^2}{T^*} \leq .0432$	$\frac{Fr_0^2}{T^*} \leq .0432$ and $T^* \geq 39$
8	$\frac{Fr_0^2}{T^*} \geq 2.85$	$\frac{Fr_0^2}{T^*} \leq .0101$	$\frac{Fr_0^2}{T^*} \leq .0101$ and $T^* \geq 154$
20	$\frac{Fr_0^2}{T^*} \geq 2.50$	$\frac{Fr_0^2}{T^*} \leq .0037$	$\frac{Fr_0^2}{T^*} \leq .0037$ and $T^* \geq 383$

4. Uniform flow

4.1 Manning's equation

From a study of viscous flow (Bedient et al., 2019; Chaudhry, 2008; Chow, 1959; White, 1999), we know that the head-loss in a system is equal to

$$h_L = f \frac{L}{D} \frac{v^2}{2g} \tag{56}$$

where f is the friction factor, L is length, D is the inside diameter, v is velocity and g is the gravity constant.

Eq. (56) is the celebrated Darcy-Weisbach equation.

And for open channel flow instead of discussing diameter it is more convenient to discuss diameter in terms of hydraulic radius, which is defined as

$$R_h = \frac{A}{P} \tag{57}$$

and diameter of a pipe in terms of R_h is given as

$$R_h = \frac{A}{P} = \frac{\pi/4 D^2}{\pi d} = \frac{D}{4} \tag{58}$$

where A is the area and P is the wetted perimeter

Additionally, the slope is defined as

$$S = \frac{h_L}{L} \tag{59}$$

Substitution of Eqs. (58), (59) into Eq. (56) for "D" and h_L/L results in

$$S = \frac{f}{8gR_h} v^2 \tag{60}$$

and solving for v

$$v = \sqrt{\frac{8g}{f}} \sqrt{SR_h} \tag{61}$$

Given that the velocity is of the order of 3 m/s, kinematic viscosity for water at 20°C is about 1e-6 m²/s, and the characteristic length of this system is the hydraulic radius, the Reynold's numbers for this system are well above 1e6 and the relative roughness is fairly large $\left(\frac{\epsilon}{R_h} > .001 \right)$ results in fully turbulent flow (see Fig. 10, Moody's Diagram) such that

$$f = g\left(\frac{\epsilon}{D}\right) \tag{62}$$

The Moody Diagram can be represented by the Colebrook-White equation

$$\frac{1}{\sqrt{f}} = -2 \, log \left[\frac{\epsilon}{3.7D} + \frac{2.51}{Re \sqrt{f}} \right] \tag{63}$$

and for fully turbulent flow (where f is independent of Re) takes on the simpler form

$$\frac{1}{\sqrt{f}} = -2 \, log \left[\frac{\epsilon}{3.7D} \right] \tag{64}$$

or

$$\frac{1}{f} = 4\left[log \left(\frac{\epsilon}{3.7D} \right) \right]^2 = 4\left[log \left(\frac{\epsilon}{14.8R_h} \right) \right]^2 \tag{65}$$

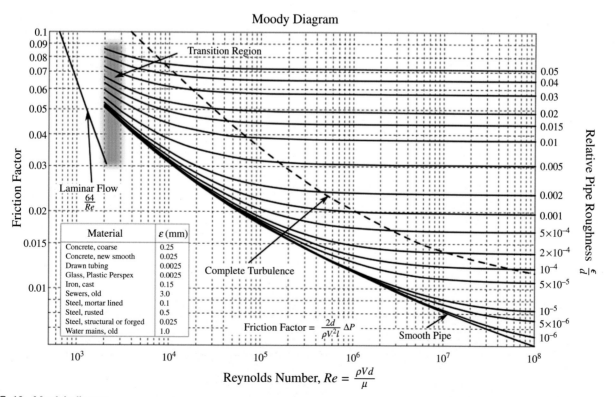

FIG. 10 Moody's diagram.

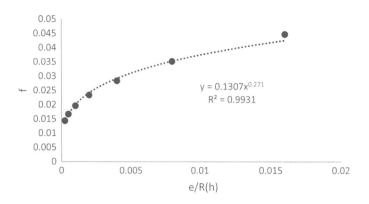

FIG. 11 f versus $e/R(h)$.

And a relationship between $\dfrac{\epsilon}{R_h}$ versus f can be established (see Fig. 11).where

$$f \approx \beta \left(\frac{\epsilon}{R_h} \right)^{.271} \tag{66}$$

Substitution of Eq. (66) into the definition of Chezy's coefficient

$$C = \sqrt{\frac{8g}{f}} \tag{67}$$

Results in

$$C \approx \frac{\sqrt{8g}}{\sqrt{\beta \left(\dfrac{\epsilon}{R_h} \right)^{.271}}} = \frac{\sqrt{8g}}{\sqrt{\beta \epsilon^{.271}}} R_h^{.271/2} \approx \frac{1}{n} R_h^{\frac{1}{6}} \tag{68}$$

It can be shown that $\dfrac{\sqrt{8g}}{\sqrt{\beta \epsilon^{.271}}}$ is roughly equal to $\frac{1}{n}$ for a range of $\epsilon \equiv (.0001, .5)$ where 0.0001 m ($n = .012$) would be related to an artificially lined canal and 0.5 m ($n = .035$) would be related to a major river (Ying et al., 2004).

Eq. (68) has been experimentally verified by Manning (Bedient et al., 2019; Chaudhry, 2008; Chow, 1959; White, 1999) and subsequent researchers. Manning's roughness coefficient, n, has been determined for various surfaces and given in Table 9.

Substitution of Eq. (68) into Eq. (61) results in

$$v = \alpha \frac{1}{n} \sqrt{S} R_h^{2/3} \tag{69}$$

where α is a coefficient with the following values

When $v[m/s] \rightarrow \alpha = 1$ and when $v[fps] \rightarrow \alpha = 1.49$

Eq. (69) is known as the Manning equation where more details about surface roughness associated with open channel flow can be found in (Yen, 2002).

For shallow, wide rivers, Manning's equation takes a simpler form. This can be shown by looking at the hydraulic radius of a wide, narrow rectangle

$$R_h = \frac{A}{P} = \frac{wy}{w + 2y} = \frac{y}{1 + \frac{2y}{w}} \tag{70}$$

and as w increases for a fixed y, then

$$R_h \approx y \tag{71}$$

where $w > 20y$ to ensure accuracy of results and the error associated with using Eq. (69) for sundry w is aptly shown in Fig. 12.

TABLE 9 Manning coefficients for various surfaces (White, 1999).

Material	Typical manning roughness coefficient
Concrete	0.012
Gravel bottom with sides—concrete	0.020
mortared stone	0.023
riprap	0.033
Natural stream channels	
Clean, straight stream	0.030
Clean, winding stream	0.040
Winding with weeds and pools	0.050
With heavy brush and timber	0.100
Flood Plains	
Pasture	0.035
Field crops	0.040
Light brush and weeds	0.050
Dense brush	0.070
Dense trees	0.100

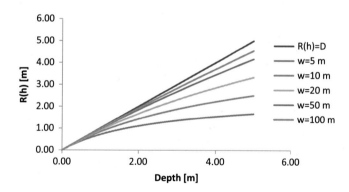

FIG. 12 $R(h)$ versus depth.

And Manning's equation for a wide, shallow river becomes

$$v = \alpha \frac{1}{n} \sqrt{S} y^{\frac{2}{3}}$$ (72)

An example of uniform flow is given next (Fig. 13).

Example 4. Uniform Flow.

Given the following cross sections

Where the main channel is 2.5 m wide by 1 m deep and the flood channel additionally includes 7.5 m wide by 2.5 m deep. The n for the main channel is 0.015 and for the flood channel it is 0.030. If the slope is 1/1000 and the Q is 3.0 cms. What's the depth of flow? As can be seen in the table below, the depth is between 1.50 and 1.75 m.

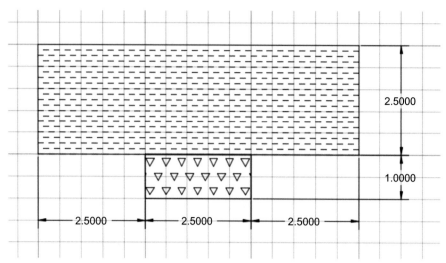

FIG. 13 Open channel flow.

4.2 Uniform flow, geometries

Three types of geometries for uniform, open channel flow will be considered in this section:

- Trapezoidal
- Rectangular
- Circular

Where a rectangular is a trapezoid with $\theta = 90$ (see Fig. 14)

The area for this cross section (White, 1999) is

$$A = by + \alpha y^2 \; where \; \alpha = y^2 \tag{73}$$

and the wetted perimeter is

$$P = b + 2W = b + 2y\sqrt{1 + \alpha^2} \tag{74}$$

and P in terms of $\{\alpha, y\}$ is

$$P = \frac{A}{y} - \alpha y + 2y\sqrt{1 + \alpha^2} \tag{75}$$

In order to maximize the flow, we need to maximize R_h, which can be done by keeping A constant and minimizing P, which can be done by

$$\frac{dP}{dy} = 0, \; \frac{-A}{y^2} - \alpha + 2\sqrt{1 + \alpha^2} = 0 \tag{76}$$

and

$$A = y^2 \left[2\sqrt{1 + \alpha^2} - \alpha \right] \tag{77}$$

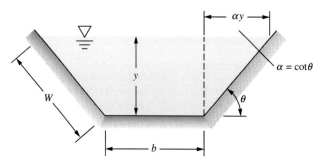

FIG. 14 Trapezoidal cross-section (Ying et al., 2004).

$$P = 4y\sqrt{1 + \alpha^2} - 2\alpha y \tag{78}$$

and

$$R_h = \frac{A}{P} = \frac{1}{2}y \tag{79}$$

What Eq. (79) states is that for any angle θ the "best" cross section is one where the hydraulic radius is half the depth of flow. It can be shown the best rectangle is one where

$$A = 2y^2, P = 4y, R_h = \frac{1}{2}y, b = 2y \tag{80}$$

and the best trapezoid is one where $\frac{dP}{d\alpha} = 0$ and

$$2\alpha = \sqrt{1 + \alpha^2}, \alpha = \cot(\theta) = \frac{1}{\sqrt{3}} \tag{81}$$

and

$$\theta = 60 \tag{82}$$

For pipe flow, the geometry gets a little dicey (see Fig. 15).

The wetted perimeter is simply the perimeter for a sector where the formula is

$$P = \frac{\theta}{2}D \tag{83}$$

and the area of the sector is

$$A_1 = \frac{\theta}{8}D^2 \tag{84}$$

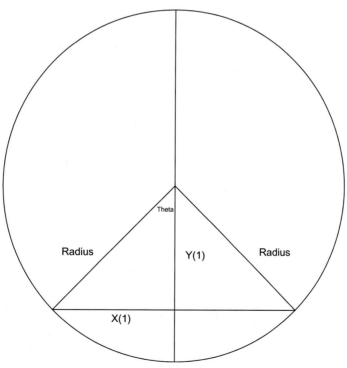

FIG. 15 Partially full pipe flow.

and the lengths of the two sides of the small triangle ABC are

$$y_1 = \frac{D}{2} \cos\left(\frac{\theta}{2}\right) \ and \ \frac{x}{2} = \frac{D}{2} \sin\left(\frac{\theta}{2}\right) \tag{85}$$

where the area of triangle ABC is

$$A_2 = \frac{1}{2}\frac{x}{2}y_1 = \frac{1}{2}\left\{\frac{D}{2} \sin\left(\frac{\theta}{2}\right)\right\}\left\{\frac{D}{2} \cos\left(\frac{\theta}{2}\right)\right\} = \frac{D^2}{8}\sin(\theta) \tag{86}$$

and the area of flow is

$$A = A_1 - A_2 = \frac{D^2}{8}[\theta - \sin(\theta)] \tag{87}$$

It can be shown how the volumetric flow rate varies with depth of flow or θ. An example of these ideas is given below.

Example 5. Partially Filled Pipe Flow.
Given a slope of 1/1000, an n of 0.03, and the diameter is 2 ft, determine how the volumetric discharge varies with theta and depth of flow. The solution is given in the table below and Figs. 16 and 17 where it is seen that the maximum discharge is not at 100% full but at 93% full as defined by y/d.

S(0)		0.001		Theta	Q	y/D	Q
n		0.03		30	0.0015	0.017	0.0015
Diameter	[feet]	2		60	0.028	0.067	0.028
y	[feet]	1.97		90	0.144	0.146	0.144
Theta	[Degrees]	330		120	0.426	0.25	0.426
Theta	[Radians]	5.76		150	0.91	0.371	0.91
A	[ft²]	3.13		180	1.55	0.5	1.55
P	[ft]	5.76		210	2.24	0.629	2.24
Rh	[ft]	0.54		240	2.83	0.75	2.83
Q	[cfs]	3.27		270	3.31	0.854	3.31
y/D		0.98		300	3.34	0.933	3.34
				330	3.27	0.983	3.27
				360	3.11	1	3.11

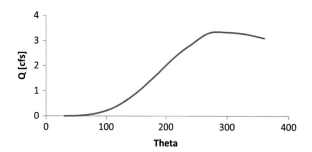

FIG. 16 Q versus theta.

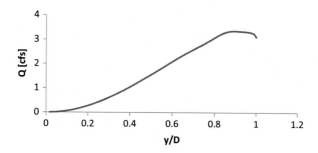

FIG. 17 Q versus y/D.

5. Specific energy

5.1 Rectangular cross-section

The specific energy for a rectangle cross section is

$$y_1 + \frac{u_1^2}{2g} = y_2 + \frac{u_2^2}{2g} \tag{88}$$

Let $q = Q/b = uy$ where b is the width and substituting this relationship into Eq. (88) results in

$$E = y + \frac{q^2}{2gy^2} \tag{89}$$

If we graph y versus E for $q = .5\frac{m^2}{s}$, then it is obvious there is a minimum in terms of specific energy at $E_c = .44\ m$ and the depth is $y_c = .29\ m$.

The minima can be determined by setting $\frac{dE}{dy}$ equal to zero and the result is

$$1 = \frac{q^2}{g}\frac{1}{y^3} \tag{90}$$

and solving for y results in

$$y_c = \sqrt[3]{\frac{q^2}{g}} \tag{91}$$

where y_c is known as the critical depth and represents an unstable open channel flow condition

And substituting Eq. (91) into Eq. (89) results in

$$E_{min} = \frac{3}{2}y_c \tag{92}$$

Associated with open channel flow is a nondimensional number known as the Froude number, which is defined as

$$Fr = \frac{u}{\sqrt{gy}} \tag{93}$$

It has been shown (Bedient et al., 2019; Chaudhry, 2008; Chow, 1959; White, 1999) that for

$$y < y_c \rightarrow Fr > 1, Super - Critical\ Flow \tag{94}$$

and

$$y > y_c \rightarrow Fr < 1, Sub - Critical\ Flow \tag{95}$$

and at the critical depth the Froude number is one (Fig. 18).

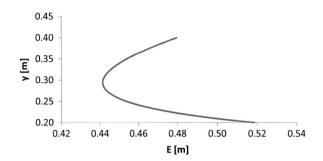

FIG. 18 y versus E.

5.2 Nonrectangular cross-section

For a nonrectangular cross-section, the specific energy is defined as

$$E = y + \frac{Q^2}{2gA^2} \tag{96}$$

and for $\frac{dE}{dy} = 0$,

$$\frac{dA}{dy} = \frac{gA^3}{Q^2} \tag{97}$$

where $dA = bdy$ and b is the width at the water surface and the critical area and velocity are

$$A_c = \sqrt[3]{\frac{bQ^2}{g}} \tag{98}$$

and

$$v_c = \sqrt{\frac{gA_c}{b}} \tag{99}$$

6. Gradually varied flow

There are situations where the slope of the surface (S_f) no longer matches the slope of the bedform (S_0) and when the differences is slight and the flow is steady, this is termed gradually, varied flow. An example of this is given as Fig. 19 where

$$S_f \neq S_0 \tag{100}$$

From Fig. 19, the energy balance from one end of the section to the other is

$$\frac{v^2}{2g} + y + S_o dx = S_f dx + \frac{v^2}{2g} + d\left(\frac{v^2}{2g}\right) + y + dy \tag{101}$$

or

$$\frac{dy}{dx} + \frac{d}{dx}\left(\frac{v^2}{2g}\right) = S_0 - S_f \tag{102}$$

Eq. (102) simply states that the change in surface elevation with distance plus the change in kinetic energy is equal to the difference between the bed slope and the water surface slope (Figs. 20 and 21; Table 10).

The change in kinetic energy can be written as

$$\frac{1}{2g}\frac{d}{dx}(v^2) = \frac{1}{g}v\frac{dv}{dx} \tag{103}$$

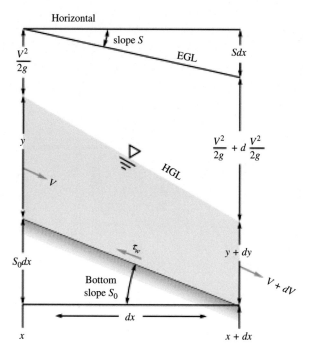

FIG. 19 Gradually, varied flow (Ying et al., 2004).

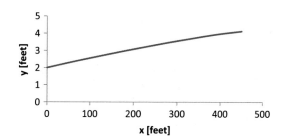

FIG. 20 Depth of flow versus distance.

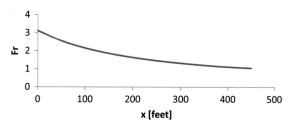

FIG. 21 Froude number versus distance.

From the conservation of mass,

$$\frac{dQ}{dx} = 0 = \frac{d}{dx}(vA) = v\frac{dA}{dx} + A\frac{dv}{dx} \tag{104}$$

or

$$\frac{dv}{dx} = -\frac{v}{A}\frac{dA}{dx} \tag{105}$$

TABLE 10 Solution for Example 4.

S	0.001	
n(1)	0.03	
n(2)	0.015	
n(3)	0.03	
b(1)	2.5	[m]
b(2)	2.5	[m]
b(3)	2.5	[m]

Depth [m]	A(1) [m²]	P(1) [m]	Rh(1) [m]	Q(1) [cms]	A(2) [m²]	P(2) [m]	Rh(2) [m]	Q(2) [cms]	A(3) [m²]	P(3) [m]	Rh(3) [m]	Q(3) [cms]	Q(total) [cms]
0.00				0.00	0.00	2.50	0.00	0.00				0.00	0.00
0.25				0.00	0.63	3.00	0.21	0.74				0.00	0.74
0.50				0.00	1.25	3.50	0.36	1.06				0.00	1.06
0.75				0.00	1.88	4.00	0.47	1.27				0.00	1.27
1.00				0.00	2.50	4.50	0.56	1.42				0.00	1.42
1.25	0.63	2.75	0.23	0.39	3.13	5.00	0.63	1.54	0.63	2.75	0.23	0.39	2.33
1.50	1.25	3.00	0.42	0.59	3.75	5.50	0.68	1.63	1.25	3.00	0.42	0.59	2.81
1.75	1.88	3.25	0.58	0.73	4.38	6.00	0.73	1.71	1.88	3.25	0.58	0.73	3.17
2.00	2.50	3.50	0.71	0.84	5.00	6.50	0.77	1.77	2.50	3.50	0.71	0.84	3.45
2.25	3.13	3.75	0.83	0.93	5.63	7.00	0.80	1.82	3.13	3.75	0.83	0.93	3.69
2.50	3.75	4.00	0.94	1.01	6.25	7.50	0.83	1.87	3.75	4.00	0.94	1.01	3.89
2.75	4.38	4.25	1.03	1.07	6.88	8.00	0.86	1.91	4.38	4.25	1.03	1.07	4.05
3.00	5.00	4.50	1.11	1.13	7.50	8.50	0.88	1.94	5.00	4.50	1.11	1.13	4.20
3.25	5.63	4.75	1.18	1.18	8.13	9.00	0.90	1.97	5.63	4.75	1.18	1.18	4.33
3.50	6.25	5.00	1.25	1.22	8.75	9.50	0.92	2.00	6.25	5.00	1.25	1.22	4.44

Substitution of Eq. (105) into Eq. (102) results in

$$\frac{1}{2g}\frac{d}{dx}(v^2) = \frac{1}{g}v\left[-\frac{v}{A}\frac{dA}{dx}\right]$$
(106)

and substitution of Eq. (106) into Eq. (102) results in

$$\frac{dy}{dx} - \frac{v^2}{gA}\frac{dA}{dx} = S_0 - S_f$$
(107)

where $dA = bdy$ such that

$$\frac{dy}{dx} - \frac{v^2}{gA}\frac{bdy}{dx} = S_0 - S_f$$
(108)

Further,

$$\frac{dy}{dx}\left[1 - b\frac{v^2}{gA}\right] = S_0 - S_f$$
(109)

and it can be shown for a nonrectangular cross-section that

$$Fr^2 = \frac{bv^2}{gA}$$
(110)

Substitution of Eq. (110) into Eq. (109) results in

$$\frac{dy}{dx} = \frac{S_0 - S_f}{1 - Fr^2}$$
(111)

And S_f can be determined from the knowledge that when the two slopes are equal the velocity of flow is equal to the normal flow velocity with the same conditions and

$$S_f = S_{0n} = \frac{n^2v^2}{\alpha^2 R_h^{4/3}}$$
(112)

Example 6. Gradually, Varied Flow.
Determine the profile of flow for a gradually, varied flow system where y_0 is 2 ft, q is 50 ft^2/s, $n = 0.022$, and $S(0)$ is 0.0048 by graphing x versus y until y is over 4.0 ft.

n		0.022	x	y	S(f)	Fr^2	dy/dx	v	Fr
S(0)		0.0048	[ft]	[ft]				[fps]	
q	[ft^2/s]	50	0	2.00	0.05	9.70	0.01	25.00	3.12
y(0)	[ft]	2	50	2.28	0.03	6.51	0.01	21.88	2.55
Y(critical)	[ft]	4.27	100	2.56	0.02	4.64	0.01	19.55	2.15
E(min)	[ft]	6.40	150	2.82	0.02	3.46	0.01	17.73	1.86
alpha		1.486	200	3.07	0.01	2.67	0.00	16.26	1.63
y(n)	[ft]	4.14	250	3.32	0.01	2.12	0.00	15.07	1.46
			300	3.55	0.01	1.73	0.00	14.08	1.32
			350	3.77	0.01	1.45	0.00	13.26	1.20
			400	3.97	0.01	1.24	0.00	12.60	1.11
			450	4.12	0.00	1.11	0.00	12.13	1.05

In the above example, a particular profile was developed for depth of flow versus distance. There are several surface profiles associated with gradually varied flow (see Fig. 22) and these various profiles can be categorized based on slope

- Mild
- Steep
- Neutral (Horizontal)
- Critical
- Adverse

Within each category above, there can be up to three possible curves based on the relationship between y (actual depth), y_N (normal depth), and y_c (critical depth), which is shown in Table 11; note that y_N is the depth of flow for a uniform flow. One challenge with solving for these profiles is determining whether the upstream condition or downstream conditions controls; more details can be found in (Bedient et al., 2019; Chaudhry, 2008; Chow, 1959; White, 1999).

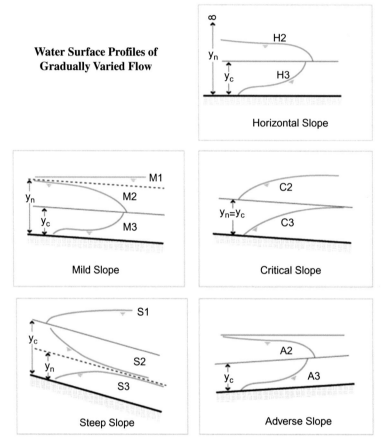

FIG. 22 Various flow profiles (White, 1999).

TABLE 11 Various flow profiles (White, 1999).

Slope class	Slope notation	Depth class	Solution curves
$S_0 > S_c$	Steep	$y_c > y_n$	S-1, S-2, S-3
$S_0 = S_c$	Critical	$y_c = y_n$	C-2, C-3
$S_0 < S_c$	Mild	$y_c < y_n$	M-1, M-2, M-3
$S_0 = 0$	Horizontal	$y_n = \infty$	H-2, H-3
$S_0 < 0$	Adverse	$y_n = $ imaginary	A-2, A-3

7.   Conclusions

This chapter has discussed both hydrological and hydraulic flood routing models. The hydrological models are developed from conservation of mass and a second relationship to relate inflow (outflow) to storage; the Muskingum and Modified Puls method were discussed. The hydraulic models were developed from a simplification of the *St.* Venant's Equation and two simplifications were amply explored. The first is the uniform flow equation for open channel flow, which is modeled using Manning's equation and the second is gradually, varied flow.

In the H&H (Hydraulics and Hydrology) community, two very prominent computer software packages are known as HEC-1 and HEC-2 and available through the US Army Corps of Engineers (https://www.hec.usace.army.mil/software/, n. d.). The hydrological and hydraulic routing models developed in this chapter make up the heart of HEC-1 and HEC-2 where HEC-1 provides for a given inflow and storage the outflow and HEC-2 for a given inflow, cross-sectional information, and friction factors provides both the outflow and depths of flow.

References

Akbari, G., Firoozi, B., 2010. Implicit and explicit numerical solution of saint Venant equations for simulating flood wave in natural Rivers. In: 5th National Congress on Civil Engineering, May 4 to 6. Ferdowsi University of Mashhad, Mashhad, Iran.

Bedient, P.B., Huber W.C. and Vieux B.E. (2019). Hydrology and Floodplain Analysis; sixth ed. Pearson, NY, USABedient, P.B., Huber, W.C., Vieux, B. E., 2019. Hydrology and Floodplain Analysis, sixth ed. Pearson, NY, USA.

Chaudhry, M.H. (2008). Open Channel Flow; Springer, second ed.Chaudhry, M.H., 2008. Open Channel Flow, second ed. Springer.

Chow, V.T., 1959. Open Channel Hydraulics. McGraw-Hill, New York, NY, USA.

https://www.hec.usace.army.mil/software/. (Accessed 28 July 2020).

Keskin, M.E., 1998. A simplified dynamic model to calculate floods. Trans. Ecol. Environ. 24, 255–263.

Moussa, R., Bocquillon, C., 1996. Criteria for the choice of flood-routing methods in natural channels. J. Hydrol. 186 (1–4), 1–30.

Moussa, R., Bocquillon, C., 2000. Approximate zones of the Saint Venant equations for flood routing with overbank flow. Hydrol. Earth Syst. Sci. 4 (2), 251–260.

USDA, 2014. Flood Routing; Part 630 Hydrology National Engineering Handbook. United States Department of Agriculture, USA.

White, F. (1999). Fluid Mechanics; McGraw-Hill, Boston, MA, USA, forth ed.White, F., 1999. Fluid Mechanics, forth ed. McGraw-Hill, Boston, MA, USA.

Yen, B.C., 2002. Open channel flow resistance. J. Hydraul. Eng. 128 (1), 20–39.

Ying, X., Khan, A., Wang, S.S.Y., 2004. Upwind conservative scheme for the Saint Venant equation. J. Hydraul. Eng. 130 (10), 977–987.

Zareeian, M.J., Eslamian, S., Gohari, A., Adamowski, J., 2017. The effect of climate change on watershed water balance. In: Furze, J.N., Swing, K., Gupta, A.K., McClatchey, R., Reynolds, D. (Eds.), Mathematical Advances Towards Sustainable Environmental Systems. Springer International Publishing, Switzerland, pp. 215–238.

Further reading

https://www.ldeo.columbia.edu/martins/climate_water/lectures/floods.htm. (Accessed 12 September 2020).

Chapter 10

Application of fuzzy logic in water resources engineering

Gokmen Tayfur

Department of Civil Engineering, Izmir Institute of Technology, Izmir, Turkey

1. Introduction

Fuzzy logic (FL) is introduced by Lotfi Ali Askerzadeh in the early 1960s (Zadeh, 1965) who had served as a scientist at UC-Berkely at the Department of Electronics. The philosophy of the FL is not immediately welcomed by the westerners, rather it has found wide application in the Eastern world, with first applications in Singapore, Korea, Malaysia, and Japan for the control of vapor engine, cement factory, and signalization (Mamdani and Assilian, 1975; Holmblad and Ostergaard, 1982). Presently, all the electronics devices such as the washing machines, vacuum cleaners, elevators, refrigerators, etc. are all controlled by the FL.

The reason that the FL has found wider and earlier applications in the Eastern world than the Western world is because the FL thinking is more embedded in the Eastern philosophy. The Western philosophy is embroiled in the Aristotelian Logic (AL) which is sharp in dividing classes; such as *"either one is a member of a set or not." If it is a member, then it is fully member.* That may be the reason that racism has been seen historically widely in the Western world, such as the slavery, segregation, and anti-Semitism. On the other hand; the Eastern thinking is more receptive of differences, and this is may be because of the harmonization of many cultures in the same pot. The famous thinker and philosopher, renowned Mavlana Celaleddeen Rumi, in his one poet briefly, says the following;

> *Come, whoever you are; wanderer, worshipper*
> *Come, even if you have broken your vow a hundred times*
> *Ours is not a Dergah (House) of despair*

The FL, as it is expressed in this poem, welcomes elements and does not perform sharp divisions, rather constitutes gray areas. Fig. 1 shows fuzzy sets for a rainfall intensity. The vertical axis stands for the degree of membership. Each set is represented using the triangle functions. For example, the *low* set includes values from 0 to 20 mm/h. 5 mm/h is the member of *very low* and *low* subsets with 50% memberships. Similarly, 15 mm/h is the member of *low* and *medium* subsets. That means; 15 mm/h can be member of both sets at the same time. This is not possible in the case of AL.

In the FL, the **fuzziness** can be thought as an expression of an **uncertainty**, which is the common case in the water resources engineering problems. For example, when we say *weather is hot*, what *hot* implies is a kind of uncertain since it can change from person to person and from location to location. For example, 20°C would be hot for a person in the North Pole but 40°C would mean the same thing for another person in the Ecuador. Since it assigns partial memberships, *hot* can be 40°C for a person in Arabia but it can be 30°C for a person in Turkey. There is no conflict between these two according to the FL. Also, the FL operates verbally and it is thus a more in line with humans. Because we think verbally not numerically. We often say, "today is hot," we do not say "today temperature is 35°C." Therefore, it can allow us to express any terminology using a fuzzy set (Eslamian and Amiri, 2011). For example; Fig. 2 presents fuzzy set for a word *young*.

First applications of the FL in the areas of water resources engineering are seen in the late 1990s for the solute transport, river basin planning, and flood control (Dou et al., 1997; Raj and Kumar, 1998; Chuntian, 1999). The application had become wider in the 2000s in the flood prediction and forecasting, rainfall-runoff, infiltration, groundwater recharge, reservoir operation, and sediment transport (See and Openshaw, 2000; Hundecha et al., 2001; Coppala et al., 2002;

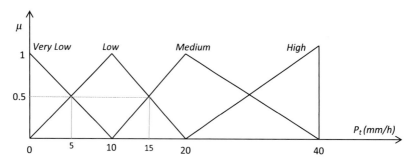

FIG. 1 Fuzzy subsets for rainfall intensity.

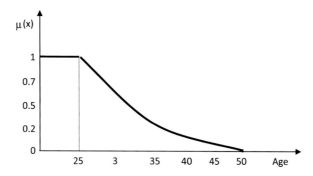

FIG. 2 Fuzzy set for *young*.

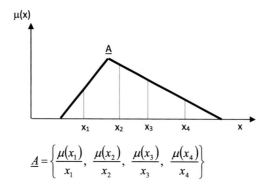

FIG. 3 Representation of a fuzzy set.

Sen and Altunkaynak, 2003; Tayfur et al., 2003; Chang et al., 2005). The applications have sparsely continued with the prediction of runoff and discharge (Tayfur and Singh, 2006, 2011; Tayfur and Brocca, 2015), urban water resources (Zhao and Chen, 2008), water quality management (Singh et al., 2019), groundwater storage and quality predictions (Masoumi et al., 2019; Selvaraj et al., 2020) and drought prediction (Malik et al., 2020; Kumar et al., 2021; Nourani et al., 2021), among many.

2. Fundamentals of fuzzy sets

2.1 Fuzzy set representation

Fuzzy sets can be represented by using many types of functions, such as the triangular, trapezoidal, Gaussian, etc. Fig. 3 shows a typical fuzzy set representation by A having triangular membership with four elements. In the set building notation, in each term, there is no division rather it is just a representation. For example $\mu(x_1)/x_1$ means; x_1 *is an element of the fuzzy set A with $\mu(\omega_1)$ degree of membership*.

2.2 Fuzzy set operations

2.2.1 Union and intersection of sets

The union of fuzzy sets is found by picking the maximum (***max***) membership degree (the bold line) of an element (see Fig. 4). The symbol for the union representation between the sets is V. By the union operation, an element can be a member of both sets. Verbally it is stated that "an element is a member of set A or set B."

The intersection of sets is shown in Fig. 5 where the intersection symbol is the upside-down V. In this case, it can be stated that "an element is a member of A **and** B" and therefore the bold line shows the intersection region.

In a set builder notation, the union of fuzzy sets can be expressed as follows:

$$\underline{A}V\underline{B} = \left\{ \frac{\max\left[m_{\underline{A}}(x),\ m_{\underline{B}}(x)\right]}{x} \right\}$$

while the intersection of the sets as;

$$\underline{A}\wedge\underline{B} = \left\{ \frac{\min\left[m_{\underline{A}}(x),\ m_{\underline{B}}(x)\right]}{x} \right\}$$

For example; given the following two fuzzy sets;

$$\underline{A} = \left\{ \frac{0.1}{a},\ \frac{0.3}{b},\ \frac{0.9}{c},\ \frac{1}{d}\ \frac{0.6}{e},\ \frac{0.2}{f} \right\}$$

$$\underline{B} = \left\{ \frac{1}{1},\ \frac{0.8}{a},\ \frac{0.6}{3},\ \frac{0.4}{b}\ \frac{0.2}{x},\ \frac{0.1}{c} \right\}$$

The union and intersections of these sets can be obtained as follows, respectively;

$$\underline{A}V\underline{B} = \left\{ \frac{0.8}{a},\ \frac{0.4}{b},\ \frac{0.9}{c},\ \frac{1}{d}\ \frac{0.6}{e},\ \frac{0.2}{f}\ \frac{1}{1},\ \frac{0.6}{3}\ \frac{0.2}{x} \right\}$$

$$\underline{A}\wedge\underline{B} = \left\{ \frac{0.1}{a},\ \frac{0.3}{b},\ \frac{0.1}{c} \right\}$$

2.2.2 Complementary sets

In a fuzzy set, the complementary set is the one that complements each element's degree to the degree one (see Fig. 6).

Using the set builder notation; $\overline{\underline{A}} = \left\{ \frac{1-(m_{\underline{A}}(x))}{x} \right\}$. The complementary set of fuzzy set $\underline{A} = \left\{ \frac{0.1}{a},\ \frac{0.3}{b},\ \frac{0.9}{c},\ \frac{0.8}{d} \right\}$ can be represented as $\overline{\underline{A}} = \left\{ \frac{0.9}{a},\ \frac{0.7}{b},\ \frac{0.1}{c},\ \frac{0.2}{d} \right\}$.

2.2.3 Unique operations peculiar to fuzzy sets

All the other operations known in the classical sets are also valid for the fuzzy sets. For details, one can see Tayfur (2012). There are, however, some unique set operations in the fuzzy sets that cannot be seen in the classical sets. These are called; the *concentration, the dilation*, the *normalization,* and the *intensification.*

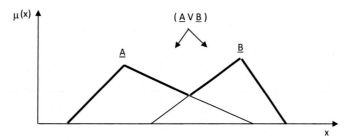

FIG. 4 Graphic representation of union of two sets.

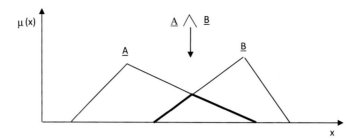

FIG. 5 Schematic representation of intersection of two fuzzy sets.

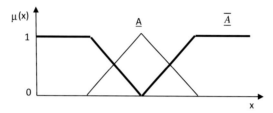

FIG. 6 Representation of a complementary fuzzy set.

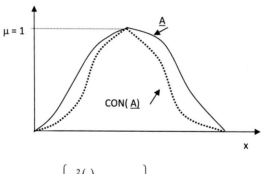

$$CON\left(\underline{A}\right) = \left\{\frac{m_{\underline{A}}^2(x)}{x}, \quad x \in U\right\}$$

FIG. 7 Representation of concentration operation in a fuzzy set.

By the *concentration operation*, the fuzzy set function is made narrower, i.e., the elements with low degrees are reduced more than the ones with high degrees (see dashed line in Fig. 7). The *concentration* is a common way of representing "*very something*" as "*CON something*," e.g., *very young*.

The *dilation* is the opposite of the concentration operation where the elements that are barely in the set increase their degree of membership tremendously (dashed line in Fig. 8). The *dilation* is a common way of representing "*slightly (or almost) something*" as "*DIL something*"; e.g., *almost* young.

The *normalization* of fuzzy set $\underline{A}$, denoted as the NORM ($\underline{A}$), normalizes the membership function in terms of the maximum membership value by dividing the degrees of $\underline{A}$ by the maximum value, resulting in at least a single value of 1.

The *intensification* operation is used to increase membership degrees of some of the elements while decreasing the degrees of others. In general, the degrees less than 0.5 are subjected to the concentration while the others are subjected to the dilation (see the dashed line in Fig. 9).

Note that in the fuzzy set operations, when dealing with the union and the intersection of the sets, we apply the **max** and min operations, respectively. The details on the fuzzy sets and fuzzy set operations can be obtained from Tayfur (2012).

3. Fuzzy logic model

There are different fuzzy systems such as the Tagaki and Sugeno (T&S), Sugeno and Kank (S&K), Tagaki-Sugeno-Kank (TSK), and Mamdani. The consequent parts of the fuzzy rules in TSK systems involve some sort of mathematical functions.

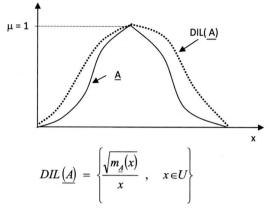

$$DIL\left(\underline{A}\right) \; = \; \left\{ \frac{\sqrt{m_{\underline{A}}(x)}}{x} \;\; , \;\; x \in U \right\}$$

FIG. 8   Dilation representation for a fuzzy set.

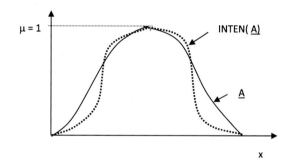

$$INTEN\left(\underline{A}\right) \; = \; \begin{cases} \dfrac{2m_{\underline{A}}^{2}(x)}{x} & 0 \le m_{\underline{A}}(x) \le 0.5 \\[3em] 1 - 2\left[1 - \left(1 - \dfrac{2m_{\underline{A}}^{2}(x)}{x}\right)\right]^{2} & 0.5 \le m_{\underline{A}}(x) \le 1.0 \end{cases}$$

FIG. 9   Representation of operation intensification.

In the case of the Mamdani fuzzy system, both the antecedent and consequent parts of the fuzzy rules are verbally expressed and therefore it is widely preferred. A general Mamdani type fuzzy model is shown in Fig. 10. The model has basically four major components: (1) Fuzzification, (2) Fuzzy Rule Base, (3) Fuzzy Inference Engine, and (4) Defuzzification.

3.1 Fuzzification

The **fuzzification component** constitutes the fuzzy sets and related membership functions for each input and output variable. The membership functions can take many forms, in practical applications triangular ones are adopted (Sen, 1998). The construction of the membership functions is both an objective and subjective process while the data distribution may help the modeler to identify obvious clusters amenable to classification. Relatively small number of extreme highs and/or lows may necessitate further classification (for example, "very very high," "very very low") to capture the system behavior under extreme conditions (Coppala et al., 2002). Fig. 11 is an example, where the clustering of the measured discharge data results in nine fuzzy subsets as formulated by Tayfur and Singh (2011) who have predicted discharge using channel cross-section data.

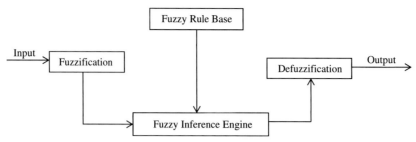

FIG. 10   Typical Mamdani type fuzzy model (Tayfur, 2012).

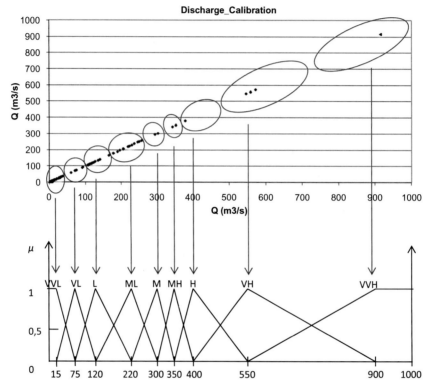

FIG. 11   Fuzzification of mean discharge.

3.2 Fuzzy rule base

The **fuzzy rule base** contains IF-THEN rules constituting possible fuzzy relations between the input and output variables. In the Mamdani rule system, both the antecedent and consequent parts of a rule are the verbal statements (Mamdani, 1977), such as the following;

IF the precipitation high THEN the runoff is high

The Mamdani rules can be intuitively produced by an expert view or they can also be constructed from available data (Sen, 2004; Casper et al., 2007; Tayfur, 2012). For example; Tayfur and Brocca (2015) have shown how to construct a fuzzy rule when predicting runoff using the rainfall and moisture data measured at an experimental basin in central Italy. According to Fig. 12, 30% soil moisture is a member of medium-high (MH) and high (H) subsets with 0.2 and 0.8 degrees of membership, respectively; 18 mm/day rainfall is a member of low (L) and medium-low (ML) subsets with 0.4 and 0.6 degrees of memberships, respectively; and 78 m^3/s discharge is a member of ML and medium (M) with 0.3 and 0.7 degrees of memberships, respectively. Hence, an observed set of 30% soil moisture, 18 mm/day rainfall and 78 m^3/s discharge would yield the following rule, always considering the sets where the variable has the maximum degree of membership (Sen, 2004; Tayfur, 2012):

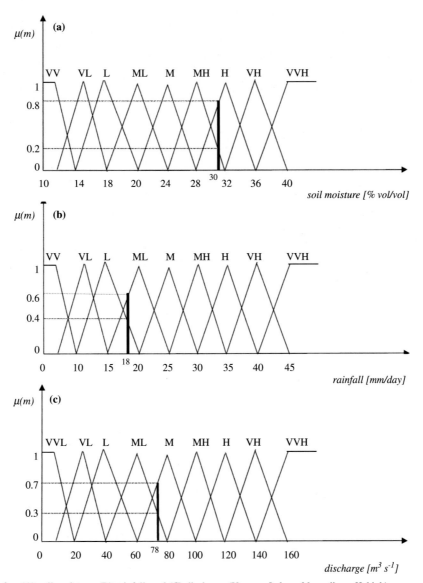

FIG. 12 Fuzzy subsets for: (A) soil moisture, (B) rainfall, and (C) discharge (*V*, very; *L*, low; *M*, medium; *H*, high).

IF the soil moisture is high and the rainfall is medium low THEN the discharge is medium

If the generated rule is not already in the rule base then it is added to the base. Note that, according to the number of fuzzy subsets of the input variables, the possible number of rules for the case in Fig. 12, can be $9 \times 9 = 81$ in order to cover every possible system state. However, working with 81 rules may be cumbersome and furthermore may not be suitable from a practical point of view. Therefore; the rules whose firing strengths are less than 10% or less can be eliminated (Coppala et al., 2002). The firing strength of a rule can be simply found by the product of the membership degrees (Coppala et al., 2002; Tayfur, 2012). The firing strength of the above rule is $0.8 \times 0.6 \times 0.7 = 0.34$. Furthermore, the rules which are not physically sound cannot be conserved in the rule base. For example; if soil moisture content is low and rainfall is low then the runoff cannot be high.

3.3 Fuzzy inference engine

The fuzzy inferencing engine performs basically two suboperations: (1) the inferencing and (2) the composition. By the inferencing, the firing strength of each triggered-rule is first computed and then applied to the consequent part of the rule to compute related the fuzzy output. For this purpose, either the min or the **prod** method can be employed. Note that the

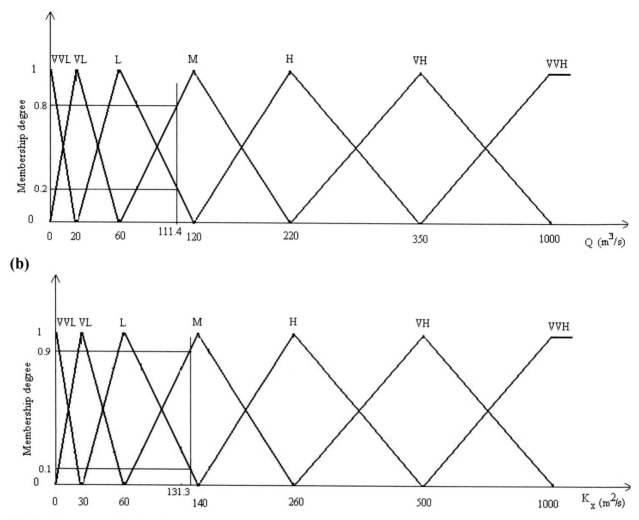

FIG. 13 Fuzzy subsets: (A) Flow discharge; (B) longitudinal dispersion coefficient (*V*, very; *L*, low; *M*, medium; *H*, high).

inferencing assigns fuzzy subsets to the output variable from each triggered rule. By the composition operation, all the fuzzy subsets obtained by the inferencing operation can be combined to form a single fuzzy output set. For this purpose, either the max or the **sum** methods can be used. The details regarding the fuzzy inferencing operations can be obtained from Tayfur (2012).

The application of the fuzzy inferencing can be shown by considering Fig. 13 where the fuzzy subsets are presented for the input variable of flow rate Q and the output variable of dispersion coefficient K_x. Also the following rules are already constructed for relating Q to K (Tayfur, 2006):

Rule 1:	IF	Q is VVL	THEN	K_x is VVL
Rule 2:	IF	Q is VL	THEN	K_x is VL
Rule 3:	IF	Q is L	THEN	K_x is L
Rule 4:	IF	Q is M	THEN	K_x is M
Rule 5:	IF	Q is H	THEN	K_x is H
Rule 6:	IF	Q is VH	THEN	K_x is H
Rule 7:	IF	Q is VVH	THEN	K_x is VVH

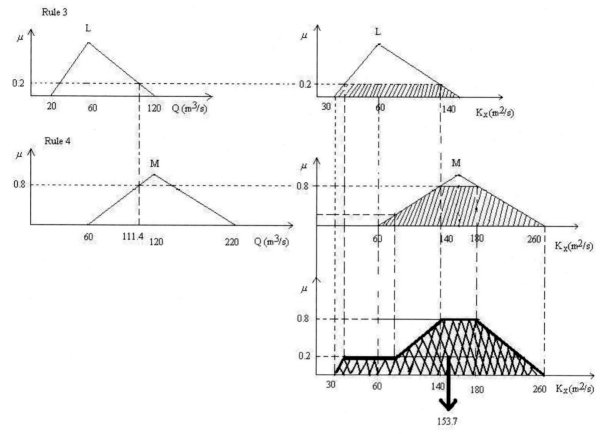

FIG. 14 Fuzzy inferencing and the **max** composition.

We can now see how the inferencing would produce the fuzzy output set for K_x given the input variable value of $Q = 111.4\,\text{m}^3/\text{s}$. As seen in Fig. 13, $Q = 111.4\,\text{m}^3/\text{s}$ triggers the low (Rule 3) and medium (Rule 4) fuzzy subsets of the flow discharge. The consequent part of each K_x variable appears, by the min operation, as the truncated trapezium for each rule, as presented in Fig. 14 where the shaded subsets show the fuzzy output subsets corresponding to 0.20 degree in the "low" and 0.80 degree in the "medium" subsets of K_x. The combination of these two truncated trapeziums by the max composition yields the single fuzzy output set (the double shaded area at the bottom of Fig. 14).

3.4 Defuzzification

The defuzzification operation converts the resulting combined fuzzy output set to a crisp value. There are many methods for this purpose such as the mean of maxima, the leftmost maximum, the right most maximum, the bisector of area, and the center of gravity (COG). One can obtain more details in Tayfur (2012). The COG is the most commonly used one; which can be expressed as (Sen and Altunkaynak, 2003):

$$\widehat{x} = \frac{\displaystyle\int_a^b x\mu(x)\,dx}{\displaystyle\int_a^b \mu(x)\,dx} \tag{1}$$

Note that for the discrete case, the integrals in Eq. (1) can be replaced by the summations. The combined single fuzzy output set given in Fig. 14 is shown in more details in Fig. 15 where the whole area is first subdivided into the triangular and rectangular subareas as A1, A2, A3, A4, A5 and A6. Then, the area and the center numerical values are computed for each subarea.

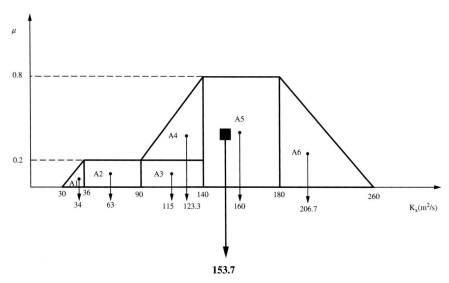

FIG. 15 Combined fuzzy output set.

Finally, by applying the COG method, as shown in Eq. (2), $K_x = 153.7\,\text{m}^2/\text{s}$ is obtained (Tayfur, 2006).

$$\widehat{K}x = \frac{\sum\limits_{i=1}^{6} K_{xi}A_i}{\sum\limits_{i=1}^{6} A_i} = \frac{0.6*34 + 10.8*63 + 10*115 + 15*123.3 + 32*160 + 32*206.7}{0.6 + 10.8 + 10 + 15 + 32 + 32} = 153.7\,m^2/s \qquad (2)$$

Note that in Eq. (2), K_{xi} stands for the centroid value of subarea A_i and $\widehat{K}x$ stands for the crisp value as a result of the COG operation.

4. Discussions

Dynamics in hydrology and water resources engineering often deal with imprecision, uncertainty, and vagueness due to lack of data, errors in measurements, and temporal and spatial variability of the hydrological and meteorological variables. Hence, handling such issues could be suitable for the fuzzy logic-based modeling (Bogardi et al., 2004). FL can be used in the analysis and decision making and it can enable a decision-maker to gain insight into the model sensitivity and the uncertainty (Ozelkan and Duckstein, 2001). Hence in many studies, the fuzzy models have shown better performance in comparison to the conventional methods (Kambalimath and Deka, 2020). Some advantages can be summarized as follows:

1. It employs IF-THEN rules (statements relating input variables to the output variable). Hence, it is more in line with human thinking since humans think verbally, not numerically.
2. These, IF-THEN statements can handle the uncertainty and vagueness within the physical dynamics. This is achieved by using the membership functions and allowing the partial belongings (memberships).
3. It can benefit from the observations, experiments and as well as experience. While dealing with irrigation, for example, it can benefit easily from the experience of a farmer who does not have to have any college degree.

On the other hand, the fuzzy modeling can be efficient if an engineer works with optimal number fuzzy rules. Especially, when there is many input variables and/or the subsets, it is possible to end up with many rules in the order of hundreds. Engineers would not welcome such irritating cases. Although the number of rules can be reduced by using the firing strength approach, this, on the other hand, may result in the loss of information. The achievement of this delicate balance is a standing issue to be resolved. Kambalimath and Deka (2020) have recently reviewed some papers dealing with fuzzy modeling applications in hydrology and water resources engineering. They have also discussed the merits and demerits of the fuzzy modeling.

5. Conclusions

This chapter has focused on the fuzzy sets and fuzzy model components. The fuzzy model components and its processes are presented using the hydrological variables that would help the readers to better visualize and understand the concepts. It has also given the fundamental background that would help some modelers who may be using FL related package programs while being unaware of the basics.

Fuzzy modeling has been applied in the field of hydrology and water resources engineering since the early 1990s. One can find more details of all these applications in Tayfur (2012) and Kambalimath and Deka (2020), among many. As pointed out above, the fuzzy modeling approach has some advantages and disadvantages. The limitations of the fuzzy modeling, some of which are pointed out above, have been overcome, in recent years, by its employment with the other expert methods, such as the ANFIS modeling. It is apparent that its employment would continue, mostly as the hybrid modeling.

References

Bogardi, I., Bardossy, A., Duckstein, L., Pongracz, R., 2004. Fuzzy Logic in Hydrology and Water Resources, Fuzzy Logic in Geology. Elsevier Science, Amsterdam, pp. 153–190.

Casper, M., Gemmar, P., Gronz, O., Johst, M. and Stüber, M. (2007). Fuzzy logic-based rainfall–runoff modelling using soil moisture measurements to represent system state. Hydrol. Sci. J., 52(3), 478–490, 2007.

Chang, L.-C., Chang, F.-J., Tsai, Y.-H., 2005. Fuzzy exampler-based interface system for flood forecasting. Water Resour. Res., W02005. https://doi.org/10.1029/2004WR003037.

Chuntian, C., 1999. Fuzzy optimal model for the flood control system of the upper and middle reaches of the Yangtze River. Hydrol. Sci. J. 44 (4), 573–582.

Coppala, E., Duckstein, L., Davis, D., 2002. Fuzzy rule-based methodology for estimating monthly groundwater recharge in temperate watershed. J. Hydraul. Eng. ASCE 7 (4), 326–335.

Dou, C., Woldt, W., Bogardi, I., Dahab, M., 1997. Numerical solute transport simulation using fuzzy set approach. J. Contam. Hydrol. 27 (1–2), 107–126.

Eslamian, S., Amiri, M.J., 2011. Estimation of daily pan evaporation using adaptive neural-based fuzzy inference system. Int. J. Hydrol. Sci. Technol. 1 (3/4), 164–175.

Holmblad, L.P., Ostergaard, J.J., 1982. Control of cement kiln by fuzzy logic. Fuzzy Set. Intell. Syst., 337–347. https://doi.org/10.1016/B978-1-4832-1450-4.50039-0.

Hundecha, Y., Bardossy, A., Theisen, H.W., 2001. Development of a fuzzy logic-based rainfall-runoff model. Hydrol. Sci. J. 46 (3), 363–376.

Kambalimath, S., Deka, P.C., 2020. A basic review of fuzzy logic applications in hydrology and water resources. Appl Water Sci 10, 191. https://doi.org/10.1007/s13201-020-01276-2.

Kumar, A.K.C., Reddy, G.P.O., Palanisamy, M., Turkar, S.Y., Sandeep, P., 2021. Integrated drought monitoring index: a tool to monitor agricultural drought by using time series datasets of space-based earth observation satellites. Adv. Space Res. 67, 298–315.

Malik, A., Kumar, A., Salih, S.Q., Kim, S., Kim, N.W., Yaseen, Z.M., Singh, V.P., 2020. Drought index prediction using advanced fuzzy logic model: regional case study over Kumaon in India. PLoS One 15 (5), e0233280. https://doi.org/10.1371/journal.pone.0233280. eCollection 2020.

Mamdani, E.H., 1977. Application of the fuzzy logic to approximate reasoning using linguistic synthesis. IEEE Trans. Comput. C-26, 1182–1191.

Mamdani, E.H., Assilian, S., 1975. An experiment in linguistic synthesis with a fuzzy logic controller. Int. J. Mach. Stud. 7 (1), 1–3.

Masoumi, Z., Rezaei, A., Maleki, J., 2019. Improvement of water table interpolation and groundwater storage volume using fuzzy computations. Environ. Monit. Assess. 191 (6), 401. https://doi.org/10.1007/s10661-019-7513-1.

Nourani, V., Najafi, H., Sharghi, E., Roushangara, K., 2021. Application of Z-numbers to monitor drought using large-scale oceanic-atmospheric parameters. J. Hydrol. 598, 126198.

Ozelkan, E.C., Duckstein, L., 2001. Fuzzy conceptual rainfall-runof models. J. Hydrol. 253, 41–68.

Raj, A.P., Kumar, D.N., 1998. Ranking multi-criterion river basin planning alternatives using fuzzy members. Fuzzy Sets Syst. 100 (1–3), 89–99.

See, L., Openshaw, S., 2000. A hybrid multi-model approach to river level forecasting. Hydrol. Sci. J. 45 (4), 523–536.

Selvaraj, A., Saravanan, S., Jennifer, J.J., 2020. Mamdani fuzzy based decision support system for prediction of groundwater quality: an application of soft computing in water resources. Environ. Sci. Pollut. Res. 27 (20), 25535–25552. https://doi.org/10.1007/s11356-020-08803-3.

Sen, Z., 1998. Fuzzy algorithm for estimation of solar irradiation from sunshine duration. Sol. Energy 63 (1), 39–49.

Sen, Z., 2004. Fuzzy Logic and System Models in Water Sciences. Turkish Water Foundation, İstanbul, Turkey.

Sen, Z., Altunkaynak, A., 2003. Fuzzy awakening in rainfall-runoff modeling. Nord. Hydrol. 35 (1), 31–43.

Singh, A.P., Dhadse, K., Ahalawat, J., 2019. Managing water quality of a river using an integrated geographically weighted regression technique with fuzzy decision-making model. Environ. Monit. Assess. 191 (6), 378. https://doi.org/10.1007/s10661-019-7487-z. 31104168.

Tayfur, G., 2006. Fuzzy, ANN, and regression models to predict longitudinal dispersion coefficient in natural streams. Nord. Hydrol. 37 (2), 143–164.

Tayfur, G., 2012. Soft Computing in Water Resources Engineering. WIT Press, Southampton, England, UK. 267 pages.

Tayfur, G., Brocca, L., 2015. Fuzzy logic for rainfall-runoff modelling considering soil moisture. Water Resour. Manag. 29, 3519–3533.

Tayfur, G., Singh, V.P., 2006. ANN and fuzzy logic models for simulating event-based rainfall-runoff. J. Hydraul. Eng. 132 (12), 1321–1330.

Tayfur, G., Singh, V.P., 2011. Predicting mean and bankfull discharge from channel cross-sectional area by expert and regression methods. Water Resour. Manag. 25, 1253–1267.

Tayfur, G., Ozdemir, S., Singh, V.P., 2003. Fuzzy logic algorithm for runoff-induced sediment transport from bare soil surfaces. Adv. Water Resour. 26, 1249–1256.

Zadeh, L.A., 1965. Fuzzy sets. Inf. Control. 8, 338–353.

Zhao, R., Chen, S.J., 2008. Fuzzy pricing for urban water resources: model construction and application. J. Environ. Manag. 88 (3), 458–466.

Chapter 11

GIS Application in floods mapping in the Ganges–Padma River basins in Bangladesh

Shafi Noor Islam[a], Sandra Reinstädtler[b,c], Mohibul Hassan Kowshik[d,*], Shammi Akther[e,*], Mohammad Nazmi Newaz[f], Albrecht Gnauck[g], and Saeid Eslamian[h,i]

[a]*Department of Geography, Environment and Development Studies, Faculty of Arts and Social Sciences, University of Brunei Darussalam, Gadong, Brunei Darussalam,* [b]*Department of Geography, Environment and Development Studies, University of Brunei Darussalam, Gadong, Brunei Darussalam,* [c]*Faculty of Environmental Sciences and Process Engineering and Faculty of Environment and Natural Sciences, Brandenburg University of Technology, Cottbus-Senftenberg, Germany,* [d]*Government Titumir College, Dhaka, Bangladesh,* [e]*Department of Geography and Environment, Jahangirnagar University, Dhaka, Bangladesh,* [f]*Management Department, Bangladesh Institute of Management (BIM), Dhaka, Bangladesh,* [g]*Brandenburg University of Technology Cottbus, Senftenbuerg, Cottbus, Germany,* [h]*Department of Water Engineering, College of Agriculture, Isfahan University of Technology, Isfahan, Iran,* [i]*Center of Excellence for Risk Management and Natural Hazards, Isfahan University of Technology, Isfahan, Iran*

1. Introduction

Bangladesh is highly floods and river bank erosion prone area in the world. There are 257 rivers in Bangladesh, and 59 of them are Transboundary Rivers which play an important role in flooding the deltaic floodplain. The excesses of water during the monsoon (June–September) cause widespread flooding which damages crops, infrastructure, *char-land* settlements, communication networks and life (Mafizuddin, 1983).The country is situation in the Ganges–Brahmaputra–Meghna drainage basin which is cover an area 1.76 million km^2 and only 7.5% lies in Bangladesh (Elahi, 1987; Elahi et al., 1998). The Ganges (Padma) is one of the potential rivers in Bangladesh which build huge *char-lands (Riverine Islands)* in the river channels. The excesses of water during the monsoon (June–September) cause widespread flooding which damages crops, infrastructure, *char-land* settlements, communication networks and life (Mafizuddin, 1983). There are many natural elements and factors are not favorable to human existence. Floods occur in the char-lands which is a potential threats to human live and resources in term of loss and damage, emerge from the intersections between extreme geophysical events and a vulnerable livelihood in the char-lands (Haque, 1997). The cause of floods in Bangladesh is based on monsoon precipitation in the GBM catchments; snow and glacial ice melt with monsoon rain in the Himalayas and runoff generated by heavy local rainfall. There are no systematic records of annual floods available until the 20th century, including 1987 and 1988, 1998, 2002, 2004, and 2008. The recent flooded land area was increased from 35% in 1974 to 71% in 2004, comparing to the flooded land area in 1954. During the flooding period, suspended sediment load reached as high as 13 million tons per day (Coleman, 1969). The newly emerged land known as *Char or Diara* is the Bengali term for the mid channel island that periodically emerges from the riverbed as accretion (Hanna, 1996; Baqee, 1993a,b, 1997, 1998a,b). This new land is fertile and a valuable natural resource. The *char-land* landscapes are of great importance for its exceptional hydro-geological setting. The physical characteristics of the geographic location, the rivers morphology and the monsoon climate render the *char-lands* highly vulnerable to natural disasters (Baqee, 1985, 1997; Coleman, 1969). Over twelve million *chaura* people are living in the *char-lands* and struggling against monsoon floods and river bank erosion.

The *char-lands* of Padma River are undergoing rapid hydro-morphological changes due to natural and anthropogenic causes (Hooper, 2001). The Padma is a meandering river and has high rate of river bank erosion and accretion in the

---

*Alumni.

Handbook of HydroInformatics. https://doi.org/10.1016/B978-0-12-821962-1.00010-6

167

channel. In the monsoon time (June–September), Brahmaputra (Jamuna) can deliver its water discharge of 100,000 m³/s (FAP, 21/22, 1993; Mofizuddin, 1983). The sediments carried by the rivers have aggraded the river bed which plays negative impacts on the floodplain and the *char-lands* (Baqee, 1993a,b, 1998a,b). The Purba Khas Bandarkhola Mouza of Char-Janajat of Sibchar Upazila of Madaripur district is part of the Ganges active delta and located in the main channel of the Ganges–Padma River (Coleman, 1969; Islam, 2014a,b, 2016a,b). The Char-Janajat is inundated by the monsoon floods every year, and in consequence people have to displace their settlements and scatter from one place to another. The agricultural crops of the *char-lands* depend on soil quality (Islam, 2016a,b; Islam and Gnauck, 2009; Islam et al., 2018a,b, 2019). The soil quality and fertility of the *char-land* relates to floods and clay siltation. Agricultural cropping pattern in the *char-land* differs from that in other places of the country because of instability of *agricultural* land (Elahi, 1991; Baqee, 1985). The *chaura* people have to relocate because of the frequent massive floods and trends of river bank and *char-land* erosion. Their dwellers are displaced from the *char* and return the native *char* when new land emerges in the river channel. The relocation distance of the *char* settlement is about 12 km range on average. The interval of displacement is about 5 years at the Purba Khas Bandarkhola Mauza. The settlements displacement and population changes are reliant on the floods occurrence and river bank erosion at the Purba Khas Bandarkhola Mouza. The outcome of the present research would be a valuable contribution to make a national plan for flood disaster management and adaptation in the *char-lands* in the Ganges delta (Islam, 2014a,b, 2016a,b).

2. Objective of this study

The objective of this study is to investigate and mapping the nature of the *char-lands* erosion, deposition pattern and settlement relocation due to floods in the Ganges–Padma River basin area and measures the vulnerability of *char* livelihoods. To investigate the *char* people cyclic dislodgment in the basin area and measures the socioeconomic impacts. Formulate some practical recommendations for sustainable livelihoods and settlements in the *char-lands* in the Ganges–Padma River basin in Bangladesh. GIS application could be a potential tool for this investigation and mapping which is advantageous for decision making for the *char-land* people, livelihoods, and settlement sustainability in the river basins in Bangladesh.

3. Geographical location and physical characteristics of the study area

The Purba Khas Bandarkhola *Mouza* is a revenue village of Char-Janajat union which is located in Padma River basin in the Ganges delta in Bangladesh. The Purba Khas Bandarkhola Mouza is under Sibchar Upazila which is situated between 23° 15′ and 23° 30′ North Latitude and 90° 05′ and 90° 18′ East Longitude (Fig. 1). The area of Purba Khas Bandarkhola Mouza is 6.73 km² (which is covering 8.00% land area of the entire Char Janajat land area), and the population of Seat no 2 is estimated 1350 in 2008.

The present study has been carried out at the seat no 2 of Purba Khas Bandarkhola Mouza. The mouza under the union of Char-Janajat (area 31.94 km²) is under the Sibchar Upazila of Madaripur District. The whole *char* (Island) is called as Char-Janajat and its area is 84.09 km² until 2003. The Case Mouza Purba Khas Bandarkhola Mouza has covered 8% Land area of the entire Char Janajat and it has covered 21.07% land area of Char Janajat Union territory (2018).

The population of the Char-Janajat union is 13,958 until 2003 and presently it is estimated over 35,000 (until 2008). The Purba Khas Bandarkhola is an area bounded by Char-Janajat Mouza, Char-Amirabad, Akater-Char, Boro Khas Bandarkhola, Bandarkhola union, Katalbari union, Panchchar union of Sibchar Upazila (Fig. 1). The two rivers are Padma and Arial Khan which are influenced by the Purba Khas Bandarkhola mouza formation, erosion and accretion (Fig. 1). The majority of the char people (55%) are involved in agricultural farming profession. Agricultural labor are 11%, wage labor are 3%, small traders 3%, transport 2%, fisher man 3%, service 4%, student 13%, and others 6%, this is the scenario of Purba Khas Bandarkhola Mouza (Fig. 1), and it is different from the main land.

4. Data and methodology

The study has been conducted based on primary and secondary data sources. The primary data on *char-land* erosion, settlement relocation, people displacement, *char* livelihoods information were collected from the local people of Seat no 2 of Purba Khas Bandarkhola mouza through PRA practice and informal interviews. Beside this a questionnaire survey on 101 families of Purba Khas Banderkhola Mouza (6.73 km²) of Char-Janajat Union of Sibchar Upazila of Madaripur District during 1997 to 2003, and a three months socioeconomic survey with *char* people in 2003 and 2018. Besides this geomorphological, anthropological village-settled study and questionnaire survey for bench mark data collection were conducted. Complete enumeration of the households covering the land holding pattern, tenancy, agricultural cropping system and

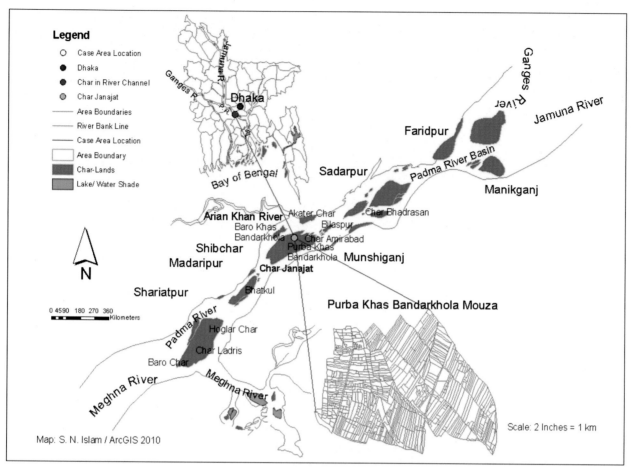

FIG. 1 The geographical location of case study area in the Padma River channel.

marking, occupation, demographic character, literacy, etc. were collected in 2003 and 2018 at the Purba Khas Banderkhola Mouza. A number of standard approaches for information and data collection and analysis were used including review literature of studies to get basic understanding of *char-land* development and rural livelihood. The secondary data inputs came from different publications of the government agencies, NGOs reports and research organizations such as BIDS, CIRDAP, and ADAB in Bangladesh. An integrated practice of PRA (Participatory Rural Appraisal) method was used to develop char erosion maps, settlement relocation pattern and people cyclic displacement in the case *char-land*. Some interpretations of life history analysis were organized with *char* people to develop the diagram of *char* people cyclic dislodgment pattern. Collected data and information was analyzed and visualized through using EXCEL, VISIO 32 and ArcGIS software. The times series Remote Sensing imageries from 1995 to 2018 have been used to investigate and compare the trends of *char-lands* erosion, settlement relocations, people displacement and land use changing pattern in the Purba Khas Bandarkhola mouza of Char-Janajat Union of Sibchar Upazila of Madaripur district. Besides two quantitative data based case studies have been arranged with two different *char* families those were living and cyclic displacing for the last 50 years in Char-Janajat island of Padma River channel. A leading person from every family was delivering the chronological historic data and recognizing the settlement relocation places in different years, cyclic displacement and *char* livelihood information, these data and information has been used to develop the people cyclic dislodgment models.

5. Geographer and anthropologist view

The problems of the *char* settlements due to monsoon floods and river bank erosion in the large rivers in Bangladesh remain under represented. Key researchers in this connection are Elahi (1987, 1989, 1991), Baqee (1993a,b, 1997), Mamun and Amin (1999). Elahi (1987, 1989, 1991) discusses the eternal struggle of the people for survival on the western bank of the Jamuna River in Rajshahi division and described how they cope up with the conditions and resettled on the banks after erosion. Wiest (1987) discussed on the point of cultural anthropology on the Brahmaputra floodplain and *char-lands* erosion in Bangladesh. Zaman (1988, 1989) in his anthropological study argued that response to natural hazards varied in accordance with the background of the family. Baqee (1993a,b) discussed that the settlements originate through the sponsorship of powerful elites and inevitably grow through a filtration process. The scenarios described the existing socio-economic relationship in the *char-lands* which become increasingly complex. Mamun and Amin (1999) discussed the perception of people in the vulnerable *char-lands* and suggested a number of strategies to reduce their sufferings. Haque (1999) provided an overview of indigenous knowledge and practices of people while coping with river erosion and floods. The earlier study on *char-lands* carried out by Adnan in 1976 pointed out the dynamics of power in remote villages of Barisal. Curry (1986, 1996) documented the overall changes in a predominantly *char*-settlements in Rangpur district in Bangladesh. Ali (1980) attempted to outline the evolution of the laws that operate in the *char-lands*. Hanna (2001) discussed in her study on indigenous and engineering knowledge in Bangladesh, especially on Jamuna River's *chars*.

Lahiri-Dutt (2000, 2015) has voiced her discomfort in imagining a river as a static physical entity, arguing that rivers are not just lines on maps. Much before the emergence of hydrological literature or mesologic and hydrocultural discourses, Lahiri-Dutt (2000) had stated that the river is "neither outside society, nor is it just a thing out there in nature" asserting how rivers keeps interacting with culture.

Zaman and Alam (2021) stated an estimated 20 million people in Bangladesh live in the char-lands in the river basins in Bangladesh. Char dwellers are extremely poor and highly vulnerable to natural hazards of floods and river bank erosion (Zaman, and Alam 2021, pp. 3–11).

The specific literature reviews suggested that, the opinion of Sociologists, Economists, Anthropologists and Geographers are almost the same. In most of the cases the Geographers have collected data and analyzed on the point of socio demographic and physical problems of river basins. The problems of *char-lands* are not solved due to lack of proper research in these areas. Hence, it is essential to carry out more in-depth applied research in terms of analysis, modeling and prediction of *char-land* people movement and settlement relocation and displacement.

6. Floods and *char-land* erosion and deposition in the river basins in Bangladesh

The primary cause of flood in Bangladesh is rainfall in the catchment areas of the rivers of Bangladesh. Situated in the monsoon belt with the Himalayas in the north, Bangladesh falls in the region of very heavy rainfall. About 80% of the rainfall occurs during the 5 months period from May to September. The *char-lands* are repeatedly affected by the massive floods, with concomitant of riverbank erosion because of the shift of river channels (Ahmed, 1965; Hassan et al., 2000). The newly formed lands (*char-lands*) (Fig. 1) and eroded riverbanks are inhabited by some of the most desperate and vulnerable people in the country. Fragile riverbanks and *char lands* have never been abandoned because of flooding (Haque, 1985). But the average rainfall in Bangladesh generates annually only 100 million acre feet of water whereas 1100 million acre feet of water comes from outside of Bangladesh. Thus, about 90% of the water carried by the river systems, namely, the Brahmaputra, the Ganges, the Meghna, and other smaller rivers, are the inflow from outside of the country (Islam, 2010a,b, 2016a,b). These rivers carry water from an area of about 960,000 km^2 of which only 7.5% lies in Bangladesh. Water enters in Bangladesh through three major channels but the discharge takes place through one major channel. The river system has evolved to carry the normal flow of water generated in the catchment area.

Tables 1 and 2 displaying the total areas of the Ganges–Brahmaputra–Meghna rivers catchment areas and its percentage covered by the countries are located in the downstream and within the catchment areas. This is the widest river system in the country flowing from north toward south. The discharge during the rainy season is enormous, the highest recorded flood was in August 1988, and the river flow discharge was 98,600 m^3/s (Islam, 1995; Islam et al., 2014a,b,c, 2017).

The reasons for massive dryness, drought, and floods in Bangladesh are to be considered based on the following evidences: Snow melting in the Himalayan region; Hydrographic and geomorphic changes in the Brahmaputra River basin;

TABLE 1 The catchment areas and the area covered by the Ganges–Brahmaputra basin.

Name of the river catchment area	Area covered	Percentage of territorial area
Brahmaputra–Jamuna River	600,000 km^2	100%
Tibet area China	293,000 km^2	48.83%
Indian Part	241,000 km^2	40.16%
Bangladesh Part	47,000 km^2	7.83%

TABLE 2 The countries covered the catchment areas and percentage of the Ganges–Brahmaputra–Meghna rivers catchment.

Country	Ganges basin		Brahmaputra basin		Meghna basin	
	Basin area (1000 km²)	Percentage of total area	Basin area (1000 km²)	Percentage of total area	Basin area (1000 km²)	Percentage of total area
China	33	3	293	50		
Nepal	140	13				
Bhutan			45	8		
India	861	80	195	34	49	58
Bangladesh	46	4	47	8	36	42
Total	1080	100	580	100	85	100

Estimated 2.4 billion tons of sediments carried by the river systems of Bangladesh by annual flood which reduces the water carrying capacity of the rivers; it is also creating massive floods (Anwar, 1988; Coleman, 1969; Islam et al. 2014a; Reinstädtler and Islam, 2017). High rate of forest depletion in the GBM catchments areas tends to aggravate the floods in the downstream areas; and Unplanned development of constructions such as roads, railways, embankments, polders, etc. also creates barrier to the flow of water and aggravates the flood (Islam, 2010a,b, 2016a,b).

In the 20th century, eighteen major floods have been recorded: 1900, 1902, 1907, 1918, 1922, 1954, 1955, 1956, 1962, 1963, 1968, 1970, 1971, 1974, 1984, 1987, 1988, 1998, and 2008, of which 3 (1987, 1988 and 1998) are catastrophic floods. According to the statistical evidence of Bangladesh both in 1987 and 1988 the country experienced disastrous floods (Islam, 2010a,b, 2016a,b). The 1987 flood is estimated a 30- to 70-year event, which affected 75,300 km², almost 40% of the total area of the country. The 1988 flood inundated about 82,000 km² and about 60% of the area and its recurrent period are estimated to be 50 to 100 years (BWDB) (Fig. 2). The damage caused by the 1987 flood was staggering, according to one estimate, the total loss was about over US $ 95 million.

No sophisticated system is so far available in Bangladesh to assess the damage due to floods. Almost all the *char-lands* in the GBM catchments are inundated by the devastating floods, so the damages of the char-lands cannot be estimated during that period of havoc, in absence of modern technology, manpower and favorable financial conditions (Islam, 2010, 1995).

In 1988 (May–September), 53 out of 64 districts was affected and 3000 km of roads, 180 bridges and culverts and 640 km of railway tracks and 2700 primary schools/madras have in the rural areas have been either been washed away or partly damaged (Islam, 1995). Considering the historical data and information it has been recognized that the most devastating floods and river bank erosions have been occurred in 1988 and 1998 (Fig. 2).

The official sources are indicating the damage of crop-acreage at TK. 4.93 million, and as a consequence rice production lowered by 2.4 million tons per year (Islam, 1995, 1993; Miah, 1988). Average annual flow of Jamuna River at Bahadurabad (the point where the Brahmaputra enters into Bangladesh) is estimated to be 501 million acre feet. August has always been the month when widespread flooding is occurred. Floods from May to July are usually high time in the catchments of Brahmaputra-Jamuna and the Meghna, from August to October due to the combined flows of that river and the Ganges (Islam and Gnauck, 2007, 2011). As a result, the flow of the Brahmaputra-Jamuna is more erratic than that of the Ganges. The gradient of the Jamuna averages 1:11,850 which is slightly more than that of the Ganges (Islam, 2010a,b, 2016a,b). During the monsoon the Jamuna widens in some parts up to 20 km and transports between 60,000 and 100,000 m³/s water (Flood Action Plan FAP 24, 1996). In the rainy season it brings down something like 13 million tons of sediment per day (Coleman, 1969; Hanna, 2001) which helps to format the *char-lands* in the Jamuna-Padma Rivers channels (Jannatul et al., 2015).

6.1 Impacts of floods on char-lands and changing rural livelihoods

Bangladesh is a country that lies mostly at the bottom of the floodplains of three major rivers, the Ganges, Brahmaputra and Meghna. Each year monsoon floodwater inundates the country leaving both adverse and favorable impacts upon the lives of its people, and particularly causing widespread damage to standing crops, livestock, infrastructure, and road network. Most of the natural hazards results from the potential for extreme geophysical events, such as floods, to create and unexpected

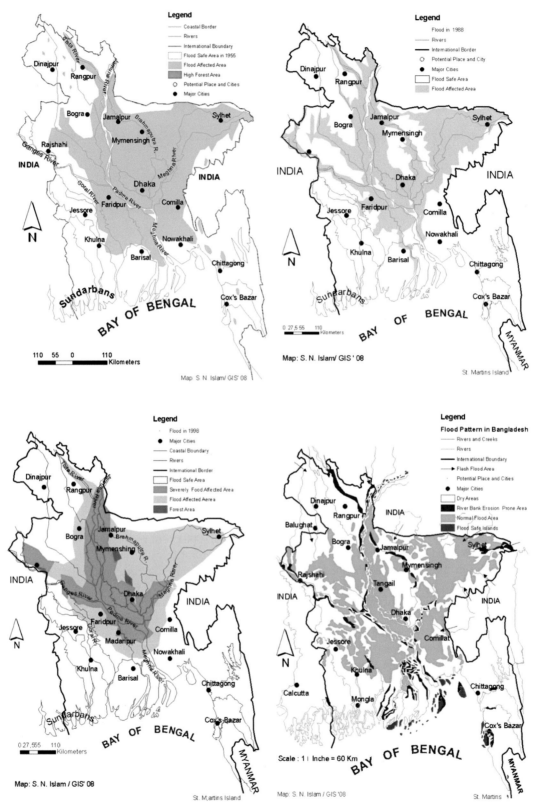

FIG. 2 Historical flood inundation and bank erosion pattern in 1955, 1988, 1998 and 2008 in the Ganges–Brahmaputra–Meghna River basins in the Bengal Delta in Bangladesh. *(Map Source: Islam, S. N. 2008.)*

threat to human life and property (Smith, 1996; Islam, 2014b; Jannatul et al., 2015). When severe floods occur in areas like *char-lands* occupied by humans, they can create natural disasters that involve the loss of human life, settlements and other property plus serious disruption to the ongoing activities of large urban and rural communities (Fig. 2). The Jamuna is braided in nature, within the braided belt of the Jamuna; there are lots of *chars* of different sizes. An assessment of the 1992 dry season Land Sat image shows that the Jamuna contained a total of 56 large island *chars*, each longer than 3.5 km. There are an additional number of 226 small island *chars*, varying in length between 0.35 and 3.5 km. This includes sandy areas as well as vegetated *chars*. In the Jamuna the period between 1973 and 2000, *chars* have consistently appeared in the reaches opposite to the Old Brahmaputra off takes, north and east of Sirajganj and in the southernmost reach above the confluence with the Ganges (EGIS, 2000; Jannatul et al., 2015; Islam and Gnauck, 2009).

In Bangladesh during 1981 to 1993, a total of about 729,000 people were displaced by river bank and *char-land* erosion. Floods are more or less a recurring phenomenon in Bangladesh and often have been within tolerable limits. But occasionally they become devastating. Each year in Bangladesh about 26,000 km^2, 18% of the country is flooded. During severe floods, the affected area may exceed 55% of the total area of the country.

In average year, 844,000 million m^3 of water flows into the country during the humid period (May to October) through the three main rivers namely the Ganges, the Brahmaputra-Jamuna and the Meghna. This volume is 95% of the total annual inflow. By comparison only about 187,000 million m^3 of stream flow is generated by rainfall inside Char-Janajat is one the large char in the Padma river basin. Fig. 2 shows the impacts of massive floods on the char-land in the Padma–Jamuna river channel in the Ganges delta. The entire char is inundated by annual floods. The inhabitants of char-land have to live with flood or to migrate in a safe place. Floods and river bank erosions are the most threatened natural phenomena for the Char-Janajat people, and almost 6000 people are facing the flood problems in the char-land. In the Purba Khas Bandarkhola mouza, there were 3000 people living in the past (Fig. 4).

But the population and settlements figure could not be estimated permanently because of random flooding, *char* land sliding and river bank erosion. Every year people have to move and adapt the strategy to live with floods. Whenever the river bank erosion creates fragile environmental situation the chaura people have to relocate from his original home. Fig. 4 shows the population and settlements increasing and decreasing trends at the Purba Khas Bandarkhola mouza in the Char-Janajat union (Fig. 1) of Madaripur district in Bangladesh. Fig. 3 shows the settlement and population increasing and decreasing tendency in the *char-land*. There a close relation of settlements and population in Char-lands. In Fig. 3, it has been stated that in 1964 there are 371 settlements and population was around 3000, and in 2000, there were only 82 settlements and population was 574, and in 2008, there were 171 settlements and the population was around 1400 (Fig. 4).

To understand char livelihoods of the riverine islands (bar) in the Padma River channel and their vulnerability to land erosion disasters, it is necessary to explore how human needs and wants interact and adapt with nature, transform nature into resources by developing its different facts. Therefore it is important to know the socioeconomic and livelihood system and resilient to the impacts of flood and char-land erosion hazards and vulnerability is determined by the interplay of a combination of several factors, such as awareness, condition of settlement stability and infrastructure (Mamun and Amin, 1999).

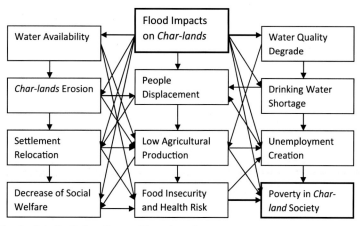

FIG. 3 Flood impacts on the *char-lands* in the Padma–Jamuna River channel.

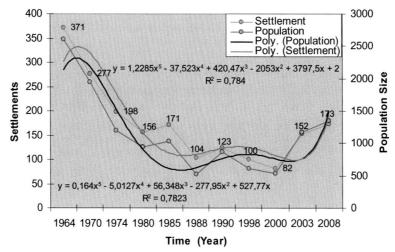

FIG. 4 Settlements and population in the Purba Khas Bandarkhola Mouza.

6.2 *Char-lands* erosion and accretion pattern in the Padma River basin

The riverine floodplain land is characterized by deltaic sediments of Quaternary formation. The combination of these sediments with high water content from the annual wet monsoon, a low degree of compaction and large amount of run off materials from water flows is the root cause to adjust their bed confutations (Hassan et al., 2000). In the Padma River channel may shift laterally by more than 300 m annually. This process makes *char-land* erosion a devastating natural hazard in Bangladesh that pushes people to displace for short time or some cases permanently (Hassan et al., 2000). The whole *char* is called Char-Janajat (Fig. 4) is unstable and prone to annual flooding and uncertain of rural Livelihoods. An estimated 35 thousand people live in the Char-Janajat and 1350 people are living in the Purba Khas Bandarkhola Mouza. In general, the population figure is also not stable figure, however, it is dependent on the stability of the land and erosional trends in the riverine channel. As the erosion and deposition process in the Padma river is a regular process in the Ganges river system. The Char-Janajat is extremely vulnerable to both erosion and flood hazards. The recent analysis of the time series Satellite images (from 1995 to 2008) indicates that over 80% of the area within the river channel had been char-lands of Char-Janajat in the Padma River Channel (Fig. 5).

The same analysis shows that about 80% of the Char-lands persisted between 1 and 10 years, while only 20% lasted for 20 years or more in Padma channel. Fig. 4 shows the large char geomorphic condition and landside pattern in 1995 (top) and in 2008 (bottom) (Fig. 5). Fig. 5 (top) shows that the whole Char (Char-Janajat) area was 84.09 km^2 and the area of Purba Khas Bandarkhola was measured 6.73 km^2 which is covering 21.07% land area of the entire mouza. The char was settled at the eastern side of the Padma River Channel which is near to Bhaggakul and Maoa Ferry Ghat of Munchiganj. The main river channel was very narrow (in 1995) in the eastern bank of the Padma river and it was wide in the western bank side. The major portion of char-land in 1995 was agricultural land and settlement land, rest of the area covered by sandy land and water body and fellow land. On the other hand, Fig. 5 (bottom) shows the Char-Janajat char and its shifting trends in 2008. The whole char size has reduced, approximately the area is 82 km^2 and the area of Purba Khas Bandarkhola is 5.70 km^2. That mean area has become smaller and land use pattern also changed. And the whole char has shifted to the west bank direction and main channel has also shifted from west bank to eastern bank side. The char-land is an unstable land and it is continuously shifting and resettles from East to west and north to south location within 20 km distance of the Padma River Channel in Madaripur and Munshigonj borders area.

Fig. 5 (bottom) shows the new formation and geomorphic pattern of Char-Janajat and the location of Purba Khas Bandarkhola mouza. The newly shifted in the west bank of the main river channel, the land is more fertile than eastern location. Almost 40% of the land was used for agricultural fertile land, 20% land is gradually developing for new agricultural land or shallow marshy land, 10% land is only sandy land, 2% land was fellow land, 5% land was used for grass land, and 3% land was used for water body, bazaars, and institution in the Purba Khas Bandarkhola mouza. The main Padma river channel which was flowing in the eastern side of river bank now it is flowing in the western bank but the wideness of the river channel is shrinking. The channel in the western bank side is almost closed in 2009 (Fig. 5). The main river channel has shifted again in the eastern bank side.

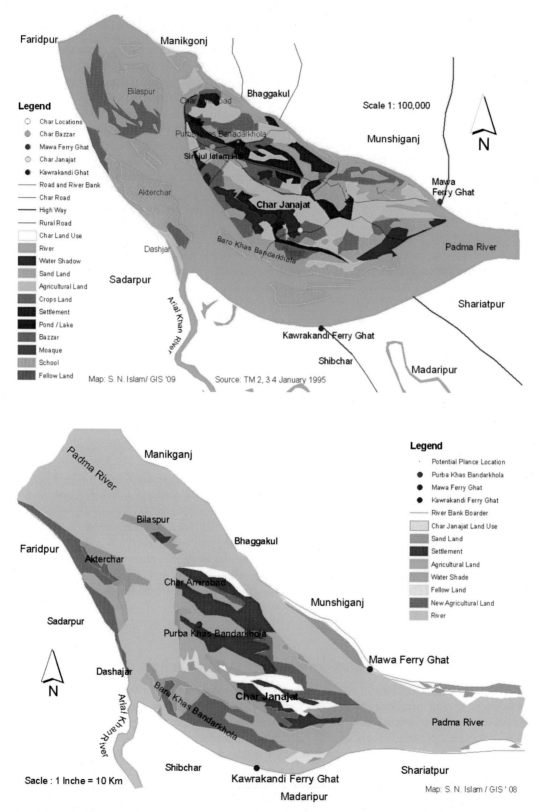

FIG. 5 GIS application in detecting of erosion and development of Char-Janajat in 1995 *(upper)* and 2008 *(bottom)* and Landuse pattern in the Ganges–Padma River channel (Jannatul et al., 2015; Islam and Gnauck, 2009).

7. Unstable settlement locations at Purba Khas Bandarkhola Mouza

The settlement relocation in the *char-land* in Bangladesh is a common feature to the people of Bangladesh and they are learnt how to survive with massive floods in the *char-lands*. The impacts of massive floods in *char-lands* are tragic and vulnerable which is demonstrating in Fig. 6. The river bank erosion and char land slide makes the Char-Janajat more vulnerable. The settlements, livelihood and the cropping systems are unstable over there. *Chaura people* always fight against poverty and food insecurity. Fig. 5 shows the quantity of settlement and relocations patterns in different times in the Purba Khas Bandarkhola Mouza of Sibchar Upazila, Madaripur district. The cyclic movement of people and their settlement relocation is occurring in the char-lands very rapidly in the Jamuna-Padma River channel. The *chaura people* of Purba Khas Bandarkhola Mouza (Fig. 6) of Char-Janajat are facing socioeconomic and habitat problems such as settlement displacement, agricultural crops production, communication and business (Fig. 2). The place where people settle down permanently or temporarily and build houses and create living environment and other facilities primarily this is called as settlement. In general it can be said that a place what is the shelter for human society and ecology is settlement.

On the other hand, a place where some community people construct houses, roads and community organization for livelihood is also called as settlement. From the time immemorial, when the culture and civilization has been developed people have tried to choose a peaceful place where they can take rest and food and can sleep which is called house or home, and gradually some homes made a settlement. Settlement is one of the fundamental basic needs of human life and gradually it has been extended to the different parts of the world. Settlement is the symbolic landmark of ancient and modern civilizations (Ahmed, 1965). The boundary lines of *char-land* villages are often of great antiquity. Whilst, the possible reasons for the origins of some nucleated designs or forms of house and homes have been postulated, there are four basic ways in which these villages could have evolved in the *char-land* (Elahi, 1987; Haq, 1981; Coleman, 1969) (Fig. 6).

- Growth from a single place,
- The agglomeration of several single place close enough to merge;
- The collapse of a pattern of dispersed settlements into one of nucleated village;
- Deliberate planning etc.

The villages of *char-lands* in Bangladesh, which occupy the central swath of planned landscapes, where rarely in existence before 1000 CE.

Village could be having developed anywhere, but were united by territorial boundaries of the land available for the subsistence of their inhabitants. The first process of development by steady growth from a single farmstead could have occurred when a family expanded and land was subdivided among the next generation (Fig. 6). The resulted form might be loose and amorphous if there were no features such as a road or road junctions along which to arrange the houses (Bhooshan, 1980). Fig. 6 shows the time series scenarios of char-land settlement relocations patterns. Fig. 6 (top left (1970) and top right (1980)) represented the time frame of 1970 to 1980 when the settlement pattern was cluster pattern. Fig. 6 (1988 middle left and 1990 middle right) shows the scenarios when the settlement pattern was mostly random pattern and cluster pattern. Fig. 6 (2000 (bottom (left) and 2008 bottom (right)) shows the time series scenarios of *char-land* settlement pattern where the settlements are distributed as random and semicluster basis, and this is one kind of mixed of random and cluster pattern (Fig. 6).

7.1 Cyclic displacement of Basir Uddin: Case analysis 1 (1960–2008)

Basir Uddin of Purba Khas Bandarkhola mouza was living with his family members in 1960. The family members of that time were 8 persons. Father, mother, brother and sisters were living together and they were leading a happy life in the char village of Padma river channel. It is time for Basir Uddin who can only realize the memory of char village life in the Padma River. He is now 78 years old; almost 50 years ago this Purba Khas Bandarkhola Mouza was a *Kaim-land* and soil was very fertile. Their house plot number was 2178 (Fig. 7) which is now under water. The river erosion and its impacts created a lot of negative attitude to the family members. The first time of *char-land* erosion in 1960 at that time they displace to the south site; it was about 2 km far from the original home. Where they displaced it was totally new place and they had no own land but fortunately they got the land for yearly rent basis of Taka 1000. After two years they had to move another new place of the same mouza its distance was more than 4 km far (Fig. 7). From the south side of that present location, they had to rent again a piece of land TK. 500 per year. After one year again, Basir Uddin moved to another new place of a new *char* of Char-Janajat in the Padma channel it was also almost 4 km far from the previous distance. They settled there only 3 years and after one year Basir Uddin decided that he should move to far distance where there is no river and river bank or *char-land* erosion and it will be safer than present horrible *char* life in char Janajat.

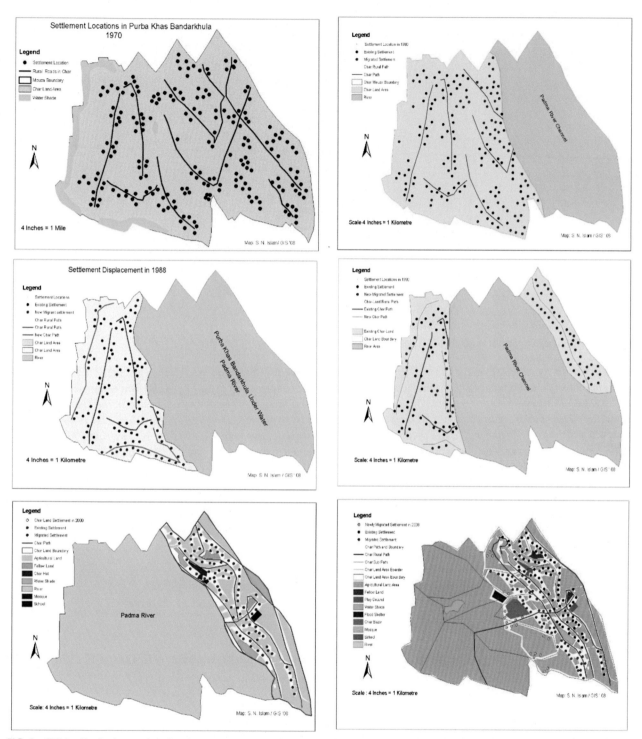

FIG. 6 GIS Application in mapping of settlement patterns in: 1970 *(top left)*; 1980 *(top right)*; 1988 *(middle left)*; 1990 *(middle right)*; 2000 *(bottom left)* and 2008 *(bottom right)* in Purba Khas Bandarkhola Mouza of Shibchar Upazila, Madaripur district in Bangladesh.

At that time Basir Uddin went to Sibchar and shifted there at the main land; its distance about 16 km from his home and he shifted in 1989 with rented land (Fig. 7). Basir Uddin and his family members were living at Sibchar for the more than 13 years after this period he again came back at his own land in 2003 in the Char-Janajat where his forefather was settled down and other family members (one brother) came back with him to the same *char*, but to another plot (Fig. 7).

The distance from Sibchar to the original living place in Char-Janajat is about 10 km. Now it is the time Basir Uddin is entirely tired for this hostile displacement of houses and life is running one after another moment. Within his long time life

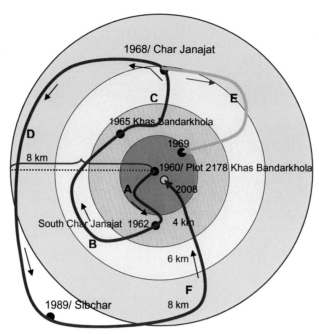

FIG. 7 People dislodgment model based on Basir Uddin's life cycle in the *Char-land*.

experience with river, river bank, *char-land* erosion and displacement in the *char-lands* he is extremely tried to find out some alternative solution to mitigate the impacts of floods and river bank erosion in the *char-land*. But he himself stated that "*There is no alternative solution to solve this natural problem, and there is no easy way to survive in the char-lands. The way we are living this is reality and we will have to live with floods, river bank and char-land erosion. This is the reality, if we would like to stay here and would like to live in the char-land we will have to face and adapt this environmental calamity within a natural process.*" The *char* people are mobile and settlements are dynamics and displacement of *char-land* is the normal process of *char* life and it is moving within a cyclic process (Fig. 7). The case study analysis and its finding shows that the people of *char-lands* are naturally habited to deal with environmental problems and they know how lead the critical *char* environment in the *char-lands*. The dislodgment model shows the cyclic displacement of Basir Uddin within the char and outside of *char-land*. Considering the model, the distance of displacement can be measured on the following formula;

Displacement Length (DL) = Starting Point + Distance of Ending Point of Displacement

A (1960–1962) = 3 km
B (1962–1965) = 4 km
C (1965–1968) = 4 km
D (1968–1969) = 16 km
E (1968–1969) = 5 km
F (1989–2008) = 14 km

Therefore Displacement Length(DL) = A(3 km) + B(4 km) + C(4 km) + D(16 km) + E(5 km) + F(14 km)
= (3 + 4 + 4 + 16 + 5 + 14) km
= 46 km

The total displacement length (DL) is 42 km

The duration of living in char-land of Basir Uddin is 48 years (1960–2008).

Yearly average dislodgment length is 46/48 = 0.976 km.

The total dislodgment time is 6 (times within 43 years), and average dislodgment trends is 43/6 = 7.16 years that means Basir Uddin displaced his home every 7 years on average and every time the average distance was 42/6 = 7 km within the Purba Khas Bandarkhola and outside of Char-Janajat.

7.2 Cyclic displacement of Omar Ali: Case analysis 2 (1945–2018)

Omar Ali Sheikh son of Abdul Gani Sheikh was living in the (Mouza Seat no 2) of Purba Khas Bandarkhola mouza of Bandarkhola union. Abdul Gani Sheikh was settled down at Bandarkhola mouza in 1945 (Fig. 8). It was good and

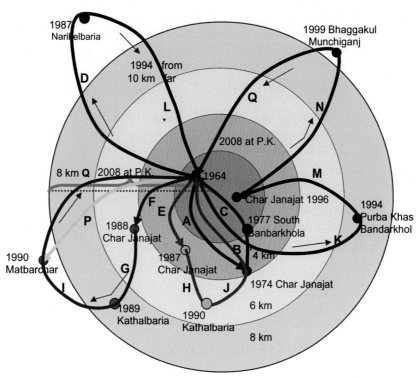

FIG. 8 The displacement model of and life cycle of Omar Ali in Char-Janajat.

comfortable life for them because they had agricultural land and he had 10 family members. They cultivated agricultural crops which was enough for the whole year. The floods and other natural calamities were not creating any massive damage of their land property and resources. Later on Omar Ali was shifted at the same char in another plot with all the 10 family members (4 brothers, 4 sisters and parents). He had 4 houses (Tin Shed), his occupation was agriculture. This land area was 3 acres; it was enough for the supply of food grain for the entire year.

In 1964, Omar Ali's houses were affected due to massive flood and river bank erosion in the Padma river channel in the Ganges delta in Bangladesh. He decided to move from his father land and gradually he shifted 5 km far from in south Char-Janajat in 1974. Only his elder stayed in their original village, but eventually moved to Char Janajat in (West) in 1987. In 1990, he relocated again to Kathalbari union and returned to Char-Janajat in 1996. He returned to his original home place in 2008. The relocation and cyclic migration were largely due to erosion and displacement at various times and available support and assistance from kin groups at different locations. This is a very common pattern associated with displacement and migration among char people (Islam, 2021).

The second brother was also relocated in 1988 at west Char-Janajat when the massive flood was occurred in Bangladesh and his house was totally damaged. He shifted to Kathalbaria union in 1989 and was living only one year and then he shifted to Madbarchar to his brother place in 1990; he was staying there only 10 months and again returned to his parent land in 2008. The second brother came back to the original plot in the Bondarkhola mouza in 1987 and the same year he again shifted to the east corner of the mouza and in 1994 he again shifted to the middle place of Purba Khas Bandarkhola mouza. The third brother also shifted in 1990 at Matbar *char* of Shibchar Upazila which is 8 km far from the original living place in Purba Khas Bandarkhola mouza. He is still living there as permanently. Omar Ali came back to Purba Khas Bandarkhola mouza in 1977 and again shifted in 1987 to Narikelbaria which is 11 km far from Purba Khas Bandarkhola mouza and it is located in Sadarpur Upazila of Faridpur district. Omar Ali was staying at Narikelbaria for the last 7 years. Due to social and economic and cultural attraction again decided to come back to the original mother land in Purba Khas Bandarkhola mouza when this char-land ridged in the Padma River basin. In 1994 Omar Ali came back to Purba Khas Bandarkhola mouza at the original place where his parents were living; and at the same year again he shifted to Bhaggakul of Munchiganj district in 1999 and at the same year he camelback to Purba Khas Bandarkhola mouza. He was living in Bhaggakul more than 8 years and returned to Purba Khas Bandarkhola mouza in 2008.

Omar Ali and his family members (10 members) have been displaced and shifted their houses 17 times (Fig. 8) within 46 years that mean every 3 years interval they had to move from one place to another. The dislodgment model which is displaying the displacement of trends of Omar Ali in the Purba Khas Bandarkhola mouza such as in Char-Janajat union of Sibchar Upazila. The length of cyclic displacement can be measured the following formula;

Displacement Length (DL) = Starting Point + Distance of Ending Point of Displacement

A (1964–1974) = 4 km
B (1974–1977) = 2 km
C (1977–1979) = 3 km
D (1979–1987) = 11 km
E (1979–1987) = 2 km
F (1987–1988) = 4 km
G (1988–1989) = 4 km
H (1987–1990) = 2 km
I (1989–1990) = 2 km
J (1990–1994) = 4 km
K (1992–1994) = 5 km
L (1987–1994) = 11 km
M (1994–1996) = 5 km
N (1996–1999) = 8 km
O (1990–1992) = 6 km
P (1987–1990) = 6 km
Q (1999–2008) = 8 km

$$\begin{aligned}
\text{Therefore Displacement Length(DL)} = {} & A(4\,km) + B(2\,km) + C(3\,km) + D(11\,km) + E(2\,km) + F(4\,km) + G(4\,km) \\
& + H(2\,km) + I(2\,km) + J(4\,km) + K(5\,km) + L(11\,km) + M(5\,km) + N(8\,km) \\
& + O(6\,km) + P(6\,km) + Q(8\,km) \\
= {} & (4\,km + 2\,km + 3\,km + 11\,km + 2\,km + 4\,km + 4\,km + 2\,km + 2\,km + 4\,km \\
& + 5\,km + 11\,km + 5\,km + 8\,km + 6\,km + 6\,km + 8\,km) = 87\,km
\end{aligned}$$

The total length of displacement is 87 km and the total displacement was 17 times.

The duration of living in *char-land* of Omar Ali is 46 years (1962–2008).

Yearly dislodgment length is 87/46 = 1.977 km.

The total dislodgment time is 17 (times within 46 years), and average dislodgment trends is 46/17 = 2.588 years; that means Omar Ali displaced his home every 3 years on average and every time the average displacement distance was 87/17 = 5.11 km within the Char-Janajat and outside of Char-Janajat. Omar Ali was living and working in the char-lands for the last 50 years and displacing one char to another place. He is an old man who is now operating a country boat and serving over there. Although it is hard life for him to maintain his family but there is a pull and pushing attraction for his motherland which is a cultural identity of *char* people. The social and cultural and rural power political attractions are also the reasons for coming back to *char-land*, and it is a cyclic process and continuing for a hundred of years in the Riverine catchments in the Padma and Jamuna river basins in Bangladesh (Islam, 2020).

7.3 Discussion on dislodgment model results

There are two case studies have been analyzed based on quantitative and qualitative data sources. The results of these two dislodgment model of char people cyclic displacement in different location within the *char* land and outside the *char-land*. Table 3 shows the model results of two case studies on Basir Uddin (Case 1) and Omar Ali (Case 2) of Purba Khas Bandarkhola mouza of Sibchar Upazila.

The result show that the people of char life of the Padma River channel is cyclic movement and unstable life. The average displacement of Basir Uddin and Omar Ali is 61 km and they were living on average for the last 45 to 50 years. The yearly average distance of displacement is 1.35 km; the interval of displacement is every 5 years. They have displaced in different mouzas and outside of the char in 10 times. The models (Figs. 7 and 8) show the cyclic displacement locational tendency of two *char* family members, the average family members are 9, it has been stated that Basir Uddin and Omar Ali have been displaced in different places with 8 and 10 family members. There are some potential factors including pull and push factors of migration and displacement of *char* people, besides this social security, *char* rural power politics, fourth-generation land property and cultural identity are the root causes to return to the native *char* after a long staying at outside of the native *char*. This is very much as usual scenarios in Char-Janajat of Padma River channel (Islam, 2021).

TABLE 3 The results of dislocation models.

Case studies	Total length km	Living time in Char (year)	Interval of displace (year)	Distance (p/y km)	Average distance (km)	Family member	Displace time
Case 1 Basir Uddin	46	48	7.16	0.976	7	8	6
Case 2 Omar Ali	87	46	2.59	1.977	5.11	10	17
Average	64.5	44.5	4.87	1.476	6.55	9	12.5

8. Conclusions

The *char* settlement relocation, people displacement, rural livelihood and annual foods are interlink age in the riverine Bangladesh. Almost every year more than 30 thousands *chaura people* of Char-Janajat are facing and struggling against the floods and *char-land* erosion. The study finding shows that the flood and *char-land* erosion is the main reason of people cyclic dislodgment and settlement relocation in the same *char* and outside of the *char* within 90 km^2 range in the Padma River channel in Sibchar Upazila of Madaripur District. The Purba Khas Bandarkhola Mouza is one of the vulnerable unstable *char-land* in Bangladesh where the *chaura people* cyclically moving and migrating because of unstable *char-land* and uncertain livelihood. The finding also shows that the mouza has been changed its shaped and size due to erosion and accretion due to devastating floods. Therefore the *char* inhabitants have been displaced maximum 17 times and minimum 6 times within 50 years of *char* life which is a threat for *char* livelihood and sustainable *char-lands* development. The floods in the *char-lands* erosion is a challenge for managing *char* settlements, cultural landscape protection, agricultural cropping systems maintain, crop biodiversity and riverine ecology of Char-Janajat island. It is necessary to find an alternative approach or adaptation strategies for sustainable livelihood in the *char-lands* of Padma River channels in the Ganges–Brahmaputra delta in Bangladesh. The findings of this study are strongly recommending the following concluding remarks that could be implemented in Purba Khas Bandarkhola Mouza as well as in Char-Janajat of Sibchar Upazila in Madaripur District as alternative approach to solve the *char-land* erosion people displacement and settlement relocation problems in the river basins in Bangladesh.

- The Government, NGOs and social organizations needs to make an effect to develop methodologies mechanism, policy making, validate tools, impact assessment indicators and instruments for flood and *char-land* erosion protection and management at micro level in the of the *char-lands* in the river channels.
- The Geographical Information Systems (GIS), Remote Sensing techniques should be introduced in floods and river bank erosion monitoring, *char-land* data analysis, visualization, mapping, planning and modeling for future flood forecasting and *char-land* erosion control and effective management measures.
- Hydro-engineering dams or embankments could be constructed to protect the *char-land* people displacement, settlements and people and their property.
- Flood monitoring, forecasting and warning systems should be developed and more transparent to the local people.
- The community participation, local adaptation strategy, awareness education, and applied research on floods, *char-land* erosion and agricultural cropping systems should incorporate in the national development agenda. Local settlers would be equal partners in the *char-land* development for safer settlement and livelihood by incorporation indigenous knowledge, skills and capacities for ecological management and a sustainable livelihood in *char-land* that could be the reality in riverine environmental context in Bangladesh.
- The practical alternative should be to promote and strongly support local, regional and national level program and initiatives to enable societies to become resilient to the negative impact of *char-land* erosion in the river channel.
- An interdisciplinary integrated *char-lands* settlement and migration policy development plan should be developed based on the result of river bank and *char-land* erosion, traditional cropping system in the Purba Khas Bandarkhola Mouza of Sibchar Upazila of Madaripur district in Bangladesh.

Acknowledgments

The chapter incorporates ideas and findings of research I have jointly carried out with my colleagues Shilpa Sing, Hasibush Shaheed, and Shouke Wei (see Islam, 2010a,b) and the work carried by me on floods, char-land erosion, and settlement displacement in the Ganges–Padma River Basin (see Islam, 2021). I acknowledge their contributions and remain thankful for their support as well as thanks to the char dwellers those gave me a lot of information concerning char settlement and livelihoods issues.

References

Ahmed, N., 1965. Rural settlement in East Pakistan. Geograph. Rev. 46, 388–398.

Anwar, J., 1988. Geology of coastal area of Bangladesh and recommendation for resource development and management. In: National Workshop on Coastal Area Resource Development and Management, Part II. Dhaka: Coastal Area Resource Development and Management Association. Academic Publishers, pp. 36–56.

Baqee, M.A., 1985. Violence and agricultural seasonality in Char-lands of Bangladesh. In: Oriental Geographer. Vol. 29-30. Bangladesh Geographical Society, Dhaka, pp. 25–36.

Baqee, M.A., 1993a. Barga and Bargadar, The case of the Char-lands of Bangladesh. In: Oriental Geographer. 37. BGA, Dhaka University, Dhaka, Bangladesh, pp. 1–17.

Baqee, M.A., 1993b. The Settlement Process in the Char-Lands. Unpublished PhD dissertation, Department of Geography, Dhaka University, Bangladesh.

Baqee, M.A., 1997. Coping with floods and erosion in Bangladesh char-lands. Asia Pacific J. Dev. 4 (2), 38–52.

Baqee, M.A., 1998a. Grameen Bashati (Bengali). Presidency Press, Dhaka, Bangladesh.

Baqee, M.A., 1998b. Peopling in the Land of Allah Jaane Power, Peopling and Environment: The Case of Char-Lands of Bangladesh. The University Press, Dhaka.

Bhooshan, B.S., 1980. Toward Alternative Settlement Strategies: The Role of Small and Intermediate Centres in the Development Process. Heritage, London.

Coleman, J.M., 1969. Brahmaputra River channel process and sedimentation. Sediment. Geol. 3, 129–239.

Coleman, J.M., 1969. Brahmaputra River channel process and sedimentation. Sediment. Geol. 3 (2–3), 129–239.

Curry, B., 1986. Changes in Chilmari: looking beyond rapid rural appraisal and farming systems research methods. In: Paper Presented in the Workshop on Water Systems July, Dhaka, Bangladesh, pp. 19–22.

Curry, B., 1996. Looking beyond rapid rural appraisal and farming systems research methods. In: Environmental Aspects of Agricultural Development in Bangladesh.

EGIS, 2000. Environmental base line of Gorai river restoration project-environment and GIS support project for water sector planning Bangladesh and the Government of the Natherlands. Environmental Geographical Information Services, pp. 1–150.

Elahi, K.M., 1987. Rural Bastees and the phenomena of rural squatting due to riverbank erosion in Bangladesh. In: Seminar Paper on Shelter for the Homeless. Urban Development Directorate, Dhaka.

Elahi, K.M., 1989. Population displacement due to river bank erosion of the Jamuna in Bangladesh. In: Clark, J.I., et al. (Eds.), Population and Disasters. Basil Blackwell, Oxford, UK.

Elahi, K.M., 1991. Riverbank erosion, flood hazards and population displacement in Bangladesh. In: Elahi, K.M., et al. (Eds.), Riverbank Erosion Impact Studies, Jahangirnagar University, Momin Offset Press, and Dhaka. Bangladesh.

Elahi, K.M., Das, S.C., Sultana, S., 1998. Geography of coastal environment: a study of selected issues. In: Bayes, A., Mahammad, A. (Eds.), Bangladesh at 25: An Analytical Discourse on Development. University Press Limited, Dhaka, pp. 336–368.

FAP, 21/22, 1993. The Dynamic Physical and Human Environment of Riverine Char-Lands. Floods Plan Coordination Organization. Ministry of Irrigation, Water Development and Flood Control, Meghna, Dhaka (Report prepared by Irrigation support Project for Asia and the Near East).

Flood Action Plan (FAP 24), 1996. River Survey project—morphology of Gorai off-take: Special Report No.: 10. WARPO, Dhaka.

Hanna, S.W., 1996. Living with the Floods Survival Strategies of Char-Dwellers in Bangladesh Asa Programme, Berlin, and Germany.

Hanna, S.W., 2001. Facing the Jamuna River-Indigenous and Engineering Knowledge in Bangladesh. Bangladesh Resource Centre for Indigenous Knowledge (BARCIK), Bersha (Pvt) Ltd, Nilkhet, Dhaka. 242 pp.

Haq, S., 1981. Rivers of Bangladesh and their floods. J. BNGA 4, 18–28. Jahangirnagar University, Dhaka.

Haque, E.C., 1997. Hazards in a Fickle Environment: Bangladesh. 1997 Kluwer Academic Publishers, The Netherlands, pp. 6–13.

Haque, M., 1985. Impact of riverbank erosion in Kazipur, an application of Landsat data. In: REIS Workshop, Jahangirnagar University, Dhaka.

Haque, M., 1999. Indigenous knowledge and practices in disaster management in Bangladesh. Vol. II (2 & 3) Grassroots Voice, Dhaka, Bangladesh.

Hassan, M., Haque, M.S., Saroar, M., 2000. Indigenous knowledge and perception of the *Charland* people in cropping with natural disasters in Bangladesh. In: Sen, S., Khan, N.A. (Eds.), Grassroots Voice – A Journal of Resources and Development. Vol. III. BARCIK, Dhaka, Bangladesh, pp. 34–44.

Hooper, A.G., 2001. Coping with river floods in Bangladesh. In: Carpenter, T.G. (Ed.), The Environmental of Constructions. Wiley Publication, UK.

Islam, A., 1995. Environment Land Use and Natural Hazards in Bangladesh. University of Dhaka. Dhanshiri Mudrayan, New Market, Dhaka, Bangladesh, pp. 227–276.

Islam, N., 1993. Rural housing in Bangladesh: An overview in search of new strategies. In: Oriental Geographer. vol. 37. BGA, Dhaka University, pp. 47–59.

Islam, S.N., 2010a. Char-lands erosion, livelihoods and cyclic displacement of people in Ganges-Padma River basin in Bangladesh. Asia-Pacific J. Rural Dev. 20 (1), 151–174.

Islam, S.N., 2010b. Threatened wetlands and ecologically sensitive ecosystems management in Bangladesh. Front. Earth Sci. China 4 (4), 438–448.

Islam, S.N., 2014a. An analysis of the damages of Chakoria Sundarban mangrove wetlands and consequences on community livelihoods in south east coast of Bangladesh. Int. J. Environ. Sustain. Dev. 13, 153. https://doi.org/10.1504/IJESD.2014.060196.

Islam, S.N., 2014b. An analysis of the damages of Chakoria Sundarbans mangrove wetlands and consequences on community livelihoods in south east coast of Bangladesh. Int. J. Environ. Dev. 13 (2), 153–171.

Islam, S.N., 2016a. Deltaic floodplains development and wetland ecosystems management in the Ganges-Brahmaputra-Meghna Rivers Delta in Bangladesh. Sustain. Water Resour. Manag. 2 (3), 237–256. https://doi.org/10.1007/s40899-016=0047-6.

Islam, S.N., 2016b. Deltaic floodplains development and wetland ecosystems Management in the Ganges-Brahmaputra-Meghna Rivers Delta in Bangladesh. Sustain. Water. Resour. Manag. 2, 237–256.

Islam, S.N., 2020. Sundarbans a dynamic ecosystem: An overview of opportunities, threats and tasks. In: Sen, H.S. (Ed.), The Sundarbans: A Disasters Prone Eco-Region, Coastal Research Library. Springer Nature, Switzerland, pp. 31–58.

Islam, S.N., 2021. Floods, Charland erosions and settlement displacement in the Ganges-Padma River basin. In: Zaman, M., Alam, M. (Eds.), Living on the Edge. Springer Geography, Springer, Cham, https://doi.org/10.1007/978-3-030-73592-0_125_18.

Islam, S.N., Gnauck, A., 2007. Effects of salinity intrusion in mangrove wetlands ecosystems in the Sundarbans: an alternative approach for sustainable management. In: Wetland: Monitoring and Management. Leider, Taylor and Francis, Rotterdam, the Netherlands, pp. 315–322.

Islam, S.N., Gnauck, A., 2009. AAPG HEDBERG CONFERENCE "Variations in fluvial-deltaic and coastal reservoirs deposited in tropical environments". In: The Coastal Mangrove Wetland Ecosystems in the Ganges Delta: A Case Study on the Sundarbans in Bangladesh.

Islam, S.N., Gnauck, A., 2009. Threats to the Sundarbans mangrove wetland ecosystems from transboundary water allocation in the Ganges basin: a preliminary problem analysis. Int. J. Ecol. Econ. Stat. 13 (09), 64–78.

Islam, S.N., Gnauck, A., 2011. Food security and ecosystem services under threat in the coastal region of Ganges Delta in Bangladesh: Preliminary result analysis. In: Gnauck, A. (Ed.), Modelling and Simulation of Ecosystems. Shaker Verlag-Aachen, Germany, pp. 158–274.

Islam, S.N., Gnauck, A., Voigt, H.-J., 2014a. Hydrological change in mangroves. In: Handbook of Engineering Hydrology. Taylors and Francis CRC Press, pp. 369–390.

Islam, S.N., Jenny, Y.S.L., Mamit, N.A.B.H., Matusin, A.M.B., Bakar, N.S.B.A., Suhaini, M.H.B., 2018a. GIS application in detecting forest and bush fire risk areas in Brunei Darussalam: case analysis on Muara and Belait districts. In: IET Conference publications. Institution of Engineering and Technology., https://doi.org/10.1049/cp.2018.1537.

Islam, S.N., Karim, R., Islam, A.N., Eslamian, S., 2014b. Wetland hydrology. In: Handbook of Engineering Hydrology: Fundamental and Application. Vol. 2. Taylors and Francis CRC Press, pp. 581–605.

Islam, S.N., Karim, R., Islam, A.N., Eslamian, S., 2014c. Wetland hydrology. In: Eslamian, S. (Ed.), Handbook of Engineering Hydrology: Fundamentals and Applications. Taylor & Francis Group, New York, pp. 581–605.

Islam, S.N., Mohamad, S.M.B.H., Azad, A.K., 2019. Acacia spp.: invasive trees along the Brunei coast, Borneo. In: Makowski, C., Finkl, C.W. (Eds.), Impacts of Invasive Species on Coastal Environments: Coasts in Crisis. Springer International Publishing AG, Cham, Switzerland, pp. 455–476, https://doi.org/10.1007/978-3-319-91382-7_14.

Islam, S.N., Reinstädtler, S., Aparecida de Sa'Xavier, M., Gnauck, A., 2017. Chapter 28: Food security and nutrition policy. In: Eslamian, S., Eslamian, F.A. (Eds.), Handbook of Drought and Water Scarcity. Vol. 1: Principles of Drought and Water Scarcity. CRC Press, Taylor & Francis Group, Geneva, New York.

Islam, S.N., Rahman, N.H.H.A., Reinstädtler, S., Aladin, M.N.A.B., 2018b. Assessment and management strategies of mangrove forests alongside the Mangsalut River basin (Brunei Darussalam, on the Island of Borneo). In: Coastal Research Library. Springer, pp. 401–417, https://doi.org/10.1007/978-3-319-73016-5_18.

Jannatul, F., Islam, S.N.R., Eslamian, S., 2015. Ethical and cultural dimensions of water reuse in global aspects. In: Urban water Reuse Handbook. Taylor and Francis CRC Press, pp. 285–296.

Lahiri-Dutt, K., 2000. Imagining rivers. Econ. Polit. Wkly. 35 (27).

Lahiri-Dutt, K., 2015. Towards a more comprehensive understanding of rivers. In: Lyer, R. (Ed.), Living Rivers. Dying Rivers. Oxford University Press, New Delhi, India.

Mafizuddin, M., 1983. The physiography of Bangladesh. In: Bhugul Patrika. Vol. 2 (in Bengali). Jahangirnagar University, Dhaka, Bangladesh.

Mamun, M.Z., Amin, A.T.M.N., 1999. Densification: A Strategic Plan to Mitigate Riverbank Erosion in Bangladesh. The University Press Limited, Dhaka, pp. 24–27.

Miah, M.M., 1988. Flood in Bangladesh: A Hydromorphologial Study of the 1987 Flood, Dhaka. Academic Publishers, Dhaka, Bangladesh.

Reinstädtler, S., Islam, S.N., 2017. Drought management for landscape and rural security. In: Eslamian, S., Eslamian, F. (Eds.), Handbook of Drought and Water Security. Management of Drought and Water Security, Vol. 3. Taylor and Francis CRC Press, pp. 195–234.

Smith, K., 1996. Natural disasters: De5nitions, data base and dilemmas. Geogr. Rev. 10, 9–12.

Wiest, R.E., 1987. Riverbank erosion impact in Bangladesh: an assessment of findings and approaches. In: South Asian Horizon. Vol. 5.

Zaman, M., Alam, M., 2021. The Delta frontiers: history and dynamics. In: Living on the Edge-Char Dwellers in Bangladesh. Springer Geography, pp. 15–24.

Zaman, M.Q., 1988. The Socioeconomic and Political Dynamics of Adjustment to Riverbank Erosion Hazard and Population Resettlement in the Brahmaputra-Jamuna Floodplain. PhD dissertation, Department of Anthropology, University of Manitoba, Canada.

Zaman, M.Q., 1989. The social and political context of adjustment to river bank erosion hazard and population resettlement in Bangladesh. Hum. Organ. 48 (3).

Chapter 12

Groundwater level forecasting using hybrid soft computing techniques

Krishnamurthy Nayak[a] and B.S. Supreetha[b]

[a]*Department of Electronics and Communication Engineering, Manipal Institute of Technology, Manipal Academy of Higher Education (MAHE), Manipal, India,* [b]*Department of Electronics and Communication, Manipal Institute of Technology, Karnataka, India*

1. Introduction

Groundwater is one of the most important sources of domestic, commercial, and agricultural supply. According to the National Groundwater Management Improvement Program (NGMIP) (National GW Management Improvement Program, 2016), seven states namely Gujarat, Maharashtra, Haryana, Karnataka, Rajasthan, Uttar Pradesh, and Madhya Pradesh have some of India's most highly developed, but most heavily exploited groundwater areas. NGMIP program reported that 25% of groundwater (GW) blocks are critical, indicating the worsening situation. In countries such as India, 85% of drinking water and 60% of irrigation needs are dependent on the GW resources. Climate change and depletion of aquifers cause depletion of groundwater. The depletion of GW may lead to serious consequences for food security and other factors that affect the country's agricultural production and economic growth. Researchers are working hard to boost the precision of Groundwater Level (GWL) forecasting prediction. With the rapid growth in hybrid soft computing technologies, demand for GWL predictions is growing on a regional scale (Dash et al., 2010). In addition, the Central Ground Water Board and State Board Department, Government of India has introduced a program aimed at developing and using various tools to forecast the GWL and interpret the economy. Long-term, comprehensive water level measurements provide critical data required to determine resource changes over time and space. These measurements are critical for developing GW models and forecasting patterns and monitoring the effectiveness of GW management and protection programs. In addition, depletion of GW sources and pollution of GW are issues, which will become increasingly relevant as further aquifer development takes place in every basin. In addition, a continuous forecast of groundwater levels is needed for an overall development of the basin in order to use any simulation model effectively for water management.

Several studies have been carried out in this direction to forecast Groundwater Level using physical models that are not only laborious but also have realistic limitations, since many interrelated variables are involved. A groundwater model is a simplified, representation of groundwater systems. The Groundwater models provide further insight into the dynamic actions of the system and may help to improve conceptual understanding. In addition, once they have been demonstrated to accurately repeat past behavior, they can predict the outcome of future groundwater behavior, improve decision making and enable alternative management strategies to be explored.

The GWL forecasting systems have proven to be successful in monitoring the effectiveness of GW management and protection programs. The basic aim of GWL forecasting system is to accurately and effectively predict the future GWL trend. Since large amount of variations is involved in GW analysis such as variation in geological formations, data scarcity, environmental condition, and so on, the GWL forecasting has become a complex task. Due to these design challenges, GWL forecasting has been an active research domain for more than 70 years. The models for forecasting the GWL may be categorized as physical or mathematical physical model is a simulation of physical processes, typically on a smaller scale than in the field.

Groundwater resources of a country constitute one of tits vital assets. The assessment of groundwater resources is carried out to determine the prevailing groundwater levels in the country. In India, the Central ground water Board (CGWB) and state governments jointly estimate the dynamic groundwater resources at periodic intervals. As per the latest assessment, the details about the dynamic groundwater resources, availability and utilization are as given in Table 1 (Groundwater Yearbook-India, 2017–18, 2018).

The major source of groundwater recharge in India is the monsoon rainfall. The overall contribution of rainfall to annual recharge of groundwater resources and its utilization details are as shown in Fig. 1.

Handbook of HydroInformatics. https://doi.org/10.1016/B978-0-12-821962-1.00001-5

TABLE 1 Details of dynamic groundwater resources.

Description	India	Karnataka	Udupi
Annual groundwater resources	447 BCM	15.93 BCM	0.53042 BCM
Net groundwater availability	411 BCM	15.30 BCM	0.45273 BCM
Annual groundwater utilization	253 BCM	10.71 BCM	0.17711 BCM
Stage of groundwater development	62%	70%	39.12%

BCM, billion cubic meter (Groundwater Yearbook-India, 2017–18, 2018).

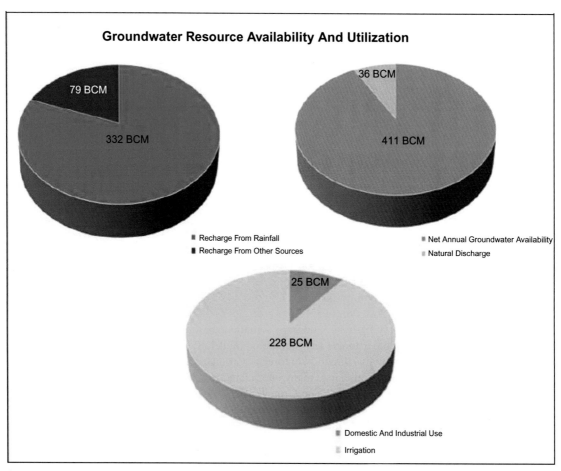

FIG. 1 Groundwater resource availability and utilization.

The overall stage of groundwater development in the country is below 70%, which implies that the annual groundwater consumption is less than recharge. The increased groundwater demands relative to groundwater availability and the need for efficient future management of groundwater resources emerges. The water availability per capita is decreasing dramatically as the population in India is expected to rise to 1.6 billion by 2051 (Groundwater Yearbook-India, 2017–18, 2018). The decline in groundwater level is a severe issue in Karnataka, and because of the growing use of water, the state faces water deficits and is unlikely to meet future agricultural, industrial and household water requirements (Raghavan et al., 2017). Thus, the estimates in the report on the dynamics of groundwater resources reveal that the demand from the agriculture sector is a significant threat to sustainability (National Compilation in Dynamic Groundwater Resources of India-2017, 2019). As per the latest report, around 1034 administrative units across India have been categorized as over-exploited, 253 as critical, 681 as semicritical and 4520 as safe. The reports estimate that the number of over-exploited and critical administrative units is significantly higher in South Indian states. The reports also reveal that 985 wells out of 1421 wells surveyed in Karnataka show a decline in GWL, and there is a continuous decline observed in the Udupi district (National Compilation in Dynamic Groundwater Resources of India-2017, 2019).

The natural GW systems are complex and have several parameters which are highly variable over time and space Researchers have been trying to explore these parameters for the creation of GW models. Therefore, advanced GW models are required to simplify this complex behavior (Pinder, 2002). With increasingly growing computing capacity, GW models have become a standard method for Skilled hydro geologists to perform most of the tasks effectively (Kumar and Singh, 2015). Today's hydrological forecasting systems are affordable and efficient as the technology Advances (Kumar, 2018). Physical hydrological models are very basic models, which require information about the physical characteristics of the study area and boundaries of a GW system using simplifying assumptions (El Alfie, 2014). The importance of these forecasting model for environment and water management is growing with urbanization and climate variability. Mathematical models are further categorized as empirical lumped conceptual and physically based models. Physically based models use physically measurable static input variables and require extensive information of the study area. Measuring the physical properties is difficult, especially for predictive models where the input values change over time (Satish and Elan Go, 2015).

2. Governing equation for groundwater flow and data driven groundwater level forecasting models

The basic model of groundwater dynamics under stationary flow is governed by Darcy's law of water motion through porous media. The flow through aquifers, most of which are natural porous media, can be expressed by Darcy's law. A primary tool for modeling water resources is an empirical relationship for liquid flow through a porous medium (Vazquez Baez et al., 2019). As per conclusions made by Darcy, the flow rate is directly proportional to the cross-sectional area (A) of the filter, and to the difference in piezo metric heads Φ_1 and $\Phi2$ inversely proportional to the length of the filter. After combining these conclusions, the Darcy's formula can be written as in Eq. (1).

$$Q = KA \left(\frac{\phi_1 - \phi_2}{L} \right) \tag{1}$$

Therefore, the discharge rate is given by Eq. (2).

$$q = \frac{Q}{A} = K \left(\frac{\phi_1 - \phi_2}{L} \right) \tag{2}$$

$$q = KJ$$

where,

$$J = \frac{\phi_1 - \phi_2}{L}, \quad \phi_1 = Z1 + \frac{P_1}{r}, \quad \phi_2 = Z2 + \frac{P_2}{r}$$

Z1 and Z2 are the datum head or elevation head and $\frac{P_1}{\gamma}, \frac{P_2}{\gamma}$, are the pressure head, ϕ_1, ϕ_2 are piezo metric heads and L is length of the filter.

The generalized three-dimensional form of the equation can be expressed as

$$q = KJ \tag{3}$$

where,

$$q = \begin{Bmatrix} q_x \\ q_y \\ q_z \end{Bmatrix} \quad J = \begin{Bmatrix} -\frac{\partial \varphi}{\partial x} \\ -\frac{\partial \varphi}{\partial y} \\ -\frac{\partial \varphi}{\partial z} \end{Bmatrix} \quad K = \begin{bmatrix} K_{xx} & K_{xy} & K_{xz} \\ K_{xy} & K_{yy} & K_{yz} \\ K_{xz} & K_{yz} & K_{zz} \end{bmatrix}$$

$$J_n = \frac{\varphi_1 - \varphi_2}{\Delta n} = -\frac{\varphi_2 - \varphi_1}{\Delta n} = -\frac{\partial \varphi}{\partial n},$$

The hydraulic gradient is negative along the direction of GW flow, since it flows from higher hydraulic head to lower hydraulic head. The above equation can also be written as Eq. (4):

$$\begin{Bmatrix} q_x \\ q_y \\ q_z \end{Bmatrix} = \begin{bmatrix} K_{xx} & K_{xy} & K_{xz} \\ K_{xy} & K_{yy} & K_{yz} \\ K_{xz} & K_{yz} & K_{zz} \end{bmatrix} \begin{Bmatrix} -\dfrac{\partial \varphi}{\partial x} \\ -\dfrac{\partial \varphi}{\partial y} \\ -\dfrac{\partial \varphi}{\partial z} \end{Bmatrix} \tag{4}$$

The flow in x, y and z direction can be written as Eq. (5)

$$q_x = \left[K_{xx}\frac{\partial \varphi}{\partial x} + K_{xy}\frac{\partial \varphi}{\partial y} + K_{xz}\frac{\partial \varphi}{\partial z} \right]$$
$$q_y = \left[K_{yx}\frac{\partial \varphi}{\partial x} + K_{yy}\frac{\partial \varphi}{\partial y} + K_{yz}\frac{\partial \varphi}{\partial z} \right] \tag{5}$$
$$q_z = \left[K_{zx}\frac{\partial \varphi}{\partial x} + K_{zy}\frac{\partial \varphi}{\partial y} + K_{zz}\frac{\partial \varphi}{\partial z} \right]$$

In case of flow through homogeneous isotropic medium, the coefficient K is a scalar constant, therefore the above equation can be written as Eq. (6)

$$q_x = -K\frac{\partial \varphi}{\partial x}$$
$$q_y = -K\frac{\partial \varphi}{\partial y} \tag{6}$$
$$q_z = -K\frac{\partial \varphi}{\partial z}$$

The above equations are also valid for flow through nonhomogeneous porous medium as long as medium is isotropic. In case of flow through homogeneous isotropic medium, the above equations are also expressed by Eq. (7).

$$q_x = -\frac{\partial}{\partial x}(K\varphi) = -\frac{\partial \phi}{\partial x}$$
$$q_y = -\frac{\partial}{\partial y}(K\varphi) = -\frac{\partial \phi}{\partial y} \tag{7}$$
$$q_z = -\frac{\partial}{\partial z}(K\varphi) = -\frac{\partial \phi}{\partial z}$$

In this case the term $\Phi = K\varphi$ is called specific discharge potential. It should be noted that the Eqs. (3)–(7) cannot be used for flow through nonhomogeneous and nonisotropic medium.

In case of anisotropic aquifer, the Darcy's law can be written as Eq. (8),

$$q_x = -K_{xx}\frac{\partial \varphi}{\partial x} - K_{xy}\frac{\partial \varphi}{\partial y} - K_{xz}\frac{\partial \varphi}{\partial z}$$
$$q_y = -K_{yx}\frac{\partial \varphi}{\partial x} - K_{yy}\frac{\partial \varphi}{\partial y} - K_{yz}\frac{\partial \varphi}{\partial z} \tag{8}$$
$$q_z = -K_{zx}\frac{\partial \varphi}{\partial x} - K_{zy}\frac{\partial \varphi}{\partial y} - K_{zz}\frac{\partial \varphi}{\partial z}$$

This can also be written as,

$$\begin{Bmatrix} q_x \\ q_y \\ q_z \end{Bmatrix} = \begin{bmatrix} K_{xx} & K_{xy} & K_{xz} \\ K_{yx} & K_{yy} & K_{yz} \\ K_{zx} & K_{zy} & K_{zz} \end{bmatrix} \begin{Bmatrix} -\dfrac{\partial \varphi}{\partial x} \\ -\dfrac{\partial \varphi}{\partial y} \\ -\dfrac{\partial \varphi}{\partial z} \end{Bmatrix}$$

The coefficients that appear in the above equations are the components of the second rank tensor of hydraulic conductivity. Since $K_{xy} = K_{yx}$, $K_{xz} = K_{zx}$, and $K_{yz} = K_{zy}$, there are six distinct components in a 3D flow. In case of 2D flow, there are only three distinct components to fully define the hydraulic conductivity as represented in Eq. (9).

$$K = \begin{bmatrix} K_{xx} & K_{xy} \\ K_{xy} & K_{yy} \end{bmatrix} \tag{9}$$

However, the component K_{ij} depends on the chosen coordinate system. These chosen directions in space are called principal directions of an anisotropic medium. As such, when principal directions are chosen, the expression is given as Eq. (10).

$$\begin{Bmatrix} q_x \\ q_y \\ q_z \end{Bmatrix} = \begin{bmatrix} K_{xx} & 0 & 0 \\ 0 & K_{yy} & 0 \\ 0 & 0 & K_{zz} \end{bmatrix} \begin{Bmatrix} -\dfrac{\partial \varphi}{\partial x} \\ -\dfrac{\partial \varphi}{\partial y} \\ -\dfrac{\partial \varphi}{\partial z} \end{Bmatrix} \tag{10}$$

In a @D flow, the equation is given as,

$$\begin{Bmatrix} q_x \\ q_y \end{Bmatrix} = \begin{bmatrix} K_{xx} & 0 \\ 0 & K_{yy} \end{bmatrix} \begin{Bmatrix} -\dfrac{\partial \varphi}{\partial x} \\ -\dfrac{\partial \varphi}{\partial y} \end{Bmatrix} \tag{11}$$

$$\left. \begin{aligned} q_x &= -K_{xx} \dfrac{\partial \varphi}{\partial x} \\ q_y &= -K_{yy} \dfrac{\partial \varphi}{\partial y} \end{aligned} \right\} \tag{12}$$

Thus, Darcy's law stated that the relation between specific discharge (q) and hydraulic gradient (J) is linear. However, in real world situations, the relation becomes nonlinear for higher values of specific discharge. As such, the Darcy's law is not valid for higher specific discharge. The relation is generally linear as long as the Reynolds number does not exceed some value between 1 and 10 (Brown, 2002). Thus, one of the traditional tools to characterize the water flow through porous media is the computational numerical model. The physical models are often given in terms of partial differential equations for specific cases of GW dynamics. These models turn out to be unfavorable when one tries to find a solution over realistic domains and conditions Alternative to the complex physical models are data driven models since physical, they are inappropriate for most of the practical problems (Tian et al., 2016). Accordingly, numerical methods of GWL forecasting were found to be less effective for GWL predictions. Though these numerical models are found to be accurate for calculation, they are less effective in predicting data patterns which vary somewhat irregularly.

The recent GWL forecasting models are implemented using a data driven approach with quantitative historical data to predict future trends (Taormina et al., 2012). Data driven models are based on existing input and output parameter relationship information. Hence, these models are region specific, with the performance values only applicable to the area where they were developed (Shareef and Abbod, 2010). Statistical, ANN, Fuzzy, and regression are typically widely used methods data- driven Models. Researchers have introduced the ANNs functionality for surface and GW quantity modeling (Wanakule and Aly, 2005). ANN are common universal approximations, used in GW studies. The ANN problem was considered as a problem of optimization. ANN's principal weakness is a lack of generalization capacity. The integration of various optimization algorithms with ANN can be achieved to improve forecast accuracy. The gradient descent training method is a local search algorithm in which the Initial weights are randomly assumed (Isaac Abiodun et al., 2018). The most frequently used algorithm for aquifer modeling in the neural network domain is the Back Propagation (BP) algorithm. It easily become stuck in local minima and is unstable in nature. To overcome the drawbacks of this algorithm, the global network parameters are searched using powerful Swarm Intelligence (SI) based algorithms.

The evolutionary algorithm and SI algorithms are the essential global met heuristic search algorithms that attracted the researchers to optimize algorithmic parameters. The Particle Optimization (PSO) which mimics flocking of birds can be used in hybrid approach for weight optimization. To address the drawbacks of traditional gradient descent-based algorithm, the parameters of the global network are searched using efficient nature inspired algorithms. The SI based algorithm called PSO that mimics bird flocking can be used for weight optimization in hybrid approach. The PSO's greater exploration capacity for discovery allows it to converge quickly (Dutta et al., 2013). The optimal Weights were defined in PSO trained ANN by training the network where the initial population of the network's synaptic weights was initialized randomly, which cause outcome fluctuations. Therefore, an algorithm is required to find the best location before updating particles. So, a new hybrid Artificial Bee Colony (ABC) controlled PSO search mechanism, was developed to solve this problem in this work. Back propagation (BP) is used widely for ANN training, but it is found that the findings obtained by using BP for ANN training are less reliable and unstable (Nourani et al., 2008). The advanced models required for GWL administration to forecast the GWL in the days ahead with the modern accessible data. Multilayer feed forward, recurrent networks, and radial base networks are the main types of ANN architectures and algorithms built in the Literature (Gao and Er, 2005). One of the commonly used representative methodologies in recent years is profound learning technique called deep learning (Shrestha and Mahmood, 2019).

The traditional ANN's cannot effectively manage sequential data which is one of the major drawbacks (Alagha et al., 2012). The Recurrent Neural Network (RNN) has been applied successfully to groundwater modeling. The standard RNN architecture has been found to have difficulty in Capturing long-term dependencies between variables due to vanishing gradient. Long Short-Term Memory (LSTM) was only recently used for the prediction function of the hydrologic time series to prevent this problem (Zidong Wang et al., 2017). Deep learning Techniques are recent contributions to GWL prediction in the field of machine learning (Glorot and Bengio, 2010). The RNNs with multiple hidden layers are commonly used in deep learning methodologies (Ruba Talal and Tariq Mohammed, 2017). Traditional ANNs cannot effectively manage sequential data which is one of the major disadvantages that can be solved by applying RNNs to GWL Predictions (Petnehazi, 2019). A computationally efficient model is needed which can forecast water levels with minimal Parameterization. At the same time, such a model should be able to address the predicted variation in the atmosphere.

The second type of soft computing approach which is statistical learning models called Support Vector Regression (SVR) are explored. The SVR based forecasting model Performances dependent on SVR based on parameter tuning (Wang and Gao, 2015). Hence, SVR with ideal parameters may be more appropriate for real-time analysis of hydro meteorological data compared to ANN. The SVR Maximizes margin, Regularizes the solution by minimizing w, so that the value of y of all examples deviates from the given regression function less than the necessary accuracy (Schoellkopf and Smola, 2002). Thus, it attempts to match a curve by deploying a nonlinear kernel, reducing cost function by using kernel trick to quantify internal products. Due to the linear and nonlinear patterns in data the underlying structure of the GWL time series cannot be easily determined. Hence, a hybrid model is a reasonable alternative to GWL forecasting rather than isolated approaches. Nature has influenced us in many ways for example many algorithms such as the evolutionary and SI algorithms are built based on these mechanisms. To boost prediction accuracy these algorithms are hybridized with machine learning algorithms. Therefore, optimize the algorithmic parameters the ANN and SVR are hybridized with global optimization algorithms.

In general, substantially large amounts of work were reported in the GWL forecasting domain (Alagha et al., 2012). It offers introductory analyses of all conventional models, valid only for linear results. Thus, it has also investigated some soft computing models such as ANN and SVR, but none of the individual models can handle them effectively. Since conventional forecasting models performance is not up to the mark till today, a hybridization need arises. The literature for GWL forecasting systems based on hybrid machine learning algorithm is very much limited. Chang et al. have (Chang et al., 2016) developed Monthly regional GWL prediction system in the study area of Taiwan, establishing monthly regional GWL prediction method using hybrid soft computing techniques. They proposed a novel and scalable soft computing technique combined with self-organized map (SOM) and NARX network using 13-year hydrologic data (2000–2013). They proposed that the hybrid SOMNARX model could predict monthly GWL's with high Correlation reasonably and appropriately. Gong et al. (2016) extended their work toward GWL prediction by comparing three ANN, SVM and ANFIS soft computing models. They used a dataset of a10 years collection from two wells located in Florida, US. They stated that model SVM was more accurate than model ANN. Jin et al. (2009) have been forecasting complex GWLs using previous GWL data from 51 wells from the Yichang irrigation district.

They used SVM with chaos optimization. They suggested the SVM model with a Limited dataset is feasible and efficient. Mohsen et al. (2010) developed data driven aquifer water level forecast where they compared SVM and ANN. In a complex GW system, they predicted transient GWLs based on pumping, temperature, and GWL data from the considered area of study. Heesung et al. (2011) have developed two types of GWL prediction models using ANN and SVM. The models were applied to forecast GWL of two wells at Korea's coastal aquifer. They used past GWL, precipitation and tide as inputs. They indicated that in mode prediction the SVM performs better than ANN, and SVM model generalization capacity is superior to an ANN model with varying lead times.

Bowes et al. (2019) proposed LSTM and RNN to forecast GW a table in the flood prone coastal town of Norfolk, Virginia. For Model and forecast GW Table response For Storm events, they explored two machine learning algorithms LSTM and RNN by using GW table, rainfall and sea level as input parameters from 2010 to 2018 train and check the models. They also indicated that LSTM networks have more predictive abilities than RNN's. They tested their approach and achieved better model efficiency, underlining LSTM's potential for applications for hydrological modeling. There are several drawbacks in using standalone LSTM network. It suffers from the lack of ability to explain the final decision that the model acquires (Hochreiter and Schmidhuber, 1997). To avoid this problem a hybrid approach of integrating global optimization algorithms with LSTM was suggested by many researchers.

In a study by Nawi et al. (2015) using the Cuckoo Search Hybrid techniques, the data classifier problem was explored using weight optimization on RNN. Using the cuckoo search algorithm, the convergence rate and the Local Minima problems are dealt with Using BPNN algorithm and other hybrid variants, the output of the proposed model is compared with ABC.

The findings show that the computational performance of conventional RNN is greatly enhanced when combined with the hybrid approaches to metaheuristics (Rashid et al., 2018). A well-structured LSTM has been developed to overcome the problems with the conventional RNN networks. They used four different optimizers based on met heuristic algorithms and the learning, speed and accuracy was explored and contrasted with RNN architecture due to long-term dependencies in LSTM. The LSTM outperforms classification accuracy compared with conventional RNN architecture. The experimental results showed that the LSTM hybrid network is outperforming the standalone approaches. To overcome the weakness and to improve the convergence rate of traditional approaches, this work proposes a new hybrid Met heuristic Approaches.

Researchers have mainly explored on GWL forecasting using dataset collected from government agencies for different parameters like rainfall, past GWL and other matereo-geological parameters. The other challenge is to develop such soft computing-based forecasting models that there are no standard database available. Therefore, the present study is focused on this research problem and has been addressed by developing different soft computing models using data from the region of interest in a hybrid system for optimal performance. The novelty of this work mainly includes building of hybrid algorithm by combining heuristic global optimization algorithms with the latest machine learning algorithm. In this research work, ANN and SVM based GWL forecasting models are explored in hybrid forms. The SI based nature inspired algorithms are integrated with neural network and SVR. The weights of ANN are optimized using a hybrid of ABC guided PSO, SI based algorithm. Another NN structure called LSTM RNN was hybridized with LA. The PSO searching strategy is a continuous process so well suited for regression problems, so it is also hybridized with SVR to tune the algorithmic parameters.

3. Soft computing based GWL forecasting model development

Soft computing is a revolutionary method for creating computationally intelligent systems that are supposed to possess human knowledge, adapt and learn to do better in evolving environments and clarify how they make decisions. Soft computing aims to combine many various paradigms of computation including ANN, fuzzy logic, and genetic algorithms.

When used together, however, the strengths of each strategy can be utilized synergistically by designing low-cost, hybrid systems. The use of soft computing in the field of hydrological forecasting is a relatively new area of research. It has already been reported that neural networks alone are effective substitutes for rainfall runoff models (Geetanjali and Sreekanth, 2005). The application of neural networks, SVR and metaheuristics as modeling methods in the field of hydrology are widely used in the literature. However, there is still to be investigated the integration of these different soft computing technologies to create a single, integrated solution for improving operational river level and flood prediction systems (Prasad et al., 2007). The Soft computing approach is becoming extremely popular, and it is getting used widely in time series forecasting. The soft computing approaches are used to model cognitive behavior of human mind.

In this work, a GWL forecasting model is considered with potential applications in regional analysis of GWL trends. The study shows that such model has been very useful in managing precious water resources/sustainability. The GWL prediction model using CGWB uses traditional tools like QGIS for analysis. At present, use of soft computing is becoming common and therefore accurate predictions of GWL trend with available data are very important. Considering all the above reasons and applications, we believe that the research investigation on GWL forecasting is very relevant and helpful.

3.1 Study area and data

One of the difficulties of GWL forecasting is the GW flow that is unique for different geological formations. Therefore, Groundwater analysis is region-specific. There is no specific benchmark for GWL's predictions to construct the model's predominance. Hence the development of regional GWL forecasting with the collection of data from the specific region for this purpose is important. In the proposed research work, the data driven GWL forecasting system with different datasets using secondary data from the government agency of Udupi, have been implemented. This requires hydro-meteorological

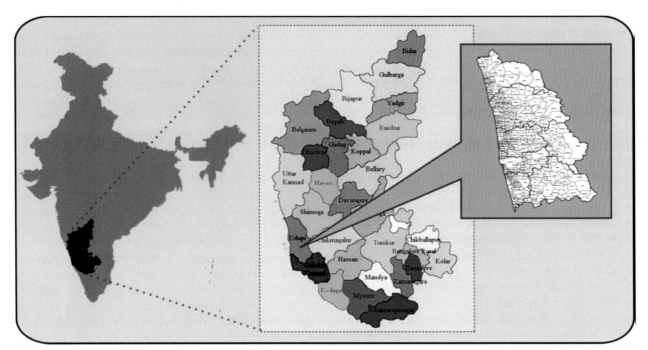

FIG. 2 Location map of Udupi District.

database with different geological formations and lead time. The location map of Udupi district is as you can see in Fig. 2. Most part of the Udupi district consists of geological formations, lateritic terrain and gneissic complex, and their features are as explained below.

The geology in its formations seems to be very complex, with varying parameters. Early researchers Radhakrishna, Vaidhyanathan, and Balasubramanyana (Radhakrishna and Vaidya Nathan, 2011) conducted the systematic work on geology and geomorphology of some parts of Dakshina Kannada district and the Udupi district. BGC and laterites cover the greater part of the Udupi district.

3.1.1 Spatial variability of groundwater level

The GWL data collected from the Department of Mines and Geology, Udupi district, Government of Karnataka were used to develop a GWL spatial map. The GWL map for premonsoon and postmonsoon period for the year 2016 and 2017 with spatial distribution of the GWL is as shown in Fig. 3.

3.2 Machine learning algorithms and metaheuristics

The machine learning algorithms is a technique that lets the computational system learn from the knowledge available. Such algorithms work by building a prediction model from a collection of training data that is used later to make predictions guided by evidence (Bateni et al., 2012). Some of the most common algorithms for machine learning include ANN feed forward and recurrent neural networks, as well as statistical models such as SVR. Traditional standalone machine learning algorithms are not efficient, so metaheuristic algorithms inspired by nature like EA and SI are used in hybrid form. The metaheuristic method applies heuristics inspired by nature in order to combine exploration and exploitation strategies. The machine learning approach and met heuristic approaches are as you can see in Fig. 4.

The key idea of applying met heuristic to neural network training is to use met heuristic algorithms instead of conventional algorithms of gradient descent to adjust weight formation. Therefore, using metaheuristic optimal global search capability to effectively train NN than conventional methods. The metaheuristic algorithms are a collective definition of a collection of algorithms like nature inspired algorithms such as Particle Swarm Optimization and Artificial Bee Colony algorithms, Lion Algorithms and their combinations of hybrids.

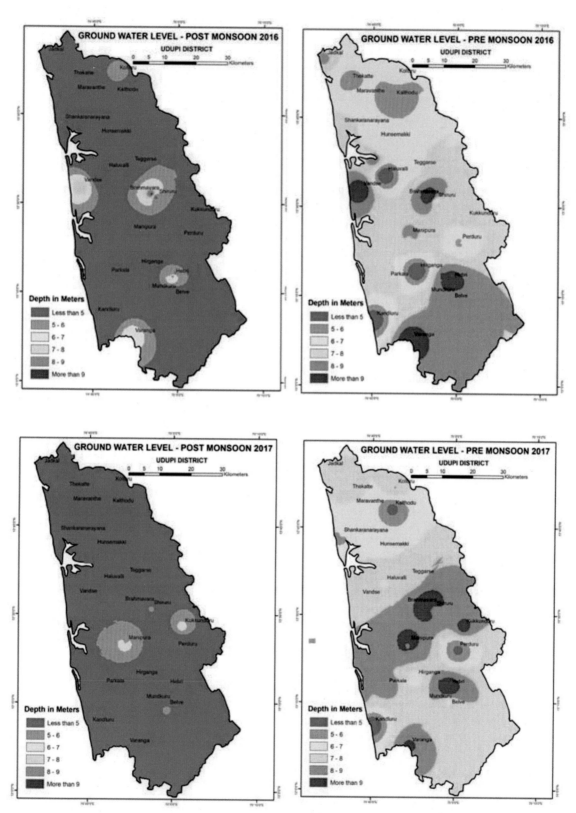

FIG. 3   Spatial groundwater level map for post and premonsoon for 2016 and 2017.

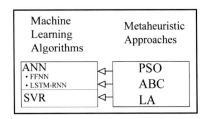

FIG. 4 Machine learning and met heuristics approaches.

3.2.1 Artificial neural networks

Artificial Neural Networks are popular universal approximations, which are applied in GW studies. The problem of ANN is formulated as an optimization problem and the main limitation with this approach is the lack of generalization capability. However due to its limitations, the metaheuristic algorithms including EA and SI are used in hybrid form. Thus the metaheuristic formulate NN component weights into an optimization problem. The conventional gradient descent-based algorithms operate on a single weight vector, whereas metaheuristics approaches use multiple weight vectors during the optimization and select best weight vector at the end of optimization iterations. Although gradient descent algorithms are computationally faster than nature inspired metaheuristic global optimization algorithms, the application of conventional algorithms are limited, because they have the tendency to fall in local minima.

The standard FFNN are straightforward networks, they allow signals to travel one way only from input to output. There are no feedback loops, so they associate inputs with outputs. Whereas RNNs are feedback networks, allow signals to travel in both directions by introducing loops in the network. These recurrent networks are powerful, where computation derived from earlier inputs are feedback into the network, which gives them a kind of memory. Thus, RNNs are dynamic where states change continuously until they reach an equilibrium point. The LSTMs, introduced by Hochreiter and Schmidhuber (1997), selectively remember patterns for long durations of time compared to traditional FFNN. LSTMs are capable of processing information to the cell state through regulated gates. In our work both standard FFNN and LSTM, a special type of RNN structures are explored. The hybrid algorithms by combining conventional and metaheuristics and two or more metaheuristic algorithms is explored. The ANN architectures FFNN and RNNs are explored in hybrid manner. The basic LSTM neuron has a separate cell state that keeps track of long-term sequential information. However, learning LSTM models for large number of memory cells becomes computationally expensive. Therefore, a hybrid LSTM-LA methodology is adopted in the current study.

3.2.2 Applying metaheuristic algorithm on NN training

Swarm intelligence on neural network

Initially GWL forecasting model was developed using hybrid ANN-PSO approach. The exploitation capability of PSO algorithm is good, therefore this algorithm is used to finding the global best solution.

For an I-H-O node topology the number of unknown weights are calculated as given by Eq. (13):

$$N_w = (I + 1) * H + H + 1) * 0 \tag{13}$$

For 2-3-1 node topology, the number of weights is calculated as 13. The weight optimization of feed forward neural network using PSO is as you can see in Fig. 5.

Thus, the neural network weights are evolved using PSO algorithm. The particles in this case are connection weights and the network is trained. The PSO was found to be poor at exploration, but better in finding a global beast solution. Therefore, to overcome this problem ABC guided PSO search mechanism was developed to enhance the exploration ability. The ABC guided PSO trained ANN was developed and tested for data from an individual well for 9 years from 2000 to 2009. For improving the accuracy, the initial population is generated by using the Artificial Bee Colony algorithm without adopting random selection. Thus, we used hybrid ABC-PSO for training Multilayer perceptron by balancing exploration and exploitation mechanisms. However, the exploitation procedure of ABC algorithm was poor, so to resolve the above problem, a new hybrid ABC approach guided by PSO search mechanism was used. In order to enhance the exploitability, ABC utilizes the search mechanism of PSO algorithm to compute new candidate solutions.

The PSO guided ABC approach was used to develop the GWL forecasting model. In the ABC algorithm, 50% of the colony comprises employed artificial bees and the remaining half comprises onlookers. One employed bee is recommended for every food source. In each cycle, the probe embraces four stages: In the first phase, the employed bees are sent out to find

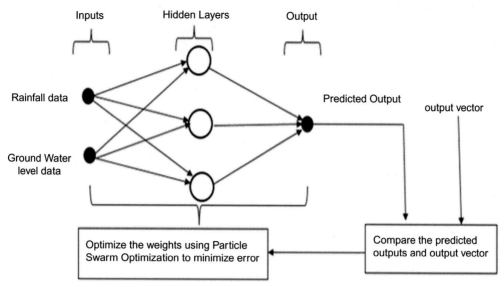

FIG. 5 Weight optimization using particle swarm optimization.

the food sources by analyzing the amount of nectar to be obtained and gives the input signal to onlooker bees. In the onlooker phase selected food sources have been identified based on the input signal from employed bees, which enhances advanced solution. If the identified nectar of the food source is rejected, the scout bees arbitrarily decide the new food sources.

In Scout bee phase, with an eye on locating the best solution Scout bee phase is substituted by PSO technique. Thus, by launching the PSO, we set out to locate the pbest and gbest, which usher in the best solution.

The particles update their position and the velocity till it arrives at its termination benchmark using following equations:

$$V_i^{(it+1)} = V_i^{(it)} + l_1 * r_1 * \left(p_{bi} - p_i^{(it)}\right) + l_2 * r_2 * \left(g_{bi} - p_i^{(it)}\right) \tag{14}$$

$$x_i^{(it+1)} = x_i^{it} + V_i^{(it+1)} \tag{15}$$

where

$V_i^{(it)}$—velocity of ith particle at iteration it
l_1, l_2—learning factors
r_1, r_2—random numbers produced lies between [0,1]
p_b—current best position
g_b—global best position
$x_i^{(it)}$—current position at iteration it

In the termination stage, the procedure gets replicated till the termination criteria is satisfied, then the process gets completed. Now, the final solution is the global best particle and it is chosen as the best solution.

At the outset, rainfall data and GWL of earlier months are employed as the input constraint and then processed by the means of the training scheme of hybrid ABC-PSO system. Here, for computing particular month GWL forecasting we used the previous month's GWL for training and testing is prepared. There is no looping's in FFNN, and the information only flows in the forward direction in every layer of the network and can predict only continuous target variables. The recurrent neural network can be used for time series prediction, irrespective of nature of data using deep learning technology. Thus, RNN is aware of time which makes them successful in time series prediction problems with sequential data.

Lion algorithm optimized long short-term memory RNN

The basic LSTM neuron has a separate cell state that keeps track of long-term sequential information. However, learning LSTM models for large number of memory cells becomes computationally expensive. Therefore, a hybrid LSTM-LA methodology is adopted in the current study is as you can see in Fig. 6

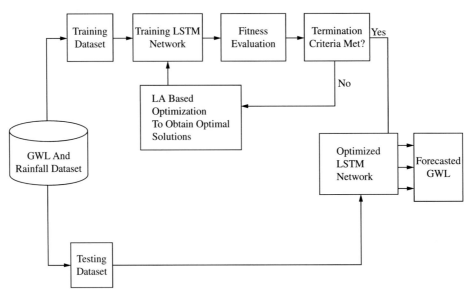

FIG. 6 Flow chart of LSTM-LA model.

A hybrid LSTM_LA model is proposed that makes use of the Lion Algorithm (LA) to optimize the weights of LSTM Network. The investigation and analysis of the proposed LSTM-LA hybrid approach is verified on a selected dataset and compared with the standard feed forward architecture. In the hybrid LSTM-LA model, the mating characteristics of lions are mathematically modeled to optimize the weights of LSTM network. The population of randomly generated set of solutions called Lions are initialized. The possible solutions are the weights and biases for the LSTM network. The population of 2n lions are assigned to two groups as the candidate population. The best weights and biases are initialized with LA in the first epoch and are passed on to LSTM network. The second step in the algorithm is mating process a platform for information exchange among different members. The important stage in LA is defense operator, which consists of defense against new matured resident males and defense against nomad lions. This defense operator plays an important role in LA by assisting it to retain powerful male lions as solutions. The nomadic lion is generated in the same way as territorial lion and new survival fight between the territorial lion and nomadic lion is performed. The male lion occupies the territory by defending and protecting the cubs and then the new solution is used to attack the male lion. If the nomadic lion is superior to the other solutions in the pride, the male lions are replaced by nomadic lions.

The territorial takeover is the last step which is same as the selection process in genetic algorithm. In this step, the optimal solution found to replace the inferior one and mating process is repeated until termination conditions are satisfied. The LA will update weights with best possible solutions in the next cycle and the searching process is continued. Thus, the weights and thresholds of all layers in the LSTM model are initialized randomly and LA searches the optimal weights. If the termination criteria, the maximum iteration number is reached, the optimal parameters are obtained, or else the optimization process is repeated until the conditions are satisfied. Then the optimized LSTM model is used to forecast the GWL. The dataset from the period year 2000–2018 was used to train and test the LSTM-LA model for different prediction horizons. The monthly forecast GWL results for year 2018 from the hybrid LSTM-LA model are compared with LSTM and FFNN approaches.

3.2.3   SVR based model

It has been found that ANN based modeling becomes inefficient, particularly for limited dataset. So, SVR based models are explored for the better results with 4 years of dataset for the study period of 2014–2018. The Optimization problem can be solved using SVR method, considering $\{(x_1, y_1), \ldots (x_m, y_m)\}$ for the datasets where each $x_i \in R^n$ which resembles the sample space for the input and target value can be considered as $y_i \in R$ for $i = 1, 2, \ldots, m$, whereas m represents size dataset to be trained.

The above problem can be formulated as the minimization problem given by Eq. (16):

$$\frac{1}{2}\|w\|^2 + C\sum_{i=0}^{m}\left(\xi + \xi^*\right) \tag{16}$$

$$\text{Subject to} \begin{cases} yi - <w, xi> -b \le \varepsilon i + \xi \\ <w, xi> +b - yi \le \varepsilon i + \xi^* . \\ \xi_i, \xi_i^* \ge 0\, i = 1, \ldots m \end{cases}$$

Where the function $\emptyset$ is used for mapping input to a higher dimensional space. Where $\boldsymbol{\xi}_i$ and $\boldsymbol{\xi}_i^*$ represents the upper and the lower training error to the ε insensitive tube $y_i - <w, x_i> -b \le \in$.

3.2.4 Applying metaheuristic algorithm on SVR training

The proposed research work implements a GWL forecasting using hybrid soft computing approach which consists of hybrid PSO based SVR with Radial Basis Function (RBF) kernel. The SVR model has three parameters represented as C, ε and γ. The global optimization algorithm PSO is helpful in determining the parameters of SVR. The accuracy of the SVR model depends on the optimal value of these parameters. The traditional cross validation method to estimate the SVR parameter leads to overfitting. Therefore, SI based optimization technique called PSO, which mimic the bird flocking mechanism was used. The flow diagram of the parameter tuning of SVR with PSO is shown is as you can see in Fig. 7.

In Particle Swarm Optimization algorithm, the particles are calculated by exploring through the search space. The upper and lower bounds of C, ε and γ are initialized and the values of these parameters are set, and the fitness is evaluated. The Normalized Mean Square Error (NMSE) serves as the fitness criteria used. Thus, the suitable parameters of the SVR algorithm were calculated using the global optimization algorithm. In this work, the monthly GWL and rainfall data of the Udupi region, were used in evaluating the performance of SVR-PSO hybrid approach. Thus, when the termination criteria is satisfied, the optimal hyper parameters are obtained else the process is repeated.

4. Results and discussion

In this section, the simulation result are presented from ABC guided PSO based FFNN structure, LA optimized LSTM RNN structure and PSO trained SVR based approach for different datasets.

4.1.1 Feedforward ANN network based GWL forecasting models

The three approaches of FFNN based GWL forecasting models ANN-ABC-PSO, LSTM-RNN-LA and SVR-PSO are tested. The details of performance analysis of all the three hybrid approaches are discussed in the following section. The ABC guided PSO was tested against data of an individual well for seasonal forecasting to improve the exploration capability.

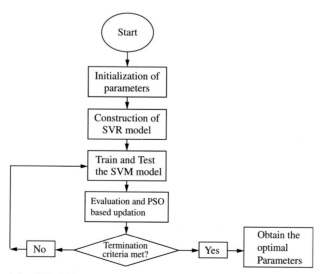

FIG. 7 Schematic representation depicting SVR-PSO methodology.

ANN-ABC-PSO GWL forecasting system

The ABC guided PSO was tested against data of an individual well for seasonal forecasting to improve the exploration capability. The ANN based ABC-PSO system is trained for a time series data of 10 years duration (2000–2009). The data incorporates rainfall data and monthly recorded GWL. The first 80% of the data was adopted for training the network and the remaining 20% for evaluating the trained network. The GWL predicted results for a small dataset of with one step ahead prediction is as you can see in Fig. 8 It is observed that in the monsoon season we observe a large difference between the expected and the predicted data. Therefore, we infer that a large dataset is required to train the ANN based models.

The GWL predicted for year 2012 is as you can see in Fig. 9. From the graph it is observed that hybrid PSO guided ABC algorithm gives satisfactory results, while the PSO trained ANN gives large errors.

The seasonal GWL is forecasted using same data set with different lead times for premonsoon and postmonsoon are as you can see in Figs. 10 and 11. It is observed that sudden change in GWL with respect to time in 2004, whereas the trend remains same up to 2002.

The yearly forecasted GWL is as you can see in Fig. 12 and it is observed that the GWL almost follows the same trend, whereas after 2009, the GWL starts decreasing.

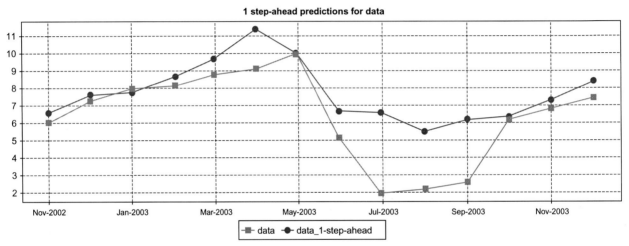

FIG. 8 GWL prediction for one step ahead.

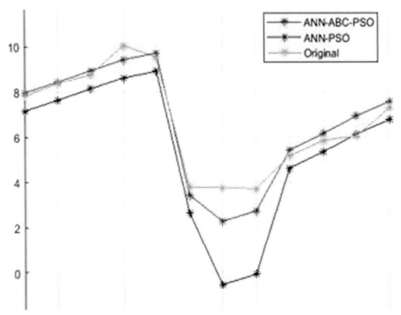

FIG. 9 GWL prediction for 2012.

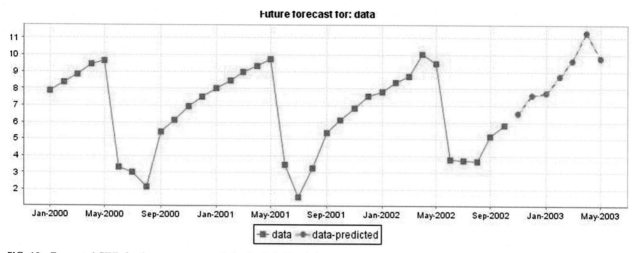

FIG. 10 Forecasted GWL for the premonsoon period using hybrid technique.

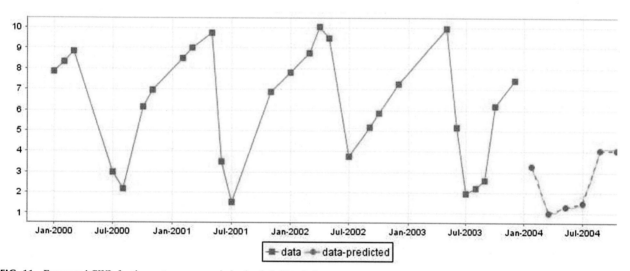

FIG. 11 Forecasted GWL for the postmonsoon period using hybrid technique.

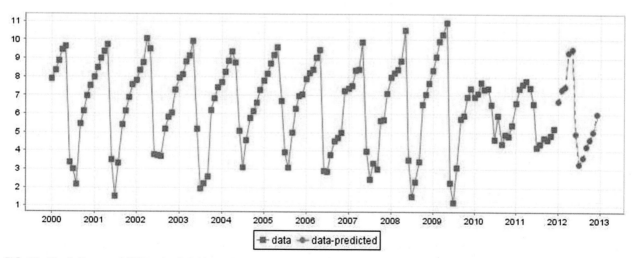

FIG. 12 Yearly Forecasted GWL using hybrid technique.

It is observed that with less data set, the output largely deviate from the desired output. It could be seen that for larger the datasets, the output almost overlaps with the expected output.

The results obtained proved to be efficient to forecast the GWL over several years. To prove the system to be effective we have compared this hybrid method with the FFBN. From this analysis we can infer that hybrid soft computing ABC-PSO trained system was found to be more accurate compared with other traditional techniques such as ABC, FFBN. Although FFNN hybrid architectures forecast the GWL accurately, they cannot process a time series sequentially, whereas RNN structures process information sequentially and exhibit temporal behavior for a time sequence. The performance analysis of a special type of RNN called LSTM RNN based GWL forecasting model is discussed in the next section.

LSTM-RNN based GWL forecasting models

The monthly forecast of GWL for the year 2019 using FFNN LSTM-LA hybrid approaches are as it can be seen in Fig. 13. The graph below shows the increasing trend irrespective of season.

The time series plot is as you can see in Fig. 14 which shows the GWL trend forecast using the hybrid LSTM-LA model.

4.1.2 Performance evaluation of SVR based GWL forecasting model

The major part of the Udupi region is covered by gneisses and laterites terrain types. The developed forecasting model was tested over the historical GWL data and rainfall for a period of 4 years collected from Udupi region for two different terrain formations, Lateritic terrain and BGC. The study was applied at the regional scale and considered two wells from two different terrains, lateritic and BGC terrain for forecasting GWL there by potentially restricting their large-scale applicability.

Lateritic terrain

The GWL of three wells identified at Perdoor and Brahmavar of Udupi district with lateritic as rock formation. The seasonal forecasted GWL using the SVR-PSO model for premonsoon period and postmonsoon period are as shown in Figs. 15–18 for above observation wells with lateritic terrain geological formations.

Well at Perdoor (well.ID: 80722) in lateritic terrain The GWL of a well1 identified at Perdoor of Udupi district with lateritic as rock formation was used to test the developed GWL forecasting model. The seasonal forecasted GWL using SVR-PSO model are as you can see in Figs. 15 and 16.

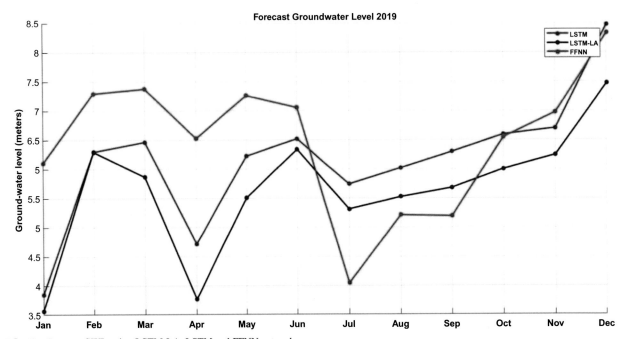

FIG. 13 Forecast GWL using LSTM-LA, LSTM and FFNN network.

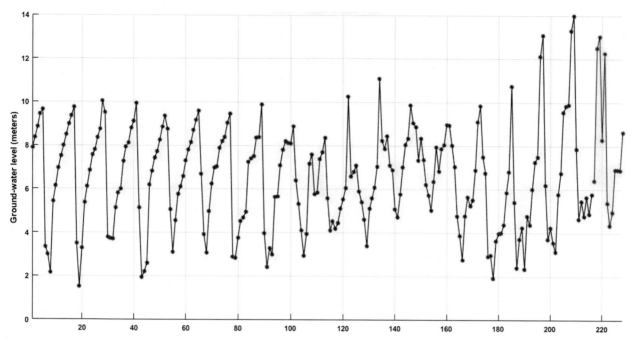

FIG. 14 Future forecast of GWL.

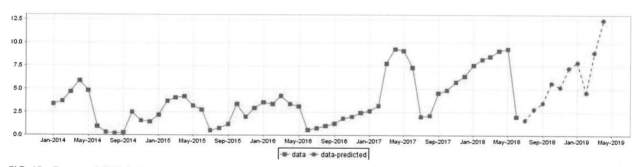

FIG. 15 Forecasted GWL in lateritic terrain for the premonsoon period for well at Perdoor.

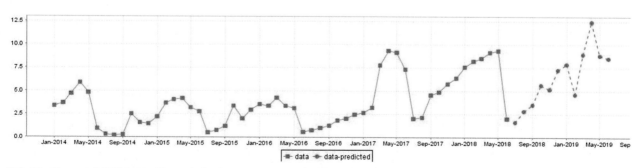

FIG. 16 Forecasted GWL in lateritic terrain for the postmonsoon period for well at Perdoor.

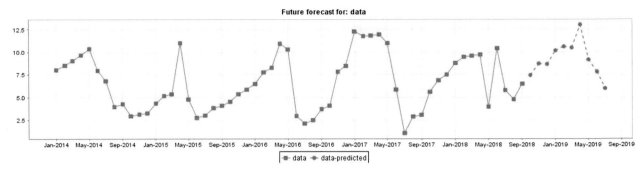

FIG. 17 Forecasted GWL in lateritic terrain for the postmonsoon period for well at Brahmavar.

FIG. 18 Forecasted GWL in lateritic terrain for the premonsoon period for well at Brahmavar.

Well at Brahmavar (well.ID: 80723) in lateritic terrain The GWL of a well1 was identified at Brahmavar of Udupi district with lateritic as rock formation was used to test the developed GWL forecasting model. The seasonal forecasted GWL using the SVR-PSO model for premonsoon and postmonsoon are as you can see in Figs. 17 and 18.

Banded gneissic complex

The GWL of a well 1 identified at Shankar Narayana of Kondapur taluk and Hiregana of Yennehole of Karkala taluk, Udupi district with BGC as rock formation was used to test the developed GWL forecasting model. The seasonal forecasted GWL using the SVR-PSO model are as you can see below.

Well at Shankar Narayana (wel.ID: 80710) in BGC terrain The GWL of a well1 2 identified at Shankar Narayana of Kondapur taluk, Udupi district with BGC (Granitic Gneiss) as rock formation was used to test the developed GWL forecasting model. The seasonal forecasted GWL using the SVR-PSO model are as you can see in Figs. 19 and 20.

Well at hiragana of Yenne hole (well.ID: 80702) in BGC terrain The GWL of a well 3 identified at Hiregana of Yennehole of Karkala taluk, Udupi district with BGC (Granitic Gneiss) as rock formation was used to test the developed GWL forecasting model. The seasonal forecasted GWL using the SVR-PSO model are as you can see in Figs. 21 and 22.

The forecasted GWL in lateritic terrain varies rapidly compared to BGC. The forecasted GWL of 2020 will be declined up to 13–14 m in a well with depth 16 m and after 2021 the GWL will be recovering as seen in above diagrams. The seasonal GWL variation in BGC is less compared to the lateritic terrain.

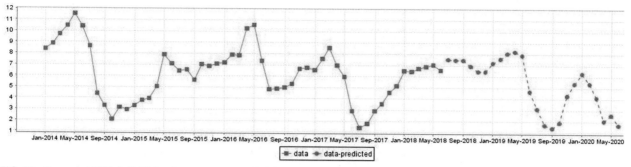

FIG. 19 Forecasted GWL in BGC for the postmonsoon period for well at Shankarnarayana.

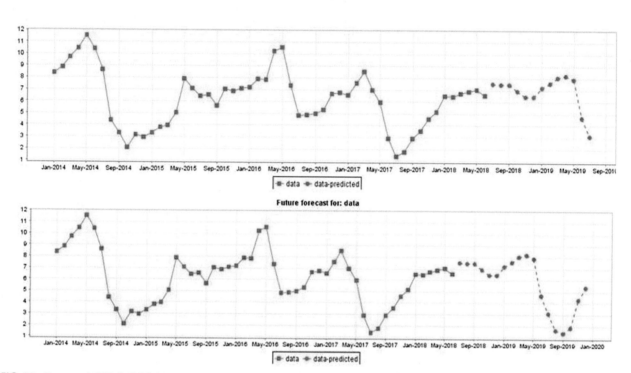

FIG. 20 Forecasted GWL in BGC for the premonsoon period for well at Shankarnarayana.

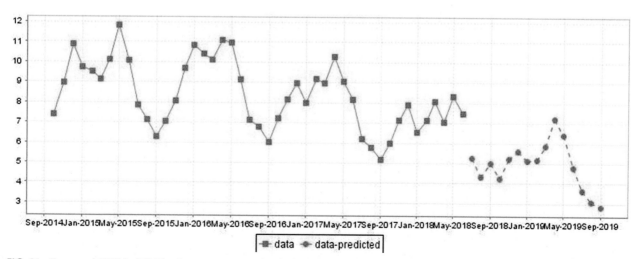

FIG. 21 Forecasted GWL in BGC for the postmonsoon period for well at Hiregana.

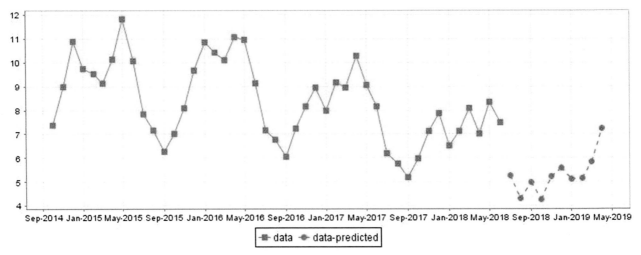

FIG. 22 Forecasted GWL in BGC for the premonsoon period for well at Hiregana.

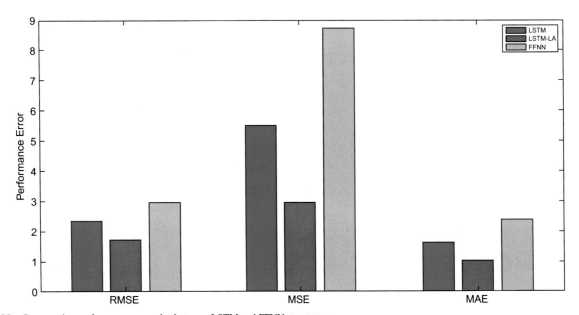

FIG. 23 Comparative performance error plot between LSTM and FFNN structures.

Gneissic granites are hard rocks from the GW occurrence point of view and are devoid of primary porosity. The occurrence, distribution and movement of GW depends on the presence of surface and deep-seated fractures and joints also on the extent of weathering. However, due to intensive weathering and presence of surface and subsurface joints and fractures, these formations are found to yield a good quantity of groundwater.

4.1.3 Comparative analysis

The performance of different models were compared using multiple performance metrics. The comparative performances error plot between LSTM and FFNN structures are presented in Fig. 23 using statistical indices.

It is found that the hybrid LSTM-LA approach gives less Mean Absolute Error (MAE) compared to FFNN. Therefore, the hybrid LSTM-LA approach outperforms the standalone FFNN approaches. The comparative plot showing the spread of the prediction accuracy scores across each validation fold for each algorithm is as shown in Fig. 24.

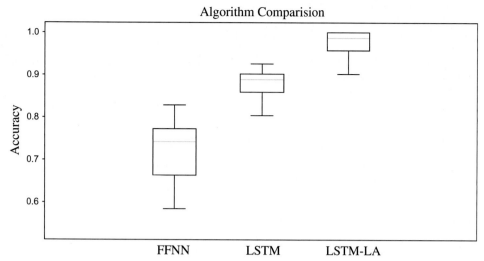

FIG. 24 Comparative plot of accuracy between LSTM and FFNN structures.

5. Conclusions

In this work, novel hybrid possibilities were investigated to forecast monthly groundwater levels using two different machine learning hybrid approaches. The three different soft computing approaches ABC guided PSO trained standard FFNN, LA optimized LSTM RNN and PSO trained SVR were implemented. The prediction capability of these models were investigated using historical monthly GWL and RF data from the Udupi region. The ANN based GWL forecasting model performance is analyzed against the SVR based model. It has been observed that the performance of SVR based models are better compared to ANN irrespective of data quality. The data used in the study is fetched from government agencies, the wells selected for this study are monitored by department of mines and geology, government of Karnataka, pumping from well is unaccounted in the study.

The developed model can be further extended by considering a greater number of input related input parameters from different geological formations in different parts of the country. The two different models ANN and SVM will be combined by ensemble multiple models using meta learning concept and also model can be extended toward hybrid framework of space-time GWL forecasting by combining soft computing tools and geostatistical models.

Acknowledgment

The authors are grateful to Nagaraj Rao, statistics department, and Mr. Niranjan and Mr. Dinakar Shetty, senior geologist, Geological Department, Government of Karnataka, for providing valuable rainfall and groundwater level data of Udupi district for this research. The authors also grateful to the Manipal Institute of Technology, Manipal Academy of Higher Education (MAHE) for their support to this research work.

References

Alagha, J.S., Md Said, M.A., Moghier, Y., 2012. Review–artificial intelligence based modelling of hydrological processes. In: 5th International Engineering Conference of 21st century, pp. 1–13.

Bateni, M., Eslamian, S.S., Mousavi, S.F., Hosseinipour, E.Z., 2012. Application of a localization scheme in estimating groundwater level using deterministic ensemble Kalman filter. In: EWRI/ASCE 10th Symposium on Groundwater Hydrology, Quality and Management, USA.

Bowes, B., Sadler, J.M., Morsy, M.M., Behl, M., Goodal, J.L., 2019. Forecasting GW table in a flood prone coastal city with long short-term memory and recurrent neural networks. Water 11, 138.

Brown, G.O., 2002. Henry Darcy and the making of a law. Water Resour. Res. 38 (7), 1–12. 1106.

Chang, F.-J., Chang, L.-C., Huang, C.-W., Kao, I.-F., 2016. Prediction of monthly regional GWLs through hybrid soft computing techniques. J. Hydrol. 541, 965–976.

Dash, N.B., Panda, S.N., Ashoo, N., Remesan, R., 2010. Hybrid neural modelling for GWL prediction. Neural Comput. Applic. 19 (8), 1251–1263.

Dutta, D., Roy, A., Choudhury, K., 2013. Training artificial neural network using particle swarm optimization. Int. J. Adv. Res. Comput. Sci. Softw. Eng. 3 (3), 430434.

El Alfie, M., 2014. Numerical GW modelling as an effective tool for management of water resources in arid areas. Hydrol. Sci. J. 59 (6), 1259–1274.

Gao, Y., Er, M.J., 2005. NARMAX time series model prediction: feedforward and recurrent fuzzy neural network approaches. Fuzzy Sets Syst. 150 (2), 331–350.

Geetanjali, N., Sreekanth, P.D., 2005. Forecasting groundwater level using artificial neural networks, National Research Centre for Cashew, Puttur. J. Hydrol. 309, 229–240.

Glorot, X., Bengio, Y., 2010. Understanding the difficulty of training deep feedforward neural networks. In: International Conference Proceedings on Artificial Intelligence and Statistics, pp. 249–256.

Gong, Y., Zhang, Y., Lan, S., Wang, H., 2016. A comparative study of ANNs, SVMs and ANFIS for forecasting GWL's near Lake Okeechobee, Florida. Water Resour. Manag. 30, 371–391.

Groundwater Yearbook-India, 2017–18, 2018. Central Ground Water Board, Ministry of Water Resources. Government of India, pp. 52–57.

Heesung, S.C.Y., JUN, H., Yunjung, G.O., Bae, K.I.K., 2011. A comparative study of Artificial Neural networks and support vector machines for predicting GWLs in a coastal aquifer. J. Hydrol. 396, 128–138.

Hochreiter, S., Schmidhuber, J., 1997. Long short-term memory. Neural Comput. 9 (8), 1735–1780.

Isaac Abiodun, O.D., Jantan, A., Omolara, A.E., Daba, K.V., Mohamed, N.A., Arshad, H., 2018. State-of-the-art in artificial neural network applications: a survey. Heliyon 4, e00938.

Jin, L., Chang, J.X., Zhang, W.G., 2009. GWL dynamic prediction based on chaos optimization and support vector machine. In: Proceedings of the 3rd International Conference on Genetic and Evolutionary Computing. IEEE Transaction, Zhengzhou, China.

Kumar, C.P., 2018. Water resources Issues and management in India. Int. J. Sci. Eng. Res. 5 (9), 137–147.

Kumar, C.P., Singh, S., 2015. Concepts and modelling of GW System. Int. J. Innov. Sci. Eng. Technol. 2 (2), 262–271.

Mohsen, B., Keyvan, A., Emery, A., Jr, C., 2010. Comparative study of SVM's and ANNs in aquifer water level prediction. J. Comput. Civ. Eng. 24 (5).

National Compilation in Dynamic Groundwater Resources of India-2017, 2019. Government of India, Ministry of Jal Shakthi. Dept. of Water Resources, CGWB, p. 87.

National GW Management Improvement Program, 2016. Environmental and Social System Assessment. Government of India, The World Bank, p. 34.

Nawi, N.M., Khan, A., Rehman, M.Z., Chiroma, H., Herawan, T., 2015. Weight optimization in recurrent neural networks with hybrid metaheuristic Cuckoo search. Math. Probl. Eng. 2015. https://doi.org/10.1155/2015/868375.

Nourani, V., Moghadam, A.A., Nadir, A.O., 2008. An ANN based model for Spatiotemporal GWL forecasting. Hydrol. Process. 22, 5054–5066.

Petnehazi, G., 2019. Recurrent Neural Networks for Time Series Forecasting, Article on Machine learning. Cornell University, USA.

Pinder, G.F., 2002. GW Modelling using Geographical Information Systems. Wiley Publication.

Prasad, R.K., Banerjee, N.C.M.P., Nandakumar, V., Singh, V.S., 2007. Deciphering Potential Groundwater Zone in Hard Rock Through the Application of GIS. Springer-Verlag.

Radhakrishna, B.P., Vaidya Nathan, R., 2011. Geology of Karnataka, Geological Society of India, Bangalore, India.

Raghavan, S., Jayaram, J., Arvind, L., Agarwal, P., Nikham, M., Nagendra, B., 2017. Report on Current Water Resources, Water Governance, Institutional Arrangements, Cross boundary issues Agreements of Karnataka State. Advanced centre for Integrated Water Resource Management, pp. 1–4.

Rashid, T.A., Fattah, P., Awla, D.K., 2018. Using accuracy measure for improving the training of LSTM with metaheuristic algorithms. Procedia Comput. Sci. 140, 324–333.

Ruba Talal, R.I., Tariq Mohammed, Z., 2017. Training RNN's by a hybrid PSO-Cuckoo search algorithm for problem optimization. Int. J. Comput. Appl. 159 (3). pp. 32–28.

Satish, S., Elan Go, L., 2015. Numerical simulation, and prediction of GW flow in coastal aquifer of Southern India. J. Water Resour. Prot. 7, 1483–1494.

Schoellkopf, B., Smola, A.J., 2002. Learning with Kernels: Support Vector Machines, Regularization, Optimization, and Beyond. MIT Press, Cambridge, p. 648. ISBN 9780262194 754.

Shareef, A.J., Abbod, M.F., 2010. Neural networks initial weights optimization. In: 12th International Conference on Computer Modelling and Simulation, pp. 57–61.

Shrestha, A., Mahmood, A., 2019. Review of Deep learning algorithms and architectures. IEEE Access 7, 53040–53065.

Taormina, R., Chau, K.W., Sethi, R., 2012. Artificial neural network simulation of hourly GWLs in a coastal aquifer system of the Venice lagoon. Eng. Appl. Artif. Intell. 25 (8), 1670–1676.

Tian, J., Li, C., Liu, J., Liang Yu, F., Chang, S., Zhao, N., Wan Jaafar, W.Z., 2016. GW depth prediction using data driven models with the assistance of gamma test. Sustainability 8 (1076), 1–17.

Vazquez Baez, V., Rubio-Arellano, A., Garcia-Toral, D., Rodriguez Mora, I., 2019. Modelling an aquifer: numerical solutions to the GW flow equations. Math. Probl. Eng., 1–9.

Wanakule, N., Aly, A., 2005. Artificial neural networks for forecasting GWLs. Comput. Civ. Eng., 1010.

Wang, X.-Q., Gao, J., 2015. Application of Particle swarm optimization for tuning the SVR parameters. In: IEEE International Conference on Advanced Intelligent Mechatronics, Korea, pp. 1173–1177.

Zidong Wang, W., Liu, X., Zeng, N., Liu, Y., Alsaadi, F.E., 2017. A Survey of deep NN architecture and their applications. Neuro Comput. 234, 11–26.

Chapter 13

Hydroinformatics methods for groundwater simulation

Nastaran Zamani, Saeid Eslamian, and Jahangir Abedi Koupai

Department of Water Engineering, College of Agriculture, Isfahan University of Technology, Isfahan, Iran

1. Introduction

Groundwater plays an important role in the semiarid region especially for water requirements in agriculture, municipal, and industrial uses (Nourani et al., 2012), and on the other hand for its exploration, prediction, and remediation (Panda and Narasimham, 2020), while this resource has been suffering from qualitative and quantitative degradation due to the effect of global warming and climate changes (Ouhamdouch et al., 2019). Groundwater modeling helps to analyze many important groundwater problems. Many hydrologists, geologists, engineers, and other researchers pay special attention to numerical-based groundwater modeling. Groundwater modeling has been performed by many computer programs. Due to the fact that a model is a simplified version of a real-world system, the first step in building a model is to build a conceptual model. After that, by the help of boundary and primary conditions, the conceptual models are translated into mathematical models in terms of flow governing equations. In this way, a problem can be solved by translating it into numerical models and writing computer programs (codes) (Chakraborty et al., 2020).

The prediction of groundwater level (GWL) has a crucial role in the management of groundwater resources (Nourani et al., 2012), at the same time it is depending on many factors for modeling such as precipitation, evapotranspiration, soil characteristics, and the topography of the watershed (Panda and Narasimham, 2020). On the other hand, due to the involvement of a large number of variables such as clogging, design, optimization, feasibility, water quality, geotechnical processes, groundwater management, recovery efficiency, saltwater intrusion, and residence time, modeling groundwater has many challenges (Diaz et al., 2020).

There are three main category models based on their physical characteristics: black-box models, conceptual models, and physical-based models. The two last ones are the main in predicting hydrological variables and understanding the system's physical processes (Nourani et al., 2012).

In spite of the development of in Artificial Intelligence application of hybrid simulation models by optimizing technique is very sparse, however, demand for water especially groundwater increase annually (Panda and Narasimham, 2020).

Recently, development of Geographical Information System (GIS) and Remote Sensing (RS) technology and the availability of spatially distributed data on climate, geomorphology, and geology make distributed watershed models useable that can be calibrated with available stream flow data. Groundwater development and management can be revolutionized by RS technology in the future due to the provision of unique and completely new hydrological and hydrogeological data (Chung et al., 2016). Each method has its own strengths and weaknesses, but by combining them, they become much stronger. So, by combining the advantages, limitations, and economics of each method, the correct solution can be obtained (Chung et al., 2016).

In simulation understanding the basic hydrologic processes is required for large area water resources development and management. Climate change, management of water supplies in arid regions, large scale flooding, and offsite impacts of land management are some concerns that are motivating the development of large area hydrologic modeling. Also advances in computer hardware and software such as: increased speed and data storage, advanced software debugging tools, and GIS/spatial analysis have allowed large area simulation more practical (Arnold et al., 1998). This chapter introduces some methods in groundwater simulation and some examples of using these methods which were applied around the world.

Handbook of HydroInformatics. https://doi.org/10.1016/B978-0-12-821962-1.00023-4

2. Methods

2.1 Time series and Markov chain methods

Markov is a tool which may describe the distribution of hydrostratigraphic units of groundwater simulation and is an approach which models the spatial variability of geologic formations. Markov based transition probability geostatistics (MTPG(for categorical variables are utilized as an efficient tool for 20 years to model the distribution of geologic facies, and aquifers' hydraulic properties for groundwater simulation (Langousis et al., 2018). This model (MTPG) was introduced by Carle and Fogg (1997). While it is conceptually simple and straightforward to implement, but its conditional simulation still has limitations.

Some factors such as anthropological effects make it so hard to see the hydrological parameters, as a result it's preferred to apply the time series model in these cases (Khorasani et al., 2016). A time series model is an empirical model for stochastic modeling and predicting the temporal behavior of hydrological system. For medium-range forecasting and artificial data generation, the stochastic time series models are popular tools. Several stochastic time-series models for these purposes are: The Markov, Box-Jenkins (BJ) Seasonal Autoregressive Integrated Moving Average (SARIMA), deseasonalized Autoregressive Moving Average (ARMA), Periodic Autoregressive (PAR), Transfer Function Noise (TFN), and Periodic Transfer Function Noise (PTFN). The Markov, ARMA and SARIMA models are considered univariate models, and also the PAR, TFN and PTFN models are considered multivariate models, and the PAR and PTFN models are periodic multivariate models (Khorasani et al., 2016).

Some factors which cause the choice of special modeling method are: required accuracy, modeling costs, simple model usage and results interpretation. Time-series modeling includes three common steps: identifying the experimental model which needs a minimum of 50 observations of the regarded series, estimating the model parameters (fitting), and verification of the model. In the first step of modeling, time series must be drowned as a time series diagram to make it easy for identification of trends, variance nonstationarity, seasonality (periodicity), and other styles of irregularities in data. Box-Cox transformation and nonstationarity in mean or variance are considered as a pretreatment step for normalizing data. These tests are divided in to two main groups: independence tests (for time) and normal distribution tests (Khorasani et al., 2016).

2.2 Geostatistics Methods

Spatial mapping and interpolation of groundwater qualitative parameters may be done using geostatistical methods as appropriate techniques. All this method may be used for simulating of depth to groundwater (Sun et al., 2009), recharge, aquifer's hydrodynamic behavior, and comparing different methods in predicting these parameters (Drias et al., 2020).

One of these techniques is easy kriging which may be a weighted linear combination that an average of definitely characterized by second-order stability and employed in the estimation process. This method requires variograms with the threshold (Maroufpoor et al., 2019). The kriging method is that the best methods among unbiased method for estimation of the value of regionalized variables in unsampled location. This method is classified into 3 groups: simple kriging (SK), ordinary kriging (OK), and universal kriging (UK) (Sun et al., 2009).

Another method is that the Co-Kriging estimator as a kriging process by secondary variables. To enhance the convenience of estimation, the state of only one secondary variable (Maroufpoor et al., 2019). To boost the transmissivity data that are insufficient for the groundwater model database for flow simulation Co-Kriging could be a good approach (Drias et al., 2020).

The inverse distance weighting (IDW) method is another interpolating method which is related to the weightings being solely a function of the gap between the point of interest and also the sampling points for $i = 1, 2, ..., n$ (Sun et al., 2009).

The radial basis function (RBF) method is based on neural networks, which features a feedforward architecture and consists of three layers: one input layer, one hidden layer, and one output layer, with a variety of neurons in each. This network has self-organizing characteristics that allow to adapt determination of the hidden neurons while the network is under training. Each input neuron is totally connected to all or any hidden neurons, and hidden neurons and output neurons also are interconnected to each other by a collection of weights. For feeding the data into the network, input neurons are utilized and transmitted to the hidden neurons; each hidden neuron then transforms the input signal employing a transfer function f. The output of hidden neurons has the form of a radial basis function (Sun et al., 2009).

The global polynomial interpolation (GPI) method could be a trend surface analysis which has been introduced into the earth sciences to analyze environments of sedimentation, contour-type map, and elevation. In trend surface analysis, a two-dimensional polynomial equation of the first, second, or a higher degree described the distribution of observational data (Yao et al., 2014).

The local polynomial interpolation (LPI) method introduces the concept of distance weight, this method combined the advantage of GPI method (reflect of tendency variation) and therefore the advantage of IDW method (reflect local characters). Generally, LPI fits different polynomials to specify overlapping surface into which the study area has been subdivided. For a defined region, the required order of the polynomial is fitted. When the regions overlap, the worth of each prediction is that the value of the fitted polynomial at the center of the region (Yao et al., 2014).

For assessing the most effective method in each interpolation, cross-validation and orthogonal-validation methods are used. By these methods, estimated values are tested with the present samples at the situation. When the estimate is calculating, it will be compared to the true sample value using statistics. To the present end, the error between the true value and also the estimated value is calculated and by compiling the root mean square error (RMSE) and therefore the correlation coefficient (R^2) the most effective method is chosen (Sun et al., 2009; Yao et al., 2014).

2.3 GIS and remote sensing

Water resource mapping may well be done by remote sensing techniques. As these techniques can make continuous and up-to-date measurements, users have been widely applied in several fields, such as hydrology, agriculture, and climate studies. One of the best known versatile tools, especially in spatial analysis, modeling, visualization, data processing, and management is GIS. Software from ArcGIS products is required to process original remote sensing images and videos (Wang and Xie, 2018). Hydrologic models can be often prepared by remote sensing and GIS, and in this way it is possible to supply a full spectrum of modeling: what happened in the past and project what is going to happen within the future, and played increasingly important roles in the hydrologic community (Wang and Xie, 2018). Water resources, quantitatively measure the hydrologic flux, and monitor the working conditions of hydraulic infrastructures, drought conditions, and flooding inundation can be done by remote sensing. Remote sensing has this ability to arrange critical data for mapping water resources (snow and glaciers, water bodies, soil moisture and groundwater), measuring hydrological fluxes (ET, precipitation and river discharge), and monitoring drought and flooding duration; while GIS is that the best tools for water resource, drought and flood risk management and for hydrologic models' setup, input data processing, output analysis and visualization. Better management of water resources, drought and flooding disasters can be done by GIS, statistics and numerical models. (Wang and Xie, 2018). GIS and RS as an integrated approach in groundwater management have this ability to support many distributed hydrological models and spatial variation in recharge (Rwanga and Ndambuki, 2017).

ArcGIS is additionally ready to link with other models, such as the WetSpass model which computes and generates flow hydrograph time series (Rwanga and Ndambuki, 2017). GIS environment also has this capability to implement models which used to predict soil loss, sediment yield, nutrient loss, pollutant transport in aquifer, and groundwater movement such as: AGNPS (agricultural nonpoint source), SWAT (soil and water assessment tool), ANSWERS (aerial nonpoint source watershed response simulation), and HSPF (hydrologic simulation program fortran). Two powerful instruments for decision makers are the integrated use of GIS and prediction models for identifying areas in danger of pesticide contamination, also they have short running time and fast in production of results. WebGIS may be a tool for integration and visualization of hydrological data within the model (Thakur et al., 2017).

2.4 Cluster analysis

For understanding the natural processes, analyzing the monitoring data of natural objects is essential. Within the process of analysis, identifying similar objects is required, observing which you can better understand the laws by which changes in these objects occur. Classification is one of the fundamental processes in science, where there is a need to classify many objects per several factors. For these multidimensional classifications, cluster analysis methods are used. Clustering is taken into account as a procedure that, starting to work with a special data type, converts them into cluster data. The most popular are hierarchical agglomerative methods and iterative grouping methods. In the case of using cluster analysis methods, it is difficult to administer unambiguous recommendations on the preference for using certain methods. Cluster analysis in conditions of the necessity for a multivariate analysis of the studied groundwater indicators and low data structure is an effective method to identify similar features (Kozhevnikova et al., 2020).

Clustering algorithms are widely used for data categorization, data comparison and model construction. In clustering, a data set is partitioned into several groups within which the similarity within the data which are in a group is much more than that of the data belonging to different groups. This sort of dividing needs a similarity metric based on which two input vectors are taken and a value reflecting their similarity is returned. Because of sensitivity of some similarity metrics such as inner products to the range of elements, each of the input variable must be normalized to lie in the unit interval [0,1], when using this metric in the clustering process (Jang et al., 1997).

Data clustering versus data classification is an unsupervised learning process that doesn't need any labeled dataset as training data. One among the foremost useful methods in clustering is C-means method which can be used in the form of classic and fuzzy models under the name of K-means (hard C-means) and fuzzy C-means (FCM), respectively (Ghosh and Dubey, 2013). K-means algorithm could be a typical dynamic clustering algorithm in which the average of objects attributes only one cluster to each object, while fuzzy C-means is a clustering method that each data belongs to two or more clusters (Yang et al., 2013).

2.5 Soft-computing methods

Artificial neural networks (ANN) is employed as a forecasting tool, especially in water resource and hydrology (Nourani et al., 2012). Artificial neural network learning processes take place in biological systems. They are composed of many artificial neurons that, according to a specific network architecture, are linked together. A neural network used for predicting future values of possibly noisy multivariate time series supported past histories. Transforming input into meaningful output is that the aim of the neural network (Adamowski and Chan, 2011). Typically, ANN uses input, hidden, and output three layers and "nodes" are artificial neurons which each layer is composed of. The model is assisted in simulations by the connection between neurons. Where each neuron in a layer is connected to any or all the neurons of the next layer, is named as feed forward architecture. Each neuron multiplies every input through its interconnection weight, sums the product, and then passes the sum by a transfer function to provide its result (Yu et al., 2018). The best structure of ANN can be obtained by a trial-and-error approach. The correlation coefficient (R), root mean square error (RMSE), standard error of estimate (SEE), coefficient of determination (R^2), and mean absolute error (MAE) are parameters that can evaluate the performance of artificial intelligence (Jeihouni et al., 2019).

The auto-regressive integrated moving average (ARIMA) method is employed to identify complex patterns in data and generate forecasts. This model has the ability to analyze and forecast univariate time series data. The ARIMA model function is represented by (p, d, q), within which p represents the amount of autoregressive terms, d the number of non-seasonal differences, and q the number of lagged forecast errors within the prediction equation. ARIMA models have three steps: identification, estimation and forecasting and defined as follows:

$$\Delta_1 Z_t = \varphi_1 Z_{t-1} + \cdots + \varphi_p Z_{t-p} + a_t - a_{t-1}\theta_1 - \cdots - a_{t-q}\theta_q$$

where $\Delta_1 Z_t$ is a differenced series (i.e., $z_t - z_{t-1}$), zt is that the set of possible observations of the time-sequenced random variable in time sequence, at is the random shock term at time t, $\varphi_1 \ldots \varphi_p$ are the autoregressive parameters of order p and $\theta_1 \ldots \theta_q$ are the moving average parameters of order q (Adamowski and Chan, 2011).

The SVM algorithm may be a machine learning theory which does not have a predetermined structure, while the training samples are judged by their contributions. Only samples that are selected contributed to the final model, which was called "support vectors." This model is known as a high-dimensional quadratic programming problem. Therefore, to avoid "dimensional disaster," high-dimensional computing is converted into low-dimensional computing by a kernel function. Generic kernel functions include linear, radial basis function (RBF), Gaussian, polynomial, and other kernel functions. Among these functions, the RBF kernel has better performance compared to the linear kernel in the case of managing complex samples; and thanks to its simple function, it is more applicable than Gaussian and polynomial kernel functions (Zhou et al., 2017).

The analytical hyporheic flux model (AHF) allows one to estimate the seepage at river banks through a riverbed employed a simplified geometry, and its numerical model was done in MODFLOW. Therefore, the model requires extensive manipulation to form the packages, by representing the river package of MODFLOW a test phase through the seepage package is completed. For developing the implementation, FloPy tool is employed due to the versatility of manipulating the packages of MODFLOW through coding. A region where surface and groundwater mix within the bed and banks of a river is that the hyporheic zone as a base of the AHF model, which could be in centimeters to meters. Estimation of the flux through this zone is complex and is simulated by numerical groundwater flow models (Diaz et al., 2020). The errors of the AHF model are significantly lower than the customarily used model, which are supported vertical water seepage through the streambed described by Darcy's law. The AHF model is considered as a suitable model due to the straightforward set of data needed to solve the problem and simple implementation in any computing environment The AHF model geometry is its limitation: a rectangular-shaped riverbed cross section which is followed by the identical shape of the sediment layer under its bottom and alongside its bank. Consequently, for small and deep rivers, neglecting the flow on the banks results in significant errors in the estimation of total flow (Nawalany et al., 2020).

As neural networks learning speed were slower than expected, a brand new learning algorithm called Extreme Learning Machine (ELM) was developed for single-hidden layer feedforward neural networks (SLFN) (Huang et al., 2006). Among soft computing techniques Extreme learning machines (ELM) has rarely used and recently gained attention than other tools, due to its less accuracy in comparison with empirical formula and few engineers familiarity therewith. For these two problems, an ELM model was developed in conjunction with db2 mother wavelet transform to reinforce the accuracy in some studies. ELM may be a training algorithm for the single layer feedforward neural network (SLFFNN). This algorithm is based on the concept which with H hidden nodes H randomly determines values of input weights and biases of the hidden layer per endless probability distribution with probability 1, so that to be able to train N separate samples (Malekzadeh et al., 2019). In theory, this algorithm has an extremely fast learning speed. The experimental results supported some artificial and real benchmark functions show that the new algorithm can produce good generalization performance in most cases and might learn thousands of times faster than popular conventional popular learning algorithms (Huang et al., 2006). In the ELM structure, the input weight and also the bias values are generated randomly. ELM calculates the output weight matrix between hidden layers and output layers analytically among the inverse operation of the hidden layer output matrix. ELM has the power of interpolating and universal approximation, which made ELM as a tool for time series prediction. The ELM models for groundwater level prediction were done in the MATLAB program which is ready to use for both a univariate and a statistical procedure. The performance of the developed models can be evaluated using several used statistical tests. The coefficient of determination (R^2) shows the degree of correlation between two variables. It indicates how well the model developed a relationship between observed and predicted variables. The root mean square error (RMSE) shows the performance of the model performance by calculating the difference between the observed and predicted values. The Nash-Sutcliffe efficiency criterion (NS) could be a parameter for assessing the predictive capacity of hydrological models. As a normalized measure, it compares the mean square error of the model to the variance of the target output (Yadav et al., 2017) (Fig. 1).

2.6 Stochastic models

Among many approaches to solving Groundwater contaminant source characterization problem, the stochastic-simulation statistic (S—S) method is extremely popular (Wang et al., 2020). In groundwater management studies there are different terms for optimization modeling, as an example, linear programing techniques are used for groundwater management due to their simple formulation and application. On the other hand, some research has been done by nonlinear programing algorithms and dynamic programing techniques. The optimization models are deterministic, but some uncertainties cause stochastic behavior in optimization modeling, such as inherent variability and fundamental lack of data about aquifer parameters (Joodavi et al., 2017).

Monte Carlo may be a stochastic optimization model during which the complete system is simulated an outsized number of times each time using a different set of random values (Joodavi et al., 2017). Mantoglou and Kourakos (2007) developed

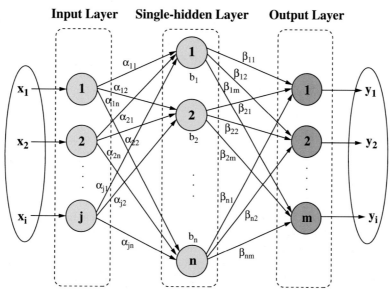

FIG. 1 The structure of ELM (Barzegar et al., 2017).

this modeling to include hydraulic conductivity uncertainty in a very multiobjective management model for optimal remediation of groundwater. This target was achieved by minimizing two objectives of the optimization model: contaminated groundwater within the aquifer and therefore the cost of remediation cost (Mantoglou and Kourakos, 2007). Although the Monte Carlo method is typically used in stochastic simulations, using them within the real world, large-scale water resource management problems can be computationally demanding (Joodavi et al., 2017).

2.7 SOM models

The self-organizing map (SOM) may be a method for clustering the homogenous monitoring piezometers within the plain by using GWL and Universal Transverse Mercator (UTM) data. For visualization of high-dimensional data, it is an efficient software tool (Nourani et al., 2012).

Wavelet transforms are a tool for the analysis, de-noising and compression of signals and pictures. Wavelets transform analysis, developed during the last 20 years, are mathematical functions that are wont to analyze time series that contain nonstationarities. Their ability to simultaneously obtain information on the time, location and frequency of a signal is their main advantage (Adamowski and Chan, 2011). Wavelet analysis may be a tool to analyze localized variations of power with the assistance of time series. The wavelet transform has been used for numerous studies like: geophysics, including tropical convection, atmospheric cold fronts, the dispersion of ocean waves, wave growth and breaking, coherent structures in turbulent flows, etc. (Torrence and Compo, 1998). Wavelet as a transformation tool is utilized for decomposition, compression, and de-noising. This method may be a time-independent spectral analysis for describing the time scale, processes and their relations. The Wavelet is in a position to rework and considers time series as a linear combination of multiple base functions. The wavelet transform has this ability to attain time, frequency, and situation data simultaneously (Malekzadeh et al., 2019). Because this method has multiresolution strength and localization capability in time and frequency domains, it is a crucial tool for the spectral analysis area (Yu et al., 2018).

2.8 Conceptual models

MODFLOW packages with independent modules in steady- and transient state conditions calibrate processes either manual or automated. This model is employed by many researchers to grasp groundwater dynamics to mend optimal pumping limits from the aquifer (Panda and Narasimham, 2020). Among its development, many packages have been created to satisfy and model some scenarios. The most challenge of modeling would be the high numerical instability, the high costs of modeling, and also the computational efficiency (Diaz et al., 2020).

This model solves the 3-D groundwater flow equation numerically using the finite-difference method in porous media:

$$\frac{\partial}{\partial x}\left(K_{xx}\frac{\partial h}{\partial x}\right) + \frac{\partial}{\partial y}\left(K_{yy}\frac{\partial h}{\partial y}\right) + \frac{\partial}{\partial x}\left(K_{zz}\frac{\partial h}{\partial z}\right) - W = \left(S_s\frac{\partial h}{\partial t}\right)$$

where x, y, z are Cartesian coordinate axes, h = potentiometric head [L], K_{xx}, K_{yy}, K_{zz} = hydraulic conductivities along the x, y, and z axes [LT^{-1}], W = volumetric flux/unit volume and represents sources and/or sinks of water [T^{-1}], S_S = specific storage of the porous material [L^{-1}] and t = time. All capabilities of MODFLOW are available in McDonald and Harbaugh (1988) (Shukla and Singh, 2018). This model has some assumption that several aquifer layer systems are met them: the flow is in saturated condition, (2) Darcy's law applies, (3) the groundwater concentration is constant, and (4) the principal direction of hydraulic conductivity or transmissivity does not change within the aquifer system (Chakraborty et al., 2020).

The SWAT (Soil and Water Assessment Tool) watershed model with 3700 published studies has been the foremost used watershed model within the world since 2000 (Bailey et al., 2020). To assist water resource managers, assess the impact of management on water supplies and nonpoint source pollution in watersheds and large river basins, this model was developed. This model can predict the impact of management on agricultural, sediment and agricultural chemical yields and also operates on a daily time step (Arnold et al., 1998). Within SWAT, the water balance equation is solved for the shallow aquifer. This model also has an updated version called SWAT+, which include MODFLOW for simulating interactions of groundwater flow and groundwater surface water. SWAT+ gives the recharge and stream stage to MODFLOW, so MODFLOW provides spatial values of groundwater and therefore the exchange rates of groundwater and surface water that are provided to the SWAT + stream channels for stream routing. SWAT+(MODFLOW) can simulate all hydrological pathways of the groundwater level, the depth of the watershed, for mapping of water table elevation, water table depth, and also the exchange rates for all grid cells within the model domain. This model is a wonderful tool for investigating water supply and also the impact of conservation practices and climate change in watersheds (Bailey et al., 2020).

GMS as a three-dimensional groundwater simulation software is employed to simulate groundwater and solute transport problems in groundwater. This software is ready to model the aquifer in both steady and unsteady conditions within the 2D and 3D modes by imposing changes in atmospheric parameters and conditions within the area for better management of groundwater in the area. Compared to other similar software, such as MODFLOW, PMWIN, and FreeWAT software, the GMS modules are better than their counterparts because of its better user efficiency and 3D visualization (Bayat et al., 2020).

3. Discussion

In Nourani et al.'s (2012) study, SOM and FFNN approaches were combined to develop a hybrid black-box model for groundwater level simulation. Their results showed that in step with the uncertainty of rainfall uncertainty and therefore the ability of fuzzy concepts, the conjunction of the ANN and FIS (Fuzzy Inference System) model as an ANFIS might be an appropriate model for model progress (Nourani et al., 2012).

In a study, wavelet transform (WA) and artificial neural networks (ANN) used for groundwater level prediction and their performance were compared to ANN and autoregressive integrated moving average (ARIMA) models for monthly groundwater level prediction. It had been revealed that the WA–ANN models are more accurate for the monthly average groundwater levels compared to the ANN and ARIMA models(Adamowski and Chan, 2011).

In a study so as to assess groundwater quality three optimization metaheuristic algorithms: imperialist competitive (ICA), election (EA), and gray wolf (GWO) were used. Ten years data were used and therefore the results indicate that GWO method with the support vector regression method (SVR) had a higher accuracy than two other methods and this method also had a higher accuracy in simulation and estimation of groundwater quality (Emami et al., 2021).

The support vector regression model (SVR) is employed to estimate heavy metals in groundwater by two input data, HCO_3, SO_4, and Pb, Zn, and Cu, as output parameter. The result showed that this model could be a reliable system for this proposal (Ghadimi, 2017).

In Africa, the estimation of the groundwater recharge rate was done using groundwater level data, the streamflow method, and therefore the water balance methods. The results indicate that the Water Table Fluctuations (WTF), Recession-Curve Displacement, and Chloride Methods may be used with certainty to boost estimation of groundwater recharge. On the opposite hand, recently developed water balance methods combined with GIS technology may well be a robust tool for estimating groundwater recharge, when spatial and temporal variability of components within the water balance is taken into consideration (Chung et al., 2016).

According to Diaz et al. (2020), Neumann boundary condition is employed for the implementation of the AHF model through the recharge package. Then, to represent it as a Cauchy boundary condition type for a river, an iterative algorithm that permits defining AHF fluxes through seepage in MODFLOW is employed. So, with assistance of the AHF model, the lateral contribution is implemented on the banks of a river, which is applied in an exceedingly real case employing a groundwater modeling tool in three dimensions as MODFLOW was done. In their study, the groundwater model is performed in several steps of fieldwork, information gathering, set-up of conceptual and numerical models, and calibration. The Biebrza River between the villages of Rogoz yn and Rogoz˙ynek villages is a study area for which the model is made for. Finally, their methodology allows the fluxes that come from the bottom and also the banks of the river to be channeled artificially rather than rivers with complex natural geometry (Diaz et al., 2020).

In a study, eight spatial interpolation methods were compared in simulating the spatial distribution of the groundwater level supported GIS in China. The results indicated that among these eight method the ordinary kriging method is that the optimal method for groundwater level interpolation based on cross-validation and orthogonal validation and orthogonal-validation (Yao et al., 2014).

In the $470 \, km^2$ Middle Bosque River Watershed (MBRW) in central Texas, USA, the SWAT+(MODFLOW) model was used. The model was tested to measure the trends of annual average hydrologic flows between annual rainfall rates and hydrologic flows, the discharge of the stream discharge at the watershed outlet, groundwater head and water table, and groundwater-surface water interactions. The simulation results were fitted well correlated with observed values (Bailey et al., 2020).

For investigating the temporal and spatial variability of nitrate leaching in watershed and modeling amount and dynamics of it the SWAT model was utilized in Hamedan-Bahar plain of Iran. The model was calibrated and validated by uncertainty analysis SUFI-2. Results of this study indicated that spatial variation of nitrate leaching were confirmed well with measured nitrate concentration in groundwater by 73% overlap. Due to the model's strong potential for considering different scenarios in reducing nitrate leaching it is suggested for best management practice in watershed (Akhavan et al., 2010).

Three models including MODFLOW, Extreme Learning Machine (ELM) and Wavelet Extreme Learning Machine (WA-ELM) were utilized during a study to simulate groundwater level. In this way, they defined 10 different models for ELM and WA-ELM models separately and examined the results of them to decide on the most effective soft computing model. The best artificial intelligence model was also compared with MODFLOW and showed that the WA-ELM model had higher accuracy in simulating groundwater level compared to comparison of MAE and RMSE (Malekzadeh et al., 2019).

In a study, AHP was accustomed determine the homogeneous recharge zones and recharge rates, modeling processes were continued in MODFLOW. During this way it is possible to resolve the physical properties of the aquifer ignoring within the resulting nonlinear equations such as soil properties, unsaturated thickness, land cover, irrigation, and precipitation. On the other hands, the results of climate change on groundwater recharge were evaluated too, and therefore the results indicated that climate change would decrease the recharge rates within the homogeneous areas (Goodarzi et al., 2016).

Hydrological and meteorological parameters such as rainfall, temperature, evapotranspiration and groundwater level monthly data were used to forecast groundwater level in two wells by ELM and SVM. The results showed that the ELM model had better ability to simulate the monthly groundwater level (Yadav et al., 2017).

In a study conducted on the Maragheh-Bonab plain in Iran, 367 months average groundwater data were used to compare the performance of different hybrid wavelet-neural networks. About 85% of the total data set were utilized for training and 15% for testing the models. The results showed that according to R^2, RMSE and NSC the ELM had better performance than other models (GMDH) (Barzegar et al., 2017).

In a study, WA-ANN and WA-SVR were developed to optimize the ANN and SVR models. To investigate the feasibility of ANN, SVR, WA-ANN and WA-SVR models in simulating groundwater depth 1, 2, and 3 months ahead in three wells. The performances of ANN, SVR, WA-ANN and WA-SVR models for 1-month-ahead forecasting were better than 2- and 3-months ahead forecasts, because as the lead time gets longer, the accuracy of simulated groundwater depth becomes worse. Also, it was observed that WA-SVR had better results than the WA-ANN model for 1, 2 and 3 months and this model has the ability to forecast groundwater under ecological water conveyance conditions (Yu et al., 2018).

The coupled wavelet support vector machine (WSVM) model was tested in a study. A partial autocorrelation function of groundwater depth time series was introduced as input variables' lag times. For preprocessing and decomposing the original groundwater level time series to four subseries by different frequencies, a three level discrete wavelet transform (DWT) was utilized. Based on the same historical data, the WSVM model was compared with ANN, SVM and WANN and the results showed that according to the RAE, r coefficient, RMSE and NSE, WSVM had more accurate results. The groundwater depth series was decomposed into four subseries with better stationary for model training by using three-level DWT, so that the extraction of mainstream components improved the prediction performance. Furthermore, the WSVM model was indicated to have a good prediction of the monthly groundwater depth (Zhou et al., 2017).

Bayat et al. (2020) used the Groundwater Modeling System (GMS software) in the 3D in Gavkhnoi Basin, Isfahan Province, Iran, to investigate the decreases of ground water and potential environmental hazards. In this study, 86 months of steady and unsteady state was utilized for calibrating and validating the model. The results showed that by using a calibrated model, reduction of almost 30% in well water harvesting leads to an increase in the mean groundwater level by 37 cm and an increase of 10% in well water harvesting; the mean groundwater level decreased by approximately 12 cm. The surface water results for this period showed that the water levels in most parts of the aquifer follow a uniform decreasing pattern, indicating that if the wells are discharged and recharged, as in the previous years, most observational wells face a drastic drop in the water level. According to scenarios, among the changes through artificial recharge, precipitation and well water harvesting, precipitation has the most effective procedure in aquifer water management (Bayat et al., 2020).

For delineating the groundwater storage and recharge potential zones in India, the RS-GIS-AHP framework was developed. Result maps are applicable for groundwater storage and facilitating aquifer recharge in the basin, which provides safe yield values for planning judicious utilization of the resource (Jena et al., 2020).

In a study which had been done by authors water table according to different scenarios was simulated by Artificial Neural Network and ArcGIS. By applying artificial neural network, water table data and cropping pattern in plain, three land use scenarios were designed: (1) existing crop pattern in study area, (2) increase fallowing every year, and (3) planting low-water use crops. The year 2008 water table was simulated with one-layer network, Levenberg–Marquart algorithm, and three functions in MATLAB R2012a (Table 1). The water table map was prepared using simulated water table in ArcGIS 10.2, the model divided data into 3 groups one for training, another for simulation variation, and the other for testing so according to covariance and RMSE the best model was selected. Zoning was performed according to the costs of water pumping. The results showed that 61% to 66% of the plain in all three scenarios had a medium limitation. Furthermore,

TABLE 1 Simulated groundwater for wells in July 2008 (m).

Well number	UTMx	UTMy	Scenario 1	Scenario 2	Scenario 3
1	575,488	3,570,835	114.88	115.31	115.5
2	566,649	3,579,262	107	107.66	108.18
3	570,186	3,579,747	118.15	118.43	118.78
4	577,276	3,577,487	157.93	158.26	158.41
5	569,258	3,575,519	93.63	92.62	92.28
6	573,653	3,578,253	121.82	122.59	122.96
7	567,753	3,586,253	66.72	66.31	66.07
8	571,769	3,584,716	106.38	106.32	106.32
9	579,173	3,574,088	138.43	138.52	138.55
10	574,258	3,575,080	115.75	115.69	115.58
11	568,927	3,584,780	88.48	89.15	89.53
12	575,963	3,578,899	158.12	158.08	158.02
13	574,223	3,580,676	140.3	141.62	141.17
14	570,638	3,586,168	99.15	99.13	99.13

the use of different management in the field like fallowing and planting low-water use crops caused an increase of 3% and 5% increase in the acreage of "without limitation" lands, respectively. Due to the results, it seems that ANN is a strong tool for simulating the water table even in case of lack of data and in managing crop planning in different regions.

4. Conclusions

For efficient management of water resources, analysis of the different hydrological parameter is needed. Groundwater systems have the dynamic and complex nature that they change in response to the climatic stresses and anthropologic activities. Also human activities induced the stresses on groundwater resources such as: population growth, increasing agricultural productions and growing of urban areas. Data scarce conditions, presence of several factors such as anthropologic effects make an accurate determination of hydrological parameters so difficult, according to the importance of groundwater especially in arid and semiarid region, it is crucial to find the best method to simulate it, therefore application of a correct modeling makes the simulation fast and accurate. Models can illustrate the simple to complicated hydrogeological system in groundwater assessment strategy. As presented before, there are many methods for modeling groundwater, selecting the best one is referring to the knowledge, skill and experience of researcher about the case, available data, dominating the methods, knowing the concepts of the method and many other factors which make the simulation accurate and close to the real world. So it is important to configure the problem and best method to solve it.

References

Adamowski, J., Chan, H.F., 2011. A wavelet neural network conjunction model for groundwater level forecasting. J. Hydrol. 407 (1–4), 28–40.
Akhavan, S., Abedi-Koupai, J., Mousavi, S.-F., Afyuni, M., Eslamian, S.-S., Abbaspour, K.C., 2010. Application of SWAT model to investigate nitrate leaching in Hamadan–Bahar Watershed Iran. Agric. Ecosyst. Environ. 139 (4), 675–688.
Arnold, J.G., Srinivasan, R., Muttiah, R.S., Williams, J., 1998. Large area hydrologic modeling and assessment part I: model development 1. J. Am. Water Resour. Assoc. 34 (1), 73–89.
Bailey, R.T., Park, S., Bieger, K., Arnold, J.G., Allen, P.M.J.E.M., Software., 2020. Enhancing SWAT+ simulation of groundwater flow and groundwater-surface water interactions using MODFLOW routines. Environ. Model. Softw. 126, 104660.
Barzegar, R., Fijani, E., Moghaddam, A.A., Tziritis, E.J.S., o. t. T. E., 2017. Forecasting of groundwater level fluctuations using ensemble hybrid multi-wavelet neural network-based models. Sci. Total Environ. 599, 20–31.

Bayat, M., Eslamian, S., Shams, G., Hajiannia, A.J.Q.G., 2020. Groundwater level prediction through gms software–case study of karvan area, Iran. Quaest. Geogr. 39 (3), 139–145.

Carle, S.F., Fogg, G.E., 1997. Modeling spatial variability with one and multidimensional continuous-lag Markov chains. Math. Geol. 29 (7), 891–918.

Chakraborty, S., Maity, P.K., Das, S.J.E., Development, and Sustainability., 2020. Investigation, simulation, identification and prediction of groundwater levels in coastal areas of Purba Midnapur, India, using MODFLOW. Environ. Dev. Sustain. 22 (4), 3805–3837.

Chung, I.-M., Sophocleous, M.A., Mitiku, D.B., Kim, N.W.J.G.J., 2016. Estimating groundwater recharge in the humid and semi-arid African regions. Geosci. J. 20 (5), 731–744.

Diaz, M., Sinicyn, G., Grodzka-Łukaszewska, M.J.W., 2020. Modelling of groundwater–surface water interaction applying the hyporheic flux model. Water 12 (12), 3303.

Drias, T., Khedidja, A., Belloula, M., Badraddine, S., Saibi, H., 2020. Groundwater modelling of the Tebessa-Morsott alluvial aquifer (northeastern Algeria): A geostatistical approach. Groundw. Sustain. Dev. 11, 100444.

Emami, S., Choopan, Y., Parsa, J., Jahandideh, O.J.A., i. E. T., 2021. Modeling groundwater quality using three novel hybrid support vector regression models. Adv. Environ. Technol. 6 (1), 99–110.

Ghadimi, F.J.J., 2017. Machine learning algorithm for prediction of heavy metal contamination in the groundwater in the Arak urban area. J. Tethys 5 (2), 115–127.

Ghosh, S., Dubey, S.K., 2013. Comparative analysis of k-means and fuzzy c-means algorithms. Int. J. Adv. Comput. Sci. Appl. 4 (4). https://doi.org/10.14569/IJACSA.2013.040406.

Goodarzi, M., Abedi-Koupai, J., Heidarpour, M., Safavi, H.R., 2016. Evaluation of the effects of climate change on groundwater recharge using a hybrid method. Water Resour. Manag. 30 (1), 133–148.

Huang, G.-B., Zhu, Q.-Y., Siew, C.-K.J.N., 2006. Extreme learning machine: theory and applications. Neurocomputing 70 (1–3), 489–501.

Jang, J.-S.R., Sun, C.-T., Mizutani, E., 1997. Neuro-fuzzy and soft computing-a computational approach to learning and machine intelligence [book review]. IEEE Trans. Autom. Control 42 (10), 1482–1484.

Jeihouni, E., Eslamian, S., Mohammadi, M., Zareian, M.J., 2019. Simulation of groundwater level fluctuations in response to main climate parameters using a wavelet–ANN hybrid technique for the Shabestar plain, Iran. Environ. Earth Sci. 78 (10), 1–9.

Jena, S., Panda, R.K., Ramadas, M., Mohanty, B.P., Pattanaik, S.K., 2020. Delineation of groundwater storage and recharge potential zones using RS-GIS-AHP: application in arable land expansion. Remote Sens. Appl.: Soc. Environ. 19, 100354.

Joodavi, A., Zare, M., Ziaei, A.N., Ferré, T.P., 2017. Groundwater management under uncertainty using a stochastic multi-cell model. J. Hydrol. 551, 265–277.

Khorasani, M., Ehteshami, M., Ghadimi, H., Salari, M., 2016. Simulation and analysis of temporal changes of groundwater depth using time series modeling. Model. Earth Syst. Environ. 2 (2), 1–10.

Kozhevnikova, T., Manzhula, I., Kondratieva, L., 2020. Simulation in the tasks of environmental monitoring of groundwater. In: Paper Presented at the IOP Conference Series: Earth and Environmental Science, p. 981.032005.

Langousis, A., Kaleris, V., Kokosi, A., Mamounakis, G., 2018. Markov based transition probability geostatistics in groundwater applications: assumptions and limitations. Stoch. Env. Res. Risk A. 32 (7), 2129–2146.

Malekzadeh, M., Kardar, S., Shabanlou, S., 2019. Simulation of groundwater level using MODFLOW, extreme learning machine and Wavelet-Extreme Learning Machine models. Groundw. Sustain. Dev. 9, 100279.

Mantoglou, A., Kourakos, G., 2007. Optimal groundwater remediation under uncertainty using multi-objective optimization. Water Resour. Manag. 21 (5), 835–847.

Maroufpoor, S., Fakheri-Fard, A., Shiri, J., 2019. Study of the spatial distribution of groundwater quality using soft computing and geostatistical models. ISH J. Hydraul. Eng. 25 (2), 232–238.

McDonald, M.G., Harbaugh, A.W., 1988. A Modular Three-Dimensional Finite-Difference Ground-Water Flow Model. US Geological Survey.

Nawalany, M., Sinicyn, G., Grodzka-Łukaszewska, M., Mirosław-Świątek, D.J.W., 2020. Groundwater–surface water interaction—analytical approach. Water 12 (6), 1792.

Nourani, V., Baghanam, A.H., Vousoughi, F.D., Alami, M.T., 2012. Classification of groundwater level data using SOM to develop ANN-based forecasting model. Int. J. Soft Comput. 2 (1), 2207–2231.

Ouhamdouch, S., Bahir, M., Ouazar, D., Carreira, P.M., Zouari, K., 2019. Evaluation of climate change impact on groundwater from semi-arid environment (Essaouira Basin, Morocco) using integrated approaches. Environ. Earth Sci. 78 (15), 1–14.

Panda, P., Narasimham, M.J.I.R., 2020. A review on modelling and simulation of ground water resources in urban regions. INFOKARA Research 9, 235–244.

Rwanga, S., Ndambuki, J., 2017. Approach to quantify groundwater recharge using gis based water balance model: a review. Int. J. Adv. Agric. Environ. Eng. 4 (1). https://doi.org/10.15242/ijrcmce.ae0317115.

Shukla, P., Singh, R.M., 2018. Groundwater system modelling and sensitivity of groundwater level prediction in indo-Gangetic Alluvial Plains. In: Groundwater. Springer, pp. 55–66.

Sun, Y., Kang, S., Li, F., Zhang, L., 2009. Comparison of interpolation methods for depth to groundwater and its temporal and spatial variations in the Minqin oasis of Northwest China. Environ. Model. Softw. 24 (10), 1163–1170.

Thakur, J.K., Singh, S.K., Ekanthalu, V.S., 2017. Integrating remote sensing, geographic information systems and global positioning system techniques with hydrological modeling. Appl Water Sci 7 (4), 1595–1608.

Torrence, C., Compo, G.P.J.B., 1998. A practical guide to wavelet analysis. Bull. Am. Meteorol. Soc. 79 (1), 61–78.

Wang, X., Xie, H., 2018. A review on applications of remote sensing and geographic information systems (GIS) in water resources and flood risk management. Water 10 (5), 608.

Wang, H., Lu, W., Li, J., 2020. Groundwater contaminant source characterization with simulation model parameter estimation utilizing a heuristic search strategy based on the stochastic-simulation statistic method. J. Contam. Hydrol. 234, 103681.

Yadav, B., Ch, S., Mathur, S., Adamowski, J., Development, L., 2017. Assessing the suitability of extreme learning machines (ELM) for groundwater level prediction. J. Water Land Dev. 32, 103–112.

Yang, H., Han, D., Yu, F., 2013. Improved fuzzy c-means clustering algorithm based on sample density. J. Theor. Appl. Inf. Technol. 48 (1), 210–214.

Yao, L., Huo, Z., Feng, S., Mao, X., Kang, S., Chen, J., et al., 2014. Evaluation of spatial interpolation methods for groundwater level in an arid inland oasis, Northwest China. Environ. Earth Sci. 71 (4), 1911–1924.

Yu, H., Wen, X., Feng, Q., Deo, R.C., Si, J., Wu, M.J.W., r. m., 2018. Comparative study of hybrid-wavelet artificial intelligence models for monthly groundwater depth forecasting in extreme arid regions, Northwest China. Water Resour. Manag. 32 (1), 301–323.

Zhou, T., Wang, F., Yang, Z.J.W., 2017. Comparative analysis of ANN and SVM models combined with wavelet preprocess for groundwater depth prediction. Water 9 (10), 781.

Chapter 14

Hydrological-Hydraulic Modeling of floodplain inundation: A case study in Bou Saâda Wadi—Subbasin_Algeria

Zohra Abdelkrim[a,b], Brahim Nouibat[a,b], and Saeid Eslamian[c,d]

[a]Institute of Management of Urban Techniques, Mohamed Boudiaf University, M'sila, Algeria, [b]Laboratory of City, Environment, Society and Sustainable Development, M'sila, Algeria, [c]Department of Water Engineering, College of Agriculture, Isfahan University of Technology, Isfahan, Iran, [d]Center of Excellence for Risk Management and Natural Hazards, Isfahan University of Technology, Isfahan, Iran

1. Introduction

Heavy rainfall events are expected, under possible climate change scenarios. This situation is likely to become worse in the future, which will lead to more frequent floods. It's known that flood risk does not differentiate between developing and developed countries, e.g., floods that have swept Germany and Belgium from 2021, at least 220 people were killed between July 12 and 15, mostly in Germany, dozens also died in Belgium and homes and other buildings were destroyed in flash flooding. Record rainfall that triggered deadly floods in Western Europe in July was made between 1.2 and 9 times more likely by human-caused climate change, according to a new study (Dewan, 2021).

Threat from floods to lives and infrastructure is increasing due to urban development, there are many effects on human settlements and economic activities (Ghazavi et al., 2010). Flood defines as high flow that exceeds or over top of the capacity either the natural or the artificial bank of the stream or the river channel (Aynalem, 2020).

Urban development and urban growth and land use changes increases flood risk in cities (Huong and Pathirana, 2013), resulting in an increase of prone areas for flood hazards over time, all this caused in hydrological regime changes. Based on facts, it is clear that intensity of land use also exerts a great impact on soils (Smith et al., 2016). The consequent impermeabilization of the soil is a current and recurrent problem, e.g., impervious areas affect rainfall runoff, flooding.

Structural and nonstructural measures are used to reduce the effects of floods and protect people; Structural measures: e.g., the construction of dams, nonstructural measures (this means nonengineering activities), like Building codes are sets of regulations, standards, and insurance. Planning and land use policies are basically nonstructural and preventive measures (Mohit and Sellu, 2013) and can help us manage risk effectively. Watershed still needs a planning and management, important feature of land-use planning and flood-risk management is its treatment of flood in watershed.

The role of flood mapping in river engineering is an important feature in planning and management (basis for managing flood plains, basis for pursuing structural and nonstructural measures), river flood mapping is the process of determining inundation extents and depth (Alaghmand et al., 2010).

Algeria is facing catastrophic floods, a significant part of its territory is exposed to flooding, all the more so the area of the Bou Saâda Wadi—Subbasin is witnessing an enormous expansion of land use, often occur caused by rapid urban growth, etc.

New data reveal May 2021 floods resulted in extreme impacts in Bou Saâda Wadi—Subbasin, causing resulting in casualties and damage, people have lost their homes, their cars are flooded: six killed and Material losses (wrecked cars). They occur within minutes to hours after a heavy rain event, and produce torrents of water, a state of emergency has been declared for Save Families from Drowning. Under extremely difficult and dangerous conditions, Bou Saâda Wadi—

Handbook of HydroInformatics. https://doi.org/10.1016/B978-0-12-821962-1.00011-8

Subbasin needs to be studied to identify flood hazard areas (depth, and extent of flooding), that the absence of such studies affects the flood risk management.

Existing literature focuses specifically on assessment and flood mapping. Most frequently used methods of floodplain modeling: integrating Geographical Information System (GIS) with hydrologic/hydraulic models (Golshan et al., 2016).

The widely used model to analyze channel flow and floodplain delineation (Romali et al., 2018) is HEC-RAS. This model is a hydraulic model; steady One-dimensional Flow. U.S. Army Corps of Engineers Hydrologic Engineering Center (HEC) developed the River Analysis System (RAS), which produced its first map of river flood hazards in 1988 (Alaghmand et al., 2010).

GIS and HEC-RAS models have been in many studies were successfully used for obtaining flood maps of Waller River in Texas (Tate et al., 2002), mid-eastern Dhaka in Bangladesh (Masood and Takeuchi, 2012). Modeling studies carried out in previous years confirm the effectiveness of Hydraulic models and GIS in developing a flood hazard map.

In this study, in order to model the floodplains of Bou Saâda Wadi—Subbasin, linked the Hydraulic model with the hydrologic model. The most important program of a hydrologic model is Hyfran_Plus; this program is essential for flood frequency analysis which is a technique to predict flow values to specific return periods.

The Hyfran_Plus includes several types of theoretical probability distributions, among them: Gumbel distribution which is a statistical method often used for predicting extreme hydrological events such as floods (Eke and Hart, 2020).

ArcGIS has a suite of extensions: HEC-GeoRAS and HEC-GeoHMS, used to extract the geometry input data and also river flood visualization (Shafroth et al., 2010).

What we need to build an engineering file for the HEC-RAS model is Hydrological and Topographic data, TIN model (Triangular Irregular Network) generated from SRTM (Shuttle Radar Topography Mission) 30 m DEM (Digital Elevation Model) to represent the topography of the study area. The DEM is a major source of topographic data for representing floodplain and river topography. SRTM DEM data is being housed on the USGS Earth Explorer.

The main objective of this study is to determine floodplain inundation for 5, 10, 50, and 100-year return period in Bou Saâda Wadi—Subbasin, using an integration of the GIS and the Hydraulic Model HEC-RAS.

2. Site of study

The Hodna basin is the fifth basin of Algeria, located at 150 km in the south of the Mediterranean coast, the Hodna basin has a drainage area of 26,000 km^2. The altitude of the Hodna summits decreases from east to west, it oscillates between 1900 and 1000 m, while to the South only a few peaks in the Saharan Atlas reach 1200 m. The situation of this basin is between two sets of mountains at the north and the south. At the center of this region, the dry salt lake named "Chott El Hodna," a center of this area is constituted by the Chott El Hodna (1150 km^2) (Hasbaia et al., 2012). According to the Algerian Agency of Water Resources called (ANRH), Hodna Basin can be divided into 23 subbasins, the 24th central one is Chott El Hodna; however these subbasins can be grouped into 8 hydrographic subbasins (Fig. 1).

The Hodna basin marks the transition between the Tellian domain in the north and the Sahara in the south. The transition between these two domains, marked by a succession of mountainous areas (the Tell in the north, the Saharan Atlas in the south) and plains, favors the diversity of bioclimatic environments. In the south, the Bou Saada wadi flows only part of the year. It is fed by springs that emerge from the Albian sandstone 7 km upstream from the town of Bou Saada.

The Bou Saâda Wadi—Subbasin is located in southwestern the Hodna basin in the highlands, located at 150 km in the south of Algeria, has a drainage area of 1008 km^2, this Subbasin is characterized by the height of a mountain, especially in the northeastern part of the basin (Fig. 2).

Bou Saâda Wadi—Subbasin is bordered to the north by Upstream Maitar—Subbasin; to the Northeast and East by Estuary Maiter—Subbasin; and to the southeast and south by Upstream Chair—Subbasin.

In Bou Saâda Subbasin, there are five lands use types such as Built Up area, Vegetation (forest, grass) and Open Land (Mountains, Soil,..) (Fig. 3). Considering the amount of damage caused by floods from the reports, it has caused more damage in Built Up areas compared to other land use types (see images 1 and 2).

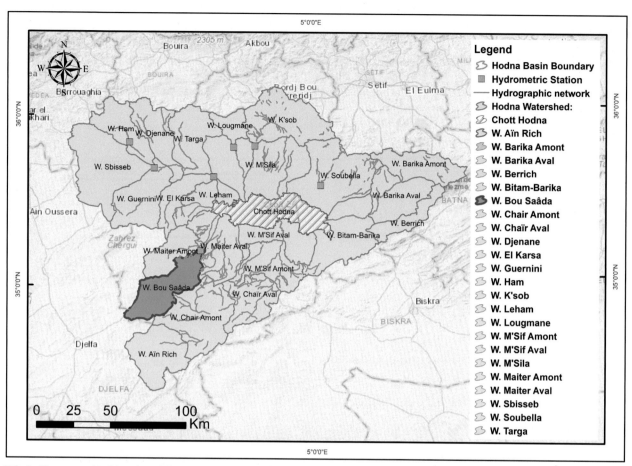

FIG. 1 The geographical location of Hodna basin.

Image.1. Bou Saâda Wadi

Image. 2. Flood damage in Bou Saâda Wadi

3. Methodology

In this study, floodplain inundation maps were obtained by using HEC-RAS, HEC-GeoRAS, and Arc-GIS. This latter, was developed by Environmental Systems Research Institute (ESRI) and it is used to create flood maps. It is an easy tool to create and use maps, and perform spatial analysis, processing, modeling, floodplain management (floodplain extents and depths). The used HEC-RAS 5.0.7 version hydraulic model was developed by the US Army Corps of Engineers and was first released in 1995 (Helwa et al., 2020). It is good for floodplain flow/flood mapping (Saad et al., 2019).

It is an integrated program and uses the following energy equation for calculating water surface profiles (Thoummalangsy et al., 2019):

$$Y_2 + Z_2 + \frac{a_2 v_2^2}{2g} = Y_1 + Z_1 + \frac{a_1 v_1^2}{2g} + h_e \quad (1)$$

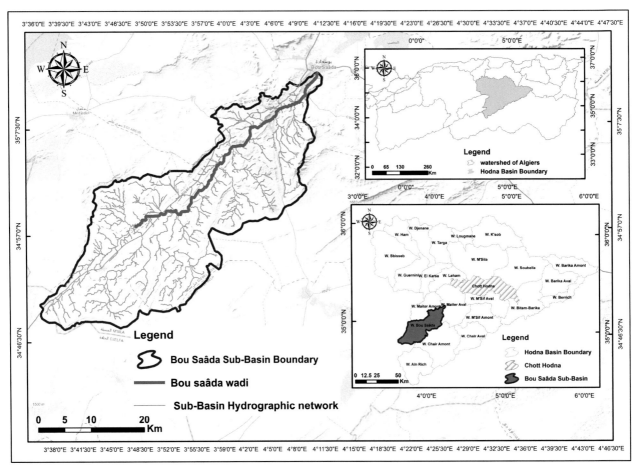

FIG. 2 The geographical location of Bou Saâda Wadi—Subbasin.

where Y_1, Y_2 is the depth of water at cross sections; Z_1, Z_2 is the elevation of the main channel inverts; v_1, v_2 is the average velocity; a_1, a_2 is the velocity weighting coefficient; g is gravitational acceleration and h_e is energy head loss.

Hydraulic analysis requires many input parameters for hydraulic analysis of the stream channel geometry and water flow, HEC-RAS uses these parameters to establish a series of cross-sections along the stream. In each cross-section, the locations of the stream banks are identified and used to divide into segments of the left floodway, main channel, and right floodway (Pallavi and Ravikumar, 2021).

HEC-GeoRAS: is an ArcGIS extension, its role is to process and transfer geospatial data, e.g.: the river cross-sections, river centerline, watercourse bank lines, flow lines. And this geospatial data are transferred between the ArcGIS and the HEC-RAS use the Geo-RAS to create floodplain inundation with depths and extents of flooding (Fig. 4).

Geo-RAS uses the terrain data: Triangular Irregular Network (TIN), which is developed from DEM. The quality of this data is based on its resolution or cell size (Desalegn and Mulu, 2021). The DEM refers to a topographic map that contains terrain elevation properties.

The methodology involved in Hydrological-Hydraulic Modeling using GIS and HEC-RAS to determine floodplain inundation for 5, 10, 50, and 100 years in Bou Saâda Wadi—Subbasin:

The first step: the hydrological analysis
    (1) Estimation of peak flow

The second step: hydraulic modeling
    (2) Preprocessing
    (3) Processing
    (4) Postprocessing

- **The first step: the hydrological analysis**
  - **Estimation of peak flow:**

Meteorological Data is considered essential to perform Flood Return Period Estimation. Meteorological Data is collected e.g., Precipitation data, from Bou saada hydraulic station. We used 41 annual flow data, for the water years between 1971 and 2012.

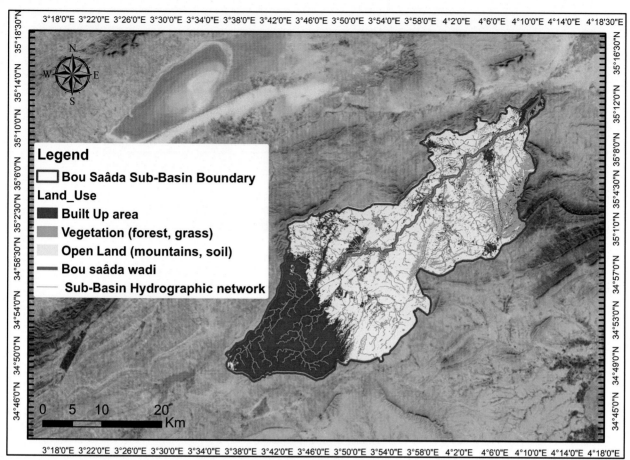

FIG. 3   Land use at Bou Saâda Wadi—Subbasin.

The Hyfran-plus software is used to select the best flood distribution model to calculate peak flows. Statistical methods were practiced to flood frequency analysis. For this study, the Gumbel plot was broadly used for the calculation of return periods.

- **The second step: hydraulic modeling**

Following stages creating a hydraulic model will allow us to determine floodplains at risk for 5, 10, 50, and 100 years in Bou Saâda Wadi—Subbasin: Preprocessing, Processing, and Postprocessing of the data.

○ **Preprocessing:**

The preprocessing stage was done within ArcMap. As Fig. 5 illustrates the preprocessing stage:
○ **Processing:**
The processing stage was done completely within HEC-RAS using the river geometry prepared in the previous stage as Fig. 6 illustrates.
   In 1D model, the roughness parameter, Manning's n value is important to the solution of the computed water surface profile (Pallavi and Ravikumar, 2021).
○ **Postprocessing:**
The Postprocessing stage was done within ArcMap, consisted of Analysis of results from the HEC-RAS model as Fig. 7 illustrates.

4. Results and discussion

4.1 Peak discharge estimation

Flood frequency analysis (using HYFRAN—Plus software) was conducted to estimate peak flow in different return periods for the hydraulic simulation. It has shown that Gumbel (Fig. 8), is the best statistical distribution.

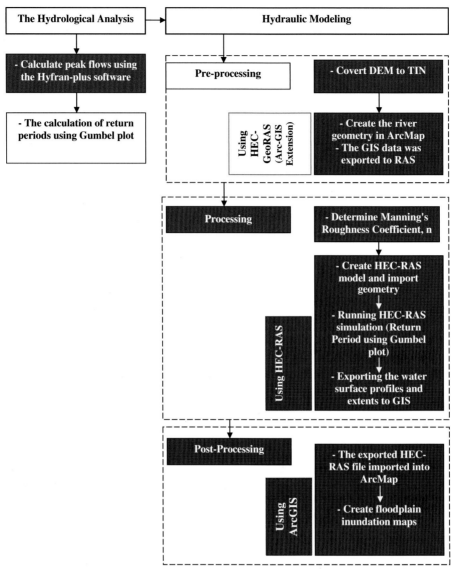

FIG. 4 Main steps in creating a Hydrological-Hydraulic Modeling with HEC-RAS and Arc-GIS.

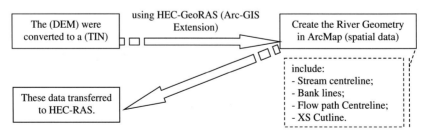

FIG. 5 Preprocessing steps.

4.2 Delineation of Bou Saâda Wadi—Subbasin

RAS geometric data was created and delineated by using TIN as base data in RAS Geometry of HEC-GeoRAS, was exported as RAS data to be used in HEC-RAS, as shown in Fig. 9.

GIS data was exported after steady flow analysis being done in HEC-RAS and imported into ArcGIS for inundation analysis using RAS Mapping.

The delineated floodplain area at different return period peak discharges is shown in the Fig. 10.

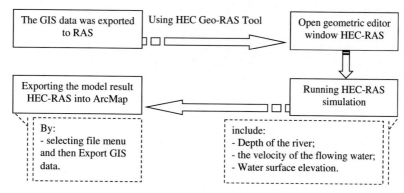

FIG. 6 Processing steps.

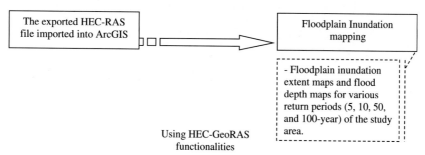

FIG. 7 Postprocessing steps.

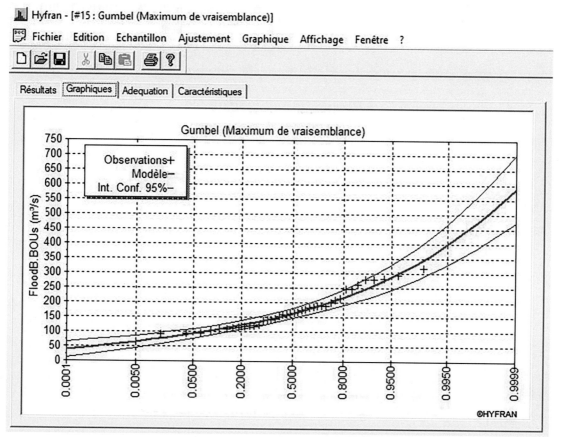

FIG. 8 Flood frequency analysis using equation Gumbel.

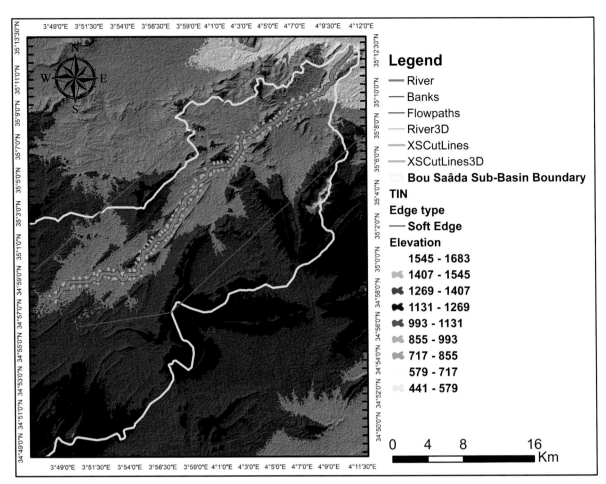

FIG. 9 Stream geometry created using TIN.

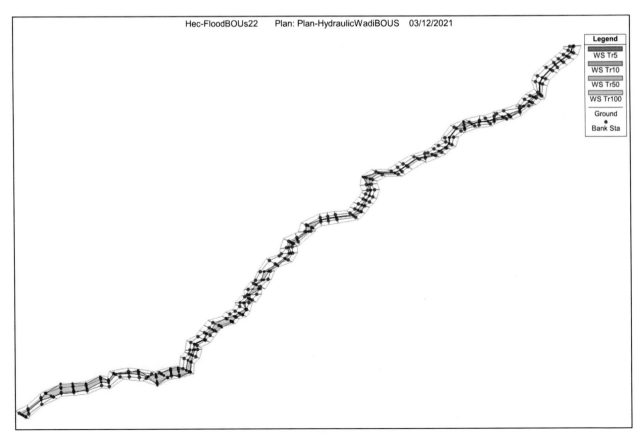

FIG. 10 3D multiple cross-section plot.

4.3 Floodplain mapping for return periods

The areas along the Bou Saâda Wadi—Subbasin simulated to be inundated for 5, 10, 50, and 100 years return periods.

Floodplain inundation analysis results in Table 1, shown that 5, 10, and 50 years return flood inundated 230.76, 239.31, and 254.52 km^2, respectively, of Built Up, Vegetation (forest, grass). And it showed that the flooded area in 100 years return period flood inundated 265.77 km^2, it can be extreme damage large area because high volume of flood.

Vegetated areas have low potential to flooding, because of the vegetation is to reduce the size of the flood peak. (Forest, grass, absorb the impact of both falling water).

Figures depict the generated river flood hazard maps based on water depth, and the intersection of land use with flood area boundaries for each flood event simulation. As shown in Figs. 11–14.

The flood inundation maps indicates a high risk to the land use and Wadi with considerable water depth.

Wadi flood hazard mapping and management water depth distribution is more essential in order to determine flooded areas. Floodplain inundation map for various return periods (5, 10, 50, and 100-year) of the study area can be mapped and overlay on Land use at Bou Saâda Wadi—Subbasin to analyze of damages caused to Built-Up area, vegetation.

The approximation of a flood-prone area on a map is shown in Figs. 11–14:

As we have seen in the results, gradually varied steady flow is characterized by minor changes in water depth from one cross-section to another.

The flooded areas along the Bou Saâda Wadi—Subbasin are 230.76 km^2, 239.31 km^2, 254.52 km^2, and 265.77 km^2 for 5, 10, 50, and 100 year return periods, respectively. The roads affected by floods have considerably affected the trade exchanges.

On the other hand, the inundation depth of the 100 years return period ranged from 0.000061035 to 5.40 m, the Built-Up areas, and the Vegetation (forest, grass) were exposed to overflows, and Whereas the impacts on Built Up areas are highly. So that the flood inundation area within 100 years return period covers 66.54% of the Built-up area, 33.45% of Vegetation (forest, grass). Similarly, the land use affected by the 5, 10, 25, and 50 years of flood frequency has different inundation depths.

People who are living near the river banks affected by catastrophic flooding are directly exposed to inundation depths of over 5.40 m. And need further considerations to flood protection.

Flood is usually a result of natural causes, it may also be caused by man-made factors. The reason lies in urbanization, density can increase risk, with the development of considered unplanned construction on flood plains, especially in high-risk areas, without proper infrastructural.

In Bou Saâda Wadi—Subbasin, flood causes can be attributed to:

1. Characteristics of the Bou Saâda Wadi, that are characterized by a gradual slope, where the floodplain is relatively lowland the hazard of the river flood is higher than the other locations.
2. Increase in Bou Saâda Subbasin land-use development causes an increase of imperviousness which leads to an increase in runoff volume.
3. Improper drainage system throughout the Bou Saâda Subbasin shares.

Based on the findings of this study, a variety of mitigation measures can be identified which will minimize the impact of flooding in Bou Saâda Wadi—Subbasin:

- Should give solutions:
- **Structural measures**.
- **Nonstructural measures:** include land use regulations, e.g., relocate the population residing along the wadi banks, and Preventing any new residential structures in the areas with high risk of flood.
- Early warning networks should be constructed in the city for the people to take the necessary precaution.
- Filed data related to channel roughness coefficient need to be collected which can help to get better results of the hydraulic model.

TABLE 1 Classification of flooded area according to land use for 5, 10, 50, and 100 years return period.

Return period	5 Years			10 Years			50 Years			100 Years		
Land use	Area (km²)	Minimum water level (m)	Maximum water level(m)	Area (km²)	Minimum water level (m)	Maximum water level(m)	Area (km²)	Minimum water level (m)	Maximum water level(m)	Area (km²)	Minimum water level (m)	Maximum water level(m)
Built Up	156.06	0.000061035	5.222117706	161.19	0.000061035	5.308575321	171	0.000061035	5.370628208	176.85	0.000061035	5.403657084
Vegetation (forest, grass)	74.7			78.12			83.52			88.92		
Total of land inundation	230.76 (km²)			239.31 (km²)			254.52 (km²)			265.77 (km²)		

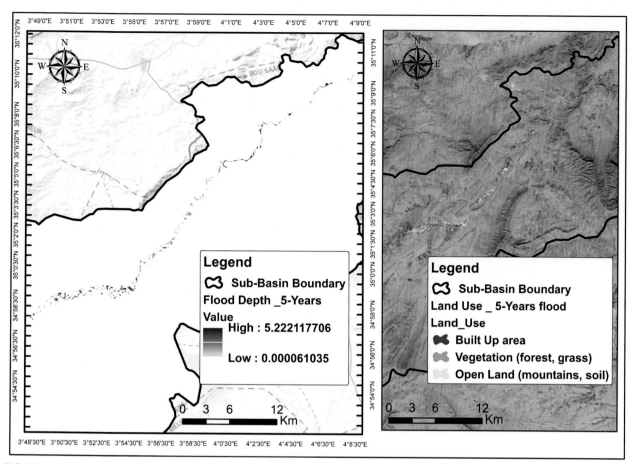

FIG. 11 Floodplain coverage extent on the land use for 5-year return period.

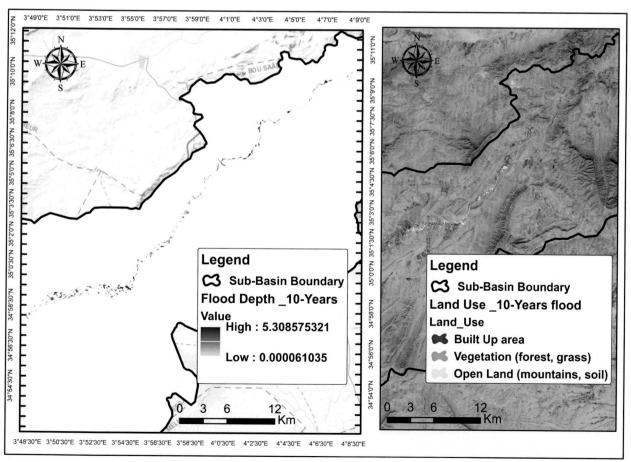

FIG. 12 Floodplain coverage extent on the land use for 10-year return period.

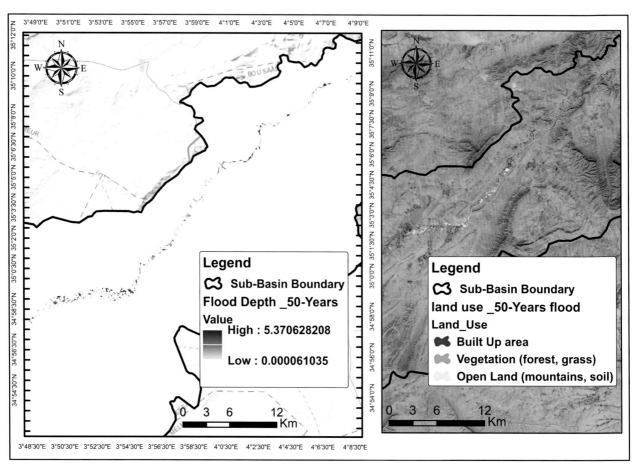

FIG. 13 Floodplain coverage extent on the land use for 50-year return period.

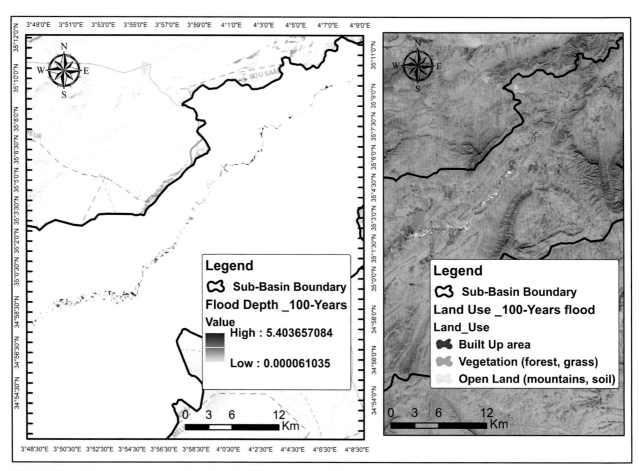

FIG. 14 Floodplain coverage extent on the land use for 100-year return period.

5. Conclusions

Floods often cause damage and suffering. Severe flooding is caused by atmospheric conditions that lead to heavy rain. Densely populated areas and the topography characteristics can also make an area more likely to flood.

This study presents a Hydrological-Hydraulic Modeling of floodplain inundation by using the one-dimensional numerical model HEC-RAS and GIS and HEC-GeoRAS extension for interfacing between HEC-RAS and GIS. Floodplain inundation along the Bou Saâda Wadi—Subbasin have been delineated based on flows for different return periods for 5, 10, 50, and 100 years.

The Bou Saâda Wadi—Subbasin is directly affected by a flood. Floodplain inundation maps for 5, 10, 50, and 100-year return period shows that $990.36\,km^2$ was the most affected. The improper use of lands contributes to an increased intensity of floods. Means, land-use intensity increases the runoff and impervious surfaces that prevent infiltration of water into soils, nearly twice as land use may be susceptible to flood damage.

From the result of floodplain inundation mapping in Bou Saâda Wadi—Subbasin, it is possible to conclude that there are severe flooding areas, the result obtained indicates that Built Up area is exposed to flooding; so that an increase in river basin land-use leads to an increase of imperviousness of the river basin. Hydrological-Hydraulic Modeling of floodplain inundation results indicate that a large percentage (33.45%–66.54%) of the inundated areas is Built Up areas and followed by Vegetation (forest, grass).

Finally, Hydrological-Hydraulic Modeling output with from this study of Bou Saâda Wadi—Subbasin can be used to implement plans to minimize the effect flood, and land use planning, and provide early warning system information for the preparedness of flood event, as well The velocity results are very important in a flood study and should be taking into consideration.

This study presents a method for improving the prediction of flood events. Furthermore, the study has limitation in accuracy valuation due to lack of field observation data. It is necessary to use the observed data to calibrate the hydraulic model. The study concludes with several recommendations for improved flood mitigation, from infrastructure to strengthening the early warning system, to prepare for disaster management.

References

Alaghmand, S., Abdullah, R.B., Abustan, I., Vosoogh, B., 2010. GIS-based river flood hazard mapping in urban area (a case study in Kayu Ara River Basin, Malaysia). Int. J. Eng. Technol. 2 (6), 488–500.

Aynalem, S.B. (2020). Flood plain mapping and hazard assessment of Muga river by using ArcGIS and HEC-RAS model Upper Blue Nile Ethiopia. LARP, 5(4), 74.Aynalem, S.B., 2020. Flood plain mapping and hazard assessment of Muga river by using ArcGIS and HEC-RAS model Upper Blue Nile Ethiopia. Landscape Architecture and Regional Planning 5 (4), 74–85.

Desalegn, H., Mulu, A., 2021. Mapping flood inundation areas using GIS and HEC-RAS model at Fetam River, Upper Abbay Basin, Ethiopia. Sci. Afr. 12, e00834.

Dewan, A., 2021. Germany's deadly floods were up to 9 times more likely because of climate change, study estimates. In: CNN Retrieved November 5, 2021, from https://edition.cnn.com/2021/08/23/europe/germany-floods-belgium-climate-change-intl/index.html.

Eke, S.N., Hart, L., 2020. Flood frequency analysis using Gumbel distribution equation in part Of Port Harcourt Metropolis. Int. J. Appl. Sci. Res. 3 (3). ISSN: 2581-7876.

Ghazavi, R., Vali, A., Eslamian, S., 2010. Impact of flood spreading on infiltration rate and soil properties in an arid environment. Water Resour. Manag. 24 (11), 2781–2793.

Golshan, M., Jahanshahi, A., Afzali, A., 2016. Flood hazard zoning using HEC-RAS in GIS environment and impact of manning roughness coefficient changes on flood zones in semi-arid climate. Desert 21 (1), 24–34.

Hasbaia, M., Seddi, A., Bournane, A., Hedjazi, A., Paquier, A., 2012. Study of The Water and Sediment Yields of Hodna Basin In The Centre of Algeria, Examination of Their Impacts. ICSE6, Aug 2012, Paris, France. pp. 103–110.

Helwa, A.M., Elgamal, M.H., Ghanem, A.H., 2020. Dam break analysis of Old Aswan Dam on Nile River using HEC-RAS. Int. J. Hydrol. Sci. Technol. 10 (6), 557–585.

Huong, H.T.L., Pathirana, A., 2013. Urbanization and climate change impacts on future urban flooding in Can Tho city, Vietnam. Hydrol. Earth Syst. Sci. 17 (1), 379–394.

Masood, M., Takeuchi, K., 2012. Assessment of flood hazard, vulnerability and risk of mid-eastern Dhaka using DEM and 1D hydrodynamic model. Nat. Hazards 61 (2), 757–770.

Mohit, M.A., Sellu, G.M., 2013. Mitigation of climate change effects through non-structural flood disaster management in Pekan Town, Malaysia. Procedia Soc. Behav. Sci. 85, 564–573.

Pallavi, H., Ravikumar, A.S., 2021. Analysis of steady flow using HEC-RAS and GIS techniques. In: International Journal of Engineering Research & Technology, NACE 2020 Conference, Bengaluru, pp. 36–41.

Romali, N.S., Yusop, Z., Ismail, A.Z., 2018. Application of HEC-RAS and ArcGIS for flood plain mapping in Segamattown, Malaysia. Int. J. GEOMATE 14 (43), 125–131.

Saad, A., Milewski, A., Benaabidate, L., El Morjani, Z., Bouchaou, L., 2019. Flood frequency analysis and urban flood modelling of Sidi Ifni Basin, Southern Morocco. Am. J. Geogr. Inf. Syst. 8 (5), 206–212.

Shafroth, P.B., Wilcox, A.C., Lytle, D.A., Hickey, J.T., Andersen, D.C., Beauchamp, V.B., Warner, A., 2010. Ecosystem effects of environmental flows: modelling and experimental floods in a dryland river. Freshw. Biol. 55 (1), 68–85.

Smith, P., House, J.I., Bustamante, M., Sobocká, J., Harper, R., Pan, G., Pugh, T.A., 2016. Global change pressures on soils from land use and management. Glob. Chang. Biol. 22 (3), 1008–1028.

Tate, E.C., Maidment, D.R., Olivera, F., Anderson, D.J., 2002. Creating a terrain model for floodplain mapping. J. Hydrol. Eng. 7 (2), 100–108.

Thoummalangsy, S., Tuankrua, V., Pukngam, S., 2019. Flood Risk Mapping Using HEC-RAS and GIS Technique: Case of the Xe Bangfai Floodplain, Khammoune Province. Lao PDR. Thai Environ. Eng. J. 33 (3), 27–38.

Chapter 15

Interoceanic waterway network system

Vladan Kuzmanović

Serbian Hydrological Association, International Association of Hydrological Sciences, Belgrade, Serbia

1. Introduction

Waterway, natural or artificial navigable inland body of water, or system of interconnected bodies of water, used for transportation, may include a lake, river, canal, or any combination of these. The existence of waterways has been an important factor in the development of regions.

In Europe, where the canal era had also started toward the end of the 17th century and continued well into the 18th, France took the lead, integrating its national waterway system further by forging the missing links. In the north the Saint-Quentin Canal, with a 3 1/2-mile tunnel, opened in 1810, linking the North Sea and the Schelde and Lys systems with the English Channel via the Somme and with Paris and Le Havre via the Oise and Seine.

In the United States, canal building began slowly; only 100 miles of canals had been built at the beginning of the 19th century; but before the end of the century more than 4000 miles were open to navigation. With wagon haulage difficult, slow, and costly for bulk commodities, water transport was the key to the opening up of the interior, but the way was barred by the Allegheny Mountains. To overcome this obstacle, it was necessary to go north by sea via the St. Lawrence River and the Great Lakes or south to the Gulf of Mexico and the Mississippi. The Erie Canal, 363 miles long with 82 locks from Albany on the Hudson to Buffalo on Lake Erie, was built by the state of New York from 1817 to 1825. Highly successful from the start, it opened up the Midwestern prairies, the produce of which could flow eastward to New York, with manufactured goods making the return journey westward, giving New York predominance over other Atlantic seaboard ports. The Champlain Canal was opened in 1823; but not until 1843, with the completion of the Chambly Canal, was access to the St. Lawrence made possible via the Richelieu River. Meanwhile, Canada had constructed the Welland Canal linking Lakes Ontario and Erie. Opened in 1829, it overcame the 327-ft difference in elevation with 40 locks, making navigation possible to Lake Michigan and Chicago. Later the St. Mary's Falls Canal connected Lake Huron and Lake Superior. To provide a southern route around the Allegheny Mountains, the Susquehanna and Ohio rivers were linked in 1834 by a 394-mile canal between Philadelphia and Pittsburgh. A unique feature of this route was the combination of water and rail transport with a 37-mile portage by rail by five inclined planes rising 1399 ft to the summit station 2334 ft above sea level and then falling 1150 ft to Johnstown on the far side of the mountains, where a 105-mile canal with 68 locks ran to Pittsburgh. By 1856 a series of canals linked this canal system to the Erie Canal.

Waterways fall into three categories, each with its particular features: natural rivers, canalized rivers, and artificial canals (Marsh and Daviess, 2018). Canalized rivers navigation is facilitated by constructing locks that create a series of steps, the length of which depends on the natural gradient of the valley and on the rise at each lock. Associated with the locks for passing vessels, weirs and sluices are required for passing surplus water; and in modern canalizations, such as the Rhône and the Rhine, hydroelectric generation has introduced deep locks with longer artificial approach channels, which require bank protection against erosion and, in some strata, bed protection against seepage losses. Artificial canals navigation can depart from natural river valleys and pass through hills and watersheds, crossing over valleys and streams along an artificial channel, the banks and sometimes the bed of which need protection against erosion and seepage. The route of an artificial canal can be selected to provide faster travel on long level pounds (stretches between locks), with necessary locks grouped either as a staircase with one chamber leading directly to another or as a flight with short intervening pounds. Where substantial differences of level arise or can be introduced, vertical lifts or inclined planes can be constructed (Marsh et al., 2018).

In the paper we will analyze, firstly, the course of the Paran river, its middle and lower flow, Pargue Nacional Iguacu, from Proto Figurueira to Guhir and Ilha Grande. The analysis of large continental rivers is made by the observation of water systems, comparisons and useful conclusions of the European, African and Asian network systems, as well as the possibilities of their further development.

Handbook of HydroInformatics. https://doi.org/10.1016/B978-0-12-821962-1.00004-0

After the analysis we will perform certain comparisons, inter-system comparisons and find certain similarities and regularities in the hydrological sense. In paleogeographic studies, we perceive certain qualities, or paleo-hydrological phenomena, unique (as geological and hydrogeological occurrences) or as specificity in classifications of hydrological dynamics, elements of paleo-propagation, that are in certain conditions, explained and defined. These paleo phenomena will be adequately registered, by mentioning the main examples, and including them in the class of significant geographical distinctions, as a sort of geographic gems in the actual construction and paleo-reconstructions. These paleo-phenomena hydrological products are valuable not only in geological, but also in overall terms, as hydro-morphological and geographical phenomena. The geo-hydrological locations are listed, as part of analyzed river flows, new terms are explained by the existing ones, casting emphasis on good examples.

The result of the observation method is demonstrated inter alia, in the notion that the interactive map is a paleo map, actual flows are here, also, actual paleo-products. Geographical observation is the main value of scientific insight. Conclusions and developments are in some cases more than obvious, on the others they represent valid data for further scientific verification. Paleo River forms a network of paleo-channels, in different periods of paleo-dynamics.

Paleo-streams are wider than the reaches of the current rivers and their recent meanders, and have not been created by the hydro-potential of the one that is presented but by the hydro-potential of the paleo river and from its paleo-cycles, as well as by other chronological factors.

2. Notable channel systems

2.1 Parana and Danube

The Parana river area consists of Pargue Nacional Iguacu, Pargue Nacional de Ilha Grande, from Proto Figurueira to Guhir, the fluvial fan of Ituzaingo (Pargue Izera), Mirinay river, the former channel Parana river and paleo influx of Parana river to Uruguay near Uruguaiana, and further Quarai river. Aguapey river has paleo channel and paleo influx at Alvear (Latrubesse, 2003; Stevaux, 2000). The Uruguay river is the paleo flow of Parana. Paleo-confluxes are the bifluxes or paleo-islands of one paleo-river. River islands could be divided in actual and paleo islands. Paleo-islands comprise the flow of two rivers, one channel that is biflux and the new channel of the older river, the older channel of the older river (biflux) through which the new river flows. In this example, it is the Latino-Iberian paleo-island (Parque Ibera): from Itaquie to Nueva Palmira (Colonia Estrella), and from Posadasa and Puerto Valle (Resistencia) to Buenos Aires (Fig. 1).

Pannonian Paleo Island is the largest continental river island in Europe as the anastomosing and distributary system with 90.000 km^2 on three paleo blocks. Iberian Paleo Insula is the largest paleo island of North America covering an area greater than 240.000 km2. Starting and ending points of paleo island are capital cities: Asuncion (distributary) and Buenos Aires (estuary), Budapest (distributary) and Belgrade (estuary). These are also the biggest river island in the world, with Marajo as an active river system (40.100 km2) and Ilha do Bananal (19.162 km2).

The Little Danube is the former paleo-Danube stream. The Danube is here making 20 km of transgression in relation to the current course. At Kolarovo paleo-conflux in Vaha river, the current Vaha from Kolarovo to Komarno is a paleo-streambed of the Danube. From Loebendorf and Kerneuburg, through Seyring, Strasshof and Schonfeld, to Marchegg in Horny Les. Horny and Dolny Les is the result of the Danube Transgression and the Moravian Meanderage. The paleo channel of the Marchegg Stadt to Devin was the former Danube flow. Zagyva is the paleo channel of the Danube, to Szolnok, as well as the present Tisza from Szolnok to Titel (supra-Pannonian). The Danube paleo-delta at Constanța is the fluvial residue of the paleo-Danube flow Cerna Voda-Constanta, with Brajla and Belem Island.

Pannonian paleo-island is a paleo-block marked by paleodynamic points Szentendrei sziget-Szolnok-Titel-Vukovar. The paleo-island is a block marked by paleodynamic points, paleo confluxes or bipaleoconfluxes, and paleo fluvial levers. Paleo river is characterized by a paleo conflux, paleo basin, paleo channel, and a paleo islet. Large paleo channels from Szeckesfehervar are also noticeable, through Nagydoreg to Szekesard (present Halasto) and Baja's wetlands. For the Pannonian paleo-island of the Quarternary Danube, two paleo confluxes are characteristic: Titel and Pančevo, and three hydrodynamic phases: Supra Pannonian, Supra Almamontic and Sub Almamontic. Upper Pannonian is described by meanders, river islands, and paleo confluxes, as specific Pannonian paleo dynamics. Pannonian lakes and river lakes produce interesting hydrological products. The basin of giant Lake Pannon in Central Europe was filled by forward accretion of sediment packages during the Late Miocene and Early Pliocene. Most sediments were supplied by the paleo-Danube. The northeastern part of Lake Pannon was filled by the paleo-Tisza system, supplying sediments from the Northeastern and Eastern Carpathians (Hickin, 1974; Holley and Jirka, 1986).

Pancevacki rit is the result of the propagation of the Danube-Tisza basin from the Begej's Perlez to the Centa, and the Tamis-Danube hydrology. The Tamisian swamp is a layered paleo-genesis: a unique European paleo transition: the

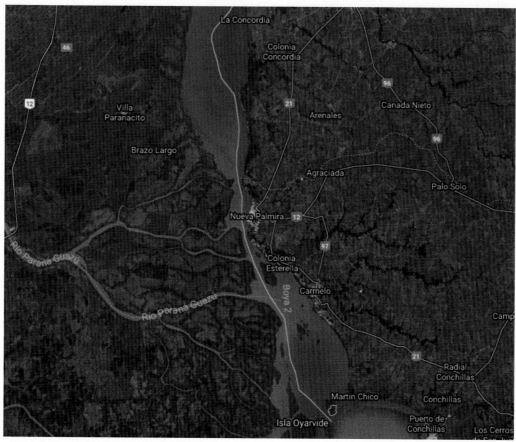

FIG. 1 Latin Iberian paleo-conflux, Colonia Estrella, Parana.

Paleodunabian, Paleotiszian (with the paleo-Danubian subphase), and the Tamisian. The Tamisian is in fact a new stream of the paleo-Danubian flow trajectory. Danubian loess deposits were initiated in response to the tectonic formation of the Pannonian basin, retreat of the large paleolakes, and increased sediment supply from the Danube (Fitzsimmons et al., 2012).

Both, double and opposite paleo flows are also recorded at different hydro chronological intervals, for example, Centa's contraflux, the flow of the river Begej into the Danube, and the Danube River to Begej (Baranda-Surduk).

3. Paleohydrography and channel systems

The Tennessee River is formed at the confluence of the Holston and French Broad rivers on the east side of present-day Knoxville, Tennessee. From Knoxville, it flows southwest through East Tennessee toward Chattanooga before crossing into Alabama. It loops through northern Alabama and eventually forms a small part of the state's border with Mississippi, before returning to Tennessee. The Divide Cut is a 29 mi (47 km) canal that makes the connection to the Tennessee River. It connects Pickwick Lake on the Tennessee to Bay Springs Lake, at Mississippi Highway 30. The cut carries the waterway between the Tennessee River watershed, which eventually empties into the Ohio River, and the Tombigbee River watershed, which eventually empties into the Gulf of Mexico (Gregory et al., 2008).

When the Tennessee Valley Act (TVA) was passed by Congress in 1933, the once free-flowing river became "less a river than a chain of lakes" by installing 25 dams along its length.

The Tennessee–Tombigbee Waterway is a 234-mile (377 km) man-made waterway that extends from the Tennessee River to the junction of the Black Warrior-Tombigbee River system near Demopolis, Alabama, United States (Green, 1985). The maximum cut through the natural divide between the Tennessee River and Tombigbee River basins is 175 ft deep and 1500 ft wide and occurs near the town of Paden, Mississippi, and 23 miles south of the waterway's northern terminus. The average cut through the divide is 50 ft. Bay Springs Lock and Dam is the only navigation feature in the Divide Cut and consists of a lock with a lift of 84 ft which is the third highest lift east of the Mississippi River and a 2750-ft-long, and 120-foot high earth and rock-filled dam. The cut through the divide required 150 million yards of excavation, or roughly one-half that required for the entire waterway. The Old River Control Structure is a floodgate system in a branch of the Mississippi River in central Louisiana. It regulates the flow of water leaving the Mississippi into the Atchafalaya River, thereby preventing the Mississippi river from changing course and was built between the Mississippi's current channel and the Atchafalaya Basin, a former channel of the Mississippi. Turnbull's Bend intercepted the Red River, which became a tributary of the Mississippi River and the Atchafalaya River was formed as a distributary of the Mississippi River. Distinguishing this hydrodynamics canal through the neck of Turnbull's Bend was made, thus shortening river disflow. Over time, the north section of Turnbull's Bend filled in with sediment. The lower half remained open and became known as Old River (Keown et al., 1986; Thorne et al., 2015).

4. Paleodynamics of large rivers, remote sensing

Paleodynamics of the middle and lower Danube provide an insight into the complex prochronology of water masses and flows with a change of influx and flow directions within the unique *Prapontic* basin. In the chronology of the Pannonian movement of the Danube, three phases are recognizable: (a) the original stream of the Danube river from Vac, with the pseudo-paleo-influx at Szolnok, the Danube flow through the Tisza channel, through Pannonia to the mouth of the Deliblatska Pescara, and Pancsevo, with suprapannonian transition I-Ia, (b) the supraalmamontic, with the paleo-Danube conflux into the Tisza at Becej, the supra-subalmamontic fluvium of the Danube and Sava (Cybalae- Ulcae, Cybalae-Syrmium) and Ulcae paleoconflux (Savus, Dravus) and Cybalae-Syrmium paleoflux of the Sava, the subalmamontic fluvium with the mediation of Slavonian Lake, (c) the supraalmamontic fluvium, with Titel and Pancsevo paleoconfluxes. Linear Vac and rectal Budim fluvium, with two subphases, sub and supraalmamonian, Danube paleoconflux, subalmamontic fluvium of Subpannonian-Carpathian paleoconflux, etc. (Pálfai, 1994; Thomas, 2000).

Some favorable examples of the incisive use of paleo hydrodynamic systems are Danube-Tisza-Danube Canal, Danube-Tiszian Supraalmamontic-Subpannonian phase, Dunav-Sava Canal Laco-Subalmamontic (laco-Savus, laco-Danubian), Euphrates-Tigris (Table 1).

Paleohydrographic observation shows us two distinct turns of the Danube, at Vukovar and Vac. Both rectangles are of neopleistocenic nature. Vac rectal is paleo-substitution for the linear Danube, the Vukovar rectal for the original savian

TABLE 1 Partial continental channels systems.

System	Permeation	Permeative distance	River	Endpoints	Permative rate	Channel	Paleo-channel
Upper Pannonian	Budapest-Szolnok	59	Danube	Baltic Sea Mediterranean Sea	0.4		y Paleo-Danubian
Lower Pannonian	Bezdan-Becej	70	Danube	Mediterranean Sea	0.7	Ferenc-csatorna	y Paleo-Tibisque Paleo-Danubian
North Asian	Maltan-Togonogh	25	Lena	Pacific Ocean	0.9		y Paleo-Lena
Mesopotamian	Albu Shejel-Taji	35	Euphrates Tiger	Indian Ocean		Haka canal	y Paleo-Euphrates
Central African	Irumu-Zekere	37	Congo	Atlantic Ocean	0.5		y Paleo-Aruwimian
South American	Puero-Valle-Itaqui	90	Parana	Atlantic Ocean		Aquapey	y Paleo-Iberian
South European	Vavousa-Mikrolivado	25	Heliacmon (Pineios) Aoos	Adriatic Sea Aegean Sea	6.5		n
South European		35	Iskar Maritsa	Danube Aegean Sea	7.8		n

semi-confluvium. For the subsupramalmontontic transition, a characteristic Paleo-Drava catabase, with angular displacement of supraalmamontic thrust. Thus, the Danube paleochannel is actually Tisza and Sava in sequences, with flows of the Sava to Slavonian laco-fluvial paleoconflux, and Paleotissiae to Szolnok paleoconflux. The length of the Pontian fluvial subtributes is consequently segmentally reduced. Paleo-Sava is a pronounced subslavonian river, with the Danube paleoinflux on the Bosnian-Slavonian limes. From the Slavonian permeation on, Bosut's channel is paleo-Danube, with a median spread from linear to quasi-flat plain expansion, in Lower Srem. The subalmamontic lever is the most remarkable example of paleo-dynamical hydro potential. Raising the river Savus flow, from Bitva river to Sremska Mitrovica was accompanied by the descent (through previous Drina paleo flow confluxial envelope) of Savus river, from the axis in direction Hrtkovci-New Belgrade to the Orid envelope.

Danubian paleo-hydro-hypocycloid (Danube H-Cycloid) consists of dozen paleo points, four confluent points with three bipaleoconfluxes and 1 contraflux: i point, Pancsevo, bipaleoconflux Danube I-II, bipaleoconflux Danube III-IIIa (Tamisian), paleoconflux Savus-Danube, ii point, Belgrade, bipaleoconflux Danube II-III, paleoconflux Savus II-Dravus, (bi)paleoconflux Savus II-III, paleoconflux Savus III-Danube III, iii point, Centa, Paleoconflux Danube III-Tamis, iv point, conflux Danube III-Tamis. (Table 2) the Tisza continues through the paleo-Danubian channel with translative paleoinflux (Kiss et al., 2015). This factor of the sub-Sirmian paleohydrographic dynamics is the main characteristic for the transition from the second to the third phase, the formation of subsirmian Sava, the alternate cata-anabasis of the sub-Almamontic Danube, with the active lever system of the sub-Sermian catabasis.

The basic sub-Sirmian stronghold initiated by the formation of the pra-Savus and the gradual disappearance of the Danube delta, was followed by the third segment of the sub-Slavonian catabasis, which is the laco-fluvial cause of sub-Sirmian dynamics. The second subalmamontal phase has characteristic linear sub-Sirmian paleo-Danubial flow, which is effectuated in the Lower Pannonian and Carpathian. The possibilities of this effectuation are different, but they are basically pre-Carpathian sediments and confluxes.

Every big river has its 100 miles event, such as proto-Nile (formerly propagation through the current Red Sea) or the paleo-Danube (and its remarkable Pannonian phases). Nasser Lake paleo channels are paleo factor of Nile's meander (Toshka-Toshka lakes to Lake Nasser by forming a circle of laco-fluvial island, Adu at Nasser Lake, Esnaq, Sohag, with an alternative system of the pool toward Fanafra). River pivots, axes and channels are representative phenomena of hydrodynamics (Liu and Zheng, 2002).

TABLE 2 Paleographic features of rivers.

Paleohydrographic feature		Paleohydrographic feature	
Bipaleoconflux[a]	Titel (Danube) Tibiscum (Danube) Nueva Palmira (Parana)	Paleo-insula	Latin-Iberean Pannonian
Counter-thread	Qena (Nile) Erdut (Danube)	Fluvial axle Flabellial axle	Panonian (Danube) Subsirmian Fanom (Nile)
Bipaleoflux Biflux	Belgrade-Tibiscum (Danube) Itaqui-Nueva Palmira (Parana Uruguay) Karada-Abadan (Euphrates Tigris)	Contra-flux Yazoo/moesian flow	Bhagdad (Euhprates) Sabac (Sava) Surduk (Danube) Pannonian-Supralmamontic Vinkovci (Danube) Moss Creek (Missouri) Jezava (Morava)
Paleodelta	Constanța (Danube) Belem (Amazon)	Fluvial lever	Subsirmian (Hrtkovci)

[a]*Bipaleoconflux is a simultaneous or chronological confluent point of two paleochannels of one river. Contra-flux and biflux are complement dynamic phenomena. Contra-flux is a paleo-dynamic phenomenon where in one channel the water stream flows in two different time sequences in both directions. Moesian stream/flow is a parallel derivative remnant of the paleo river. Many Yazoo streams are actually paleo remnants of only one, original, parallel river. Counter-thread is a meander, accompanied by simultaneous cut-bank confluxes.There is a major difference between fluvial lever and meander, whereas lever is a major river occurrence followed by a series of smaller meanders, and a unique river axis, that initiates adequate downstream adjusting. A fluvial lever is a major river turn forming paleo loop and is usually found at confluent points of paleo-islands. Paleo-island is a block framed by an actual or historical flow of one paleo river. Paleo-insula is a river island formed partially or completely at any chronological point.*

To such are added polygeohydrographic phenomena as the results of a complex synergy of two or even more factors such as hydro entities, most frequently flows, faded currents, paleo lagoons or poly-, bi- to mono- paleo-confluxes.

5. Integrated waterways systems

The channel system, integrated waterways network systems, interoceanic waterways system network (WSN), or simply the network system (NS) is an interoceanic and transcontinental entity (Fig. 2). The WSN paradigmatically consists of (a) river–river channel, (b) lake–river channel, (c) sea–river channel, (d) sea-lake–river channel system (Brookes, 1968; Gilvear, 1999). A typical example of a river basin system is the transoceanic river channel system, the system of the diversions of river basins watersheds, whereas watersheds is in the form of river permeations. Permeations are understood as spans within the integrated networks of the transcontinental system. Classical river channel systems (CS) are at the beginning, permeation of intercontinental basins, i.e., connecting family of functions of related river systems, river systems of the same sea or oceanic basin (Danube-Tisza-Danube, Danube-Rhine-Main). Intercontinental networks are the lake–river-sea-ocean traffic trajectories (as is the case in the Great Lakes-Illinois River-Mississippi-Mexico Bay system, Great Loop) (Fisk, 1951; Rahman et al., 2015). The WSN is an engineering permeation of interoceanic water channels, a system of manageable channels and long-distance navigation lines. The hydrologic analysis has yet to decide which of the available geomorphological structures is efficient, in which way some paleohydrological outlets are more favorable than alternative ones, which permeations are functional and communicatively effective, how to design and use the hydrological potential of flows (Szostak, 1996; Zhang et al., 2012).

WNS is an integrated river marine system. In relation to the transoceanic, it covers much larger territory providing a holistic approach to overall river resources. An interoceanic system is a connection or conjunction between the transoceanic systems. The integrated system is cost-effective considering the percentage of navigable accessibility. IS costs are composed of various factors: the cost of reducing interoceanic distances, indirect and circumventive navigation, better approximations of territorial distances between transport points, availability of alternatives and choices, less isolation, interconnection of the interior with the ocean (territorial compactness), overcoming the natural and communication barriers, reduction of intermodal traffic costs, and harmonization of intramodal traffic (ship-land, ocean-river), reduction of differences in intramodal traffic and water system integration, international cooperation, spatial organization and territorial unification, intersystem distribution, long distances equilibrium.

FIG. 2 Aimchan-Aldoma, Aldan Maya-Aldoma permeation (Middle Asian b).

Disequilibrium of long distances is the consequence of the unavailability of favorable alternatives and the impossibility of making an adequate cost–benefit analysis (Szostak, 1996).

An interoceanic channel system is an integrated channel system of transoceanic and interoceanic distances composed of subsystems and partial network systems. The integral interoceanic system is composed of two interoceanic systems: the inland system and the marine interoceanic system. Without aquatic, the interoceanic channel system would in fact only be a system of continental networks. The integrated system is comprised of an integrated continental network and an interconnected oceanic system. Thus, the integral waterways system is the optimal interpretation of the integrated network theorem. The waterways system is not composed of double-ended points but rather as a global integrated poly-oriented network. According to this, the integrated network theorem is applied within the system of integrated and equally distributed nodes. Continental knots of the waterways network are parts of the one integrated system. The water channel system is global, covering the entire territory in terms of logistical prevalence.

Permeative basin transjection distance refers to the optimal channel connections of large rivers (river systems and optimal subbasins interoceanic transgressions). An additional problem is that a dividing line is defined as a line between the two peaks, without taking into account basins capacities, divide portages, and that transversions and permeation lines are affected in optimal physical communication. The divide distance is unfavorable due to the fact that the mountain is a two-sided drainage system, having one peak as the source of two divides. The divide distance refers to divides nonselectively, while the permeative distance refers to the most important strategic directions and communications, as effective communicative transversions. On the other hand, the permeation lines are not linear but are selective given the uniqueness and strategic importance of such geographical phenomena.

Permeation lines are draft maps, or mapping system templates for the elaboration of specific diversion maps, having diversion lines and distances as their precise technical tool. In practice, one divide line has one or two favorable strategic communications, here defined as permeation points. Permeation zones are the most optimal divide zones or communicatively oriented intercontinental exchanges.

Every continent has its own integrated network. North American, European, and European-Asian network, and further, South-American, African and Asian (South-Asian) (Fig. 3, Table 3). The degree of exploitation of the network system capacity is not satisfactory, yet nor negative. Presently, a low percentage of the network river system is used as a completely integrated waterway system (Szostak, 1996). Existing continental coverage can also be improved, partially and integrally; the North American network system should be enhanced from the Transoceanic to the Atlantic-Pacific system.

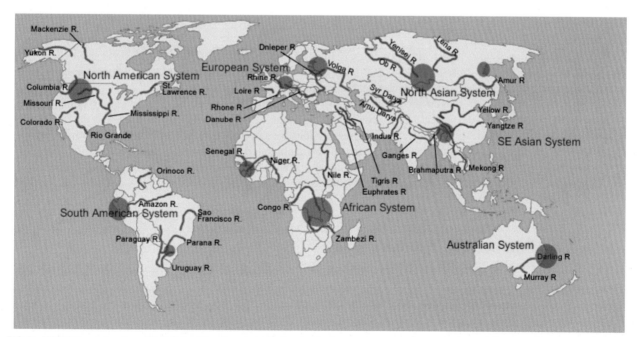

FIG. 3 IWSN Permeation Areas World Map. List of continental systems: North American System, South American System: South American a and South American b, European System: European and Euro-Asian, Asian System: North Asian b - Middle Lower Asian, Lower Asian, South Asian, Australian System, African System: West African and South African.

TABLE 3 Great interoceanic network systems.

System	Permeation	Permeative distances	Rivers, lakes	Endpoints	Territories of permeation	Permeative rates
North American	Tennessee-Tombigbee	234	Great Lakes, Mississippi, Tennessee-Tombigbee Waterway	Gulf of Mexico, Great Lakes	USA	9.1
Middle Lower Asian	Aldan Maya-Aldoma Batomga-Aimchan-Aldoma	20	Ob, Yenysey, Lena, Amur	Arctic Ocean North Pacific Ocean	Russia	0.7
Lower Asian	Junnan Xiangyun-Nanjian -Mawan (Three Parallel Rivers)	120	Yangtse (Jinsha), Mekong (Lancang)	North China Sea Golf of Thailand Andaman Sea	China	2.1
South American	Peruvian Iquitos (Huancajo) -Lima	125	Amazon	Atlantic Ocean Pacific Ocean	Peru	2.8
African	Irumu-Zekere Aruwimi (Congo)	37	Lake Albert Nile, Zambezi, Congo	Mediterranean Sea Atlantic Ocean	Congo	0.5

The challenge of the actual channel system is the transition to the integrated water system (integrated network system), trans and interoceanic, which would connect the Arctic, Atlantic and Pacific oceans (Fig. 3, Table 4). The synergy of these interactions is substantial. The important point of integrated systems is to increase coverage of the territory. For integrated water system engineering is therefore to further develop the integrated channel traffic system. The network systems are characteristic of large continental spaces.

IWSN is a promising field of water management. Continental systems are territorial networks and in this sense, we divide them into two categories: partial systems (as unintegrated river systems) and prospective integral systems (river systems integrated into the network of intercontinental channels) (Fig. 3).

River systems of natural and integrated coverage. It must be further assumed that the management system should be conceived in correspondence to the natural system (Hey, 1994; Sihn et al., 2015). Continents are covered by natural, not yet, manageable systems, integratively. Certainly, there are possibilities for a holistic approach in terms of integrative communicative systems (Gore and Petts, 1989). In future, we can talk about smart, sustainable, continental systems. Integrated traffic networks are such sustainable systems. The fine-tuned use and efficient choice between the offered alternatives is the framework modality in integral management. Management has a choice between options, systems, or elements of systems, such as the management of the holistic networks.

6. Integrated interoceanic channel systems

The African system consists of river and lake–river alternatives, in an integrated interoceanic channel system (Congo-Zambezi, Congo-Nile). Asian system is divided into North and South West Asian: the permeation of the lower flows Ob, Yenisei, Lena, Amur and their integration into the lake and marine system. Permeations are the traces of former Siberian river traffic routes. The North Asian network is interoceanic (connecting the Kara and Laptev sea to the Pacific). The option of the North Asian system is (a) river–river channel system (Ob-Yenisey, Lena-Yensey-Lena, Lena-Amur), (b) river–river-sea channel system (Lena-Okhotsk sea, Okhotsk-Amur-Lena-Lake Baikal). The ratio between the length of the channel's fissure and the length of the river is high. (The projected extension rate is significant.) The projected

TABLE 4 Interoceanic and transoceanic water system permeations.

System	Permeation	Permeative distance	Rivers, lakes	Endpoints	Territories of permeation	Permeative rates
North American	Teton Wilderness (Atlantic Creek-Pacific Creek)	0.5	Yellowstone (Mississippi) Snake (Columbia)	Atlantic Sea Pacific Sea	USA	0.001
West African	Siguiri-Bafing	45	Senegal Niger	Atlantic Ocean	Guinea	0.01
Euro-Iberian	Cuenca-Montalbo	35	Jucar Gudiana (Ciguela)	Mediterranean Sea Atlantic Sea	Spain	3.8
Euro-Iberian	Cuenca-Culebras (Guadamejud-Chillaron)	20	Jucar Tagus	Mediterranean Sea Atlantic Sea	Spain	2.1
South Asian	Apti-Khandaj (Nira-Dhavali)	4	Krishna	Bay of Bengal Arabian Sea	India	0.5
Euro Iberian	Reinosa-Bilbao (Trueba-Cadagua)	60	Ebro	Mediterranean Sea Atlantic Sea	Spain	9.4
Euro Apennine	Ceva-Savona	18.6	Po Tanaro	Adriatic Sea Ligurian Sea	Italy	4.5
Euro Apennine	Perugia-Ancona	25	Tiber Esino, Metauro	Adriatic Sea Tyrrhenian Sea	Italy	8.7

extensibility rate is in correlation with the permeative rate. This is best interpreted on the permeations of Nile-Congo, Lena-Amur (Fig. 2) (Table 5).

The Lower Asian system covers the territories of China, Laos, Thailand, Cambodia (Yangtse, Mekong), Thailand and Myanmar (Salween-Mekong), and is a river–river-sea system. The river waterways system has been successfully applied in the case of the Yangtse-Yellow River channel system (Table 3) (Join-Lambert and Devisme, 2018). Yangtse-Mekong permeation connects the Yangtse river system (the great Asian spaces of the Yangtse River), the East China Sea with the

TABLE 5 North Asian integrated networks with permeations.

System	Permeation	Length	Rivers, lakes	Endpoints	Territories of permeation	Permeative rates
North Asian	Ust-Kut-river Angara	80	Yenisey Lena	Arctic Ocean Laptev Sea		1.6
North Asian	Narim-Yeniseysk		Ob Yenisey	Arctic Ocean		
North Asian	Okhotsk-Adan	250	Lena	Laptev Sea		9.3
North Asian	Nerchinsk-Chita-Ulan-Ude	300	Amur	Pacific Ocean		6.2
North Asian	Maltan-Togonogh	25	Lena	Arctic Sea Pacific Ocean		0.9

TABLE 6 Some representative transoceanic canal systems.

Waterway	Start-end point	Length	Confluences	Country
Main-Danube	Bamberg-Kelheim	106	North Sea Black Sea	Germany
White Sea-Baltic Sea	Povenets-Belomorsk	141	White Sea Baltic Sea	Russia
Volga-Baltic Sea	Cherepovets-Lake Onega	229	Black Sea Baltic Sea	Russia
Volga-Don	Sarepta-Kalach na Donu	63	Caspian Sea Black Sea	Russia
Kiel	Brunsbuttel-Kiel	31	North Sea Baltic Sea	Germany
Oswego	Oswego-Erie	23.7	Great Lakes Atlantic Ocean	United States
Canal du Midi	Toulouse-Etang de Thau	150	Atlantic Ocean Mediterranean Sea	France

Thailand Bay and the South China Sea, and is in interaction with the Yangtse-Yellow River channel system (Table 6). The Yangtse-Salween permeation connects the Chinese Sea to the Andaman Sea, and the Pacific Ocean to the Indian Ocean. Lower Asia has a complex and diverse system of permeable river flows. These flows have great international geohydrological, geopolitical and communicative significance. (Table 7) The North Asian and South East Asian systems complete the image of the Asian continental network system (see Fig. 2) (Latrubesse, 2003).

The Amazon channel with its most important tributaries covers the entire longitude of the South American continent. 700 km of Amazon River is separated by 150 km long Lima's permeation. Amazon is navigable to Iquitos. The geohydrological insight recognizes the great potencies for the Peruvian channel (Iquitos-Lima), the Iquiots-Lima navigational shaft of Peruvian permeation. The South American network is an ocean-river–river-ocean interoceanic system of Atlantic-Pacific Iquitos-Lima permeation. The transoceanic shaft has permeable points, and the connection of two oceans at low distances of a low permeation rate is more than attractive. There is a geomorphological, touristic, logistic potential of the Peruvian, Pacific Amazonian channels. The system is also interesting as a downstream link between the Amazon and the interior of the continent and the somewhat closed West Coast of the Pacific. The partial systems should be added to the intraregional systems. We emphasize only some of the continental systems: the Mediterranean-Atlantic (Euro Iberian

TABLE 7 Transoceanic channels.

Canal	Length	Place	Status	Endpoints	Notes
Grand Canal	1.115	Chinese subcontinent	Construction began 15th c. BC First major section completed in 6th c. BC	South China North China Plain	
Rhine–Main–Danube Canal	2.177		Initially proposed by Charlemagne Post-War WWII reconstruction completed 1992	North Sea Black Sea	Channelizing the Rhine, the Main and the Danube, and connecting with a canal crossing the European Continental Divide, it traverses Europe. When combined with the Marne–Rhine Canal, it connects to the English Channel. With the addition of the proposed Danube–Oder Canal, the waterway system would also access the Baltic Sea.

TABLE 7 Transoceanic channels—cont'd

Canal	Length	Place	Status	Endpoints	Notes
Mississippi River System	2320	Central United States		Gulf of Mexico Great Lakes	Channelizing the Mississippi River and major tributaries/distributaries, it accesses central North America. With the St. Lawrence Seaway it reaches the Gulf of Saint Lawrence, and turns the Midwest and East Coast of the United States into a virtual island.
Unified Deep Water System of European Russia		Russia		Baltic Sea Caspian Sea White Sea	Traversing across European Russia, the shipping waterway accesses the Atlantic and Arctic Ocean basins, and combined with the Suez Canal, accesses the Indian Ocean as well.
Suez Canal	102	Suez Isthmus		Atlantic Ocean (Mediterranean Sea) Indian Ocean (Red Sea)	The level (lock-less) canal follows the Suez Rift Valley, from the Gulf of Suez to the Mediterranean. Completed 1869
Panama Canal	51	Panamanian Isthmus		Atlantic Ocean (Caribbean Sea) Pacific Ocean	This lock-encumbered canal takes advantage of the Chagres River, used to create Gatun Lake, over the western side of the continental divide

channel), The Adriatic-The Mediterranean Sea (Euro Apennine), Hindu (sub-Asian, the Bengal Bay-the Arabian Sea) as well as many other, communicative and internal geo-logistic systems (Table 3).

7. Conclusions

Networks are an opportunity for the more efficient disposal of water resources. The use of the net in the management of water systems, river and sea-river channels, emphasizes the efficiency of integrated water management. The conclusions of geomorphologic and paleohydrologic research can provide a more complete picture of hydrological potentials for advanced hydrological management. Analysis of river flows uses premise and paleodynamic capacities, such as paleo-flows, hydro-whorls, fluvial blocks, for environmental and ecological projects. Paleohydrology is the great presumption factor of a sustainable water economy. Long-term effectiveness is in correlation with environmental and hydrodynamic conditions.

The paper recognizes permeations as the basis for designing river channel systems; planning and implementating the river, lake–river and sea-river canals to better exploit water resources and connect river flows. Diverting permeations are hidden hydro-spatial potentials in terms of continental water coverage and continuous net of river communication opts. This are prerequisites for water management of the construction of systemic river traffic distances to completely regulated or at least controllable river systems. The water system includes two or more river basins. The use of natural water resources in integrated systems is a great advantage of interoceanic water system communications.

References

Brookes, A., 1968. Channelized rivers: perspectives for environmental management. Vol. 659 Wiley, Chichester.

Fisk, H.N., 1951. Mississippi River valley geology relation to river regime. Proc. ASCE 77 (7), 1–16.

Fitzsimmons, K.E., Marković, S.B., Hambach, U., 2012. Pleistocene environmental dynamics recorded in the loess of the middle and lower Danube basin. Quat. Sci. Rev. 41, 104–118. https://doi.org/10.1016/j.quascirev.2012.03.002.

Gilvear, D.J., 1999. Fluvial geomorphology and river engineering: future roles utilizing a fluvial hydrosystems framework. Geomorphology 31 (1–4), 229–245. https://doi.org/10.1016/S0169-555X(99)00086-0.

Gore, J.A., Petts, G.E., 1989. Alternatives in Regulated River Management. CRC Press.

Green, S.R., 1985. An overview of the Tennessee-Tombigbee waterway. Environ. Geol. Water Sci. 7, 9–13.

Gregory, K.J., Benito, G., Downs, P.W., 2008. Applying fluvial geomorphology to river channel management: background for progress towards a palaeo-hydrology protocol. Geomorphology 98 (1–2), 153–172. https://doi.org/10.1016/j.geomorph.2007.02.031.

Hey, R.D., 1994. Environmentally sensitive river engineering. In: The Rivers Handbook: Hydrological and Ecological Principles, pp. 337–362.

Hickin, E.J., 1974. The development of meanders in natural river-channels. Am. J. Sci. 174,274 (4), 414–442. https://doi.org/10.2475/ajs.274.4.414.

Holley, E.R., Jirka, G.H., 1986. Mixing in Rivers. US Army Engineer Waterways Experiment Station, USA.

Join-Lambert, P., Devisme, P., 2018. Voies Navigables France Itinéraires Fluviaux. Editions De L'Ecluse.

Keown, M.P., Dardeau, E.A., Causey, E.M., 1986. Historic trends in the sediment flow regime of the Mississippi River. Water Resour. Res. 22 (11), 1555–1564. https://doi.org/10.1029/WR022i011p01555.

Kiss, T., Hernesz, P., Sümeghy, B., Györgyövics, K., Sipos, S., 2015. The evolution of the Great Hungarian Plain fluvial system–Fluvial processes in a subsiding area from the beginning of the Weichselian. Quat. Int. 388, 142–155. https://doi.org/10.1016/j.quaint.2014.05.050.

Latrubesse, E.M., 2003. The late-quaternary palaeohydrology of large South American Fluvial systems. In: Palaeohydrology: Understanding Global Change, p. 193.

Liu, C., Zheng, H., 2002. South-to-north water transfer schemes for China. Int. J. Water Resour. D 18 (3), 453–471.

Marsh, C.M., Daviess, E.A.J., 2018. Canals and inland waterways. In: Modern Waterway Engineering. Encyclopedia Britannica.

Pálfai, I., 1994. Summary of the causes of the groundwater-sinking process in the Danube–Tisza Interfluve and the possible solutions for the water-shortage. A Nagyalföld Alapítvány Kötetei. 3 Nagyalföld Alapítvány, pp. 111–126.

Rahman, M.A., Jaumann, L., Lerche, N., Renatus, F., Buchs, A.K., Gade, R., Geldermann, J., Sauter, M., 2015. Selection of the best inland waterway structure: a multicriteria decision analysis approach. Water Resour. Manag. 29 (8), 2733–2749. https://doi.org/10.1007/s11269-015-0967-1.

Sihn, W., Pascher, H., Ott, K., Stein, S., Schumacher, A., Mascolo, G., 2015. A green and economic future of inland waterway shipping. Procedia CIRP 29, 317–322.

Stevaux, J.C., 2000. Climatic events during the late Pleistocene and Holocene in the upper Parana River: correlation with NE Argentina and South-Central Brazil. Quat. Int. 72 (1), 73–85.

Szostak, R., 1996. Economic impacts of road and waterway improvements. Transp. Quat. 50 (4).

Thomas, M.F., 2000. Late Quaternary environmental changes and the alluvial record in humid tropical environments. Quat. Int. 72 (1), 23–36. https://doi.org/10.1016/S1040-6182(00)00018-5.

Thorne, C., Hey, R., Newson, N., 2015. Applied Fluvial Geomorphology for River Engineering and Management. John Wiley & Sons Ltd.

Zhang, C., Wang, G., Peng, Y., Tang, G., Liang, G., 2012. A negotiation-based multi-objective, multi-party decision-making model for inter-basin water transfer scheme optimization. Water Resour. Manag. 26 (14). https://doi.org/10.1007/s11269-012-0127-9.

Further reading

Fisk, H.N., 1951. Mississippi River valley geology relation to river regime. In Proc. ASCE 77 (7), 1–16.

Fitzsimmons, K.E., Marković, S.B., Hambach, U., 2012. Pleistocene environmental dynamics recorded in the loess of the middle and lower Danube basin. Quat. Sci. Rev. 41, 104–118. https://doi.org/10.1016/j.quascirev.2012.03.002.

Keown, M.P., Dardeau, E.A., Causey, E.M., 1986. Historic trends in the sediment flow regime of the Mississippi River. Water Resour. Res. 22 (11), 1555–1564. https://doi.org/10.1029/WR022i011p01555.

Liu, C., Zheng, H., 2002. South-to-north water transfer schemes for China. Int. J. Water Resour. D. 18 (3), 453–471. https://doi.org/10.1080/079006202200000693.

Piesse, M., 2016. Water Governance in the Tigris-Euphrates Basin.

Rahman, M.A., Jaumann, L., Lerche, N., Renatus, F., Buchs, A.K., Gade, R., Geldermann, J., Sauter, M., 2015. Selection of the best inland waterway structure: a multicriteria decision analysis approach. Water Resour. Manag. 29 (8), 2733–2749. https://doi.org/10.1007/s11269-015-0967-1.

Zhang, C., Wang, G., Peng, Y., Tang, G., Liang, G., 2012. A negotiation-based multi-objective, multi-party decision-making model for inter-basin water transfer scheme optimization. Water Resour. Manag. 26 (14), 4029–4038. https://doi.org/10.1007/s11269-012-0127-9.

Chapter 16

Lattice Boltzmann models for hydraulic engineering problems

Ayurzana Badarch[a] and Hosoyamada Tokuzo[b]

[a]*School of Civil Engineering and Architecture, Mongolian University of Science and Technology, Ulan Bator, Mongolia,* [b]*Graduate School of Engineering, Nagaoka University of Technology, Nagaoka, Japan*

1. Introduction

Since their original introduction, enormous numbers of lattice Boltzmann (LB) models have been used in each field of fluid science. The type generally depends on the model formulation and the purpose of applications. Numerical instability and the inaptness for turbulent flows (Nathen et al., 2017) experienced in the widely used LB models with a single-relaxation time have been led to the introduction of more stable and appropriate formulations with two-relaxation time (Ginzburg et al., 2008), multirelaxation time (d'Humieres, 2002) and other models such as cascaded LB models (Asinari, 2008), the simplified LB method (Chen et al., 2018) and entropic LB models (Ansumali and Karlin, 2002). Different types of fluid problems from single-phase to multiphase flows involving heat transfer and fluid–solid interactions have led to different types of LB models such as the phase-field LB model (Wang et al., 2019), the color-gradient LB model (Wen et al., 2019), the free surface LB model (Körner et al., 2005), the immersed-boundary LB model (Noble and Torczynski, 1998) and many more, where one of the above formulations became the base. At each development of those, the LB models presented new possibilities to solve crucial complex fluid problems in computational fluid dynamics (Qian et al., 1995; Aidun and Clausen, 2010) based on conventional computational knowledge. The LB models become powerful alternative computational tools because of their simple nature of implementation and suitability for solving flows in complex configurations.

The LB models can be successfully applied in hydraulic engineering problems without acquiring knowledge on the background theory. Complex problems such as water hammer in a pipe system, particulate flows or sediment transportation, free-surface flow dynamics and turbulent flows are extensively solved and studied by the LB models. Depending on the relevant macroscopic governing equations, two types of LB models have been used in the hydraulic engineering field namely shallow water equation based LB model (Zhou, 2002) and traditional LB models with numerical ingredients which reduces the Navier–Stokes equation via the Chapman–Enskog expansion (Halliday et al., 2001). In this chapter some of the traditional LB models for hydraulic engineering problems will be discussed to show the applicability of the LB models in the field. The brief review and outlook of the recent research trends on the selected topics with the LB models are given at the end of each section.

2. Lattice Boltzmann models for closed conduit hydraulics

Unlike conventional numerical methods, the main variable of the LB models is the density distribution function which is the probability of finding a particle in given space and time (Chen and Doolen, 1998). The time advance of the distribution function can be governed by the dimensionless discrete-velocity Boltzmann equation discretized in space and time (Frisch et al., 1987) as follow:

$$f_i(\mathbf{x} + \mathbf{c}_i \delta x, t + \delta t) - f_i(\mathbf{x}, t) = \frac{\delta t}{\tau_v}\left(f_i^{eq} - f_i\right) + \delta t A_i, \tag{1}$$

where f_i is the density distribution function in the i-th direction of the lattice, $\mathbf{c}_i$ is the discrete set of velocities, τ_v is the dimensionless relaxation time, f_i^{eq} is the equilibrium density distribution function and A_i is the body force term. Eq. (1), referred to as the single-relaxation time LB model, adapts the Bhatnagar–Gross–Krook model (BGK) proposed

Handbook of HydroInformatics. https://doi.org/10.1016/B978-0-12-821962-1.00008-8

independently by Bhatnagar's group (Bhatnagar et al., 1954) and Welander (1954). The equilibrium distribution function implicating the limiting form of the any distribution function expressing fluid flow (Kerson, 1987) can be found in the discrete form of the Maxwell-Boltzmann distribution function:

$$f_i^{eq} = w_i \rho \left[1 + \frac{\mathbf{u}\mathbf{c}_i}{c_s^2} - \frac{\mathbf{u}^2}{2c_s^2} + \frac{\mathbf{u}^2 \mathbf{c}_i^2}{2c_s^4} \right]. \tag{2}$$

where w_i is the weighting function of the selected lattices for the model, $\rho(\mathbf{x}, t)$ is the macroscopic dimensionless fluid density, $\mathbf{u}(\mathbf{x}, t)$ is the macroscopic velocity vector and $c_s^2 (=RT)$ is the lattice speed of sound, in which R is the universal gas constant, T is the system temperature. The macroscopic variables are defined by the ensemble average of the distribution function along the lattice and can be expressed in a different form depending on the choice of the body force scheme. In the model the discrete force scheme proposed in Guo and Zhao (2002) can be applied, so that the macroscopic variables are given as

$$\rho = \sum_i^n f_i, \quad \rho\mathbf{u} = \sum_i^n \mathbf{c}_i f_i + \frac{\delta t}{2}\mathbf{F}, \tag{3}$$

where n is the number of the chosen lattice directions, and $\mathbf{F}$ is the dimensionless body force vector.

A numerical procedure to solve the discretized Boltzmann equation in Eq. (1) is split into two numerical steps called streaming and collision. Those steps are seen from a general formulation of the discretized Boltzmann equation and are suitable to handle boundary conditions for a complicated geometry (Chen and Doolen, 1998). Detailed discussions of the initial and boundary conditions which are typically realized in terms of the distribution function can be found in Maier et al., (1996) and Guo and Shu (2013). This intrinsic formulation of the LB models is straightforward for any applications to closed conduit flows. In this section, two selected problem representing the closed conduit hydraulics will be presented with their solutions with the LB model compared to analytical and theoretical ones to examine the validity of the LB model described above. The problems are very simple examples of the hydraulics in a closed conduit.

2.1 LB solutions to selected problems

The first problem is the Stokes second problem. Consider two parallel infinite plates that bound a fluid between them and the upper plate moves with a sinusoidal oscillation as shown in Fig. 1A. The oscillating plate generates laminar flow between the two infinite plates (Batchelor, 1967; Acheson, 1991) and the analytical solution to the problem is given as

$$u(y, t) = Ue^{-ky}\sin(ky - wt), \text{where } k = \sqrt{\frac{w}{2D}} \tag{4}$$

in which, $u(y,t)$ is the velocity distribution between two plates, $A(=1.0 \text{ ms}^{-1}$ in the numerical example of Fig. 1Aa) is amplitude of the moving plate, $w(=5 \text{ s})$ is the period, $D(=20 \text{ m}^2\text{s}^{-1})$ is the diffusivity or viscosity of the fluid and t is time.

The LB model on a 1D lattices (Qian et al., 1992) is sufficient for this problem. Eq. (1) for this problem reduces to

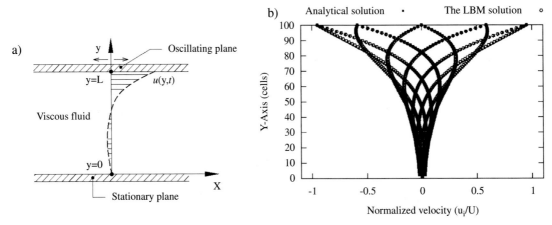

FIG. 1 (A) Schematic illustration of the Stokes second problem and (B) its solutions by the analytical equation and 1D LB models.

$$f_i^{n+1}\left(j + c_{yi}\delta t\right) = (1 - w_v)f_i^n(j) + f_i^{eq}(j)w_v, \tag{5}$$

where $f_i^{n+1}(j)$ is the distribution function in time advance at j cell, c_{yi} is the y-component of discrete velocity vector of the lattice, δt is the lattice time step, w_v is the relaxation parameter. The time step δt for the lattice is assumed to be unity to maintain the unit discrete velocity as well as unit lattice spacing. Time step and spacing for the lattice are unit forms of the computational time step, $\triangle t$, and grid spacing, $\triangle y$. The equilibrium distribution function in Eq. (2) for this problem becomes

$$f_i^{eq}(j) = w_i u_L(j), \tag{6}$$

where $u_L(j)$ is the velocity defined by Eq. (3). The relaxation parameter can be calculated from the diffusivity D as

$$w_v = \frac{1}{c_s^2 D \frac{\triangle t}{\triangle y^2} + 0.5}, x \tag{7}$$

where $c_s^2 = 2$ and $\triangle y = 0.2$ m. The time step $\triangle t$ for the LBM can be defined from a velocity scaling between the physical and the LB variables as

$$U_R = U_L \frac{\triangle y}{\triangle t}, \tag{8}$$

where U_L is the dimensionless boundary velocity at the upper plate given as

$$U(t) = A \sin\left(\frac{2\pi}{w}t\right)$$

As seen so far, all variables in the LBM must be dimensionless and after the simulation, they must be scaled appropriately just like Eq. (8).

Solutions of the spanwise velocity profiles at time intervals of $0.1\,T$ by the LB model are plotted in Fig. 1B against the analytical solution. Very good agreement proves that the LB model is assessed as a potential numerical method to solve simple fluid flows.

The second problem is to examine hydrodynamics entrance lengths at different Reynolds number in pipe flows. The pipe is represented by two infinite plates so that it can be assumed to be an infinitely-wide rectangular duct. The problem is solved by the LB model on a D2Q9 lattice arrangement (Chen and Doolen, 1998). The velocity boundary condition referred as the Zou/He boundary condition (Zou and He, 1997) is assigned to give a uniform unit velocity distribution at the pipe inlet. At the two plates, the no-slip boundary condition is realized by the bounce-back rule (Ziegler, 1993). The computational domain is formed by 40 cells in the vertical direction and 1000 cells in the streamwise direction to have sufficient length for the fully developed region.

Fig. 2 shows the numerically defined hydrodynamic entrance region and velocity boundary layers at different Reynolds numbers (Re). The flows with lower Re number have a smaller entrance region than the flows with a higher Re number. The separation line between core and wall layers at each case was defined by $\partial u(j)/\partial x = 0$. Theoretically the hydrodynamic entrance length to attain fully developed region at the center is specified by the characteristic length (D_e) and Reynolds number yielding

$$L_e = \Phi\left(D_e Re\right) \tag{9}$$

where Φ is the dimensionless entrance length. For the rectangular duct including two parallel plates, numbers of researchers have proposed different dimensionless entrance lengths in laminar flow (Han, 1960). The numerically defined

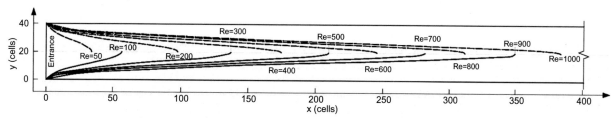

FIG. 2 Hydrodynamic entrance region at different Re numbers simulated by the 2D LB model.

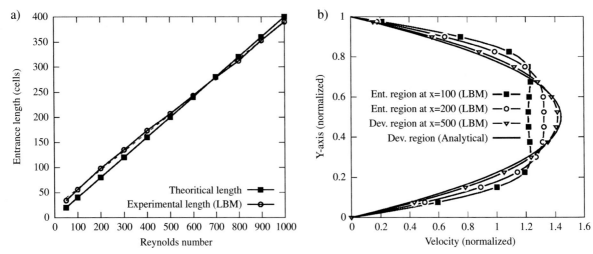

FIG. 3 Comparison of the hydrodynamic entrance lengths and the spanwise velocity profiles of $Re = 1000$ case in hydrodynamic entrance region and fully-developed region.

hydrodynamic entrance length as a function of Re numbers together with the theoretical entrance length of $L_e = 0.01(D_e Re)$ proposed by Schlichting in 1934 (Schlichting and Gersten, 2016) are given in Fig. 3A. Except for the nonlinearity and an intercept of linear fit observed in the numerical results, comparatively good agreement is found. The deviation observed becomes large with decreasing Re, which might be an effect of the entrance condition and grid dependency in the simulation. The small deviation in Fig. 3B was also observed when comparing the numerical velocity profile with the analytical velocity profile in the fully developed region. Nevertheless, the LB model provides good agreement of the hydrodynamic entrance and fully developed region and their velocity distributions.

2.2 Brief review on recent trends of pipe flows by the LB models

Pipe flow studies with different LB models are emerging including the topics of turbulence (Hou et al., 1996; Jahanshaloo et al., 2013), sediment transport in pipe flow (Dolanský et al., 2017), water hammer or transient flow (Budinski, 2016) and complex pipe network solutions (Meng et al., 2019).

Among the various types of the conventional turbulence models, the large eddy simulation (Kang and Hassan, 2013) and k-epsilon models (Teixeira, 1998; Bartlett et al., 2013) are successfully incorporated in the framework of the LB models. More attention is being paid to large eddy simulations and direct numerical simulations in LB models. There is a significant statement reported by Peng et al., (2018) on the direct numerical simulations with the LB model which proves that the stable LB models like the multirelaxation time model on the resolvable smallest length of scale can provide greater details of the turbulent statistics and valuable insight into practical applications. Not only conventional turbulence models provide opportunities for the LB turbulent modeling, but also enhanced LB models on the revised kinetic theory such as cascaded LB models (Asinari, 2008) and entropic LB models (Boghosian et al., 2001; Ansumali and Karlin, 2002; Badarch et al., 2017) are of great importance.

For sediment transport in fluid flows, correct evaluations of the mutual interactions of the fluid and particle, particle and wall effect are challenging the existing conventional models (Feng et al., 2007). Implementation of the solid boundaries in the LB models and coupling superiority of the LB model with other Lagrangian models such as the discrete element model (Galindo-Torres, 2013) enables the capture of the correct physics of both carrier fluid and sediment particles. Among the numerical schemes, the immersed boundary method incorporated with the LB model (Feng and Michaelides, 2004) is promising and is rather simple than the other models such as the momentum exchange and multiphase-based models (He ct al., 2019) in the LB framework. It has been repeatedly claimed by the many studies (a short review can be found in Aidun and Clausen, 2010) that the LB models for particulate flows can provide efficient solutions and better understandings for industrial problems.

The transient situation, so-called water hammer which causes the intensive pressure and velocity fluctuation in a pipe system, is of the great importance of practical applications. Major drawbacks on the conventional numerical methods for the water hammer are the numerical instability, efficiency and inaccuracy when the problem configuration is featured by the long and complicated pipe network (Ghidaoui et al., 2005). The LB models are considered as more efficient method to the water hammer than the traditional methods of it (Wu et al., 2008). To overcome the current issues and to improve the performance of the LB model to the water hammer problem, Budinski (2016) is modified the existing LB models (Cheng et al., 1998) with the adaptive grid technique which successfully eliminates the restriction of the Courant number and grid spacing and enables flexibility on the boundaries such as pumps and valves on the pipe networks. Louati et al., (2019) stated that the LB models are capable of modeling water hammer for low frequency cases. So that the more research on the dissipative mechanism of wave propagation are needed with the application of the LB models.

3. Lattice Boltzmann models for open channel hydraulics

Problems of fluid flows with a free surface which is the main characteristic of open channel hydraulics often occur in nature and engineering. In terms of fluids involved in the modeling, the free surface modeling can be divided into single- and multiphase modeling. For single-phase free surface modeling, the primary fluid, e.g., the water, is considered and the secondary fluid, e.g., the air, is neglected with the existence of the precise boundary condition on the interface between primary and secondary fluids. Examples of the single-phase free-surface numerical methods are the mesh-free methods, such as smoothed particle hydrodynamics (SPH) (Gingold and Monaghan, 1977; Monaghan, 1994) and the element-free Galerkin method (EFG) (Belytschko et al., 1994), and fixed grid methods, such as, to mention few, the shallow water approximation (Sielecki and Wurtele, 1970), and the free-surface LB model (Körner et al., 2005). In the multiphase flow approach, two or more fluids are considered in a model and interfaces of those flows are determined as a free surface under the influences of those fluids. Famous representatives of multiphase models are the Volume of Fluid (VoF) (Hirt and Nichols, 1981) in Eulerian approach and multiphase particle methods (Monaghan and Kocharyan, 1995) in Lagrangian approach. Alternatively to other numerical models for free surface flows, extensive researches reveal a variety of LB models. Among those the free-surface LB model developed by Körner et al., 2005 is widely used. Other applicable LB models for free surface flows are multiphase models (Shan and Hudong, 1993). Similar to the LB models by Körner et al., (2005) another free-surface LB model coupled with the VoF technique which was introduced by Janssen and Krafczyk (2010) using an additional advection equation solved with a classical finite volume method, while the free surface is reconstructed by a piecewise linear interface reconstruction. In this chapter we employ the free-surface LB model introduced by Körner et al., (2005) and later modified and improved by Thürey et al., (2006). Details of the model and its validations can be found in the original papers and some other applications for hydraulic engineering can be found in Badarch et al., (2016), Ayurzana and Hosoyamada (2017), Badarch, (2017) and Diao et al., (2018). The free-surface LB model applies implementation of the mass advection equation in a macroscopic way in which a flux-based advection scheme is used to track the movement of single-layered interface cells (depicted in Fig. 4) which is the representation of free surface (Badarch, 2017).

In this section the 2D free-surface LB models is used to derive solutions to the theoretically important transcritical flow over a broad-crested weir.

3.1 Transcritical flow over a weir

Finite slopes and flow curvature caused by vertical acceleration dominates the change of the subcritical flow from upstream to the super-critical flow downstream. The same geometrical configuration and three cases of the flow rates over a trapezoidal weir considered for the experimentation in Darvishi et al., (2017) is adapted for the LB computation. The computational domain formed with 120 cells in the y-axis, 400 cells in the x-axis. At the inlet the velocity boundary condition is given. The broad-crested weir is modeled with the slip boundary condition. Water surface profiles defined by the free-surface

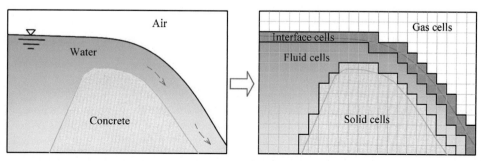

FIG. 4 Geometric discretization of the free-surface LB model.

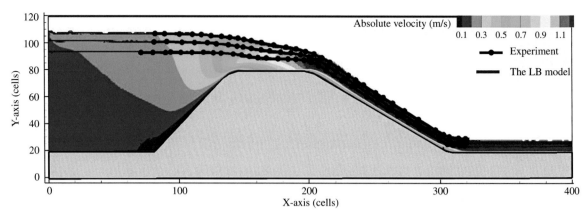

FIG. 5 Comparison of the free surface profiles from Darvishi et al., (2017) and from the simulations by the free-surface LB model. The absolute velocity field shown is of the case with highest discharge.

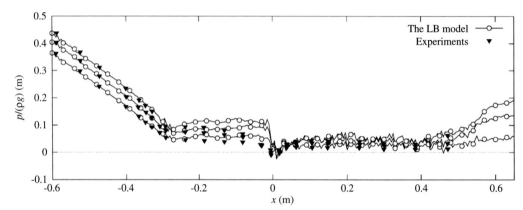

FIG. 6 Comparison of the pressure profiles at the surface of the broad-crested weir.

LB model agree well with the experimental data of all cases in Darvishi et al., (2017). It can be stated that the free-surface LB model describes the transition from the subcritical flow to the supercritical flow on top of the weir very well. The water surface curvature accounted by the free-surface LB model guarantee that the internal flow structure in transcritical flow is attained to a certain degree of accuracy.

The effects of finite slope on resistance and on pressure in the fluid as obtained by the free-surface LB model are seen in Fig. 6. The free-surface LB model predicts the nonhydrostatic pressure distribution along the transcritical region confirmed by the comparison with the experimental pressure data given in Darvishi et al., (2017). The lowest curve and associated points in Figs. 5 and 6 correspond to the smallest flow case and the highest to the largest flow case. Some slight over-estimation of the pressure at the lower edge of the crest is found by the free-surface LB model. The stair-case discretization of the boundary can be attributed as a cause to the pressure overestimation and fluctuation along the pressure profiles. This could be fixed when finer grid gradually representing the finite slope condition is used. Nevertheless, the sudden drop and spike of the pressure profile is attained by the free-surface LB model. The free-surface LB model is thus evaluated as a promising numerical method for other purpose of the study in the free surface flows over hydraulic structures.

3.2 Brief review of recent trends in open channel flows with the LB model

In the past many efforts on mathematical and numerical modeling of free surface flows had been made because of the practical need for accurate solution of the freely moving interface between two immiscible fluids. When problems such as sediment transport or forces acting on a structure or on a solid body, two-way coupling interactions need to be considered in the free surface flow modeling. To fulfill this necessity the LB models are extensively studied in two procedures, namely the free-surface LB models and shallow water LB models with and without the presence of the solid effect. The free-surface

LB model corresponds to the Navier–Stokes equations, while the shallow water LB models provide solutions to the shallow water equations with different formulations which are generally suitable for the problem where the horizontal length scale is much larger than the flow depth.

Earliest work on the shallow water LB model was introduced for ocean circulation (Salmon, 1999). Later improvements were made by Zhou (2002) which enabled the solution of both unsteady and steady flow problems. Differences between the free-surface LB model and the shallow water LB model exist in the formulation of the local equilibrium distribution function and the source term where the bed slope and bed shear stress are defined (Zhou, 2002). Shallow water LB models with a different collision operator have been successfully incorporated with the existing turbulent models (Liu et al., 2012). Single and multilayer shallow water flows on rectangular and curvilinear coordinates have been proposed (Budinski, 2014). Shallow water LB models applied to solve 1D channel networks (Thang et al., 2010) and operations of cascades of the hydropower stations (Zhang et al., 2016) show good agreement with analytical solutions. The 2D shallow water LB models are proven to be useful, not only for the dynamics, but also for the ecological assessment of the large-scale lakes (Ding et al., 2019) and other water bodies (Meng et al., 2020). Recently LB models of the Saint–Venant equations have been developed using the variation of discharge and cross-sectional area and removing the bed slope term which make the model unconditionally suitable for arbitrary cross-sectional geometry. All of those and other affords on the shallow water LB model provides alterative and powerful background for the solutions of open channel problems.

Fluid–structure interactions using the models discussed are under consideration in the research community. The framework of the LB model provides several treatments for solid boundaries (Guo and Shu, 2013). LB models have good prospects for problems studied by notable solutions using conventional methods, such as the immersed boundary method, including fluid–structure interactions. Both the free-surface LB models and shallow water LB models (Peng et al., 2013) are being used to study the fluid–solid interactions. Two general type of the immersed boundary LB models were introduced by Noble and Torczynski (1998) and Feng and Michaelides (2004), respectively. Badarch and Tokuzo (2018) studied fluid–solid interactions in free surface flows by slightly modifying the immersed boundary model of Noble and Torczynski (1998) to account for the free surface effect on the immersed solid. The so-called free-surface immersed-boundary LB model has been successfully applied to moving body simulation in free surface flows (Badarch and Tokuzo, 2018), gravity wave propagation (Ayurzana et al., 2018), wave run-up on slopes and hydrodynamics force estimation of waves on vertical structures (Badarch et al., 2020b). Diao et al., (2018) improved the accuracy of the coupled scheme of the free-surface LB model and conventional immersed boundary method (Dupuis et al., 2008) by correcting velocity and applying an iterative force correction and applied it to the performance study of overflow gates. To date, applications to the hydraulic jump and multiphase sediment transportation using the LB model remain of great interest.

4. Lattice Boltzmann models for seepage flows

Flows through porous media, which can be referred to as seepage flow in hydraulic engineering, are complex physical problems involving a scattered solid structure such as a porous medium and flows through it. The flows can be classified into pre-Darcy, Darcy, and Forchheimer zones, depending on the magnitude of flow velocity through the porous medium (Scheidegger, 1958). In hydraulics, seepage flow is often described by Darcy's law in both practical application and numerical study (Bardet and Tobita, 2002). Weak treatment of the conventional numerical models to the complex nature of a geometry is effectively solved when the LB model is applied to seepage flows. Guo and Zhao (2002) introduced the Brinkman-Forchheimer-extended Darcy model into the LB models by adding new equilibrium distribution functions and forcing terms enabling both linear and nonlinear drag effects of the porous media. Their model, celebrated as the generalized LB model for the flows through porous media, has been used to solve problems at a relatively larger scale. Based on the bounce back rules of the LB model, several schemes with an additional collision step for the treatments of porous media have been introduced to simulate pore scale porous media flows (Dardis and McCloskey, 1998; Walsh et al., 2009). Badarch et al., (2020a) believed the immersed boundary LB model proposed by Noble and Torczynski (1998) is applicable to porous media flow since the basic concept of the immersed-boundary model is based on the probabilistic nature of the bounce back rule. They questioningly found the connections between modeled permeability and solid fractions of the cell in the computational domain and the scaling of the physical permeability to modeled permeability as $k \cong 1.618 K_p / \triangle x^2$, where k is the modeled permeability, which is a function of the solid fraction of the cell and lattice fluid viscosity (Walsh et al., 2009). This model is exactly same as the free-surface immersed-boundary LB model used in Badarch et al., (2020b) and the sole validations are provided in Badarch et al., (2020a). In this section we will directly apply the free-surface immersed-boundary LB model to the seepage flow through a simple earth dam to demonstrate the applicability of the model.

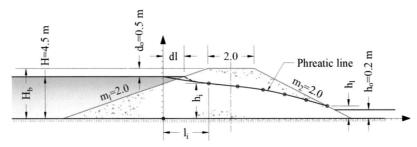

FIG. 7 Schematics of the analytical solution for the seepage flow through homogeneous dam.

4.1 Seepage flow through an earth dam

A simple homogenous trapezoidal dam with a height of 5.0 m, headwater depth of $H = 4.5$ m, and upstream and downstream inverse slopes of $m_1 = m_2 = 2$ were considered as computational domain, as shown in Fig. 7. The analytical solutions at steady state seepage flows expressed by the Darcy law yields (Pavlowsky, 1956)

$$\frac{q}{k_s} = \frac{H - h_o}{m_1} \bullet 2.3 \log \left(\frac{H_b - d_o}{H_b - d_o - h_1} \right), \tag{10}$$

where q is the discharge per unit width, H_b is the dam height, d_o is the freeboard, $k_s (=0.1$ m/day) is the hydraulic conductivity of earth material and $h_o (=0.2$ m) is the tailwater level. The following Dupuit parabola can be used to define the phreatic line if tailwater exists (Pavlowsky, 1956):

$$h_i^2 = H^2 - 2\frac{q}{k_s} \bullet l_i. \tag{11}$$

The origin of the vertical co-ordinate of the analytical phreatic line is located at $dl = m_1 H / (2m_1 + 1)$.

The computational domain of this example dam was formed with 300 cells in the x-axis and 90 cells in the y-axis. To ensure steady state condition, a uniform distribution of velocity defined from the seepage discharge obtained by the analytical solution, is given by the Zou/He boundary condition (Zou and He, 1997). No boundary condition at the surface of the dam body is needed, since the immersed boundary modification in the LB model automatically solves the boundary as the porous media while partially bounces back fluid momentum. Attention must be paid to define correctly control parameters in computations such as the solid fraction value along the dam body and the computational time step. The solid fraction value is defined by the modeled permeability which can be scaled down from the physical permeability of earth material (Badarch et al., 2020a, b).

Very good agreement between the analytical and the computational phreatic lines compared in Fig. 8 confirming the immersed-boundary LB model satisfying the Darcy law. The exit height of the phreatic line defined by the LB model was 0.42 m which also agrees with the analytical solution. In Fig. 9A the absolute velocity distribution through the homogeneous seepage region is shown. The velocities are uniformly distributed through the seepage region except the exit region near the toe and near the free surface where the nonuniform local velocities emerged due to the free exit and the free surface gradient, respectively. The exit velocity defined by the analytical solution was 0.054 m/s while

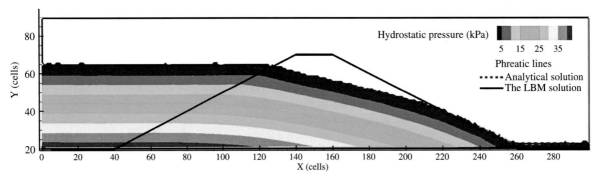

FIG. 8 Comparison of the phreatic lines defined by the analytical solutions and computed by the LB model. The background color map is the hydrostatic pressure distribution when the simulation reaches the steady state case.

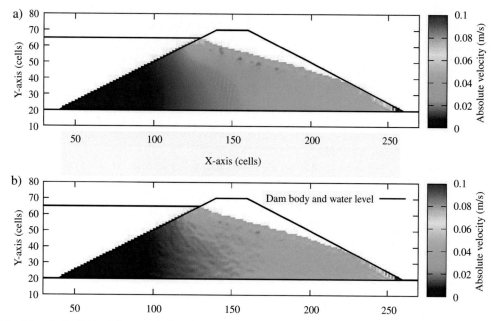

FIG. 9 Instant absolute velocity fields of the seepage flow through (A) homogenous and (B) heterogeneous dams.

the free-surface immersed-boundary LB model gives the velocity in the range of 0.04–0.08 m/s. Numerical tests were continued with a heterogeneous dam whose ensemble average of random permeability matrix is the same as that of the homogenous dam. The seepage velocity distributions in homo- and heterogeneous cases shown in Fig. 9B reveal more local properties of the seepage velocity. The velocity near the free surface of both cases was of a similar pattern. Since the solid fraction related to the permeability is a function of space, the free-surface immersed-boundary LB model is applicable for both flows through homogeneous and heterogeneous porous media.

4.2 Future outlook on the seepage flow modeling with the LB models

The generalized LB models and other LB models based on the probabilistic nature of the bounce back rule including the free-surface immersed-boundary LB model for the flow through porous media might be limited by the cohesiveness and water saturation of the earth material. Volz et al., (2017) proposed the procedure to solve Richards' equation in terms of the LB model for the unsaturated seepage flows. Consequently, the LB model solving the Richards' equation reflects the effect of saturation degree of the soil and apparent cohesion effects which are difficult to be attained by the other LB models for porous media flows. The internal erosion and the following fractures and deformations due to seepage flows are crucial topics where the LB models might provide new insight. The coupling of the Lagrangian methods with the LB models is one of the possibilities to model seepage and internal erosion. However computational expenses of this model are high. Besides the generalized LB models with Brinkman-Forchheimer-extended Darcy equation and LB models based on the probabilistic rules of bounce back, Xing et al., (2022) introduced novel LB models for seepage flow incorporating friction forces produces by porous media in terms of force term. They successfully demonstrated the model applicability for rather complicated interactions of waves and porous structures in 3D space which shows there exist more research activities with fruitful discoveries in framework of the LB models.

5. Conclusions

The basic formulations of the single-relaxation time LB model is used to solve simple fluid flows in 1D and 2D space to provide validity of the models for various hydraulic engineering problems. Good agreements between analytical solutions or experimental data and the LB results are found for the selected problems. Together with the other validations and the applications of the different LB models being used for hydraulics research in the literature and the results presented in this chapter prove that the LB models are alternative and promising tools to understand or solve crucial hydraulics problems. A brief review and outlook for each research topic among the limited literature show that there exists an enormous number of research possibilities and breakthrough advances in both fields of LB methods and hydraulic engineering. For instance, difficulties encountered with the cumbersome treatments in the current conventional numerical models could be overcame by replacing them with the newly emerging or existing LB models. Also, the LB models are well-known due to its simplicity in implementation and parallelization in computation. So that the LB models could bring fast and simple solutions to the problems. It is important that while leaning on the advantage of the LB models which could also be useful for the researching problem in soft term, overcoming the difficulties of the method at same time would bring fruitful results.

References

Acheson, D.J., 1991. Elementary Fluid Dynamics., p. 3020.

Aidun, C.K., Clausen, J.R., 2010. Lattice-Boltzmann method for complex flows. Annu. Rev. Fluid Mech. 42, 439–472.

Ansumali, S., Karlin, I.V., 2002. Single relaxation time model for entropic lattice Boltzmann methods. Phys. Rev. E 65 (5), 056312.

Asinari, P., 2008. Generalized local equilibrium in the cascaded lattice Boltzmann method. Phys. Rev. E 78 (1), 016701.

Ayurzana, B., Hosoyamada, T., 2017. Application of the lattice Boltzmann method to liquid-solid phase change in free surface flow: an example of Mongolian small hydropower. Journal of Japan J. Jpn. Soc. Civ. Eng 73 (4). pp. I_607-I_612.

Ayurzana, B., Tokuzo, H., Erdenebayar, T., 2018. Application of a free-surface immersed boundary-lattice Boltzmann modeling to wave forces acting on a breakwater. In: Aachen, Germany, 7th IAHR International Symposium on Hydraulic Structures.

Badarch, A., 2017. Application of macro and mesoscopic numerical models to hydraulic problems with solid substances. PhD diss, Nagaoka University of Technology.

Badarch, A., Tokuzo, H., 2018. Lattice Boltzmann method for the numerical simulations of the melting and floating of ice. In: Free Surface Flows and Transport Processes. Springer, pp. 143–154.

Badarch, A., Tokuzo, H., Narantsogt, N., 2016. Hydraulics application of the free-surface lattice Boltzmann method. In: 11th International Forum on Strategic Technology (IFOST) IEEE, pp. 195–199.

Badarch, A., Khenmedekh, L., Hosoyamada, T., 2017. Parallel implementation of Entropic lattice Boltzmann method for flow past a circular cylinder at high Reynolds number. Trans. GIGAKU 4 (1), 04006.

Badarch, A., Fenton, J.D., Hosoyamada, T., 2020a. Application of free-surface immersed-boundary lattice Boltzmann method to waves acting on coastal structures. J. Hydraul. Eng. 146 (2), 04019062.

Badarch, A., Fenton, J.D., Tokuzo, H., 2020b. A free-surface immersed-boundary lattice Boltzmann method for flows in porous media. In: Recent Trends in Environmental Hydraulics. Springer, pp. 23–31.

Bardet, J.-P., Tobita, T., 2002. A practical method for solving free-surface seepage problems. Comput. Geotech 29 (6), 451–475.

Bartlett, C., Chen, H., Staroselsky, I., Wanderer, J., Yakhot, V., 2013. Lattice Boltzmann two-equation model for turbulence simulations: high-Reynolds number flow past circular cylinder. Int. J. Heat Fluid Flow 42, 1–9.

G.K. Batchelor, F., 1967. An Introduction to Fluid Dynamics. Cambridge: Cambridge University Press.

Belytschko, T., Yun, Y.L., Lei, G., 1994. Element-free Galerkin methods. Int. J. Numer. Methods Eng. 37 (2), 229–256.

Bhatnagar, P., Gross, E., Krook, M., 1954. A model for collision processes in gases. I. Small amplitude processes in charged and neutral one-component systems. Phys. Rev. 94 (3), 511.

Boghosian, B.M., Yepez, J., Coveney, P.V., Wager, A., 2001. Entropic lattice Boltzmann methods. Proc. R. Soc. A: Math. Phys. Eng. Sci. 457 (2007), 717–766.

Budinski, L., 2014. MRT lattice Boltzmann method for 2D flows in curvilinear coordinates. Comput. Fluids 96, 288–301.

Budinski, L., 2016. Application of the LBM with adaptive grid on water hammer simulation. J. Hydroinformatics 18 (4), 687–701.

Chen, S., Doolen, G.D., 1998. Lattice Boltzmann method for fluid flows. Annu. Rev. Fluid Mech. 30 (1), 329–364.

Chen, Z., Shu, C., Tan, D., Wu, C., 2018. On improvements of simplified and highly stable lattice Boltzmann method: formulations, boundary treatment, and stability analysis. Int. J. Numer. Methods Fluids 87 (4), 161–179.

Dardis, O., McCloskey, J., 1998. Lattice Boltzmann scheme with real numbered solid density for the simulation of flow in porous media. Phys. Rev. E 57 (4), 4834.

Darvishi, E., Fenton, J.D., Kouchakzadeh, S., 2017. Boussinesq equations for flows over steep slopes and structures. J. Hydraul. Res. 55 (3), 324–337.

Cheng, Y.G., Zhang, S.H., Chen, J.Z., 1998. Water hammer simulation by the lattice Boltzmann method. Transactions of the Chinese Hydraulic Engineering. J. Hydraul. Eng. 6, 25–31.

d'Humieres, D., 2002. Multiple–relaxation–time lattice Boltzmann models in three dimensions. Philosophical Transactions of the Royal Society of London. Proc. R. Soc. A: Math. Phys. Eng. Sci. 360 (1792), 437–451.

Diao, W., Cheng, Y., Xu, M., Wu, J., Zhang, C., Zhou, W., 2018. Simulation of hydraulic characteristics of an inclined overflow gate by the free-surface lattice Boltzmann-immersed boundary coupling scheme. Eng. Appl. Comput. Fluid Mech. 12 (1), 250–260.

Ding, Y., Liu, H., Yang, W., Xing, L., Tu, G., Ru, Z., Xu, Z., 2019. The assessment of ecological water replenishment scheme based on the two-dimensional lattice-Boltzmann water age theory. J. Hydro-Environ. Res. 25, 25–34.

Dolanský, J., Chára, Z., Vlasák, P., Kysela, B., 2017. Lattice Boltzmann method used to simulate particle motion in a conduit. J. Hydrol. Hydromech. 65 (2), 105–113.

Dupuis, A., Chatelain, P., Koumoutsakos, P., 2008. An immersed boundary–lattice-Boltzmann method for the simulation of the flow past an impulsively started cylinder. J. Comput. Phys. 227 (98), 4486–4498.

Feng, Z.-G., Michaelides, E.E., 2004. The immersed boundary-lattice Boltzmann method for solving fluid–particles interaction problems. J. Comput. Phys. 195 (2), 602–628.

Feng, Y.T., Han, K., Owen, D.R.J., 2007. Coupled lattice Boltzmann method and discrete element modelling of particle transport in turbulent fluid flows: computational issues. Int. J. Numer. Methods Eng. 72 (9), 1111–1134.

Frisch, U., d'Humieres, D., Hasslacher, B., Lallemand, P., Pomeau, Y., Rivet, J.P., 1987. Lattice gas hydrodynamics in two and three dimensions. Complex Syst. 1 (4), 647–707.

Galindo-Torres, S.A., 2013. A coupled discrete element lattice Boltzmann method for the simulation of fluid–solid interaction with particles of general shapes. Comput. Methods Appl. Mech. Eng. 265, 107–119.

Ghidaoui, M.S., Zhao, M., McInnis, D.A., Axworthy, D.H., 2005. A review of water hammer theory and practice. Appl. Mech. Rev. 58 (1), 49–76.

Gingold, R.A., Monaghan, J.J., 1977. Smoothed particle hydrodynamics: theory and application to non-spherical stars. Mon. Notices Royal Astr. Soc. 181 (3), 375–389.

Ginzburg, I., Verhaeghe, F., d'Humieres, D., 2008. Two-relaxation-time lattice boltzmann scheme: about parametrization, velocity, pressure and mixed boundary conditions. Commun. Comput. Phys. 3 (2), 427–478.

Guo, Z., Shu, C., 2013. Lattice Boltzmann Method and its Application in engineering. World scientific Publishing, Singapore.

Guo, Z., Zhao, T.S., 2002. Lattice Boltzmann model for incompressible flows through porous media. Phys. Rev. E 66 (3), 036304.

Halliday, I., Hammond, L.A., Care, C.M., Good, K., Stevens, A., 2001. Lattice Boltzmann equation hydrodynamics. Phys. Rev. E 64 (1), 011208.

Han, L.S., 1960. Hydrodynamic entrance lengths for incompressible laminar flow in rectangular ducts. J. Appl. Mech., 403–409.

He, Q., Li, Y., Huang, W., Hu, Y., Wang, Y., 2019. Phase-field-based lattice Boltzmann model for liquid-gas-solid flow. Phys. Rev. E 100 (3), 033314.

Hirt, C.W., Nichols, B.D., 1981. Volume of fluid (VOF) method for the dynamics of free boundaries. J. Comput. Phys. 39 (01), 201–225.

Hou, S., Sterling, J., Chen, S., Doolen, G.D., 1996. A Lattice Boltzmann Subgrid Model for High Reynolds Number Flows. Fields Institute Communications.

Jahanshaloo, L., Pouryazdanpanah, E., Sidik, N.A.C., 2013. A review on the application of the lattice Boltzmann method for turbulent flow simulation. Numer. Heat Transf. A: Appl. 64 (11), 938–953.

Janssen, C., Krafczyk, M., 2010. A lattice Boltzmann approach for free-surface-flow simulations on non-uniform block-structured grids. Comput. Math. Appl. 59 (7), 2215–2235.

Kang, S.K., Hassan, Y.A., 2013. The effect of lattice models within the lattice Boltzmann method in the simulation of wall-bounded turbulent flows. J. Comput. Phys. 232 (1), 100–117.

Kerson, H., 1987. Statistical mechanics. John Willey and Sons, Massachusetts.

Körner, C., Thies, M., Hofmann, T., Thürey, N., Rüde, U., 2005. Lattice Boltzmann model for free surface flow for modeling foaming. J. Stat. Phys. 121 (1), 179–196.

Liu, H., Li, M., Shu, A., 2012. Large eddy simulation of turbulent shallow water flows using multi-relaxation-time lattice Boltzmann model. Int. J. Numer. Methods Fluids 70 (12), 1573–1589.

Louati, M., Tekitek, M.M., Ghidaoui, M.S., 2019. On the dissipation mechanism of lattice Boltzmann method when modeling 1-d and 2-d water hammer flows. Comput. Fluids 193, 103996.

Maier, R.S., Bernard, R.S., Grunau, D.W., 1996. Boundary conditions for the lattice Boltzmann method. Phys. Fluids 8 (7), 1788–1801.

Meng, W., Cheng, Y., Wu, J., Yang, Z., Zhu, Y., Shang, S., 2019. GPU acceleration of hydraulic transient simulations of large-scale water supply systems. Appl. Sci. 9 (1), 91.

Meng, W., Cheng, Y., Wu, J., Zhang, C., Xia, L., 2020. A 1D–2D coupled lattice Boltzmann model for shallow water flows in large scale River-lake systems. Appl. Sci. 10 (1), 108.

Monaghan, J.J., 1994. Simulating free surface flows with SPH. J. Comput. Phys. 110 (2), 399–406.

Monaghan, J.J., Kocharyan, A., 1995. SPH simulation of multi-phase flow. Comput. Phys. Commun. 87 (1–2), 225–235.

Nathen, P., Gaudlitz, D., Krause, M.J., Adams, N.A., 2017. On the stability and accuracy of the BGK, MRT and RLB Boltzmann schemes for the simulation of turbulent flows. J. Commun. Comput. Phys. 23, 846–876.

Noble, D.R., Torczynski, J.R., 1998. A lattice-Boltzmann method for partially saturated computational cells. Int. J. Mod. Phys. C 9 (08), 1189–1201.

Pavlowsky, N., 1956. Collected Studies (in Russian), Moscow, Leningrad.

Peng, Y., Zhou, J.G., Zhang, J.M., Burrows, R., 2013. Modeling moving boundary in shallow water by LBM. Int. J. Mod. Phys. C 24 (01), 1250094.

Peng, C., Geneva, N., Guo, Z., Wang, L.-P., 2018. Direct numerical simulation of turbulent pipe flow using the lattice Boltzmann method. J. Comput. Phys. 357, 16–42.

Qian, Y.-H., d'Humières, D., Lallemand, P., 1992. Lattice BGK models for Navier-Stokes equation. Europhys. Lett. 17 (6), 179.

Qian, Y.-H., Succi, S., Orszag, S.A., 1995. Recent advances in lattice Boltzmann computing. Annu. Rev. Compu. Phy. III, 195–242.

Salmon, R., 1999. The lattice Boltzmann method as a basis for ocean circulation modeling. J. Mar. Res. 57 (3), 503–535.

Scheidegger, A., 1958. The Physics of Flow Through Porous Media. University Of Toronto Press, London.

Schlichting, H., Gersten, K., 2016. Boundary-Layer Theory. Springer.

Shan, X., Hudong, C., 1993. Lattice Boltzmann model for simulating flows with multiple phases and components. Phys. Rev. E 47 (3), 1815.

Sielecki, A., Wurtele, M.G., 1970. The numerical integration of the nonlinear shallow-water equations with sloping boundaries. J. Comput. Phys. 6 (2), 219–236.

Teixeira, C.M., 1998. Incorporating turbulence models into the lattice-Boltzmann method. Int. J. Mod. Phys. C 9 (08), 1159–1175.

Thürey, N., Pohl, T., Rüde, U., Oechsner, M., Körner, C., 2006. Optimization and stabilization of LBM free surface flow simulations using adaptive parameterization. Comput. Fluids 35 (8–9), 934–939.

Van Thang, P., Chopard, B., Lefèvre, L., Ondo, D.A., Mendes, E., 2010. Study of the 1D lattice Boltzmann shallow water equation and its coupling to build a canal network. J. Comput. Phys. 229 (19), 7373–7400.

Volz, C., Frank, P.J., Vetsch, D.F., Hager, W.H., Boes, R.M., 2017. Numerical embankment breach modelling including seepage flow effects. J. Hydraul. Res. 55 (4), 480–490.

Walsh, S.D., Burwinkle, H., Saar, M.O., 2009. A new partial-bounceback lattice-Boltzmann method for fluid flow through heterogeneous media. Comput. Geosci. 35 (6), 1186–1193.

Wang, H., Yuan, X., Liang, H., Chai, Z., Shi, B., 2019. A brief review of the phase-field-based lattice Boltzmann method for multiphase flows. Capillarity 2 (3), 33–52.

Welander, P., 1954. On the temperature jump in a rarefied gas. Ark. Fys. 7, 507–553.

Wen, Z.X., Li, Q., Yu, Y., Luo, K.H., 2019. Improved three-dimensional color-gradient lattice Boltzmann model for immiscible two-phase flows. Phys. Rev. E 100 (2), 023301.

Wu, Y., Chi, L., Zhang, H., 2008. Study of resistance distribution and numerical modeling of water hammer in a long-distance water supply pipeline. Water Distribut. Syst. Anal. 2008, 1–10.

Xing, E., Zhang, Q., Liu, G., Zhang, J., Ji, C., 2022. A three-dimensional model of wave interactions with permeable structures using the lattice Boltzmann method. App. Math. Model. 104, 67–95. https://doi.org/10.1016/j.apm.2021.11.018.

Zhang, C.Z., Cheng, Y.G., Wu, J.Y., Diao, W., 2016. Lattice Boltzmann simulation of the open channel flow connecting two cascaded hydropower stations. J. Hydrodyn. 28 (3), 400–410.

Zhou, J.G., 2002. A lattice Boltzmann model for the shallow water equations. Comput. Methods Appl. Mech. Eng. 191 (32), 3527–3539.

Ziegler, D.P., 1993. Boundary conditions for lattice Boltzmann simulations. J. Stat. Phys. 71 (5–6), 1171–1177.

Zou, Q., He, X., 1997. On pressure and velocity boundary conditions for the lattice Boltzmann BGK model. Phys. Fluids 9 (6), 1591–1598.

Chapter 17

Developments in sediment transport modeling in alluvial channels

Gokmen Tayfur

Department of Civil Engineering, Izmir Institute of Technology, Izmir, Turkey

1. Introduction

Sediment transport is one of vital importance for regulation, training, and restoration of rivers, design of hydraulic structures, ecosystem maintenance, and coastal restoration. To ensure safe and viable navigation routes, the dredging of river sediments is carried out systematically and periodically. River navigation, like river restoration works, requires information and data on river sedimentation. Dam reservoirs are filled with sediment, adversely impacting their economic life expectancy. Therefore, sediment transport dynamics entail economic, environmental, and safety concerns. Hence, sediment transport has long been studied both experimentally and mathematically.

Experimental studies on understanding the mechanisms and frameworks for sediment transport had started in the late 19th century with the works of DuBoys (1879) and continued up to the first half of the 20th century (Schoklitsch, 1934; Shields, 1936; Kalinske, 1947; Meyer-Peter and Muller, 1948). Experimental investigations continued with the laboratory experiments of Guy et al. (1966), Soni (1981), and Bombar et al. (2011), to name but a few. Many laboratory flume flow and sediment transport experiments are summarized by Lisle et al. (1997). Field experiments had also been carried out (Langbein and Leopold, 1968; Wathen and Hoey, 1998; Lisle et al., 2001; among others). Experimental studies have sought to understand the basic mechanisms, of important issues such as incipient motion, particle fall velocity, particle velocity, and sediment rate transport. These studies also tried to develop equations for these issues and empirical equations to predict sediment rates. The developed equations have however shown a wide range of variability, since each had been developed, based on experimental conditions.

Considerable effort has also been devoted to the development of theoretical prediction equations. Einstein (1942, 1950) introduced a probabilistic approach, while Yang and Sayre (1971) employed a stochastic one, and Rottner (1959) used a regression approach to develop empirical equations for bed load rates. Lane and Kalinske (1941), Brooks (1963), and Chang et al. (1965) developed equations for the prediction of suspended sediment rate, based on the balance of forces and assumed different velocity profiles. Developments of fundamental equations for prediction of total load are also seen (Velikanov, 1954; Bagnold, 1966; Toffaleti, 1969; Yang, 1972, 1979; Shen and Hung, 1972; Ackers and White, 1973; Simons and Senturk, 1977; Karim and Kennedy, 1990).

With the advent of computers, sediment transport has been treated comprehensively, resulting in equations based on the conservation of mass, energy, and momentum. The effects of turbulence have also been recently taken into account, together with sediment mixtures.

2. Approaches for predicting sediment transport

2.1 Empirical approaches

The first studies are, in essence, experimental, mostly involving laboratory flume experiments of bed load transport. DuBoys (1879) is the first one who introduced the concept of critical shear stress, while Shields (1936) developed a method to quantify it. DuBoys (1879) developed one of the first equations to estimate sediment rate by carrying out experiments on a small experimental flume. Later, Shields (1936), based on his experimental works, introduced a new equation and Shields diagram to quantify the critical shear stress. Meyer-Peter et al. (1934) conducted extensive laboratory experiments on bed load transport and developed empirical equations based on the energy-slope approach. Then, they revised their original

Handbook of HydroInformatics. https://doi.org/10.1016/B978-0-12-821962-1.00022-2

equation in 1947 (Meyer-Peter and Muller, 1948). Schoklitsch (1949) proposed an empirical bed load equation based on the discharge approach.

The sediment transport capacity of flow is generally expressed as follows:

$$T_c = \alpha(D - D_c)^\beta \tag{1}$$

where T_c is the sediment transport capacity of flow; α and β are the parameters related to flow and sediment conditions; D is the dominant variable; and D_c is the critical condition of dominant variable at the incipient motion. The dominant variable can be shear stress (τ), velocity (v), discharge (Q), stream power (τv), and unit stream power (vS).

Rouse (1937) worked on the suspended sediment based on the exchange theory under equilibrium conditions. Lane and Kalinske (1941) revised Rouse's equation. Einstein (1950) employed a probabilistic approach for quantifying suspended sediment. Brooks (1963) and Chang et al. (1965) developed equations based on the Einstein approach.

Based on the concepts of Einstein (1950) and Einstein and Chien (1954), Toffaleti (1969) developed a procedure to compute the total sediment load. Velikanov (1954) derived a transport equation based on the gravitational power theory. Bagnold (1966) developed a sediment transport function using the stream power concept. The stream power is expressed as the product of flow velocity and shear stress (τV). Yang (1972) developed an empirical total load equation based on the unit stream power concept. Using 587 sets of field data, the unit stream power is expressed as the product of velocity and slope (vS). Shen and Hung (1972), proposed a regression equation. In a similar fashion, employing 947 sets of field data, Karim and Kennedy (1990) proposed an equation, based on the nonlinear multiple regression analysis. Details of all these approaches and applications are discussed by Yang (1996).

2.2 Physics-based approaches

With the advent of the developments in computer technology, sediment transport has been treated as physics-based. In the early 1970s, differential equations representing the processes treated sediment dynamics in one-dimension under steady flow conditions (de Vries, 1965; de Vries, 1973; Mahmood, 1975). Later, equations were solved in one-dimension under unsteady flow (de Vries, 1975; Pianese, 1994; Cao and Carling, 2003; Wu et al., 2004). Tayfur and Singh (2006) treated the dynamics in one-dimension under unsteady flow and equilibrium conditions. These equations in one-dimension can be represented as follows (Tayfur and Singh, 2006):

$$\frac{\partial h(1-c)}{\partial t} + \frac{\partial(hu(1-c))}{\partial x} + p\frac{\partial z}{\partial t} = 0 \tag{2}$$

$$\frac{\partial hc}{\partial t} + \frac{\partial(huc)}{\partial x} + (1-p)\frac{\partial z}{\partial t} + \frac{\partial q_{bs}}{\partial x} = 0 \tag{3}$$

where h is the flow depth (L); u is the flow velocity (L/T); c is the volumetric sediment concentration in suspension (L^3/L^3); p is the bed layer sediment porosity (L^3/L^3); z is the mobile bed layer elevation (L); and q_{bs} is the sediment rate in the movable bed layer (L^2/T).

Eqs. (2), (3) stand for the conservation of water and sediment in both the layers (the bed layer phase, and the suspension layer phase (water flow phase)), respectively, under the equilibrium conditions where the entrainment rate (E_z) is equal to the deposition rate (D_c) (i.e., $E_z = D_c$). That is, the amounts of sediment exchange between the two phases are equal.

Under nonequilibrium conditions ($E_z \neq D_c$), there are different approaches in the literature. Pianese (1994), in addition to the main equation, employed the adaptation equation (or lag equation):

$$(1-p)\frac{\partial z}{\partial t} = \frac{uh}{\lambda}\left(c_{eq} - c\right) \tag{4}$$

where λ is called the adaptation length and c_{eq} is the equilibrium suspended sediment concentration. The right hand side of Eq. (4) represents the deposition rate (positive) or detachment rate (negative) (Pianese, 1994).

Mohammadian et al. (2004) employed one equation for the conservation of water and one equation for the conservation of suspended sediment in the water flow layer as:

$$\frac{\partial hc}{\partial t} + \frac{\partial huc}{\partial x} = \frac{\partial}{\partial x}\left(V_x h\frac{\partial c}{\partial x}\right) + \frac{v_f}{\eta}\left(c_{eq} - c\right) \tag{5}$$

where V_x = the sediment mixing coefficient; v_f = the sediment particle fall velocity; and η is a coefficient. The last term on the right hand side of Eq. (5) represents the deposition rate (negative) or detachment rate (positive). In addition, they employed the following equation for relating the change in bed level in time to the particle fall velocity, equilibrium suspended sediment concentration, and suspended sediment concentration as:

$$(1-p)\frac{\partial z}{\partial t} = \frac{v_f}{\eta}\left(c_{eq} - c\right) \tag{6}$$

Tayfur and Singh (2007), on the other hand, employed a different approach. They wrote the conservation of mass for suspended sediment in the water flow layer and conservation of mass for bed sediment in the movable bed layer separately, considering the exchange of sediment due to the detachment and deposition between the two layers. They, thus, expressed the main sediment transport equations, in addition to the water flow dynamics, as follows:

$$\frac{\partial hc}{\partial t} + \frac{\partial(huc)}{\partial x} = \frac{1}{\rho_s}(E_z - D_c) \tag{7}$$

$$(1-p)\frac{\partial z}{\partial t} + \frac{\partial q_{bs}}{\partial x} = \frac{1}{\rho_s}(D_c - E_z) \tag{8}$$

where E_z is the entrainment rate (detachment rate) $(M/L^2/T)$; and D_c is the deposition rate $(M/L^2/T)$. Eqs. (2), (7), and (8) are for the conservation of mass for water in both the layers, suspended sediment in the water flow layer, and sediment in the movable bed layer, respectively, and constitute the basic equations for modeling transient bed profiles under equilibrium $(E_z = D_c)$ and nonequilibrium conditions $(E_z \neq D_c)$.

According to Eqs. (2), (7), and (8), there are 5 unknowns, namely, h, u, c, z, and q_{bs}. In order to close the system, two more equations are needed. One equation can be obtained using the momentum equation for water flow as follows:

$$\frac{\partial u}{\partial t} + u\frac{\partial u}{\partial x} + g\frac{\partial h}{\partial x} = g\left(S_o - S_f\right) \tag{9}$$

where g is the gravitational acceleration, S_o is the channel bed slope, and S_f is the friction slope which can be quantified using Manning's equation.

The fifth equation can be obtained by relating the sediment transport rate (sediment flux) to sediment concentration in the movable bed layer. Previous researchers (Velikanov, 1954; Ching and Cheng, 1964; de Vries, 1965; Lai, 1991; Pianese, 1994) had related sediment flux in the movable bed layer to the flow variables in the water flow layer. Tayfur and Singh (2006) had, on the other hand, proposed a new equation for relating sediment flux to the sediment concentration in the bed layer, as follows, following the fundamental work of Langbein and Leopold (1968);

$$q_{bs} = (1-p)v_s z\left[1 - \frac{z}{z_{max}}\right] \tag{10}$$

where v_s is the velocity of sediment particles as concentration approaches zero (L/T); z_{max} is the maximum bed elevation (L). Eqs. (2) and (7)–(10) form the system of five equations for modeling the sediment transport in alluvial channels under nonequilibrium conditions.

2.3 Advanced approaches

Advanced approaches treat the dynamics as having the processes of three phases: water, air, and sediment. The related equations (Reynolds Averaged Navier-Stokes (RANS)) can be found in Marsooli and Wu (2015):

$$\frac{\partial u_j}{\partial x_j} = 0 \tag{11}$$

$$\frac{\partial u_i}{\partial t} + \frac{\partial u_i u_j}{\partial x_j} = g - \frac{1}{\rho}\frac{\partial P}{\partial x_i} + \frac{1}{\rho}\frac{\partial}{\partial x_j}\left[(\mu + \mu_t)\left(\frac{\partial u_i}{\partial x_j} + \frac{\partial u_j}{\partial x_i}\right)\right] \tag{12}$$

$$\frac{\partial \chi}{\partial t} + u_j\frac{\partial \chi}{\partial x_j} = 0 \tag{13}$$

where u is the flow velocity, g is the gravitational acceleration, P is the pressure, ρ is the fluid density, μ is the dynamic viscosity, μ_t is the eddy viscosity; χ is the phase characteristic, $\chi = \alpha_q \rho_q$ (where α_q is the volume fraction of the qth phase (can be fluid, air, or sediment) in the cell, ρ_q is the density of the qth phase, and i and j stand for the x-, y-, and z directions.

The phase characteristic (χ) is the fundamental parameter in the volume of fluid (VOF) formulation when solving the multiphase fluid flow transport. In each control volume, the volume fractions of all phases sum to the unity. The fields for all variables and properties are shared by the phases and represented by the volume-averaged values, as long as the volume fraction of each phase is known at each location. Thus, the variables and properties in any given cell are either purely representative of one of the phases or representative of a combination of the phases, depending upon the volume fraction values. In other words, if the qth fluid-phase volume fraction in the cell is denoted as α_q, then three conditions are possible: $\alpha_q = 0$; the cell is empty of the qth fluid, $\alpha_q = 1$; the cell is full of the qth fluid; $0 < \alpha_q < 1$, the cell contains the interface between the qth fluid and one or more other phases. Based on the local value of α_q the appropriate properties and variables are assigned to each control volume within the domain. The volume fraction equation is not solved for the primary phase; the primary-phase volume fraction is computed, based on the constraint of $\sum_{q=1}^{n} \alpha_q = 1$.

The tensor stress can be expressed as (Landau and Lifshitz, 1997; Issakhov et al., 2018):

$$\tau = \mu \left(\frac{\partial u_i}{\partial x_j} + \frac{\partial u_j}{\partial x_i} \right) \tag{14}$$

To close the RANS equations, the k-ω SST (shear stress transport) turbulent model with two additional equations for the two variables k and ω is employed. Issakhov et al. (2018) employed several methods to close the RANS equations and concluded that the k-ω was the better one, especially with respect to the CPU time.

3. Issues under considerations

In the physics-based approaches, there remain three issues: (1) particle fall velocity, (2) particle velocity, and (3) sediment rate function. In the literature, it is possible to find different approaches to each of these issues. The following section briefly summarizes the common approaches.

3.1 Particle fall velocity

The fall velocity is directly related to flow conditions and reflects the integrated effects of particle shape, size, surface roughness, and fluid density and viscosity. Therefore, it is difficult to quantify this parameter even for steady uniform flow (Yang, 1996). It is possible to find several approaches, starting with Rubey (1933) who developed a relation for particles whose diameter greater than 1 mm and 2 mm under theoretical considerations. The fall velocity (v_f) of a particle can be expressed as (Yang, 1996):

$$v_f = \begin{cases} \dfrac{1}{18} \dfrac{(\gamma_s - \gamma_w)}{\gamma_w} \dfrac{g d_s^2}{\upsilon} & d_s \leq 0.1\,\text{mm} \\[2ex] F \left[\dfrac{g d_s (\gamma_s - \gamma_w)}{\gamma_w} \right]^{0.5} & 0.1\,\text{mm} < d_s \leq 2.0\,\text{mm} \\[2ex] 3.32 \sqrt{d_s} & d_s > 2.0\,\text{mm} \end{cases} \tag{15}$$

where

$$F = \begin{cases} \left[\dfrac{2}{3} + \dfrac{36\upsilon^2 \gamma_w}{g d_s^3 (\gamma_s - \gamma_w)} \right]^{0.5} - \left[\dfrac{36\upsilon^2 \gamma_w}{g d_s^3 (\gamma_s - \gamma_w)} \right]^{0.5} & 0.1\,\text{mm} < d_s \leq 1.0\,\text{mm} \\[2ex] 0.79 & 1.0\,\text{mm} < d_s \leq 2.0\,\text{mm} \end{cases} \tag{16}$$

where d_s is the particle diameter, g is the gravitational acceleration, γ_s is the specific weight of sediment, and γ_w is the specific weight of water.

When the particle Reynolds number is greater than 2.0, the particle fall velocity can be determined experimentally. Yang (1996) gives a figure summarizing the particle fall velocity values depending on the sieve diameter and the shape factor. For most natural sands, the shape factor is 0.7 and Rouse (1938) gives $v_f = 0.024$ m/s for $d_s = 0.2$ mm.

Analyzing a wide range of empirical data, Dietrich (1982) developed the following equation for the dimensionless particle fall velocity (W_*):

$$W_* = R_3 10^{(R_1 + R_2)} \tag{17}$$

where

$$R_1 = -3.767 + 1.929(\log D_*) - 0.0982(\log D_*)^2 - 0.00575(\log D_*)^3 + 0.00056(\log D_*)^4 \tag{18}$$

$$R_2 = \log\left[1 - \frac{(1 - CSF)}{0.85}\right] - (1 - CSF)^{2.3} \tanh\left[\log D_* - 4.6\right] + 0.3(0.5 - CSF)(1 - CSF)^2(\log D_* - 4.6) \tag{19}$$

$$R_3 = \left[0.65 - \left(\frac{CSF}{2.83} \tanh\left[\log D_* - 4.6\right]\right)\right]^{\left[1 + \frac{(3.5 - P)}{2.5}\right]} \tag{20}$$

where CSF is called the Corey shape factor and defined as (Dietrich, 1982):

$$CSF = \frac{c}{(ab)^{0.5}} \tag{21}$$

where a, b, and c are the longest, intermediate, and shortest mutually perpendicular axes of the sediment particle, respectively, and P is the value of roundness.

The dimensionless settling (fall) velocity of particles (W_*) is related to the particle fall velocity as (Dietrich, 1982):

$$W_* = \frac{\rho v_f^3}{(\rho_s - \rho)gv} \tag{22}$$

where ρ is the fluid (water) density. The dimensionless particle size (D_*) is related to the particle diameter as (Dietrich, 1982):

$$D_* = \frac{(\rho_s - \rho)gd_s^3}{\rho v^2} \tag{23}$$

The numerical study of Bor (2008) showed that the approaches given by Rouse (1938), Dietrich (1982), and Yang (1996) have produced the similar results. These relations have been developed under steady conditions. According to Haushild et al. (1961), the fall velocity decreases with the increase in fine particles within the system. Therefore, there are many conditions that can affect the fall velocity, such as nonuniformity, unsteadiness, presence of fine particles, turbulence effects, fluid and sediment characteristics, etc. Under such general conditions, quantifying the fall velocity would be challenging and remains to be an outstanding issue for the scientific community.

3.2 Sediment particle velocity

The particle velocity in a flow is another important issue for the predication of sediment rates in alluvial rivers. As in the case of particle fall velocity, herein too this parameter is a function of many variables, such as particle and fluid characteristics and flow conditions. It is possible to find different approaches and thus formulations to quantify this parameter in the literature (Levy, 1957; Sharmov, 1959).

Kalinske (1947) proposed the following equation for the particle velocity (v_s):

$$v_s = \beta(u - u_c) \tag{24}$$

where u is the flow velocity, u_c is the flow velocity at the incipient motion, and β is a constant, almost unity.

Chien and Wan (1999) suggested the following equation for sediment particle velocity:

$$v_s = u - \frac{(u_c/1.4)^3}{u^2} \tag{25}$$

where u_c is the critical flow velocity at the incipient sediment motion, expressed as a function of particle fall velocity (v_f) and the shear velocity Reynolds number (R^*), for which details can be found in Yang (1996).

Based on theoretical considerations of the dynamics of bed load motion, Bridge and Dominic (1984) developed the following expression for the particle velocity:

$$v_s = \delta(u_* - u_{*_c}) \qquad (26)$$

where δ is defined as:

$$\delta = \frac{1}{K} \ln\left(\frac{y_n}{y_1}\right) \qquad (27)$$

where K is the von Karman constant; y_n is the distance from the boundary of effective fluid thrust on bed load grain; and y_1 is the roughness height. The average value of δ is between 8 and 12 (Bridge and Dominic, 1984). u_{*_c} is the critical shear velocity at the incipient motion and is defined as (Bridge and Dominic, 1984):

$$u_{*_c} = \frac{v_f(\tan\varphi)^2}{\delta} \qquad (28)$$

where $\tan\phi$ is the dynamic friction coefficient and has an average value between 0.48 and 0.58 (Bridge and Dominic, 1984). As seen in Eq. (28), the critical shear velocity is a function of the particle fall velocity.

In a numerical study, Bor (2008) showed that the equations proposed by Kalinske (1947) and Bridge and Dominic (1984) behaved similarly, as opposed to that of Chien and Wan (1999). As in the case of fall velocity, sediment particle velocity is also a function of many parameters, such as fluid and particle characteristics, flow conditions, particle fall velocity, and the presence of fine particles. Development of a relation for particle velocity under a comprehensive case remains to be a challenge for the researchers.

3.3 Sediment rate function

DuBoys (1879) employed the shear stress approach for a sediment rate function as follows (Straub, 1935):

$$q_{bs} = \frac{0.172}{d_s^{0.75}} \tau(\tau - \tau_c) \qquad (29)$$

where τ is the shear stress, τ_c is the critical shear stress, and d_s is the particle diameter.

Conducting series of laboratory experiments, Shields (1936), proposed the following;

$$\frac{q_{bs}\gamma_s}{q\gamma_w S} = \frac{10(\tau - \tau_c)}{(\gamma_s - \gamma_w)d_s} \qquad (30)$$

where γ_s and γ_w are the specific weights of sediment and water, respectively; S is the bed slope, and q is the flow rate.

Meyer-Peter and Muller (1948) proposed the following;

$$\gamma_w RS = 0.047(\gamma_s - \gamma_w)d_s + 0.25\rho^{0.33}q_{bs}^{0.67} \qquad (31)$$

where R is the hydraulic radius.

Chang et al. (1965) proposed an approach based on the shear stress as follows:

$$q_{bs} = \frac{K_b\gamma_s u(\tau - \tau_c)}{(\gamma_s - \gamma_w)\tan(\theta)} \qquad (32)$$

where K_b is a constant, and θ is the angle of repose of submerged material.

Parker et al. (1982) proposed a different approach, based on the hypothesis of equal mobility, which requires cumbersome computations of many parameters. Finally, Tayfur and Singh (2006) proposed the following:

$$q_{bs} = (1-p)v_s z\left[1 - \frac{z}{z_{max}}\right] \qquad (33)$$

Note that the proposed equation by Tayfur and Singh (2006) is for the case when there is only aggradation within an alluvial channel. The above approaches point out that the suitable formulation for the sediment rate function under comprehensive flow and sediment transport conditions is still under investigation.

4. Conclusions

As this chapter briefly summarizes, for the last 150 years, there have been many studies to model sediment transport in alluvial rivers. The first studies had been devoted to the understanding of the basic mechanisms in the sediment dynamics by through flume experiments. Later, based on experimental data and theoretical studies, several empirical equations had been developed for bed load, suspended load, and total load predictions. The developments had involved basic laws of physics and as well as stochastic, physically-based regression approaches.

While the earlier studies had been carried out in steady and uniform flows, later ones were extended to nonuniform and unsteady flows. With the advent of computers, sediment transport was first treated in one-dimension in nonuniform and steady flows. Later, it was treated in two dimensions under equilibrium and nonequilibrium conditions in unsteady flows. More recently, more advanced approaches have been employed by treating the flow as a multiphase transport in three dimensions under turbulence effects.

No matter which approach is employed there remains several issues that are still under investigation:

1. The sediment transport capacity can be based on one of the several variables, such as discharge, velocity, stream power, and the unit stream power.
2. The particle fall velocity depends on fluid, sediment, and flow characteristics that are hard to quantify and therefore many approaches have been developed.
3. In a similar fashion, the particle velocity depends on the fluid, particle, flow characteristics and even fall velocity.
4. There are different approaches to sediment rate, based on shear stress, velocity, shear velocity, and sediment concentration.

Sediment transport in a river depends on the dynamics not only in the river but also on the wash load from the upstream fields and therefore it can be a function of precipitation, geography, geology, surface cover, vegetation, soil type, climate, soil and fluid characteristics, and flow dynamics. Therefore, there are more ways to go, and there are many fields that can be the topics of scientific research.

References

Ackers, P., White, W.R., 1973. Sediment transport: new approach and analysis. J. Hydraul. Div. ASCE 99 (11), 2041–2060.

Bagnold, R.A., 1966. An Approach to Sediment Transport Problem from General Physics. US Geological Survey Professional Paper 422-J, USA.

Bombar, G., Elci, S., Tayfur, G., Guney, S., Bor, A., 2011. Experimental and numerical investigation of bed-load transport under unsteady flows. J. Hydraul. Eng. 137 (10), 1276–1282.

Bor, A., 2008. Numerical Modeling of Unsteady and Non Equilibrium Sediment Transport in Rivers (M.Sc thesis). Department of Civil Engineering, Izmir Institute of Technology, Turkey.

Bridge, J.S., Dominic, D.F., 1984. Bed load grain velocities and sediment transport rates. Water Resour. Res. 20 (4), 476–490.

Brooks, N.H., 1963. Calculation of suspended sediment load discharge from velocity concentration parameters. In: Proceed. Federal Interagency Sedimentation Conference, US-DA, Publication Number 970, USA.

Cao, Z., Carling, P.A., 2003. On evolution of bed material waves in alluvial rivers. Earth Surf. Process. Landf. 28, 437–441.

Chang, F.M., Simons, D.B., Richardson, E.V., 1965. Total Bed Material Discharge in Alluvial Channels. USGS Water Supply Paper 1498-I, USA.

Chien, N., Wan, Z.H., 1999. Mechanics of Sediment Transport. ASCE Press, USA.

Ching, H.H., Cheng, C.P., 1964. Study of river bed degradation and aggradation by the method of characteristics. Chin. J. Hydraul. Eng. 5. 41 pp.

de Vries, M., 1965. Consideration about non-steady bed-load transport in open channels. In: Proc. XI Congress, IAHR, Vol. 3, paper 3.8, Leningrad, Russia.

de Vries, M., 1973. River bed variation—aggradation and degradation. In: International Seminar on Hydraulics of Alluvial Streams, IAHR, New Delhi (Also available as Delft Hydraulic Lab., Delft, 1973, Pub. No. 107.

de Vries, M., 1975. A morphological time-scale for rivers. In: Proceedings of the XVI Congress, IAHR, vol. 2, Sao Paulo, Brasil.

Dietrich, W.E., 1982. Settling velocity of natural particles. Water Resour. Res. 18 (6), 1615–1626.

DuBoys, M.P., 1879. Le rhone et les rivieres a lit affouillable. Ann. Ponts Chausses 5 (18), 141–195.

Einstein, H.A., 1942. Formula for the transportation of bed load. Trans. ASCE 107.

Einstein, H.A., 1950. The Bed-Load Function for Sediment Transportation in Open Channel Flows. USDA Soil Conservation Service, Technical Bulletin, No. 1026, USA.

Einstein, H.A., Chien, N., 1954. Second Approximation to the Solution of Suspended Load Theory. University of California, Institute of Engineering Research, No. 3, USA.

Guy, H.P., Simons, D.B., Richardson, E.V., 1966. Summary of Alluvial Channel Data from Flume Experiments, 1956-1961. U.S. Geological Survey Professional Paper, 462-I. 96 pp.

Haushild, W.L., Simons, D.B., Richardson, E.V., 1961. The Significance of the Fall Velocity and Effective Fall Dimatetr of Bed Materials. USGS Professional Papers, 424-D, USA.

Issakhov, A., Zhandaulet, Y., Nogaeva, A., 2018. Numerical simulation of dam break flow for various forms of the obstacle by VOF method. Int. J. Multiphase Flow 109, 191–206.

Kalinske, A.A., 1947. Movement of sediment as bed-load in rivers. Trans. AGU 28 (4).

Karim, M.F., Kennedy, J.F., 1990. Menu of coupled velocity and sediment discharge relations for rivers. J. Hydraul. Eng. 116 (8), 973–996.

Lai, C.-T., 1991. Modeling alluvial-channel flow by multimode characteristic method. J. Eng. Mech., 32–53.

Landau, L.D., Lifshitz, E.M., 1997. Fluid Mechanics. Translated by Sykes, J. B.; Reid, W. H. second ed. Butterworth Heinemann, ISBN: 0-7506-2767-0.

Lane, E.W., Kalinske, A.A., 1941. Engineering calculation of suspended sediment. Trans. AGU 20 (3), 603–607.

Langbein, W.B., Leopold, L.B., 1968. River channel Bars and Dunes—Theory of Kinematic Waves. U.S. Geological Survey Professional Paper, 422-L, USA. 20 pp.

Levy, I.I., 1957. River Dynamics. National Energy Resources Press, Moscow, Russia.

Lisle, T.E., Cui, Y.T., Parker, G., Pizzuto, J.E., Dodd, A.M., 2001. The dominance of dispersion in the evolution of bed material waves in gravel-bed rivers. Earth Surf. Process. Landf. 26, 1409–1420.

Lisle, T.E., Pizzuto, J.E., Ikeda, H., Iseya, F., Kodama, Y., 1997. Evolution of a sediment wave in an experimental channel. Water Resour. Res. 33, 1971–1981.

Mahmood, K., 1975. Mathematical modeling of morphological transients in sandbed canals. In: Proc. 16th IAHR Congress, vol. 2, Paper BB, pp. 57–64.

Marsooli, R., Wu, W., 2015. Three-dimensional numerical modeling of dam-break flows with sediment transport over movable beds. J. Hydraul. Eng. 141 (1), 04014066.

Meyer-Peter, E., Favre, H., Einstein, A., 1934. Neuere versuchssresultate uber den geschiebetrieb. Schweiz Bauzellung 103 (13) (in German).

Meyer-Peter, E., Muller, 1948. Formula for bed-load transport. In: Proceedings of International Association for Hydraulic Research, 2nd Meeting, Stockholm, Sweden.

Mohammadian, A., Tajrishi, M., Azad, F.L., 2004. Two-dimenrsional numerical simulation of flow and geo-morphological processes near headlands by using unstructured grid. Int. J. Sediment Res. 19 (4), 258–277.

Parker, G., Klingeman, P.C., McLean, D.G., 1982. Bed load and size distribution in paved gravel bed streams. J. Hydraul. Eng. ASCE 108 (HY4), 544–571.

Pianese, D., 1994. Comparison of different mathematical models for river dynamics analysis. In: International Workshop on Floods and Inundations related to Large Earth Movements-Trent, Italy, October 4–7, Paper No. 782. 24 pp.

Rottner, J., 1959. A formula for bed-load transportation. La Houille Blanche 14 (3), 285–307.

Rouse, H., 1937. Modern conceptions of the mechanics of turbulence. Trans. ASCE 102.

Rouse, H., 1938. Fluid Mechanics for Hydraulic Engineers. Hunter Rouse, Dover, New York, USA (Chapter XI).

Rubey, W.W., 1933. Settling velocities of gravel, sand and silt particles. Am. J. Sci. 25, 325–338.

Schoklitsch, A., 1934. Der geschiebetrieb und die geschiebefracht. Wasserwirtsch. Wassertech. 29 (4), 37–43.

Schoklitsch, A., 1949. 'Berechnung der geschiebefracht', Wasser und Energiewirischaft, No. 1. Wasserwirtsch. Wassertech. 29 (4), 37–43.

Sharmov, G.I., 1959. River Sedimentation. Hydrology and Morphology Press, Leningrad, Russia.

Shen, H.W., Hung, C.S., 1972. An engineering approach to total bed material load by regression analysis. In: Proceedings Sedimentation Symposium, pp. 1–17 (Chapter 14).

Shields, A., 1936. Application of Similarity Principles and Turbulence Research to Bed Laod Movement. California Institute of Technology, Pasadena (translated from German).

Simons, D.B., Senturk, F., 1977. Sediment Transport Technology. Water Resources Publications, Fort Collins, Colorado, USA.

Soni, J.P., 1981. Laboratory study of aggradation in alluvial channels. J. Hydrol. 49, 87–106.

Straub, L.G., 1935. Missouri River Report'. In: In_house Document 238, 73rd Congress, 2nd Session. Washington DC, USA, US Government Printing Office, p. 1135.

Tayfur, G., Singh, V.P., 2006. Kinematic wave model of bed profiles in alluvial channels. Water Resour. Res. 42 (6), W06414.

Tayfur, G., Singh, V.P., 2007. Kinematic wave model for transient bed profiles in alluvial channels under nonequilibrium conditions. Water Resour. Res. 43, W12412. https://doi.org/10.1029/2006WR005681.

Toffaleti, F.B., 1969. Definitive computations of sand discharge in rivers. J. Hydraul. Div. ASCE 95 (1), 225–246.

Velikanov, M.A., 1954. Gravitational theory of sediment transport. J. Sci. Sov. Union Geophys. 4 (in Russian).

Wathen, S.J., Hoey, T.B., 1998. Morphological controls on the downstream passage of a sediment wave in a gravel-bed stream. Earth Surf. Process. Landf. 23, 715–730.

Wu, W., Vieria, D.A., Wang, S.Y.S., 2004. One-dimensional numerical model for nonuniform sediment transport under unsteady flows in channel networks. J. Hydraul. Eng. 130 (9), 914–923.

Yang, C.T., 1972. Unit stream power and sediment trasnport. J. Hydraul. Div. ASCE 98 (10), 1805–1826.

Yang, C.T., 1979. Unit stream power equations for total load. J. Hydrol. 40, 123–138.

Yang, C.T., 1996. Sediment Transport Theory and Practice. McGraw-Hill, New York, USA.

Yang, C.T., Sayre, W.W., 1971. Stochastic model for sand deposition. J. Hydraul. Div. ASCE 97 (2), 265–288.

Chapter 18

Modeling approaches for simulating the processes of wetland ecosystems

Shahid Ahmad Dar[a], Sajad Ahmad Dar[b], Sami Ullah Bhat[a], Irfan Rashid[c], and Saeid Eslamian[d,e]

[a]*Department of Environmental Science, University of Kashmir, Srinagar, Jammu and Kashmir, India,* [b]*Department of Environmental Science, Uttarakhand Technical University, Uttarakhand, India,* [c]*Department of Geoinformatics, University of Kashmir, Srinagar, Jammu and Kashmir, India,* [d]*Department of Water Engineering, College of Agriculture, Isfahan University of Technology, Isfahan, Iran,* [e]*Center of Excellence for Risk Management and Natural Hazards, Isfahan University of Technology, Isfahan, Iran*

1. Introduction

Wetlands are complex biogeochemical systems that are currently recognized as eco-friendly systems for the treatment of contaminants due to their easy handling, low cost, and high-performance efficiency (Vymazal, 2002; Nyquist and Greger, 2009; Rai et al., 2013; Dar et al., 2021a). They have been widely used for the treatment of domestic, agricultural, landfill leachate, and industrial wastewaters (Yalcuk and Ugurlu, 2009; Corbella and Puigagut, 2018; Tan et al., 2019). The increase in human population, urbanization, and land system changes has increased the pollution of wetlands (Dar et al., 2020a). Despite the increasing applications and several research studies on wetlands all over the world, the mechanism of decontamination of contaminants and pollutants is still troublesome to simulate as the removal of contaminants in wetlands is ascribed to an assortment of physical, chemical, and biological transformations in the tissues of macrophytes (Yuan et al., 2020). The mechanisms of pollution decontamination become more difficult to describe due to the combined effects between plants, microorganisms, water, and porous media (Pálfy et al., 2017).

In recent years, wetland models have been regarded as a promising tool to intensify the knowledge about the various physical, biological, and chemical processes governing the health of wetlands (Garcia et al., 2010). This led to the heavy development and utilization of models in understanding ecosystem processes that translated into an increased number of research publications and policy briefs over a short period (Samsó et al., 2015; Defo et al., 2017). Furthermore, the large diversity of processes occurring among different types of wetlands has formed an extensive range of models based on the processes and types they anticipate to define. In the progressively diversified arena of works related to the biogeochemistry of wetland ecosystems, the main focus is to understand the chemical, physical, and biological processes and to identify their basic mechanisms (Reddy and DeLaune, 2008). The rationale behind the use of modeling approach is to better understand the process and perhaps to change the processes to get some sort of anticipated results (Rezanezhad et al., 2020). Nowadays, most of the research studies are concerned with the change in wetland processes resulted due to increasing human activities like the huge inflows of sediments and nutrients to wetlands and to know the remedial measures to cope up with these changes (Okhravi et al., 2017). For wetland authorities to find the best management practices for the conservation of wetlands and monitoring their effects (Rashid et al., 2022), various gears like the analytical models have been developed to best explain the biological, chemical, and physical processes in wetlands (Chouinard et al., 2014).

Nowadays various studies have been published to review the state of knowledge of wetland modeling (Langergraber, 2008). During the last few decades, there has been large-scale development of wetland models, but it is an uphill task to choose the most suitable model for every research question we are interested in. This is the outcome of the circumstances that the previous knowledge generally comprises of descriptive features of models, and no in-depth studies were made between different models. Additionally, during the last ten years, numerous related publications on wetland modeling have been published that remained untouched in the preceding reviews. Against this backdrop, an attempt is made to offer an elaborative look on wetland modeling by providing an in-depth analysis of these models.

Handbook of HydroInformatics. https://doi.org/10.1016/B978-0-12-821962-1.00026-X

2. Types of models

Employing our knowledge and understanding of processes of wetland ecosystems in a simulation model, whether deterministic or stochastic, synthesizes knowledge and understanding in a coherent and explicit form. For a better understanding of wetland ecosystem functions, the development of models is regarded as a promising tool in defining the process of biogeochemical transformations (Kumar and Zhao, 2011; Meng et al., 2014). This led to a large number of models being developed for simulating the processes of wetlands (Stein et al., 2006; Langergraber et al., 2009). The various wetland models are broadly categorized into 2 groups (i) black box (statistical/empirical), and (ii) process-based models.

2.1 Black box models

Black box models are built on several simple and innovative regression modeling approaches to envisage the discharge concentration of constructed wetlands. Black box models are based on a specific functioning of their hydraulic loading rates and inflow concentrations. The main characteristic and limitations of various black box models are described in Table 1. Based on inputs, several Black box models have found their applications in various research studies for designing and predicting the removal efficacy of wetland ecosystems (Toscano et al., 2009). Using Black box models, many researchers have testified satisfactory results. However, few studies have highlighted that although these tools can envisage outcomes based on certain input information with a high (70%) coefficient of variation, the major limitation with the Black box models being that they are incompetent to explain the inner processes of pollution removal (Defo et al., 2017).

2.1.1 Stochastic models

Stochastic models are the numerous methods of computing the changing associations of wetland processes and communities (Daniel et al., 2019). These models perform a significant part in explaining the natural processes of wetland biogeochemistry. Stochastic models are preferably used for the prediction of several possible results of various wetland processes weighted by their probability and likelihoods (Webster et al., 2009). The principle of the long-run relative frequency interpretation of probability is the main building block in modern stochastic modeling, made accurate and acceptable within the manifested structure by the law of large numbers. The recent method of stochastic modeling is to split up the explanation of possibility from any specific category of application. Stochastic models are based on a correlation between macrophytes

TABLE 1 An overview of Black box models used in wetlands.

Black box model	Main features	Limitations	References
Artificial Neural Networks	- Simulates the structural aspects of biological neural networks - Simulates the removal of nitrogen, suspended solids, heavy metals, BOD, and COD from wastewaters	Regression coefficient for removal of ammonia is low	Akratos et al. (2009)
First order models	- Suitable non-linear deterministic method is suitable for sizing the systems	Incompetent to simulate random events, variations in concentration and flow of inputs	Kadlec (2000); Rousseau et al. (2004)
Regression models	- Agreements with observed investigation of relations among inlet and outlet concentrations	Focusses on input and output information	Rousseau et al. (2004); Tang et al. (2009)
Monod models	- Characterizes zero order reactions for higher concentration and first-order for lower concentrations	Averts total degradation of contaminants	Langergraber and Šimůnek (2005)
Time dependent retardation models	- Simulates elimination rates decline during a time reference, contaminants that are biodegradable are eliminated fast, Highly efficient for removal of COD at various loadings and depths in wetlands	Involves tracer studies for calculating these elimination percentage constants	Shepherd et al. (2001)

and abiotic parameters and contain one or more variables (Baird and Mehta, 2011). Stochastic models are based on the rules or expert knowledge or may be regression models built on field observation and calculations.

The stochastic model Dose-Effect Model for terrestrial Nature (DEMNAT-2.1) has been used for the study of dryness of wetland ecosystems (van Ek et al., 2000). The DEMNAT-2.1 model has 3 units a geographical database of ecosystems, dose-effect function, and a nature estimation module (Claessen et al., 1994). The Stochastic modeling was used to illustrate its general utility, and role of hydrological inconsistency, bathymetry, groundwater, and hydro-climatic forcing of geographically remote wetland systems (Park et al., 2014).

The stochastic water balance equation for a wetland is described as

$$\frac{dW}{dt} = -\rho(W) + \xi_t \tag{1}$$

Here, $W(t)$ [L^3] is the volume of water deposited in the wetland, t [T] is time, and $\rho[W(t)]$ [L^3/T] is the deterministic function of loss of water.

Multivariate regression statistics permit modeling of multifaceted performance and interface between various physicochemical parameters of wetland ecosystems (Tomenko et al., 2007). Cluster plot and ordination procedures define the structural variability within the multivariate datasets by decreasing the dimensionality of several parameters to an underlying smaller number of parameters (Zhou et al., 2012). The clustering technique is normally used in case of bigger ecosystem characteristics in identifying basic ecosystem attributes such as types of habitats or damage by the influx of nutrients. Principal Component Analysis (PCA) is the most widely used approach belonging to this category (de Almeida et al., 2015). The uncorrelated variables created through PCA have been used to scrutinize the structural attributes of biological populations, insect assemblages, physical, and chemical characteristics in numerous wetland ecosystems.

2.1.2 Empirical statistical models

Empirical-statistical tools envisage the possibility of the existence of a final climax community of plants in response to the changing ecosystem. As compared to the process-based models, statistical models are less versatile as they link only 2 time-steps. An overview of statistical models in the Netherlands was given by Venterink and Wassen (1997), by comparing 6 statistical models to predict the response of plant communities to the changes in hydrogeological elements. Models based on expertise such as the DEMNAT-2 model uses different parameters such as moisture, pH, cover for determining the response of plant communities are not as robust as the regression models (ITORS and ICHORS) that work with continuous parameters (Witte et al., 1993). A spatially unambiguous logistic regression model based on hydrometeorological data defines the possibility of occurrence of plant communities along with a riparian ecosystem (Toner and Keddy, 1997). Forecasts of climax communities based on empirical-statistical models are far away from actual communities and very difficult to understand and this has been attributed to the lack of explanatory variables or due to mathematical errors (Ersten, 1998). In addition, several abiotic factors such as the behavior of a particular plant community during successional stages are not given consideration (Timmermann, 1999a) as a result the climax community in wetland ecosystems takes hundreds of years in an intermediate stage (Timmermann, 1999a,b).

2.2 Process-based models

An overview of process-based models used in the simulation of wetland processes is provided in Table 2. Process-based/ mechanistic models enhance our knowledge of ecosystem characteristics and processes (Meyer et al., 2015). Mechanistic models have been developed on the principle that basic information of biogeochemical processes, their mechanisms, and exchanges are easily available, and algebraic procedures are then employed to define the basic chemistry behind the biogeochemical transformations (Langergraber, 2011). These models are temporally and spatially unambiguous and permit modeling of the wetland ecosystem mechanisms over longer periods. The main constraint of mechanistic models is the huge data requirement like inputs to test, run, evaluate, and validation of the results of the model (Kadlec, 2003; Rousseau et al., 2004). Due to our incomplete knowledge of biogeochemical mechanisms, the deterministic method is mostly used through experimental procedures. Nowadays, the quasi-physically–based process models are widely in use to determine the transport of nutrients, flux of water, and conversions, often encompassing modest kinetics to define the conversion of nutrients (Flynn, 2001; Chavan and Dennett, 2008; Hantush et al., 2013; Langergraber, 2017). Further inspiring is to model biogeochemical processes tempted by microorganisms that are extremely active, changing, and ensue at different geographical scales. Numerous mechanical simulation eco-hydrological tools for wetland ecosystems have been developed such as wide-ranging Wetland-DNDC biogeochemical model, for assessment of biogeochemical cycles (carbon) and

TABLE 2 Software's and models for simulation of processes in wetland ecosystems.

Software	Sub-models linked to processes			Reference
	Hydraulic and hydrodynamic processes	Biokinetic process	Physicochemical process	
FITOVERT	• Richards/transport equations • Evapotranspiration • Clogging process • Surface flow	• C, N, O • Bacterial growth • Biomass description • Monod type	• Atmospheric oxygen transfer • Gas, particulates transport • Filtration	Brovelli et al. (2007); Giraldi et al. (2010)
PHREEQC	–	• Aerobic processes based on CW2D, • C, P, N • Monod type	• Sediment water interactions • Mineral and gas transfer	Claveau-Mallet et al. (2012)
PHWAT	• Darcy's equations • Clogging process	• C, N, O, P • Monod type (CW2D) • 3 functional bacterial groups • Bacterial growth, biomass description	• Transfer of atmospheric oxygen • Transport of gas, particle, and particulates • Chemical equilibrium pH, redox	Brovelli et al. (2009)
BIO_PORE	• Darcy's equations	• C, N, S, O • Monod type (CWM1) • 6 functional bacterial groups • Bacterial growth, biomass description	• Atmospheric oxygen transfer • Filtration • Transport of particulates components	Samsó and Garcia (2013)
STELLA	• Variably saturated settings	• Organics removal • Nutrient elimination • Microbial degradation	• Adsorption • Desorption	Ouyang et al. (2010)
AQUASIM-CWM1	• TIS with saturated water	• C, N, O, P, S • Monod type (CWM1) • 6 functional bacterial groups • Bacterial growth, biomass description (suspended)	• Atmospheric oxygen transfer • Gas transport • Adsorption/desorption	Llorens et al. (2011)
Dual-porosity model in HYDRUS-1D	• Richard's equation • Dual porosity)	–	• Atmospheric O_2 transfer • Gas transport • Sorption processes	Mornnavou et al. (2013)
Model				
Activated sludge model	–	• Biokinetic process	–	Henze et al. (2000)
CMW1-RETRASO model	Darcy's equations	• C, O, P, and N • Monod type (CWM1) • 6 functional bacterial groups • Biomass description (suspended)	• Atmospheric oxygen transfer • Gas, particles and particulates transport	Langergraber et al. (2009)

TABLE 2 Software's and models for simulation of processes in wetland ecosystems—cont'd

Software	Sub-models linked to processes			Reference
	Hydraulic and hydrodynamic processes	Biokinetic process	Physicochemical process	
CWM1 (Constructed Wetland Model No. 1)	No	• Biochemical transformations • Organic matter degradation • N and S in subsurface flow	• No	Langergraber et al. (2009)
Constructed wetland Two-dimensional (CW2D) model	–	• Biokinetic process	–	Langergraber and Šimůnek (2005)
2 D HYDRUS/ CWM1	• Richard's equations • Transpiration • Surface flow	• C, O, N, P • Monod type (CWM1) • 3 functional bacterial groups • Bacterial growth, biomass description	• Atmospheric oxygen transfer • Gas transport • Sorption processes	Pálfy and Langergraber (2014); Rizzo et al. (2014)
HYDRUS/CW2D (HYDRUS Wetland Module)	• Richards/ Transport equations • Evapotranspiration • Surface flow	• C, N, P, O • Monod type (CW2D) • 3 functional bacterial groups • Growth of bacteria, description of biomass	• Atmospheric oxygen transfer • Gas transport Sorption processes	Langergraber and Šimůnek (2005); Mornnavou et al. (2013)
Retraso Code Bright (RCB) model	• Fick's/Darcy's law	• Degradation of N, S, and Organic matter	• Diffusion • Advection • Transport of gaseous and inorganic dissolved substances • Dispersion	Ojeda et al. (2008)
Wang-Scholz-Model (COMSOL)	• Darcy's law • Clogging process	–	• Liquid-solid • Diffusion • Sedimentation	Sani et al. (2013)
RTD/GPS-X model	• TIS having inconstant H_2O content	• 12 species considered • COD, N (only soluble), Interaction with biofilm growth	–	Zeng et al. (2013)
RSF_Sim	• TIS having adjustable water content	• C, N, P • Sediment accumulation description	• Particles and particulates transport • Filtration/ Sedimentation • Sorption processes	Meyer et al. (2013)

hydrological processes (Zhang et al., 2002; Cui et al., 2005); the Ecological Dynamic Simulation Model (EDYS) that describes water, soil, nutrient dynamics amid numerous other processes (Childress et al., 2002); a geographically diverse ecological model casing diverse aquatic ecosystems that predicts energy flux, the composition of species, and patterns of biomass and the Everglades Landscape Model (ELM) for production of biomass, water-flux, transport of phosphorus, and decomposition (Voinov et al., 1998; Fitz et al., 2004). Simulation models related to watersheds model uplands and wetlands, thus are perfectly suitable to evaluate the nutrient inputs, effects of land-use management practices, and other impacts on wetlands. Watershed simulation models were used by (Rahman et al. (2016) and Evenson et al. (2018)) who applied the soil and water assessment tool (SWAT) to predict the transportation of nutrient materials across various categories of ecosystems. The watershed ecosystem nutrient dynamics-phosphorus (WEND-P) model by Aschmann et al. (1999); Cassell et al. (2001), and Kort et al. (2007), and the model MIKE SHE (Waseem et al., 2020) are deterministic, spatially disseminated, physical models for simulating various processes related to the hydrology. Several process-based simulation tools have been employed to define the detention time distribution (DTD) in wetlands like the plug-flow (PF) and Tanks-In-Series (TIS) model (Merriman et al., 2017). Despite describing the TIS mixing through the gamma-distribution function, the model doesn't describe the mixing process in wetland ecosystems. The TIS model rather adhering to the internal processing focuses only on the inputs and outputs. The gamma function of the TIS model is described as:

$$g(t) = \frac{1}{(N-1)!t_1} \left(\frac{t}{t_1}\right)^{N-1} \left(\exp^{\frac{t}{t_1}}\right) \tag{2}$$

where "t" is the detention time, "t_1" is the average detention time in one tank, and "N" is the number of tanks.

2.2.1 Geospatial models

Location explicit biogeochemical studies on wetlands have produced widespread information on discrete fragments and developments of the functioning of environments (Rebelo et al., 2010). Consideration as to how several factors cooperate as a total needs a complete viewpoint that reflects the distribution, and interface of all mechanisms of an ecosystem. Geographical Information System (GIS) offers the context to create physicochemical and biological information and scale up the data to ecosystem level (Ozesmi and Bauer, 2002). To advance wetland geospatial tools, it is important to map the fundamental three-dimensional inconsistency and dispersal of assets. Even though biogeochemical goods are estimated at several temporal and spatial scales, measurements are restricted to few samples at select sites to examine processes. Converting this process-based information into a spatial clear framework has been hindered by the intricacy of ecosystems, numerous nested levels of interrelated biogeochemical processes, and the absence of adequate numerical data. Nowadays, wetland biogeochemistry and geoscience have interconnected to explore features of wetlands in a geographically clear context (Rains et al., 2016). Geostatistics permits the enumeration of spatial autocorrelation forms of various physicochemical and biological parameters of wetlands and their geographical relations with other parameters of ecological landscapes.

2.2.2 Geostatistical techniques

Geostatistical analysis of wetland data removes many corresponding flaws and shortcomings compared to traditional statistical analysis (Setianto and Triandini, 2013). The information on the autocorrelation of geographical entities is imperative for the interpolation of biogeochemical parameters across the ecosystems.

Inverse distance weighted (IDW)

IDW has been widely used for producing seamless water quality and bathymetric maps of wetlands (Dar et al., 2021b,c,d). IDW interpolation is based on the assumption that the locations close to one another have more similarities and correlations as compared to points that are away from one another. In IDW interpolation, it is supposed significantly that the proportion of relationships and likenesses among different neighbors is proportionate to the remoteness, and is described by the function of the reverse distance of every location from neighboring locations. To foresee the value of an unsampled location, IDW assumes that every single measured location has an adjacent impact on the unmeasured point that decreases with distance from the measured points. The IDW algorithm is expressed as:

$$Z_0 = \frac{\sum_{i=1}^{n} z_i . d_i^{-N}}{\sum_{i=1}^{n} d_i^{-N}} \tag{3}$$

where "Z_0" is the assessment value of parameter z at point i, "z_i" is the value of the sample at point i, "d_i" is the distance of an estimated point to sample point, "n" is the coefficient determining the value based on a distance, and "N" is the total number of estimates for each endorsement case.

Kriging

Kriging is an interpolation technique widely used in wetland modeling. Kriging and its variants such as kriging with external drift, regression kriging, and co-kriging is a subjective geostatistical technique that accounts for the 3-D edifice and inconsistency of wetlands, and has been used in various studies of wetlands (Van Horssen et al., 1999). Based on the prejudiced mean of neighboring observed locations in the sampling area, Kriging delivers an estimation at an unsampled point of parameter z. An estimate of the weighted average given by Kriging is given as

$$Z(s_0) = \sum_{i=1}^{n} \lambda_i z(si) \tag{4}$$

where λ are the values allotted to detected samples. These values amount to unity so that the analyst delivers an impartial estimate:

$$\sum_{i=1}^{n} \lambda_i = 1 \tag{5}$$

The values are estimated from the matrix as

$$C = A^{-1} \times b \tag{6}$$

where "A" is the medium of semi variances between numbers, "c" is the resulting value, and "b" is the path of assessed discrepancies among various value locations and the locations at which the variable z is to be assessed.

Natural neighbor interpolation (NNI)

Natural Neighbor has been widely used in the modeling of wetland ecosystems with irregularly spaced and scattered data (Dar et al., 2020b). NNI is an average-weighted technique developed by Robin Sibson (Sibson, 1981). NNI is used for the estimation of the value of unknown points. The locations used to calculate the value of a variable at point x are the accepted neighbors of x, and the value of neighbor is equal to the natural neighbor coordinate of x concerning this neighbor. If we consider that each data point in S has an attribute a_i (a scalar value), the natural neighbor interpolation is

$$f(x) = \sum_{i=1}^{n} w_i \times a_i \tag{7}$$

where $f(x)$ is the intercalated function value at point x.

3. Discussion

To provide a comprehensive description of the pollution degradation process, various mathematical models based on numerous treatments were developed. The main characteristics and limitations of some process-based models are provided in Table 3. A 2D mechanistic model to define the main biological and chemical processes linked to the transformation of nutrients and organic compounds in wetlands was developed (Ojeda et al., 2008). Significant involvement in mechanistic/process-based models has been the development of CWM1 and CW2D tools that are built on the mathematical formulation for the activated sludge model, these models have been used to describe the degradation mechanisms of nutrients, organic compounds, and other contaminants in wetlands (Langergraber et al., 2009). The process-based models generally employ numerous equations or additional methods to define processes in wetlands such as biokinetic, diffusion in gases, the saturated flow of water, and transformation of pollutants (Garcia et al., 2010). Richard's equation, Fick's law, and Darcy's law

TABLE 3 Evaluation of mechanistic models for wetland processes.

Model	Main features	Limitations	References
Activated Sludge Model (ASM)	Simulation of nitrogen, phosphorus, and chemical oxygen demand Elimination built on Monod kinetics	Location specific and appropriate only for the conditions for which developed	Henze et al. (2000)
CW2D model	Subsurface flow is vertical Transformation of C, N, and P	Appropriate only for dissolved materials	Langergraber and Šimůnek (2005)
CWM1 (Constructed Wetland Model No.1)	Degradation processes for organic matter, N, and S Simulation of porous media Transportation of suspended solids	High cost Computing time lengthy	Langergraber et al. (2009)
2 D HYDRUS/ CWM1	Saturated and unsaturated conditions Elimination of N, S, and COD Adsorption of ammonium	Few organic pollutants	Rizzo et al. (2014)
RCB flow model	Simulates S, N, and organic matter	–	Ojeda et al. (2008)
Wang-Scholz-Model (COMSOL)	Uniform and vertical flow Clogging process	No biochemical model	Sani et al. (2013)
Dual-porosity model (DPM) in HYDRUS-1D	Unreliably saturated settings	Appropriate only for nonreactive tracer transportation	Mornnavou et al. (2013)
CMW1-RETRASO model	Nineteen reactions instead of seventeen as in CWM1 Transfer of oxygen Development of biofilms and clogging, plant uptake	Growth of bacterial populations is not involved	Langergraber et al. (2009)
RTD/GPS-X model	TIS, removal of soluble nitrogen and COD, Growth of biofilms, Kinetic and hydraulic modeling on fixed bed aerated bioreactors	Verified for very few numbers of contaminants	Zeng et al. (2013)

were used in hydrological models to define the horizontal subsurface and vertical subsurface flow over a diversified multi-dimensional and allotropic permeable substrate in wetlands (Samsó and Garcia, 2013). The sub-model reactive transport built on the concept of convection dispersal was created for simulation of particulates and passage of solutes and degradation of pollutants in wetland ecosystems (Llorens et al., 2011; Samsó et al., 2015).

Statistical tools include arbitrary forcing (i.e., variation or random error) and their purpose is to recognize the functional relationships. Statistical models sometimes are used in the simulation of a deterministic phenomenon that in practice can't be exactly detected or exhibited (Chase and Myers, 2011). In wetland science, to link physical, chemical, and biological elements in wetland processes, various stochastic tools, regression, correlation, and analysis of variance have been used. Recent procedures have permitted to assembly of wide-ranging information and data on numerous biogeochemical parameters that are supposed to be related. These datasets are accompanied by solid information on additional ecosystem characteristics like depth analysis, water level fluctuations, and vegetation analysis through high-resolution remote sensing.

4. Modeling of emerging contaminants

The emerging contaminants such as microplastics, perfluoroalkyl constituents, pesticides, fertilizers, and antibiotics are believed to be a hot topic in modeling present and future generation and treatment of wastewaters (Ji et al., 2020). Nowadays, emerging pollutants are viewed as a global environmental problem due to their difficulty in removal from wastewaters, and most of them have been listed in the European Union legislation (Gorito et al., 2017). Presently the wetlands are being increasingly used in the treatment processes of emerging pollutants. In a study, emerging pollutants were significantly removed in constructed wetlands with Ciprofloxacin achieving a removal efficiency of 93.9% and Bisphenol-A achieving a removal efficiency of 76.2% (Christofilopoulos et al., 2019). However, because of the different properties of emerging pollutants, different elimination processes are preferred and this also has been a burning issue in this field (Chen et al., 2019). Therefore, the emerging contaminants should be given thorough consideration in the forthcoming course of the development of wetland models. Hence, for future studies, the advance in wetland modeling must include (i) innovative monitoring techniques or computation methods to quantitatively govern the variables in wetland models, (ii) correction and adjustment of errors produced due to supposition, (iii) simplifying the generalized model, and (iv) modeling of emerging contaminants in the wetland ecosystems.

5. Conclusions

Wetland modeling has emerged as a promising tool for simulating and describing the processes of degradation and transformations of contaminants in wetland ecosystems. Simulation models have provided better insights into the Black box and have increased the knowledge and understanding of multifaceted processes in wetlands. During the previous decades, various wetland-based mechanistic/process-based, stochastic, and geospatial models have been developed to predict the mechanisms of degradation and transformations of contaminants. As an efficient investigation tool, wetland models can largely decrease the period and workload of experimentation research. Previous decades have seen a continuous rise in the development of wetland simulation models for the depiction of biogeochemical processes governing the fate of contaminants. Almost all the process-based, stochastic, and geospatial models have attained best-fit simulation results, however, there is a long way to go before simulation tools can direct the plan in constructed wetlands. The full understanding of the decontamination mechanisms and transformations of pollutants in wetlands demands the further development of simulation models. A single or a compound model joining few numbers of sub-models can't completely depict the decontamination processes in wetlands. In addition, an inclusive model together with all sub-models of present understanding includes several factors most of which are interactive and can't be determined in quantitative terms; therefore, it makes the model complicated and interactions in a diffuse way. Therefore, for future studies, the advance in wetland modeling must include innovative monitoring techniques, computation methods, adjustment of simulation errors, simplification of generalized models, and modeling of emerging contaminants.

References

Akratos, C.S., Papaspyros, J.N., Tsihrintzis, V.A., 2009. Total nitrogen and ammonia removal prediction in horizontal subsurface flow constructed wetlands: use of artificial neural networks and development of a design equation. Bioresour. Technol. 100 (2), 586–596.

Aschmann, S.G., Anderson, D.P., Croft, R.J., Cassell, E.A., 1999. Using a watershed nutrient dynamics model, WEND, to address watershed-scale nutrient management challenges. J. Soil Water Conserv. 54 (4), 630–635.

Baird, D., Mehta, A.J., 2011. Estuarine and coastal ecosystem modeling. In: Wolanski, E., McLusky, D.S. (Eds.), Treatise on Estuarine and Coastal Science. Academic Press.

Brovelli, A., Baechler, S., Rossi, L., Langergraber, G., Barry, D.A., 2007. Coupled flow and hydro-geochemical modelling for design and optimization of horizontal flow constructed wetlands. In: Mander, U., Koiv, M., Vohla, C. (Eds.), Second International Symposium on "Wetland Pollutant Dynamics and Control WETPOL, pp. 393–395.

Brovelli, A., Malaguerra, F., Barry, D.A., 2009. Bioclogging in porous media: model development and sensitivity to initial conditions. Environ. Model Softw. 24 (5), 611–626.

Cassell, E.A., Kort, R.L., Meals, D.W., Aschmann, S.G., Dorioz, J.M., Anderson, D.P., 2001. Dynamic phosphorus mass balance modeling of large watersheds: long-term implications of management strategies. Water Sci. Technol. 43 (5), 153–162.

Chase, J.M., Myers, J.A., 2011. Disentangling the importance of ecological niches from stochastic processes across scales. Philos. Trans. R. Soc. B: Biol. Sci. 366 (1576), 2351–2363.

Chavan, P.V., Dennett, K.E., 2008. Wetland simulation model for nitrogen, phosphorus, and sediments retention in constructed wetlands. Water Air Soil Pollut. 187 (1), 109–118.

Chen, J., Liu, Y.S., Deng, W.J., Ying, G.G., 2019. Removal of steroid hormones and biocides from rural wastewater by an integrated constructed wetland. Sci. Total Environ. 660, 358–365.

Childress, W.M., Coldren, C.L., McLendon, T., 2002. Applying a complex, general ecosystem model (EDYS) in large-scale land management. Ecol. Model. 153 (1–2), 97–108.

Chouinard, A., Balch, G.C., Wootton, B.C., Jørgensen, S.E., Anderson, B.C., 2014. SubWet 2.0. Modeling the performance of treatment wetlands. Dev. Environ. Modell. 26, 519–537.

Christofilopoulos, S., Kaliakatsos, A., Triantafyllou, K., Gounaki, I., Venieri, D., Kalogerakis, N., 2019. Evaluation of a constructed wetland for wastewater treatment: addressing emerging organic contaminants and antibiotic resistant bacteria. New Biotechnol. 52, 94–103.

Claessen, F.A., Klijn, F., Witte, J.F.P., Nienhuis, J.G., 1994. Ecosystem classification and hydro-ecological modelling for national water management. In: Ecosystem Classification for Environmental Management. Springer, Dordrecht, The Netherlands, pp. 199–222.

Claveau-Mallet, D., Wallace, S., Comeau, Y., 2012. Model of phosphorus precipitation and crystal formation in electric arc furnace steel slag filters. Environ. Sci. Technol. 46 (3), 1465–1470.

Corbella, C., Puigagut, J., 2018. Improving domestic wastewater treatment efficiency with constructed wetland microbial fuel cells: Influence of anode material and external resistance. Sci. Total Environ. 631, 1406–1414.

Cui, J., Li, C., Sun, G., Trettin, C., 2005. Linkage of MIKE SHE to Wetland-DNDC for carbon budgeting and anaerobic biogeochemistry simulation. Biogeochemistry 72 (2), 147–167.

Daniel, J., Gleason, J.E., Cottenie, K., Rooney, R.C., 2019. Stochastic and deterministic processes drive wetland community assembly across a gradient of environmental filtering. Oikos 128 (8), 1158–1169.

Dar, S.A., Bhat, S.U., Aneaus, S., Rashid, I., 2020b. A geospatial approach for limnological characterization of Nigeen Lake, Kashmir Himalaya. Environ. Monit. Assess. 192 (2), 1–18.

Dar, S.A., Bhat, S.U., Rashid, I., 2021a. The status of current knowledge, distribution, and conservation challenges of wetland ecosystems in Kashmir Himalaya, India. In: Sharma, S., Singh, P. (Eds.), Wetlands Conservation. Wiley-Blackwell.

Dar, S.A., Bhat, S.U., Rashid, I., 2021d. Landscape transformations, morphometry, and trophic status of Anchar wetland in Kashmir Himalaya: implications for urban wetland management. Water Air Soil Pollut. 232 (462), 1–19. https://doi.org/10.1007/s11270-021-05416-5.

Dar, S.A., Bhat, S.U., Rashid, I., Dar, S.A., 2020a. Current status of Wetlands in Srinagar City: threats, management strategies, and future perspectives. Front. Environ. Sci. 7, 1–11.

Dar, S.A., Rashid, I., Bhat, S.U., 2021b. Linking land system changes (1980–2017) with the trophic status of an urban wetland: implications for Wetland Management. Environ. Monit. Assess. 193 (710), 1–17.

Dar, S.A., Rashid, I., Bhat, S.U., 2021c. Land system transformations govern the trophic status of an urban wetland ecosystem: Perspectives from remote sensing and water quality analysis. Land Degrad. Dev. 32 (14), 4087–4104.

de Almeida, T.I.R., Penatti, N.C., Ferreira, L.G., Arantes, A.E., do Amaral, C.H., 2015. Principal component analysis applied to a time series of MODIS images: the spatio-temporal variability of the Pantanal wetland, Brazil. Wetlands Ecol. Manage. 23 (4), 737–748.

Defo, C., Kaur, R., Bharadwaj, A., Lal, K., Kumar, P., 2017. Modelling approaches for simulating wetland pollutant dynamics. Crit. Rev. Environ. Sci. Technol. 47 (15), 1371–1408.

Ersten, A.C.D., 1998. Ecohydrological Response Modelling. PhD Thesis, University of Utrecht, The Netherlands. 143 pp.

Evenson, G.R., Jones, C.N., McLaughlin, D.L., Golden, H.E., Lane, C.R., DeVries, B., Alexander, L.C., Lang, M.W., McCarty, G.W., Sharifi, A., 2018. A watershed-scale model for depressional wetland-rich landscapes. J. Hydrol. X 1, 100002.

Fitz, C., Sklar, F., Waring, T., Voinov, A., Costanza, R., Maxwell, T., 2004. Development and application of the everglades landscape model. In: Costanza, R., Voinov, A. (Eds.), Landscape Simulation Modeling. Springer, New York, NY, USA, pp. 143–171.

Flynn, K.J., 2001. A mechanistic model for describing dynamic multi-nutrient, light, temperature interactions in phytoplankton. J. Plankton Res. 23 (9), 977–997.

Garcia, J., Rousseau, D.P., Morato, J., Lesage, E.L.S., Matamoros, V., Bayona, J.M., 2010. Contaminant removal processes in subsurface-flow constructed wetlands: a review. Crit. Rev. Environ. Sci. Technol. 40 (7), 561–661.

Giraldi, D., de Michieli Vitturi, M., Iannelli, R., 2010. FITOVERT: a dynamic numerical model of subsurface vertical flow constructed wetlands. Environ. Model. Softw. 25 (5), 633–640.

Gorito, A.M., Ribeiro, A.R., Almeida, C.M.R., Silva, A.M., 2017. A review on the application of constructed wetlands for the removal of priority substances and contaminants of emerging concern listed in recently launched EU legislation. Environ. Pollut. 227, 428–443.

Hantush, M.M., Kalin, L., Isik, S., Yucekaya, A., 2013. Nutrient dynamics in flooded wetlands. I: Model development. J. Hydrol. Eng. 18 (12), 1709–1723.

Henze, M., Gujer, W., Mino, T., van Loosdrecht, M.C., 2000. Activated Sludge Models ASM1, ASM2, ASM2d and ASM3. IWA Publishing.

Ji, B., Kang, P., Wei, T., Zhao, Y., 2020. Challenges of aqueous per-and polyfluoroalkyl substances (PFASs) and their foreseeable removal strategies. Chemosphere 250, 126316.

Kadlec, R.H., 2000. The inadequacy of first-order treatment wetland models. Ecol. Eng. 15 (1–2), 105–119.

Kadlec, R.H., 2003. Effects of pollutant speciation in treatment wetlands design. Ecol. Eng. 20 (1), 1–16.

Kort, R.L., Cassell, E.A., Aschmann, S.G., 2007. Watershed ecosystem nutrient dynamics-phosphorus (WEND-P models). In: Radcliffe, D.E., Cabrera, M.L.(Eds.), Modeling Phosphorus in the Environment. CRC Press, USA, pp. 261–276.

Kumar, J.L.G., Zhao, Y.Q., 2011. A review on numerous modeling approaches for effective, economical and ecological treatment wetlands. J. Environ. Manag. 92 (3), 400–406.

Langergraber, G., 2008. Modeling of processes in subsurface flow constructed wetlands: a review. Vadose Zone J. 7 (2), 830–842.

Langergraber, G., 2011. Numerical modelling: a tool for better constructed wetland design? Water Sci. Technol. 64 (1), 14–21.

Langergraber, G., 2017. Applying process-based models for subsurface flow treatment wetlands: recent Developments and Challenges. Water 9 (1), 1–5.

Langergraber, G., Rousseau, D.P., García, J., Mena, J., 2009. CWM1: a general model to describe biokinetic processes in subsurface flow constructed wetlands. Water Sci. Technol. 59 (9), 1687–1697.

Langergraber, G., Šimůnek, J., 2005. Modeling variably saturated water flow and multicomponent reactive transport in constructed wetlands. Vadose Zone J. 4 (4), 924–938.

Llorens, E., Saaltink, M.W., Poch, M., García, J., 2011. Bacterial transformation and biodegradation processes simulation in horizontal subsurface flow constructed wetlands using CWM1-RETRASO. Bioresour. Technol. 102 (2), 928–936.

Meng, P., Pei, H., Hu, W., Shao, Y., Li, Z., 2014. How to increase microbial degradation in constructed wetlands: influencing factors and improvement measures. Bioresour. Technol. 157, 316–326.

Merriman, L.S., Hathaway, J.M., Burchell, M.R., Hunt, W.F., 2017. Adapting the relaxed tanks-in-series model for stormwater wetland water quality performance. Water 9 (9), 691.

Meyer, D., Chazarenc, F., Claveau-Mallet, D., Dittmer, U., Forquet, N., Molle, P., Morvannou, A., Pálfy, T., Petitjean, A., Rizzo, A., Campà, R.S., 2015. Modelling constructed wetlands: Scopes and aims – a comparative review. Ecol. Eng. 80, 205–213.

Meyer, D., Molle, P., Esser, D., Troesch, S., Masi, F., Dittmer, U., 2013. Constructed wetlands for combined sewer overflow treatment—Comparison of German, French and Italian approaches. Water 5 (1), 1–12.

Mornnavou, A., Forquet, N., Vanclooster, M., Molle, P., 2013. Which hydraulic model to use in vertical flow constructed wetlands. In: Simunek, J., Kodesova, R. (Eds.), Proceedings of the 4th International Conference HYDRUS Software Applications to Subsurface Flow and Contaminant Transport, Czech University of Life Sciences, Prague, Czech Republic.

Nyquist, J., Greger, M., 2009. A field study of constructed wetlands for preventing and treating acid mine drainage. Ecol. Eng. 35 (5), 630–642.

Ojeda, E., Caldentey, J., Saaltink, M.W., García, J., 2008. Evaluation of relative importance of different microbial reactions on organic matter removal in horizontal subsurface-flow constructed wetlands using a 2D simulation model. Ecol. Eng. 34 (1), 65–75.

Okhravi, S.S., Eslamian, S.S., Fathianpour, N., 2017. Assessing the effects of flow distribution on the internal hydraulic behaviour of a constructed horizontal sub-surface flow wetland using a numerical model and a tracer study. Ecohydrol. Hydrobiol. 17 (4), 264–273.

Ouyang, Y., Zhang, J.E., Lin, D., Liu, G.D., 2010. A STELLA model for the estimation of atrazine runoff, leaching, adsorption, and degradation from an agricultural land. J. Soils Sediments 10 (2), 263–271.

Ozesmi, S.L., Bauer, M.E., 2002. Satellite remote sensing of wetlands. Wetl. Ecol. Manag. 10 (5), 381–402.

Pálfy, T.G., Gourdon, R., Meyer, D., Troesch, S., Molle, P., 2017. Model-based optimization of constructed wetlands treating combined sewer overflow. Ecol. Eng. 101, 261–267.

Pálfy, T.G., Langergraber, G., 2014. The verification of the Constructed Wetland Model No. 1 implementation in HYDRUS using column experiment data. Ecol. Eng. 68, 105–115.

Park, J., Botter, G., Jawitz, J.W., Rao, P.S.C., 2014. Stochastic modeling of hydrologic variability of geographically isolated wetlands: Effects of hydro-climatic forcing and wetland bathymetry. Adv. Water Resour. 69, 38–48.

Rahman, M.M., Thompson, J.R., Flower, R.J., 2016. An enhanced SWAT wetland module to quantify hydraulic interactions between riparian depressional wetlands, rivers and aquifers. Environ. Model Softw. 84, 263–289.

Rai, U.N., Tripathi, R.D., Singh, N.K., Upadhyay, A.K., Dwivedi, S., Shukla, M.K., Mallick, S., Nautiyal, C.S., 2013. Constructed wetland as an eco-technological tool for pollution treatment for conservation of Ganga river. Bioresour. Technol. 148, 535–541.

Rains, M.C., Leibowitz, S.G., Cohen, M.J., Creed, I.F., Golden, H.E., Jawitz, J.W., Kalla, P., Lane, C.R., Lang, M.W., McLaughlin, D.L., 2016. Geographically isolated wetlands are part of the hydrological landscape. Hydrol. Process. 30, 153–160. https://doi.org/10.1002/hyp.10610.

Rashid, I., Dar, S.A., Bhat, S.U., 2022. Modelling the hydrological response to urban land-use changes in three wetland catchments of the Western Himalayan Region. Wetlands 42 (64). https://doi.org/10.1007/s13157-022-01593-z.

Rebelo, L.M., McCartney, M.P., Finlayson, C.M., 2010. Wetlands of Sub-Saharan Africa: distribution and contribution of agriculture to livelihoods. Wetl. Ecol. Manag. 18 (5), 557–572.

Reddy, K.R., DeLaune, R.D., 2008. Biogeochemistry of Wetlands: Science and Applications. CRC Press, USA.

Rezanezhad, F., McCarter, C.P.R., Lennartz, B., 2020. Editorial: wetland biogeochemistry: response to environmental change. Front. Environ. Sci. 8 (55), 1–3. https://doi.org/10.3389/fenvs.2020.00055.

Rizzo, A., Langergraber, G., Galvão, A., Boano, F., Revelli, R., Ridolfi, L., 2014. Modelling the response of laboratory horizontal flow constructed wetlands to unsteady organic loads with HYDRUS-CWM1. Ecol. Eng. 68, 209–213.

Rousseau, D.P., Vanrolleghem, P.A., De Pauw, N., 2004. Model-based design of horizontal subsurface flow constructed treatment wetlands: a review. Water Res. 38 (6), 1484–1493.

Samsó, R., Garcia, J., 2013. BIO_PORE, a mathematical model to simulate biofilm growth and water quality improvement in porous media: application and calibration for constructed wetlands. Ecol. Eng. 54, 116–127.

Samsó, R., Meyer, D., García, J., 2015. Subsurface flow constructed wetland models: review and prospects. In: The Role of Natural and Constructed Wetlands in Nutrient Cycling and Retention on the Landscape, pp. 149–174.

Sani, A., Scholz, M., Babatunde, A., Wang, Y., 2013. Impact of water quality parameters on the clogging of vertical-flow constructed wetlands treating urban wastewater. Water Air Soil Pollut. 224 (3), 1–18.

Setianto, A., Triandini, T., 2013. Comparison of kriging and inverse distance weighted (IDW) interpolation methods in lineament extraction and analysis. J. Southeast Asian Appl. Geol. 5 (1), 21–29.

Shepherd, H.L., Tchobanoglous, G., Grismer, M.E., 2001. Time-dependent retardation model for chemical oxygen demand removal in a subsurface-flow constructed wetland for winery wastewater treatment. Water Environ. Res. 73 (5), 597–606.

Sibson, R., 1981. A brief description of natural neighbour interpolation. In: Barnett, V. (Ed.), Interpreting Multivariate Data. Wiley, New York, USA, pp. 21–36.

Stein, O.R., Biederman, J.A., Hook, P.B., Allen, W.C., 2006. Plant species and temperature effects on the k–C* first-order model for COD removal in batch-loaded SSF wetlands. Ecol. Eng. 26 (2), 100–112.

Tan, X., Yang, Y., Liu, Y., Li, X., Fan, X., Zhou, Z., Liu, C., Yin, W., 2019. Enhanced simultaneous organics and nutrients removal in tidal flow constructed wetland using activated alumina as substrate treating domestic wastewater. Bioresour. Technol. 280, 441–446.

Tang, X., Eke, P.E., Scholz, M., Huang, S., 2009. Processes impacting on benzene removal in vertical-flow constructed wetlands. Bioresour. Technol. 100 (1), 227–234.

Timmermann, T., 1999a. Anbau von Schilf (*Phragmites australis*) als ein Weg zur Sanierung von Niedermooren -Eine Fallstudie zu Etablierungs-methoden, Vegetationsentwicklung und Konsequenzen für die Praxis. Arch. Nat. Conserv. Landsc. Res. 38, 111–143.

Timmermann, T., 1999b. Sphagnum-Moore in Nordostbrandenburg: Stratigraphisch-hydrodynamische Typisierung und Vegetationswandel seit 1923. Dissertationes Botanicae 305, 1–175.

Tomenko, V., Ahmed, S., Popov, V., 2007. Modelling constructed wetland treatment system performance. Ecol. Model. 205 (3–4), 355–364.

Toner, M., Keddy, P., 1997. River hydrology and riparian wetlands: a predictive model for ecological assembly. Ecol. Appl. 7, 236–246.

Toscano, A., Langergraber, G., Consoli, S., Cirelli, G.L., 2009. Modelling pollutant removal in a pilot-scale two-stage subsurface flow constructed wet-lands. Ecol. Eng. 35 (2), 281–289.

van Ek, R., Witte, J.P.M., Runhaar, H., Klijn, F., 2000. Ecological effects of water management in the Netherlands: the model DEMNAT. Ecol. Eng. 16 (1), 127–141.

Van Horssen, P.W., Schot, P.P., Barendregt, A., 1999. A GIS-based plant prediction model for wetland ecosystems. Landsc. Ecol. 14 (3), 253–265.

Venterink, H.O., Wassen, M.J., 1997. A comparison of six models predicting vegetation response to hydrological habitat change. Ecol. Model. 101 (2–3), 347–361.

Voinov, A.A., Fitz, H.C., Costanza, R., 1998. Surface water flow in landscape models: 1. Everglades case study. Ecol. Model. 108 (1–3), 131–144.

Vymazal, J., 2002. The use of sub-surface constructed wetlands for wastewater treatment in the Czech Republic: 10 years experience. Ecol. Eng. 18 (5), 633–646.

Waseem, M., Kachholz, F., Klehr, W., Traenckner, J., 2020. Suitability of a coupled hydrologic and hydraulic model to simulate surface water and ground-water hydrology in a typical North-Eastern Germany lowland catchment. Appl. Sci. 10 (4), 1281.

Webster, K.L., Creed, I.F., Skowronski, M.D., Kaheil, Y.H., 2009. Comparison of the performance of statistical models that predict soil respiration from forests. Soil Sci. Soc. Am. J. 73 (4), 1157–1167.

Witte, J.P.M., Groen, C.L.G., van der Meijden, R., Nienhuis, J.G., 1993. DEMNAT: a national model for the effects of water management on the veg-etation. In: Hooghart, J.C., Posthumus, C.W.S. (Eds.), The Use of hydro-ecological Models in the Netherlands, Proceedings and Information: TNO Committee on Hydrological Research, No. 47, The Netherlands, Delft, pp. 31–51.

Yalcuk, A., Ugurlu, A., 2009. Comparison of horizontal and vertical constructed wetland systems for landfill leachate treatment. Bioresour. Technol. 100 (9), 2521–2526.

Yuan, C., Huang, T., Zhao, X., Zhao, Y., 2020. Numerical models of subsurface flow constructed wetlands: review and future development. Sustainability 12 (8), 3498.

Zeng, M., Soric, A., Ferrasse, J.H., Roche, N., 2013. Interpreting hydrodynamic behaviour by the model of stirred tanks in series with exchanged zones: preliminary study in lab-scale trickling filters. Environ. Technol. 34 (18), 2571–2578.

Zhang, Y., Li, C., Trettin, C.C., Li, H., Sun, G., 2002. An integrated model of soil, hydrology, and vegetation for carbon dynamics in wetland ecosystems. Glob. Biogeochem. Cycles 16 (4), 1061. https://doi.org/10.1029/2001GB001838.

Zhou, D., Luan, Z., Guo, X., Lou, Y., 2012. Spatial distribution patterns of wetland plants in relation to environmental gradient in the Honghe National Nature Reserve, Northeast China. J. Geogr. Sci. 22 (1), 57–70.

Chapter 19

Multivariate linear modeling for the application in the field of hydrological engineering

María C. Patino-Alonso[a], Jose-Luis Molina[b], and S. Zazo[b]

[a]IGA Research Group, Department of Statistics, University of Salamanca, Campus Miguel de Unamuno, Salamanca, Spain, [b]High Polytechnic School of Engineering, University of Salamanca, Ávila, Spain

1. Introduction

Researchers interested in studying change in behavior over time must use a longitudinal research design where behavior in the same subjects over time is examined. Those studies may be conducted over the short term or much longer duration. However, longitudinal research also has limitations, such as they are expensive and sometimes, all parts of the study are not completed. In addition, the results of longitudinal studies may be affected by repeated assessments.

Linear models are increasingly used for the analysis in a variety of scientific disciplines, including hydrology, ecology, economics, agronomy, medicine, etc. (Cook et al., 2017; Molina et al., 2021) and the work of many applied researchers in demonstrating the utility of the linear model, has resulted in the widespread use of these methods. The representation and measurement of change is both an interest and a fundamental concern to almost all scientific disciplines. Traditional approaches to studying change have used ANOVA and Multiple Regression. Linear regressions are designed to measure one specific type of relationship between variables: those that take linear form. Hydrological processes' variability is increasing on a global and local scale (Pfahl et al., 2017). Consequently, those events, with values and trends far from the historical average behavior, are more frequent (Chang et al., 2015). To address this new hydrological reality and aim to anticipate and forecast it, new approaches, methodologies, and techniques have emerged, incorporating this changing behavior. However, to build tools for the present future, not all the reasons that explain this increasing variability are brand new (climate change and projections to the future), but also, the historical behavior should be more deeply understood (Li et al., 2017; Molina and Zazo, 2018).

One of the ·aims of science is to describe and predict events in the world in which we live. Statistical data analysis includes a set of univariate and multivariate methods that allow being studied together, or in a group of individuals. The variables can be quantitative, qualitative, or a mixture of both. One way this is accomplished is by finding a formula or equation that relates quantities in the real world (Bazrkar et al., 2017). Statisticians often use linear models for data analysis and for developing new statistical methods. Linear models are fundamental to the practice of statistics. These are widely used as part of this learning process and they have been widely used in the field of hydrology (Tasker and Stedinger, 1989; Verdin et al., 2015).

When there are too many variables in sample factorial methods, some of the information collected may be excessive. In this case, multivariate methods of dimension in reduction, such as Principal Component Analysis (PCA, several quantitative variables), Correspondence Analysis (CA, two qualitative variables), Multiple Correspondence Analysis (MCA, many variables qualitative) can be applied. On other occasions, individuals may present certain common characteristics in their responses, which allow us to try to classify their groups of a certain homogeneity. Classification techniques or three-way methods (Decision Trees; Cluster Analysis; Discriminant Analysis, Biplot, Statis, etc. seek to analyze the relationships between variables to see if individuals can be separated into posterior clusters (Carrasco et al., 2019; González-Narváez et al., 2021). And finally, researchers want to know how an independent variable affects several distinct dependent variables. This includes techniques such as simple and multiple linear regression, nonlinear regression, logistic regression, simple and multiple analysis of variance, time series analysis, etc. Data tables from the design of the experimental studies with multiple response variables are becoming more common in hydrology, water resources, hydrogeochemical groundwater studies, among others (Molina et al., 2020; Tahroudi et al., 2020).

Handbook of HydroInformatics. https://doi.org/10.1016/B978-0-12-821962-1.00014-3

However, before carrying out the analyses described above, it is necessary to the stages of analysis. In the first stage, researchers use univariate analyses to understand structures and distributions of variables, as these can affect the choice of what types of models to create and how to interpret the output of those models. In the second stage, researchers use bivariate analyses to understand relationships between variable pairs that can sometimes indicate important findings.

This chapter aims to provide a conceptual overview of univariate and multivariate linear model statistical techniques of data analysis, to examine some of the most powerful modeling approaches, and to discuss issues related to data structures and model building. The fundamental tools examined are linear structures for modeling data.

2. General linear model

The classical general linear model (GLM) has been the most research setting. Cohen (Cohen, 1968) presented Multiple Linear Regression (MLR) as the univariate general linear model. This comprises just two variables and it is modeled as a linear relationship with an error term. Multivariate linear models may be represented using the general form:

$$Y = (y_1, \ldots, y_n)' = X\beta + e; e = (e_1, e_2, \ldots, e_n)' \tag{1}$$

where the dependent variable Y can be described as a linear combination of independent variables, y_n are independent random p-vectors, $X = (X_1, \ldots, X_n)$ matrix of known constants, $\beta = (\beta_1, \ldots, \beta_n)$ is a (r x **n**) matrix of unknown parameters and e_i are independent distributed random vectors $e \sim N(0, \sigma^2)$.

The GLM comprises highly flexible statistical techniques allowing to combine of categorical and continuous variables, which can be used for testing multiple hypotheses. Those represent a broad class of different statistical models that include linear regression, ANOVA for one dependent variable by one or more factors and/or variables, ANCOVA (with fixed effects only). In these cases, the GLM may predict one response variable from one or more other predictor variables. Fig. 1 shows a general outline of the models.

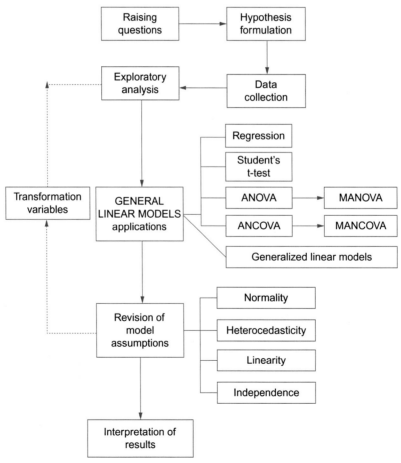

FIG. 1 Conceptual scheme for fitting a model (Cayuela, 2014).

GLM is long established in the statistical literature, provides a flexible and rigorous formal framework and, can deal with high levels of variability, such as those typically associated with landslides susceptibility mapping (Chandler and Wheater, 2002).

2.1 Simple linear regression

Linear regression is one of the most widely used statistical techniques. It is used to model the relationship between a response Y and one or more and one or several predictor variables X_i. For example, linear regression could use to test whether annual rainfall (the explanatory variable) is a good predictor of runoff (the response variable). Regression models attempt to model the relationship between variables by fitting a line to the observed data, while logistic and nonlinear regression models use a curved line. Regression allows you to estimate how the dependent variable changes as the independent variable(s) change.

Simple linear regression (SLR) is a statistical method that allows us to study and estimate relationships between two quantitative variables. SLR can use when the objective is to know how strong the relationship is between two variables (e.g., the relationship between rainfall and soil erosion).

For a linear relationship, the model has the following form:

$$Y = \beta_0 + \beta_1 X + e \tag{2}$$

where Y is regarded as the response, outcome, or dependent variable (it is the predicted value of the dependent variable (Y) for any given value of the independent variable X); β_0 is the intercept, the predicted value of Y when the X is 0; β_1 is the regression coefficient, is that how much we expect Y to change as X increases; X is regarded as the predictor, explanatory, or independent variable (the variable influencing the Y); e is the error of the estimate, or how much variation there is in our estimate of the regression coefficient.

Linear regression finds the line of the best fit line through your data by searching for the regression coefficient β_1 that minimizes the total error (e) of the model. SLR is a parametric test, meaning that it makes certain assumptions about the data. These assumptions are:

1. Homogeneity of variance (homoscedasticity). The equal variances assume that the variance of the residuals is the same for all values of X. The size of the error in the prediction does not change significantly across the values of the independent variable.
2. Independence of errors. Y is independent of errors.
3. Normality of errors. The residuals must be normally distributed.
4. Linearity. The relationship between the independent and dependent variable is linear: the line of best fit through the data points is straight (rather than a curve or others).

In many cases, hydrological models have some parameters, such as infiltration or runoff coefficient, among others, that need to be adjusted to obtain a satisfactory simulation performance, and they are usually related to specific basin characteristics. In this regard, extensive tests are typically performed between the sensitive parameters of the hydrological model and all available basin characteristics to analyze their possible relationships and then adopt a linear regression model to fit a reliable relation for each sensitive parameter, respectively (Huang et al., 2015).

2.2 Multiple linear regression model

Multiple linear regression (MLR) is simply an extended form of simple regression in which two or more independent variables are used. The response is usually influenced by more than one predictor variable. For example, Ewaid and colleagues were interested in modeling the most important parameters responsible for variation in water quality (Ewaid et al., 2018). They considered that the water quality index depends on the physicochemical. Patel et al. (2016) used the technique based on multiple linear regression for the determination of rainfall-runoff relationships because the use of this technique provided them with a fast and straightforward way to determine runoff.

The mathematical expression of the model can be expressed as:

$$Y = \beta_0 + \beta_1 X_1 + \ldots + \beta_i X_i + e = \beta_0 + \sum_{i=1}^{n} \beta_i X_i + e \tag{3}$$

MLR estimates Y as a linear function of predictors $X_1, ..., X_n$, where $\beta_0, ..., \beta_i$ are the linear coefficients for the MLR which are estimated here using the standard least square method. β_0 is the y-intercept, $\beta_1, \beta_2, ..., \beta_i$ are the slope coefficients of the first, second, and nth independent variables, and e is the error of the residuals.

The assumptions are essentially the same as those for simple linear regression, although there are also some important issues to be considered:

- A1.-Dependent variable should be measured on a continuous scale.
- A2.-There must be two or more independent variables, which can be either continuous (an interval or ratio variable) or categorical (an ordinal or nominal variable).
- A3.-Linear relationship. The model needs the relationship between the independent and dependent variables to be linear.
- A4.-Homocedasticity. The variance of Y should be constant and therefore does not depend on the X's.
- A5.-Normally distributed residuals: the residuals should follow a normal distribution. Those are the differences between the observed value of the dependent variable and the predicted value. Common methods to check this assumption are: (i) the histogram, (ii) a Normal P-P Plot, and (iii) a normal Q-Q Plot of the studentized residuals.
- A6.-No autocorrelation. Autocorrelation occurs when the error terms violate the ordinary least squares assumption that the error terms are uncorrelated (Uyanto, 2020). Durbin-Watson's d tests the null hypothesis that the residuals are not linearly auto-correlated. While d can assume values between 0 and 4, values around 2 indicate no autocorrelation. As a rule of thumb values of 1.5 are normal (Durbin and Watson, 1950).
- A7.-Outliers cases. It is important to consider the extreme cases that may influence the regression model.
- A8.-No or little multicollinearity. Multicollinearity exists when two or more of the independent variables are highly correlated with each other. It also suggests that the two variables may represent the same underlying factor, and this can cause problems when the results are interpreted. Multicollinearity is checked against four criteria:
  - **(a)** Correlation matrix: when computing the matrix of Pearson's Bivariate Correlation among all independent variables, the correlation coefficients need to be smaller than 0.08.
  - **(b)** Tolerance: it measures the influence of one independent variable on all other independent variables. Tolerance is defined as $= 1 - R_j^2$ ($j = 1, ..., k$. A tolerance of less than 0.20 could be contributing to multicollinearity problems (O'Brien, 2007).
  - **(c)** Variance Inflation Factor (VIF): it is defined as $VIF = 1 / T$. Similarly, with $VIF > 10$ there is an indication for multicollinearity to be present.
  - **(d)** Condition Index: the condition index is calculated using factor analysis on the independent variables. Values between 10 and 30 indicate mediocre multicollinearity in the regression variables, values >30 indicate strong multicollinearity.

If one or more assumptions do not hold, the estimators may be poor. The researcher has to try several procedures to obtain the best fit model as are to adding additional data, dropping variable(s), or transformation the variables. Finally, one of the great advantages of multiple regression models is that they allow for the inclusion of control variables. Control variables not only help researchers account for spurious relationships, but also measure the impact of any given variable above and beyond the effects of other variables.

Accurate prediction of runoff signatures is important for numerous hydrological and water resources applications. Zhang et al. (2018), for the first time, introduces regression tree ensemble approach and compares it with other three widely used approaches (multiple linear regression, multiple log-transformed linear regression, and hydrological modeling) for assessing prediction accuracy of 13 runoff characteristics or signatures which were grouped into three categories: low flow (e.g., zero flow ratio, daily flow at 10th and 50th percentile), high flow (e.g., mean daily flow, mean log-transformed daily flow, mean daily flow during winter and summer), and flow dynamics (e.g., coefficient of variation of daily streamflow, standard deviation of daily streamflow, interquartile range), using an extensive data set from 605 catchments across Australia.

2.2.1 Logarithmic transformations of variables

Data do not always come in a form that is suitable for analysis. It often becomes necessary to fit a linear regression model to the transformed rather than the original variables. There are two kinds of transformations: linear and nonlinear. A linear transformation preserves linear relationships between variables, whereas a nonlinear transformation changes a linear relationship. The most frequently used transformation is logarithmic.

Logarithmically transforming variables in a regression model are often recommended for skewed data and a nonlinear relationship exists between the independent and dependent variables. In these cases, using the logarithm of one or more variables makes the effective relationship nonlinear, while still preserving the linear model, it improves the fit of the model by transforming the distribution of the features to a more normally shaped bell curve. A regression model has the unit changes between the X and Y variables. Taking the log of one or both variables will effectively change the case from a unit change to a percent change.

The log-transformation is widely used in biomedical, social sciences, and hydrology research to deal with skewed data. There are several possible combinations of transformations involving logarithms:

1. Linear log-model. The only predictor variable is log-transformed

$$Y = \beta_0 + \beta \log (X_i) + e_i \tag{4}$$

The interpretation of the coefficient β is that a one-unit increase in $logX$, produces an expected increase Y of β units

2. Log-linear model. A linear relationship is hypothesized between a log-transformed outcome variable, and a group of predictor variables is that only the dependent/response variable is log-transformed. It is assumed that log-normal conditional on all the covariates

$$\log (Y) = \beta_0 + \beta X_i + e_i \tag{5}$$

The interpretation of the coefficient β is that a one-unit increase in X, produce an expected increase $logY$ of β (units)

3. Log-log model. Both response variables and predictor variable(s) are log-transformed

$$\log (Y) = \beta_0 + \beta \log (X_i) + e_i \tag{6}$$

2.2.2 Method of moments estimation in linear regression

The problem of fitting a straight line to bivariate (X,Y) data where the data are scattered about the line is a fundamental one in statistics. The usual way of fitting a line of best fit for a set of data is to use the principle of least squares. This line is called the regression line of Y on X. Another method of estimation that has been used in errors in variables regression is the method of moments.

The method of moments is a statistical technique for constructing point estimators of the parameters in a statistical model (Lindsay, 2014). The researchers rarely obtain information on an entire population therefore, they use a sample from the population to estimate population moments. Suppose that the observed sample is $\{y_i; i = 1, \ldots, n\}$ with mean μ (population average), where y is a random variable describing the population of interest. The population means can be written as E(y), is that the expected value or mean of y. The mean of y is called the first moment of y and the population variance Var (y) is defined as the second moment of y: $Var(y) = E((y - \mu)^2)$. The method of moments estimator of μ is just the sample average (Wooldridge, 2001).

The method of moments has been widely applied in soil science (Kerry and Oliver, 2007), process-based rainfall models (Kaczmarska et al., 2015), simulation of solute transport through heterogeneous networks (e.g., mixing of fresh and salt water in coastal aquifers) (Li et al., 2018), etc.

2.3 Analysis of variance (ANOVA) vs. analysis of covariance (ANCOVA)

Gosset (1908) developed the Student's t-test to compare the means of two small sets for quantitative data when simples are collected independently of one another of two experimental conditions. Whereas there are more than two conditions in an experiment is needed to compare all of the means. The problem of comparing more than two conditions when more than one Student's t-test is applied results from the increase in Type I error which occurs when the null hypothesis is rejecting but it is true. ANOVA is a statistical technique that has three or more levels for mean differences based on a continuous dependent variable. Therefore Student's t-test is used to compare the means between two groups, whereas analysis of variance (ANOVA) extends the two-group t-test to several groups. When one categorical is used, the method is called one-way ANOVA; for two categorical independent variables, it is called two-way ANOVA (Mishra et al., 2019).

ANOVA was developed by Fisher in the 1920's. ANOVA models have become widely used tools and play a fundamental role in much of the application of statistics. It is a technique for testing causality, which has been used in a variety of fields including, hydrology, biology data, for example for uncertainty assessment in climate change studies (Bosshard et al., 2013).

One-way ANOVA compares levels of a single factor based on the single continuous response variable. The null hypothesis tested by single-factor ANOVA is that two or more means are equal:

$$H_0 = u_1 = u_2 = \ldots = u_i \tag{7}$$

The research hypothesis is that the means are not all equal. The test statistic F for testing the null hypothesis assumes equal variability in the i populations, and it is defined as follows:

$$F = \frac{\sum (\overline{X_i} - \overline{X})n_i/(k-1)}{\sum \sum (X - \overline{X_i})/(N-k)} \tag{8}$$

where i represents the number of independent groups and N represents the total number of observations in the analysis (N does not refer to population size). n_i is the sample size in the ith group, $\overline{X_i}$ is the sample mean and $\overline{X}$ is the overall mean. The *F-statistic* is the ratio where the numerator captures between groups variability, is that for each data value, look at the difference between its group means and the overall mean. The denominator contains an estimate of the variability within the groups, it is searched for each data value, the difference between that value, and the mean of its group.

In ANOVA, the variable in question is often called the response or dependent variable. It is a univariate method that is applied to each variable separately, which involves a large number of tests and requires some form of multiple testing correction. When the null hypothesis H_0 is rejected after ANOVA, the researcher does not know how one group differs from a certain group, and they test pairwise differences. To this end, it is necessary to apply the multiple comparison test (MCT), called post hoc tests. The most common uses MCT include the following tests: Bonferroni, Tukey, Scheffee and Dunnett (Lee and Lee, 2018).

ANOVA requires the following assumptions:

- Random, independent sampling from the i groups.
- Normal population distributions for each of the i groups.
- Equal variances within the i groups (homoscedasticity). When the data do not meet this assumption, it may be able to use a nonparametric alternative.

Researchers use ANOVA to explain variation in the magnitude of a response variable of interest. For example, Zeigler and Whitledge (2011) used single-factor ANOVA to assess differences in individual water and otolith mean elemental concentrations among river segments.

An extension to the one-way ANOVA is the two-way ANOVA, also known as two-factor ANOVA. In this case, there are two independent variables which are called factors that affect the dependent variable. It compares levels of two or more factors for mean differences on a single continuous response variable. These factors can take different values known as levels, and the groups must have the same sample size. The following equation describes two factors ANOVA:

$$Y_{ijk} = \mu + \alpha_j + \beta_k + (\alpha\beta)_{jk} + e_{ijk} \tag{9}$$

In two-way ANOVA with interaction, the factors can again be considered fixed or random. If both factors have nonrandom levels, it is called a fixed-effects design. However, if the levels are chosen at random, then it is called a random effects design. A mixed-effects design is a combination of fixed and random-effects models.

ANCOVA is an extension of ANOVA. It compares the adjusted mean scores of two or more independent groups. The rationale underlying ANCOVA is that the effect of the independent variables on the dependent variable is revealed more accurately when the influence on the dependent variable represented by the covariate is equal across the experimental conditions (Cox and McCullagh, 1982). The covariates are uncontrolled variables that are used to reduce variance due to error from outside influences from error variance.

The expression of the simple analysis of covariance model is as follows:

$$Y = F(x_1, x_2, \ldots, x_n) \tag{10}$$

The dependent variable Y is metric and the independent variables some metric and nonmetric. The specific ANCOVA assumptions are:

- The slope of the line relating the dependent variable to the covariate does not differ across the different conditions in the experiment.
- The relationship between the covariate and the dependent variable is linear.

ANOVA and ANCOVA analysis have been used in numerous hydrological studies to test for changes in the rainfall-soil moisture relationship in the treated catchment (Page et al., 2015), to quantify the sources of uncertainty in an ensemble of hydrological climate-impact projections (Aryal et al., 2019), sensitivity analysis for water resource and environmental models (Wang et al., 2020). In the research carried out by Verma et al. (2018) about the role of large river flow events in annual load, they conducted an ANCOVA analysis to describe and quantify the relationships between large flow events and nutrient loadings.

Sensitivity analysis is an important component for modeling water resource and environmental processes. Analysis of Variance (ANOVA), has been widely used for global sensitivity analysis for various models (Wang et al., 2020).

2.4 Multivariate analysis of variance (MANOVA) vs. multivariate analysis of covariance (MANCOVA)

Empirical research in nearly every discipline is rarely confined to the study of a single response variable, a characteristic, or attribute, and it may be desirable to consider all the variables simultaneously. Data analysis methods used to conduct such analyses are in the general domain of multivariate statistical methods, such as MANOVA.

The MANOVA in multivariate GLM extends the ideas and methods of univariate by considering multiple continuous dependent variables. MANOVA is ANOVA in which the single response variable is replaced by several variables. The procedure groups these variables into a weighted linear combination.

A MANOVA is a statistical technique for the simultaneous analysis of several response variables measured on individual members of several groups of subjects. It is an extension of the ANOVA, that is, an ANOVA with two or more continuous response variables, which tests for the difference in two or more vectors of means (Hair et al., 1987). It is used to test the significance of the effects of one or more independent variables on two or more dependent to investigate the interrelationships between them.

The main objective in using MANOVA is to analyze if the dependent variables are altered by multiple levels of independent variables. If multiple ANOVA's were conducted independently might occur Type I errors so it is more advantageous to use MANOVA, and this may reveal differences not found by ANOVA tests. Therefore, when one independent variable affects many dependent variables, the multivariate test gives better results than separate univariate tests because it controls errors in the type-I and type-II sum of squares which may lead to false rejection of the hypothesis. It was developed for experiments where the number of observations (n) is larger than the number of variables (p), and it assumes that the observations in the experimental design are statistically independent. The null hypothesis (H_0) is defined:

$$\left.\begin{array}{c} u_{11} = u_{12} = \cdots = u_{1k} \\ u_{21} = u_{22} = \cdots = u_{2k} \\ . \\ . \\ . \\ u_{p1} = u_{p2} = \cdots = u_{pk} \end{array}\right\} \quad u_j = \begin{bmatrix} u_{1j} \\ u_{2j} \\ \vdots \\ u_{pj} \end{bmatrix} \Rightarrow H_0 : u_1 = u_2 = \cdots = u_k \tag{11}$$

where p is the number of dependent variables and u_{mj} is the mean on the variable for group j ($m = 1...p, j = 1...k$). The alternative hypothesis (H_1) is that at least one of the groups has a different mean.

MANOVA has been applied in various domains of science, e.g., hydrology, chemistry, ecology, etc. In a hydrological study, Shukla and Gedam (2018) performed the MANOVA analysis between spatial urbanization level and hydrological variables to study the influence of spatial urbanization on the hydrological components for comprehending the existing relationships between them.

MANCOVA is an extension of the analysis of covariance (ANCOVA) (Cohen, 2013). While a MANOVA can include only factors, MANCOVA controls a variable called the covariate, and one or more covariates are added to the mix. The objective of MANCOVA is to investigate whether the combination of dependent variables varies concerning the independent variable.

The hypothesis is to determine if there are statistically reliable mean differences that can be demonstrated among groups after adjusting the dependent variable for differences on one or more covariates (Brown, 2012).

The expression of the MANCOVA model is as follows:

$$G(y_1, y_2, ..., y_m) = F(x_1, x_2, ..., x_n) \tag{12}$$

The matrix of total variances and covariances are compared via multivariate F tests to determine any significant differences regarding all variables.

MANCOVA and MANOVA models have been used in several publications in the field of Hydrology, for example, to evaluate the principal factors and mechanisms governing the spatial variations, sensitivity, and response of water resources to climatic, land use, and environmental changes (Nathan et al., 2017; Shukla and Gedam, 2018; Vystavna et al., 2020) and environmental sciences (Zheng et al., 2018).

2.5 Overview of generalized linear models (GzLM)

The generalized linear model (GzLM) was first presented by Nelder and Wedderbum (1972) and developed by McCullagh and Nelder (1989). It is a systematic extension of familiar regression models such as the linear models for a continuous response variable given continuous and/or categorical predictors. They have been developed in a wide range of methods for analyzing data from nonnormal distributions, including linear, logistic, and Poisson regression (McCullagh, 1984).

GzLM can be also expressed as:

$$g(y) = X\beta + e; e \sim N(0, \sigma^2 I) \rightarrow y \sim N(X\beta, \sigma^2 I) \tag{13}$$

where $g(y)$ is the $n \times 1$ data vector, X is an $n \times p$ matrix of exploratory variables or a design matrix, β is a $p \times 1$ vector of coefficients. A link function $g(.)$ describes how the mean depends on the linear predictor and a variance function that describes how the variance depends on the mean. Therefore, a GLM specifies a nonlinear link function and variance function to allow, while maintaining the simple interpretation of the linear model.

GLMs extend standard linear regression models to encompass nonnormal response distributions and possibly nonlinear functions of the mean. They have three components (Agresti, 2015):

1. Random component. This specifies the response variable y and its probability distribution.
2. Linear predictor. The systematic component specifies the explanatory variables. For a parameter vector $\beta = (\beta_1, \beta_2, ..., \beta_p)$ and a $n \times p$ model matrix X that contains values of p explanatory variables for the n observations, the linear predictor is $X\beta$.
3. Link function. This is a function g applied to each component of $E(y)$ that relates it to the linear predictor:

$$g(E(y)) = X\beta \tag{14}$$

GzLM is merely a linear model when it uses the identity link function. Therefore, a GzLM with Identity Link Function is a "Linear Model. Ordinary linear models equate the linear predictor directly to the mean of a response variable y and assume constant variance for that response. The normal linear model also assumes normality. By contrast, a GzLM is an extension that equates the linear predictor to a link-function-transformed mean of y, and assumes a distribution for y that need not be normal but is in the exponential family (Agresti, 2015).

GzLM procedure has been applied in the field of hydrological forecasting due to its flexibility in modeling simulation of daily rainfall series based on atmospheric predictors and historical data, model and simulates rainfall data (Tosunoğlu and Regional, 2017). Moreover, GzLM is one of the models that have been widely used in project meteorological series, rainfall modeling (Kenabatho et al., 2012; Tosunoğlu and Regional, 2017).

3. Hybrid causal-multivariate linear modeling (H_C-MLM)

Hybrid Causal-Multivariate Linear Modeling (H_C-MLM) method was proposed by Molina et al. (2021). It describes the joint development of two different methods for temporal riveŕs runoff assessment. This is performed through a hybrid approach by means of Multivariate General Linear Models (MGLM; inspired by MLR as a statistical method), and Causal Reasoning (CR; as nonlinear ones).

Causality, addressed from the perspective of Causal Reasoning, is one of the innovative methods based on Artificial Intelligence (AI) (Molina et al., 2016) and is very useful for analyzing rivers runoff dynamics. This has addressed through a hybrid approach between traditional and novel methodologies by means of an ARMA model (a stochastic hydrological parametric model) and Bayesian Causal modeling (a powerful AI technique based on a Probabilistic Graphical Model). CR based on BNs presents several advantages into the framework in which this research is performed. Firstly, relations and connections of complex models can be automatically identified and characterized (Molina et al., 2010). This allows quantifying the variables' relationship strength (Molina et al., 2019; Zazo et al., 2020). Secondly, they enable a compact representation of the joint probability distribution over sets of random variables (Castelletti and Soncini-Sessa, 2007). According to Castelletti and Soncini-Sessa (2007), it is formally expressed as:

$$P(B|A) = \frac{P(A|B)P(B)}{P(A)} \tag{15}$$

where $P(B|A)$ is the conditional probability of B for a given state of variable A, $P(A|B)$ the other-way-round conditional probability, $P(B)$ the probability of B, and $P(A)$ the probability of A. And thirdly, they compute inference omnidirectionally (Molina and Zazo, 2017).

H-CMLM is mainly aimed to empower the analysis of temporal hydrological records behavior. This method comprises the integration and hybridization of two modules: CR and MLM Modeling. Both modules are integrated into a final interface (Fig. 2). For this, variables considered for both developments are equivalent, but they are treated from a different approach. In this sense, CR considers each variable as a probabilistic distribution while MLM comprises pure deterministic management of them.

H-CMLM was applied to three different Spanish basins (Adaja, Mijares, and Porma), which were chosen due to their disparate features. All these case studies are defined by gauging stations that are located upstream of the first regulating reservoirs, and therefore they comprise unregulated river stretches. For comparing temporal dependence propagation for the period 1945/46 to 2013/2014, multivariate general linear models were conducted in which each period was covaried. All periods 1945/46 to 2013/ 2014 were used as dependent variables. For each covariate assessment approach, covariates for each time period were adjusted. Sixty-seventh models were run to look at the overall relationship between the outcome and the covariate. Numerical results show a very high level of equivalence between the average value of temporal dependence provided by MLM module and the continuous behavior of temporal dependence computed by CR module and visualized through Dependence Mitigation Graph (DMG). This high coherent outcome from both modules makes the analysis much more robust from a stochastic hydrology point of view. The results, especially the average temporal dependence value, are very useful for the optimal dimensioning of hydraulic infrastructures like reservoirs.

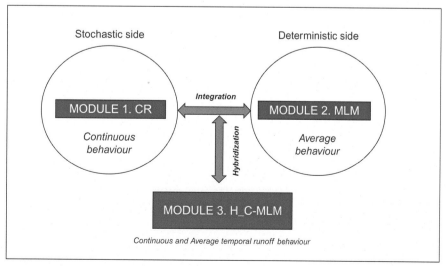

FIG. 2 Methodology for the building of H_C-MLM (Molina et al., 2021).

4. Conclusions

Like almost all fields of science, hydrology has benefited to a large extent from the tremendous improvements in statistical methods. The General Linear Model is a statistical theory which underpins many parametric analytic techniques. The general aim of methods is to determine whether the independent variable(s) affect or relate to the dependent variable(s).

These models are proposed in numerous investigations for certain hydrological applications in which they produce satisfactory results. Sometimes these techniques are combined with other statistical analyses to improve performance. The combined use of stochastic models for generating the inputs and deterministic models are found to be a powerful approach for providing realistic and statistically consistent simulations.

Developed statistical methods such as the H_C-MLM can be embedded as a tool/module for a wider Decision Support System aimed to reach a sustainable and efficient integrated management of water systems.

Table 1 shows some of the main applications of the methodologies presented in the field of hydrology.

TABLE 1 Main applications of shown mathematical-statistical models in the field of hydrology.

Mathematical/statistical methods	Applications	Authors
Simple Linear Regression	Simplification of the parameter calibration procedure and estimation of parameters in ungauged basins.	Huang et al. (2015)
Multiple Linear Regression	Assessment prediction accuracy of runoff characteristics or signatures (zero flow ratio, mean daily flow, standard deviation of daily streamflow).	Zhang et al. (2018)
Moments Estimation in Linear Regression	Application in soil science at four field sites in the United Kingdom: Shuttleworth Clay (SC), Shuttleworth Sand (SS), Wallingford and Yattendon. Fitting point process-based rainfall models and artificial rainfall simulations.	Kerry and Oliver (2007) Kaczmarska et al. (2015)
ANOVA and ANCOVA	To test for changes in the rainfall-soil moisture relationship. To quantify the sources of uncertainty in an ensemble of hydrological climate-impact projections. Sensitivity analysis for water resource and environmental models.	Page et al. (2015) Aryal et al. (2019) Wang et al. (2020)
MANOVA and MANCOVA	Study between spatial urbanization level and hydrological variables to study the influence of spatial urbanization on the hydrological components. Evaluation of the availability of primary and secondary food resources, and allochthonous particulate carbon to aquatic consumers in the Irtysh River	Shukla and Gedam (2018) Zheng et al. (2018)
Generalized Liner Models	Modeling and simulation of rainfall data. To provide a stochastic weather generator that generates daily weather ensembles where the minimum and maximum temperatures are conditioned on precipitation occurrence. Analysis of infilling historic records and climate change.	Tosunoğlu and Regional (2017) Verdin et al. (2015) Kenabatho et al. (2012)
Hybrid causal multivariate linear modeling (H_CMLM)	It describes the joint development of two different methods for temporal rivers´ runoff assessment	Molina et al. (2021)
Predictive Runoff Methods	Extensive analysis of the most important methods for the hydrological understanding and prediction of rivers´ runoff behaviors.	Molina et al. (2020)

TABLE 1 Main applications of shown mathematical-statistical models in the field of hydrology—cont'd

Mathematical/statistical methods	Applications	Authors
Causal Reasoning	To identify, characterize and quantify the influence of time (dependence) for each time step in annual run-off series in five Spanish River basins. To determine and quantify two opposite temporal-fractions within runoff.	Molina and Zazo (2017); Molina et al. (2019)
MixSTATICO	It examines the relationship between phytoplankton and environmental parameters of the eastern equatorial Pacific	González-Narváez et al. (2021)
HJ-Biplot	Water quality evaluation at the sampling sites, Gamboa and Paraiso, located at Gatun Lake (Panama)	Carrasco et al. (2019)
Nayesian networks	To analyze the temporal dependence of an annual runoff series dynamically. Integrated water management and used as a Decision Support System.	Molina et al. (2016) Molina et al. (2010)

References

Agresti, A., 2015. Foundations of Linear and Generalized Linear Models. John Wiley & Sons, Inc.

Aryal, A., Shrestha, S., Babel, M.S., 2019. Quantifying the sources of uncertainty in an ensemble of hydrological climate-impact projections. Theor. Appl. Climatol. 135, 193–209. https://doi.org/10.1007/s00704-017-2359-3.

Bazrkar, M.H., Adamowski, J., Eslamian, S., 2017. Water system modeling. In: Furze, J.N., Swing, K., Gupta, A.K., McClatchey, R., Reynolds, D.M. (Eds.), Mathematical Advances Towards Sustainable Environmental Systems. Springer International Publishing, Switzerland, pp. 61–88.

Bosshard, T., Carambia, M., Goergen, K., Kotlarski, S., Krahe, P., Zappa, M., Schär, C., 2013. Quantifying uncertainty sources in an ensemble of hydrological climate-impact projections. Water Resour. Res. 49, 1523–1536.

Brown, C.E., 2012. Applied Multivariate Statistics in Geohydrology and Related Sciences. Springer Science & Business Media.

Carrasco, G., Molina, J.L., Patino-Alonso, M.C., Castillo, M.D.C., Vicente-Galindo, M.P., Galindo-Villardón, M.P., 2019. Water quality evaluation through a multivariate statistical HJ-Biplot approach. J. Hydrol. 577, 123993. https://doi.org/10.1016/j.jhydrol.2019.123993.

Castelletti, A., Soncini-Sessa, R., 2007. Bayesian Networks and participatory modelling in water resource management. Environ. Model. Softw. 22 (8), 1075–1088. https://doi.org/10.1016/j.envsoft.2006.06.003.

Cayuela, L., 2014. Modelos Lineales: Regresión. ANOVA y ANCOVA.

Chandler, R.E., Wheater, H.S., 2002. Analysis of rainfall variability using generalized linear models: a case study from the west of Ireland. Water Resour. Res. 38, 1–11. https://doi.org/10.1029/2001wr000906.

Chang, B., Guan, J., Aral, M.M., 2015. Scientific discourse: climate change and sea-level rise. J. Hydrol. Eng. 20, 1–14. https://doi.org/10.1061/(asce)he.1943-5584.0000860.

Cohen, J., 1968. Multiple regression as a general data-analytic system. Psychol. Bull. 70, 426–433. https://doi.org/10.1037/h0026714.

Cohen, J., 2013. Statistical Power Analysis for the Behavioral Sciences. Academic Press.

Cook, J.P., Mahajan, A., Morris, A.P., 2017. Guidance for the utility of linear models in meta-analysis of genetic association studies of binary phenotypes. Eur. J. Hum. Genet. 25, 240–245. https://doi.org/10.1038/ejhg.2016.150.

Cox, D., McCullagh, P., 1982. Some aspects of analysis of covariance. Biometrics, 1–10.

Durbin, J., Watson, G., 1950. Testing for serial correlation in least squares regression: I. Biometrika 37 (3/4), 409–428.

Ewaid, S.H., Abed, S.A., Kadhum, S.A., 2018. Predicting the Tigris River water quality within Baghdad, Iraq by using water quality index and regression analysis. Environ. Technol. Innov. 11, 390–398. https://doi.org/10.1016/j.eti.2018.06.013.

González-Narváez, M., Fernández-Gómez, M.J., Mendes, S., Molina, J.L., Ruiz-Barzola, O., Galindo-Villardón, P., 2021. Study of temporal variations in species–environment association through an innovative multivariate method: Mixstatico. Sustainability 13 (11), 5924. https://doi.org/10.3390/su13115924.

Gosset, W., 1908. The probable error of a mean. Biometrika 6, 1–25.

Hair, J.F., Anderson, R.E., Tatham, R.L., Black, W.C., 1987. Multivariate Data Analysis with Readings. Macmillan, New York, USA.

Huang, C., Wang, G., Zheng, X., Yu, J., Xu, X., 2015. Simple linear modeling approach for linking hydrological model parameters to the physical features of a river basin. Water Resour. Manag. 29, 3265–3289. https://doi.org/10.1007/s11269-015-0996-9.

Kaczmarska, J.M., Isham, V.S., Northrop, P., 2015. Local generalised method of moments: an application to point process-based rainfall models. Environmetrics 26, 312–325. https://doi.org/10.1002/env.2338.

Kenabatho, P.K., McIntyre, N.R., Chandler, R.E., Wheater, H.S., 2012. Stochastic simulation of rainfall in the semi-arid Limpopo basin, Botswana. Int. J. Climatol. 32, 1113–1127. https://doi.org/10.1002/joc.2323.

Kerry, R., Oliver, M.A., 2007. Comparing sampling needs for variograms of soil properties computed by the method of moments and residual maximum likelihood. Geoderma 140, 383–396. https://doi.org/10.1016/j.geoderma.2007.04.019.

Lee, S., Lee, D.K., 2018. What is the proper way to apply the multiple comparison test? Korean J. Anesthesiol. 71, 353–360. https://doi.org/10.4097/kja.d.18.00242.

Li, J., Alvarez, B., Siwabessy, J., Tran, M., Huang, Z., Przeslawski, R., Radke, L., Howard, F., Nichol, S., 2017. Application of random forest and generalised linear model and their hybrid methods with geostatistical techniques to count data: predicting sponge species richness. Environ. Model. Softw. 97, 112–129. https://doi.org/10.1016/j.envsoft.2017.07.016.

Li, M., Qi, T., Bernabé, Y., Zhao, J., Wang, Y., Wang, D., Wang, Z., 2018. Simulation of solute transport through heterogeneous networks: analysis using the method of moments and the statistics of local transport characteristics. Sci. Rep. 8, 1–15. https://doi.org/10.1038/s41598-018-22224-w.

Lindsay, B., 2014. Method of Moments. Wiley StatsRef: Statistics Reference Online.

McCullagh, P., 1984. Generalized linear-models. Eur. J. Oper. Res. 16, 285–292.

McCullagh, P., Nelder, J., 1989. Generalized Linear Models. Chapman & Hall, New York, UK.

Mishra, P., Singh, U., Pandey, C., Mishra, P., Pandey, G., 2019. Application of student's t-test, analysis of variance, and covariance. Ann. Card. Anaesth. 22, 407. https://doi.org/10.4103/aca.aca_94_19.

Molina, J., Patino-alonso, C., Zazo, S., 2021. Hybrid causal multivariate linear modelling (H_CMLM) method for the analysis of temporal rivers runoff. J. Hydrol. 599, 126501. https://doi.org/10.1016/j.jhydrol.2021.126501.

Molina, J.L., Bromley, J., García-Aróstegui, J.L., Sullivan, C., Benavente, J., 2010. Integrated water resources management of overexploited hydrogeological systems using Object-Oriented Bayesian Networks. Environ. Model. Softw. 25, 383–397. https://doi.org/10.1016/j.envsoft.2009.10.007.

Molina, J.L., Zazo, S., 2017. Causal reasoning for the analysis of rivers runoff temporal behavior. Water Resour. Manag. 31 (14), 4669–4681.

Molina, J.L., Zazo, S., 2018. Assessment of temporally conditioned runoff fractions in unregulated rivers. J. Hydrol. Eng. 23, 1–15. https://doi.org/10.1061/(ASCE)HE.1943-5584.0001645.

Molina, J.L., Zazo, S., Martín, A.M., 2019. Causal reasoning: towards dynamic predictive models for runoff temporal behavior of high dependence rivers. Water 11 (5), 877. https://doi.org/10.3390/w11050877.

Molina, J.L., Zazo, S., Martín-Casado, A.M., Patino-Alonso, M.C., 2020. Rivers' temporal sustainability through the evaluation of predictive runoff methods. Sustainability 12 (5), 1720. https://doi.org/10.3390/su12051720.

Molina, J.L., Zazo, S., Rodríguez-González, P., González-Aguilera, D., 2016. Innovative analysis of runoff temporal behavior through Bayesian networks. Water (Switzerland) 8, 1–21. https://doi.org/10.3390/w8110484.

Nathan, N.S., Saravanane, R., Sundararajan, T., 2017. Spatial variability of ground water quality using HCA, PCA and MANOVA at Lawspet, Puducherry in India. Comput. Water Energy Environ. Eng. 06, 243–268. https://doi.org/10.4236/cweee.2017.63017.

Nelder, J., Wedderbum, R., 1972. Generalized linear models. J. Roy. Stat. Soc. Ser. A 135 (3), 370–384.

O'Brien, R.M., 2007. A caution regarding rules of thumb for variance inflation factors. Qual. Quant. 41, 673–690. https://doi.org/10.1007/s11135-006-9018-6.

Page, J., Winston, R., Mayes, D., Perrin, C., Hunt III, W., 2015. Retrofitting with innovative stormwater control measures: hydrologic mitigation of impervious cover in the municipal right-of-way. J. Hydrol. 527, 923–932.

Patel, S., Hardaha, M.K., Seetpal, M.K., Madankar, K.K., 2016. Multiple linear regression model for stream flow estimation of Wainganga river. Am. J. Water Sci. Eng. 2, 1–5. https://doi.org/10.11648/j.ajwse.20160201.11.

Pfahl, S., O'Gorman, P.A., Fischer, E.M., 2017. Understanding the regional pattern of projected future changes in extreme precipitation. Nat. Clim. Chang. 7, 423–427. https://doi.org/10.1038/nclimate3287.

Shukla, S., Gedam, S., 2018. Assessing the impacts of urbanization on hydrological processes in a semi-arid river basin of Maharashtra, India. Model. Earth Syst. Environ. 4, 699–728.

Tahroudi, M.N., Ramezani, Y., De Michele, C., Mirabbasi, R., 2020. Analyzing the conditional behavior of rainfall deficiency and groundwater level deficiency signatures by using copula functions. Hydrol. Res. 51, 1332–1348. https://doi.org/10.2166/nh.2020.036.

Tasker, G.D., Stedinger, J.R., 1989. An operational GLS model for hydrologic regression. J. Hydrol. 111, 361–375. https://doi.org/10.1016/0022-1694(89)90268-0.

Tosunoğlu, F., Regional, C., 2017. Joint modelling of drought characteristics derived from historical and synthetic rainfalls: application of generalized linear models and copulas. J. Hydrol. Region. Stud. 14, 167–181.

Uyanto, S.S., 2020. Power comparisons of five most commonly used autocorrelation tests. Pak.j.stat.oper.res 16, 119–130. https://doi.org/10.18187/pjsor.v16i1.2691.

Verdin, A., Rajagopalan, B., Kleiber, W., Katz, R.W., 2015. Coupled stochastic weather generation using spatial and generalized linear models. Stoch. Env. Res. Risk A. 29, 347–356. https://doi.org/10.1007/s00477-014-0911-6.

Verma, S., Bartosova, A., Markus, M., Cooke, R., Um, M.J., Park, D., 2018. Quantifying the role of large floods in riverine nutrient loadings using linear regression and analysis of covariance. Sustainability (Switzerland) 10. https://doi.org/10.3390/su10082876.

Vystavna, Y., Schmidt, S., Kopáček, J., Hejzlar, J., Holko, L., Matiatos, I., Wassenaar, L.I., Persoiu, A., Badaluta, C.A., Huneau, F., 2020. Small-scale chemical and isotopic variability of hydrological pathways in a mountain lake catchment. J. Hydrol. 585, 124834.

Wang, F., Huang, G.H., Fan, Y., Li, Y.P., 2020. Robust subsampling ANOVA methods for sensitivity analysis of water resource and environmental models. Water Resour. Manag. 34, 3199–3217. https://doi.org/10.1007/s11269-020-02608-2.

Wooldridge, J.M., 2001. Applications of generalized method of moments estimation. J. Econ. Perspect. 15, 87–100. https://doi.org/10.1257/jep.15.4.87.

Zazo, S., Molina, J., Ruiz-Ortiz, V., Velez-Nicolas, M., García-Lopez, S., 2020. Modeling river runoff temporal behavior through a hybrid causal-hydrological (HCH). Water 12, 3137.

Zeigler, J.M., Whitledge, G.W., 2011. Otolith trace element and stable isotopic compositions differentiate fishes from the Middle Mississippi River, its tributaries, and floodplain lakes. Hidrobiologia 661, 289–302. https://doi.org/10.1007/s10750-010-0538-7.

Zhang, Y., Chiew, F.H.S., Li, M., Post, D., 2018. Predicting runoff signatures using regression and hydrological modeling approaches. Water Resour. Res. 54, 7859–7878. https://doi.org/10.1029/2018WR023325.

Zheng, Y., Niu, J., Zhou, Q., Xie, C., Ke, Z., Li, D., Gao, Y., 2018. Effects of resource availability and hydrological regime on autochthonous and allochthonous carbon in the food web of a large cross-border river (China). Sci. Total Environ. 612, 501–512.

Chapter 20

Ontology-based knowledge management framework: Toward CBR-supported risk response to hydrological cascading disasters

Feng Yu[a,b] and Yubo Guo[c]

[a]*School of International and Public Affairs, Shanghai Jiao Tong University, Shanghai, China,* [b]*School of Emergency Management, Shanghai Jiao Tong University, Shanghai, China,* [c]*School of Design, Shanghai Jiao Tong University, Shanghai, China*

1. Introduction

Hydrological cascading disaster risk (HCDR) is the possibility of successive losses caused by a hydrological disaster chain. Clarifying the characteristics of cascading disaster risk helps build a resilient city (Pescaroli and Alexander, 2015, 2018). As a typical class of complex risk, coping with HCDR more effectively deserves to be studied. HCDR response refers to the decision and implementation of measures to avoid, transfer, and mitigate sequential risks triggered by a typhoon, rainstorm, or dam break (Yu et al., 2018). The complexity of HCDR makes response difficult and unpredictable. Uncertain and incomplete disaster information generally results in knowledge scarcity (Eslamian and Eslamian (2021). Once response fails, it quickly evolves as cascading disasters with incremental negative impacts on critical infrastructures and urban communities (Pescaroli and Alexander, 2015, 2016). In August 2018, the Minister of China's Emergency Management visited Shanghai and pointed out: "the more modern the city, the more concentrated the risk and the greater the vulnerability, thus it is necessary to strengthen the strategic planning of risk management, deepen the implementation of risk prevention and control, and focus on building a full-time and systematic modernized urban public safety system. Emergency management's focus is gradually transitioning from mid-disaster response to predisaster risk response, and HCDR response is one of the essential components.

In the face of HCDR, emergency decisions to enhance urban resilience still need to rely on historical knowledge. Historical cases can provide many risk response knowledge with a high-value density (Fan et al., 2015; Yu et al., 2018, 2020). Effective use of historical cases to support HCDR response is a crucial practical need. A standard knowledge management framework needs to address the combined influence of time window constraint and psychological pressure. A well-designed framework should involve knowledge creation, acquisition, sharing, utilization, and evaluation (uit Beijerse, 2000; Zack, 2003). Hence, this paper considers three research demands: (1) How to refine common and unique risk response knowledge from historical cases and manage them with a well-structured base? (2) How to quickly match and adapt similar cases to estimate target risk evolution and generate the corresponding response plan? (3) How to evaluate the effectiveness of historical experience for avoiding response failures?

In social sciences, the risk is understood as the uncertainty and severity of consequences of human-related activities (Aven and Renn, 2009). As urbanization accelerates, the links between nature, society, economy, politics, management, and technology become closer. Disaster risk no longer appears to be homogeneous but changes dynamically with the change of risk context and tends to be more complex (Helbing, 2013). Pescaroli and Alexander proposed that cascading disaster risk is one of the most typical complex risks, with risk evolution and incremental negative impacts (Pescaroli and Alexander, 2015). Cascading disaster risk is related to the vulnerability of risk carriers, e.g., the interdependence between critical infrastructures tends to cause cascading effects (Pescaroli and Alexander, 2016, 2018). Besides, integrated perception of disaster information from physical space, information space, and psychological space can facilitate the HCDR scenario cognition (Li et al., 2019; Office of Force Transformation, 2005).

Handbook of HydroInformatics. https://doi.org/10.1016/B978-0-12-821962-1.00016-7

Scenario-dependent demand drives the "scenario-response" decision-making to become the dominant paradigm for HCDR response (Pang et al., 2011; Wang et al., 2020). The US Department of Homeland Security detailed 15 major disaster risk scenarios as the response target, including natural disasters, chemical attacks, infectious disease outbreaks, etc. (US Department of Homeland Security, 2006). Alexander explained the scenario to answer the "what if…" question (Alexander, 2013). In the field of emergency management, case-based reasoning (CBR) is a helpful tool to support emergency decision-making by identifying the evolutionary pattern of disaster risk and developing a response plan (Amailef and Lu, 2013; Yu et al., 2018, 2020). CBR consists of case retrieval, reuse, revision, and retention (i.e., the 4R model) (Aamodt and Plaza, 1994). It also involves case representation (Lu et al., 2013) and repartition (Finnie and Sun, 2003). Therefore, scenario-based CBR is a suitable knowledge management framework for HCDR response.

In order to structure response knowledge, ontology can be adopted to enhance CBR (Amailef and Lu, 2013; Yu et al., 2018, 2020). An ontology is "an explicit specification of a conceptualization" (Gruber, 1993). There are five primitives to express knowledge: concept, relation, function, axiom, and instance (Gómez-Pérez and Benjamins, 1999; Gruber, 1993; Yang et al., 2009). Amailef and Lu explained three advantages of ontology-based CBR (Amailef and Lu, 2013): (1) Ontology could model case knowledge in a semantic network, (2) Ontology defines the unified concepts to facilitate query setting. (3) Ontology provides a comparable way to calculate the similarity between the target and source cases. At present, ontology-based CBR has been widely used in risk response. For example, a risk case ontology model of urban water supply network was built to generate risk response strategies (Yu et al., 2018). Subway risk case can be modeled by the ontologies of risk precursor, safety risk, and safety measures to find similar solutions (Lu et al., 2013). Furthermore, an earthquake case ontology model was presented to serve CBR (Yang et al., 2009).

To sum up, HCDR response needs to build an ontology-based knowledge management framework. CBR can provide a suitable logic process for this framework and operate in a complex disaster risk situation. However, the current ontology-based CBR cannot adapt to the HCDR response well due to the complexity of risk. To improve the preparedness of HCDR response, we propose a knowledge management framework that consists of scenario modeling, layout, reuse, and capability assessment. It attempts to deal with a series of problems, i.e., how to model, store, and reuse HCDR response knowledge and how well it works.

2. Ontology modeling for hydrological cascading disaster risk

Scenario modeling for HCDR is crucial for structuring risk response knowledge. This section will integrate gene theory and ontology model (Yu and Li, 2018) to represent HCDR with a multiphase nested genetic model, which focuses on the design of risk structure and elements.

2.1 Hydrological cascading disaster risk identification

As the preparatory work of HCDR modeling, HCDR identification should first preprocess multisource heterogeneous mass data collected from historical cases. A rough data warehouse is built with many texts, images, audio, and videos. As mentioned above, ontology modeling should consider the basic concepts and relations. Concepts correspond to the elements of HCDR. Based on disaster system theory (Shi et al., 2010) and public safety triangle model (Wu et al., 2020), risk identification needs to discover risk evolution, risk situation elements (hazard, risk carrier, and risk environment), response task elements (strategy, responder, action, and resource), and task effect elements (evaluation, failure, and improvement). All of the common features are clarified, and special features are marked for the historical cases. Relations represent the interaction between two or more elements. The relation between the same elements in different cases should be unified, which lays the foundation of ontology base building.

2.2 Risk "context-scenario" nested model building

The prominent feature of HCDR is the evolution of disaster risks. Based on the "context-scenario" framework of complex disasters (Yu and Li, 2017, 2018), the ontology model of risk evolution can be built in a nested structure. A disaster risk scenario describes the state and trend of risk with response tasks determined by the risk carrier. It is a detailed scene due to the existence of a specific hazard-affect object (e.g., critical infrastructure, urban community, and citizen). A disaster risk context is a type of circumstance that involves a set of disaster risk scenarios. Hazard determines the disaster risk context, which can be generated due to the effect of the previous hazard or the transformation from a hazard-affected object (Yu et al., 2020). Assumed that $f_{DRC}(H)$ and $g_{DRS}(RC)$ denote disaster risk context and scenario respectively, there is the following nested structure:

$$f_{DRC_1}(H) \xrightarrow{\text{triggers}} f_{DRC_2}(H) \ldots \xrightarrow{\text{triggers}} f_{DRC_i}(H) \ldots$$

$$\downarrow \text{involves} \qquad\qquad \downarrow \text{involves} \qquad\qquad \downarrow \text{involves}$$

$$\left\{ g_{DRS_{1j}}(RC) \right\} \quad \left\{ g_{DRS_{2j}}(RC) \right\} \qquad \left\{ g_{DRS_{ij}}(RC) \right\}$$

Specifically, as mentioned above, the relation "triggers" is established when a subsequent hazard is generated (e.g., typhoon triggers rainstorm) or the damaged object transforms to a new hazard (e.g., disabled power grid triggers power outage). Besides, risk context involves one or multiple risk scenarios when the hazard-affected object is under threat.

2.3 Genetic model of disaster risk scenario

Disaster risk scenario consists of risk situation, response task, and task effect. The genetic structure can be introduced to build a complete scenario ontology model with leader region, control region, transcription region, and tail region (Yu and Li, 2018). It makes up for the traditional ontology model's lack of a fixed structure. All the information is arranged in each region using base pairs (see Fig. 1). The setting of four regions is consistent with the logic of CBR.

Specifically, (1) Leader region is comprised of spatiotemporal information and risk information (e.g., the type and possibility of risk). It describes the basic information of a disaster risk scenario. (2) Control region activates and enhances the knowledge retrieval with a sequence of risk situation elements and their attributes. Promoter and enhancer constitute the central part of this region. Promoter is used to judge whether scenario retrieval can be started. Once all the types of risk situation elements are matched, a quantitative similarity measure should be conducted. Enhancer refines the process of scenario retrieval for finding out similar cases. Attribute comparison is executed in this stage. Risk situation elements and the corresponding attributes form the pairs with their weights. (3) Transcription region is the most critical part that executes the reuse of past knowledge. Response tasks are listed in the form of "problem-solution" pair. The transcription

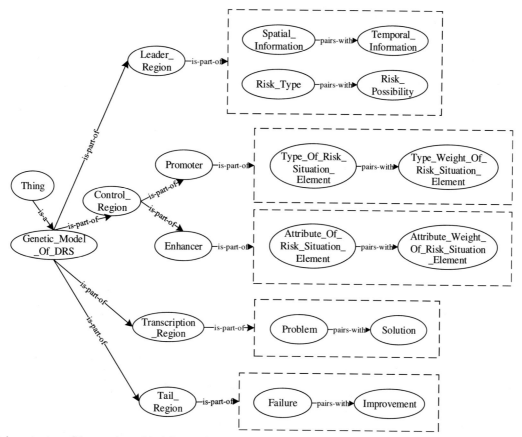

FIG. 1 Ontology structure of the genetic model of disaster risk scenario.

regions of similar cases can reference the target case if the problem is matched. (4) Tail region ends the transcription based on the acceptability of past failures. If a failure in the past solution cannot be tolerated, scenario reuse stops. The terminator is set in this region as a "failure-improvement" pair. In particular, historical after-action reports provide suggestions for avoiding solution failure. These comments can be adopted to adapt the incorrect response tasks.

3. Scenario layout with ontology base

This section presents a picturesque scenario layout called evolutionary risk forest based on biological evolutionary theory. Here, the "layout" means the reorganization of HCDR ontologies. These ontologies are contained in historical cases. Practical scenario layout is the precondition of scenario reuse, which involves hierarchical and classified knowledge storage.

3.1 Planning criteria of scenario layout

The traditional knowledge base is generally partitioned based on the type of knowledge. When facing a continuous knowledge requirement, fragmented knowledge cannot be integrated well to cope with a holistic problem. Knowledge acquisition will become inefficient, and the incompatibility between knowledge is difficult to be avoided. In order to serve rapid knowledge reuse, it is necessary to build a three-level base structure that consists of an ontology base, knowledge base, and case base (see Fig. 2). Ontology base comprises upper, domain, task, and application ontologies (Yu et al., 2020). In detail, upper ontologies restrain the logic scope of the general concepts as the basis of domain ontologies. Domain ontologies set a naming standard for concepts related to a risk situation, and task ontologies clarify the terms of response tasks and task effect. It is helpful to provide an understandable communication way. Application ontologies mean the applied ontologies and their instances in historical cases. These ontologies representing response knowledge are stored in the knowledge base in three categories, i.e., risk situation knowledge, response task knowledge, and task effect knowledge. Considering the analogy between risk and biological evolution, we use the evolutionary tree to build the case base. An evolutionary risk forest can be generated to express an intricate structure of HCDR.

Practical evolutionary risk forest planning determines the efficiency of the knowledge management framework. Here, we give three criteria for managing these knowledge sets (i.e., historical cases): (1) The collection scope of historical cases should be clearly defined. The minimal jurisdiction that a case involves is the urban district or town, i.e., the spatial scale of a disaster risk context is limited to a specific region. For a disaster risk scenario, we focus on a particular risk carrier. (2) Each Case should be represented as an evolutionary tree. The initial root node denotes the primary disaster risk,

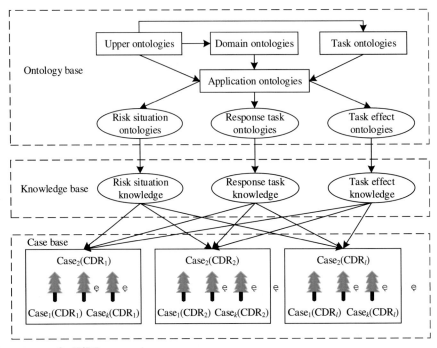

FIG. 2 Three-level base structure for HCDR response.

and the secondary disaster risks are regarded as the leaf nodes. Every node means a disaster risk context. (3) Case base should be divided into multiple case groups based on the type of trigger hazards, such as typhoon-triggered HCDR, rainstorm-triggered HCDR, and dam break-triggered HCDR.

3.2 Structure design of scenario layout

A well-structured case base is helpful to improve the efficiency of knowledge reuse. As mentioned above, evolutionary risk forest is decomposed into three distinct groups based on the type of primary disaster risk, including typhoon group, rainstorm group, and dam break group. Each group consists of abundant evolutionary risk trees with different luxuriance. Each evolutionary risk tree represents a complete HCDR case. Here, luxuriance means the complexity and richness of disaster risk evolution, which can be evaluated by depth and density. Depth refers to the number of generations. For example, the risk chain "typhoon storm surge urban waterlogging power outage water supply interruption" has five generations. Furthermore, disaster risk is likely to cause adverse public opinion risk in social networks and further lead to social safety incident risk. The evolutionary risk tree is decomposed into three layers: physical, information, and psychological space. Density means the number of average branches, which is the ratio of the sum of branches to the number of generations.

According to the phylogenetic tree (Mayr and Bock, 2002), a coding system can be designed to abstract the evolutionary risk tree. The trigger condition between two risk contexts and the involvement condition between risk context and scenario should also be defined. Besides, the length of cascading path is assigned to reflect the association between risk contexts. Therefore, the historical case can be transformed into a code text for convenient case storage. These code texts can be used to reason the code text of the target case and transform it into an evolutionary risk tree in turn. In order to guarantee reliable knowledge reuse, a substitution mechanism is designed. The disaster risk scenarios with the same hazard, risk carrier, and risk environment should be classified into a partition. Once the most similar scenario is not practical, the replaceable knowledge from the subordinate partition could be found.

3.3 Expansion of scenario layout

Since the number of HCDR cases may not be enough to achieve high CBR performance, it is necessary to expand the capacity of the case base to avoid unsuccessful reuse. There are two expansion ways for scenario layout, including transplant expansion and replay expansion. If an HCDR case (**Case A**) partly overlaps with another one (**Case B**) and the two cases are triggered by different hazards, transplant expansion can be used to cut out part of **Case B** and joint it with **Case A**. Hence, a new case (**Case C**) that belongs to the same group as **Case A** can be obtained. Indeed, some scenario parameters need to be modified. For example, a partial risk chain "urban waterlogging power outage water supply interruption" can be cut out from a rainstorm-triggered HCDR case. Typhoons can lead to the same risk chain, and thus the extracted part can be used to generate a new typhoon-triggered HCDR case.

On the other hand, if we need to add complex cases with greater depth, we can assume a complicated situation that can further drive risk evolution. Scenario simulation and expert evaluation can be integrated to generate a deduced case. For example, a case that contains a risk chain of "typhoon urban waterlogging power outage" can be used to deduce a more complex case with a risk chain of "typhoon urban waterlogging power outage water supply interruption".

4. Ontology-supported four-stage scenario reuse

Scenario reuse is the primary function of CBR-supported HCDR response based on ontologies. This section integrates the traditional case retrieval into case reuse to propose a four-stage scenario reuse process (i.e., "Filtration-Deduction-Copy-Adaptation," FDCA).

4.1 Scenario filtration

Scenario filtration, as the first step of scenario reuse, is to compare the scenario structure and elements. First, the specific partition can be matched in the evolutionary risk forest according to the primary disaster risk scenario. An initial evolutionary risk tree set can be obtained. Second, the attributes of risk situation elements can be divided into essential attributes and precursory attributes. Essential attribute means the background feature of a risk situation. Precursory attribute represents the evidence of imminent risks, such as overflowing groundwater to indicate landslide risk. Hence, a two-phase matching can be executed. One compares the essential attributes based on a hybrid similarity measure (Fan et al., 2014).

Another compares the precursory attributes based on a dot matrix diagram (Gibbs and McIntyre, 1970). Afterward, a similar evolutionary risk tree set can be generated, and the potential of triggering secondary risk can be judged.

4.2 Scenario deduction

We need to describe the secondary disaster risks based on a similar tree set. First, any similar case can be decomposed into multiple scenarios and transformed into an instantiated ontology model. Then, the risk evolution rules can be acquired based on the relations between instances, which involve the trigger factor, trigger probability, and trigger threshold. Decision-makers can deduce the possible cascading path if the target case meets the conditions. Once the deduced scenario evolutionary path is no longer applicable as the disaster evolves, we can adopt the Bayesian network to modify the trigger probability and update the cascading path.

4.3 Scenario copy

After scenario deduction, it is necessary to integrate similar cases' risk response tasks to support the target HCDR response. In this step, the past solutions can be transcribed when the new problems are matched with the old ones. In order to address knowledge scarcity, if there are two or more different old solutions for a new problem, the law of independent assortment can be introduced by recombining the past response task elements. Hence, more feasible solutions can be generated. Finally, by considering each solution as a set of response tasks, an evaluation criterion system should be proposed to compute the utility of each solution. The utility ranking can be used to select satisfying solutions.

4.4 Scenario adaptation

Scenario adaptation is the final but most crucial step due to the need to fit the target situation. First, the copied response solutions may still not adapt to the target case. The transformation operations (Aamodt and Plaza, 1994; Fan et al., 2015) can be employed to manually remedy the genes of HCDR, including addition, deletion, and modification of the "problem-solution" pair. The comprehensive risk similarity can be used to generate a predicted interval value for quantitative task elements (e.g., the number of emergency resources). Then, soft computing, such as evolutionary algorithms, artificial neural networks, and fuzzy analysis, can be adapted to optimize the solutions. For example, fuzzy analysis can judge whether the acceptable task elements meet decision-makers' preferences. The evolutionary algorithm can provide a way to find out the optimal predicted value of the task element with the highest confidence level. Finally, HCDR has the characteristic of dynamic change. With the continuous import of disaster information, the original optimal solution may cause deviations when there are new requirements. In such a situation, an evidence combination model (Yu et al., 2018) can be built to compute the optimal solution's credibility and evaluate whether it is against expertise.

5. Gap analysis on ontology-based CBR from a failure perspective

The feasibility and effectiveness of CBR-supported capability need to be evaluated whether it is suitable for HCDR response. Therefore, this section provides a failure analysis framework for risk evolution analysis, response plan generation, and emergency decision cognition. The "genetic disease" (i.e., maladaptation to the target context) inherited from similar cases can be identified based on gap analysis.

5.1 Preparation for gap analysis

Decision-makers need to set the trial HCDR case in the selected city. Primary hazard, urban characteristics, meteorological conditions, geographic environment, critical infrastructure status, and emergency capability reserve should be considered and structured using the proposed genetic model. We can select a local historical case as the target case in the case study. Hence, we can obtain the estimated and actual results of risk evolution analysis, response plan generation, and emergency decision cognition. In particular, the historical cases in the same city and other cities should be collected together.

5.2 Failure analysis on CBR-supported HCDR response

Based on the failure analysis process (Jackson, 2008; Jackson et al., 2010, 2011), we emphasize verifying the three core functions of CBR-supported HCDR response (i.e., risk evolution analysis, response plan generation, and emergency decision cognition). First, risk deduction relates to the effect of risk avoidance, transfer, and mitigation. If risk response successes, the cascading path will be cut off. Hence, the difference between deduced and actual risk evolution can be confirmed. If there is no difference, the proposed method is considered adequate. If a difference exists, decision-makers should determine why scenario deduction fails and evaluate the potential impact. Because emergency capability will gradually improve as emergency management investment increases, the only acceptable state is that the estimated cascading path can cover the actual cascading path. Otherwise, we deem that the function of scenario deduction is unreliable and needs to enhance the consideration in the worst-case situation.

Second, the differences can be identified by comparing the estimated and actual risk response strategies in different risk phases. In particular, the actual risk response strategies are modified based on the after-action report. Decision-makers should analyze whether these differences can lead to risk response failure. If confirmed, scenario filtration and adaptation need to be improved, such as algorithm substitution, weight adjustment, and goal enhancement. Indeed, if the emergent consequences of failures can be tolerated by prepared backup, the capability gap can be accepted.

Third, HCDR response decision-making relies on the risk preference of decision-makers. Cognition toward the risk of a CBR-generated response plan needs to be considered. The relevant behavior experiments can be conducted to determine the association between personality traits and risk preference. Personality traits of decision-makers originate from the educational background, individual psychological experience, and HCDR knowledge reserve. These factors co-determine whether and how much decision-makers are willing to undertake the negative influence by learning from the historical cases. Hence, we can confirm whether CBR fits the risk preference of decision-makers in the target case.

6. Conclusions

This paper builds an ontology-based knowledge management framework for HCDR response and adopts CBR to design an operational process of knowledge reasoning. The proposed framework enables decision-makers to model the knowledge involved in HCDR cases and reuse them with a four-stage process. Besides, we present a way to organize and manage ontologies from a biological perspective and discuss how to identify the failures of ontology-based CBR for HCDR response. There are three contributions: (1) A comprehensive knowledge representation model integrating genetic structure and ontologies is presented, which can be used to build a complete HCDR case quickly. (2) Concept of the evolutionary tree is introduced to plan the HCDR case base and enhance the capability of complex case storage. (3) A systematic scenario reuse is reconstructed to promote similar scenarios discovery, and the applicability of ontology-based CBR is also explained. In addition, the proposed framework can also be extended and applied in other domains, such as meteorological, geological, and earthquake-triggered cascading disaster risk. It is worth mentioning that the CBR-supported model needs to be improved to adapt to the emerging risks and new changes of the typical cascading disaster risk.

Acknowledgments

This work was supported by the National Natural Science Foundation of China [Grant number 71904121], Science and Technology Commission of Shanghai Municipality [Grant number 22692195800], Shanghai Jiao Tong University [Grant number JCZXZGB-03], China Postdoctoral Science Foundation [Grant number 2019 M661531], and Shanghai Planning Office of Philosophy and Social Science [Grant number 2020EGL001]. We also thank Prof. Saeid Eslamian for his kind helps.

References

Aamodt, A., Plaza, E., 1994. Case-based reasoning: foundational issues, methodological variations, and system approaches. AI Commun. 7 (1), 39–59.

Alexander, D.E., 2013. Emergency and disaster planning. In: López-Carresi, A., Fordham, M., Wisner, B., et al. (Eds.), Disaster Management: International Lessons in Risk Reduction, Response and Recovery. Routledge, UK, pp. 125–141.

Amailef, K., Lu, J., 2013. Ontology-supported case-based reasoning approach for intelligent m-Government emergency response services. Decis. Support Syst. 55 (1), 79–97.

Aven, T., Renn, O., 2009. On risk defined as an event where the outcome is uncertain. J. Risk Res. 12 (1), 1–11.

Eslamian, S., Eslamian, F., 2021. Disaster Risk Reduction for Resilience: New Frameworks for Building Resilience to Disasters, Springer Nature Switzerland. 487 Pages.

Fan, Z.P., Li, Y.H., Wang, X.H., et al., 2014. Hybrid similarity measure for case retrieval in CBR and its application to emergency response towards gas explosion. Expert Syst. Appl. 41 (5), 2526–2534.

Fan, Z.P., Li, Y.H., Zhang, Y., 2015. Generating project risk response strategies based on CBR: a case study. Expert Syst. Appl. 42 (6), 2870–2883.

Finnie, G., Sun, Z.H., 2003. R5 model for case-based reasoning. Knowl.-Based Syst. 16 (1), 59–65.

Gibbs, A.J., McIntyre, G.A., 1970. The diagram, a method for comparing sequences: its use with amino acid and nucleotide sequences. Eur. J. Biochem. 16 (1), 1–11.

Gómez-Pérez, A., Benjamins, R., 1999. Overview of knowledge sharing and reuse components: Ontologies and problem-solving methods. In: Proceedings of the 16th International Joint Conference on Artificial Intelligence, pp. 1–15.

Gruber, T.R., 1993. A translation approach to portable ontology specifications. Knowl. Acquis. 5 (2), 199–220.

Helbing, D., 2013. Globally networked risks and how to respond. Nature 497 (7447), 51–59.

Jackson, B.A., 2008. The Problem of Measuring Emergency Preparedness: The Need for Assessing "Response Reliability" as Part of Homeland Security Planning. RAND Corporate, USA.

Jackson, B.A., Sullivan Faith, K., Willis, H.H., 2010. Evaluating the Reliability of Emergency Response Systems for Large-Scale Incident Operations. RAND Corporate, USA.

Jackson, B.A., Sullivan Faith, K., Willis, H.H., 2011. Are we prepared? Using reliability analysis to evaluate emergency response systems. J. Conting. Crisis Manag. 19 (3), 147–157.

Li, S., Du, W.W., Wang, Z.S., 2019. Properties of the physical space, the information space and out-structure space of information. J. Phys. Conf. Ser. 1168 (3), 032054.

Lu, Y., Li, Q.M., Xiao, W.J., 2013. Case-based reasoning for automated safety risk analysis on subway operation: case representation and retrieval. Saf. Sci. 57, 75–81.

Mayr, E., Bock, W.J., 2002. Classifications and other ordering systems. J. Zoolog. Syst. Evol. Res. 40 (4), 169–194.

Office of Force Transformation, 2005. The Implementation of Network-Centric Warfare, Washington DC, USA.

Pang, J.J., Liu, L., Li, S.M., 2011. A comparative study between "prediction-response" and "scenario-response" in unconventional emergency decision-making management. In: Proceedings of the 4th International Joint Conference on Computational Sciences and Optimization, pp. 649–652.

Pescaroli, G., Alexander, D.E., 2015. A definition of cascading disasters and cascading effects: going beyond the "toppling dominos" metaphor. GRF Davos Planet@Risk 3 (1), 58–67.

Pescaroli, G., Alexander, D.E., 2016. Critical infrastructure, panarchies and the vulnerability paths of cascading disasters. Nat. Hazards 82 (1), 175–192.

Pescaroli, G., Alexander, D.E., 2018. Understanding compound, interconnected, interacting, and cascading risks: a holistic framework. Risk Anal. 38 (11), 2245–2257.

Shi, P.J., Shuai, J.B., Chen, W.F., et al., 2010. Study on large-scale disaster risk assessment and risk transfer models. Int. J. Disaster Risk Sci. 1 (2), 1–8.

US Department of Homeland Security, 2006. National Planning Scenarios. USDHS, USA.

uit Beijerse, R.P., 2000. Knowledge management in small and medium-sized companies: knowledge management for entrepreneurs. J. Knowl. Manag. 4 (2), 162–179.

Wang, D.L., Wan, K.D., Ma, W.X., 2020. Emergency decision-making model of environmental emergencies based on case-based reasoning method. J. Environ. Manage. 262, 110382.

Wu, L., Li, J., Ruan, Y.Z., et al., 2020. Study on risk assessment of typhoon storm surge based on public safety triangle theory. IOP Conf. Ser. Earth Environ. Sci. 526, 012055.

Yang, P., Wang, W.J., Dong, C.X., 2009. Application of emergency case ontology model in earthquake. In: Proceedings of the 2009 International Conference on Management and Service Science, China, pp. 1–5.

Yu, F., Fan, B., Li, X.Y., 2020. Improving emergency preparedness to cascading disasters: a case-driven risk ontology modelling. J. Conting. Crisis Manag. 28 (3), 194–214.

Yu, F., Li, X.Y., 2017. Complex emergency case representation method based on genetic structure. Syst. Eng. Theory Pract. 37 (3), 677–690.

Yu, F., Li, X.Y., 2018. Improving emergency response to cascading disasters: applying case-based reasoning towards urban critical infrastructure. Int. J. Disaster Risk Reduct. 30, 244–256.

Yu, F., Li, X.Y., Han, X.S., 2018. Risk response for urban water supply network using case-based reasoning during a natural disaster. Saf. Sci. 106, 121–139.

Zack, M.H., 2003. Rethinking the knowledge-based organization. MIT Sloan Manag. Rev. 44 (4), 67–71.

Chapter 21

Optimally pruned extreme learning machine: A new nontuned machine learning model for predicting chlorophyll concentration

Salim Heddam

Faculty of Science, Agronomy Department, Hydraulics Division, Laboratory of Research in Biodiversity Interaction Ecosystem and Biotechnology, Skikda, Algeria

1. Introduction

During the last few years, study and control of water pollution and its direct relationship to the aquatic and human's biology are well documented (ALabdeh et al., 2020; Munyebvu et al., 2020). For a long period of time, chlorophyll-a pigment (Chl-a) is used for the quantification of phytoplankton biomass (Blix et al., 2019) and is considered a primary cause of eutrophication (Millette et al., 2019). Monitoring water quality of freshwater ecosystems can be achieved through different methods: (i) traditional in situ monitoring methods using water quality instruments, (ii) laboratory analysis of water quality variables, and (iii) mechanistic empirical and semiempirical models which have filled the gaps encountered using traditional methods (Zhang et al., 2020b). In response to the continued increase in the proliferation of nutrients causing the eutrophication, and the high level of concentration of phytoplankton biomass in many regions of the world, efficient and robust computational algorithms need to be introduced to help in improving the early warning and to qualitatively and quantitatively predicting the amount of water quality variables, e.g., chlorophyll-a, biochemical oxygen demand (BOD), chemical oxygen demand (COD), among others (Zhang et al., 2020a). In order to avoid rather common difficulties encountered in the application of the mechanistic models of eutrophication, especially due to the lack of sufficient and accurate measured data (Park et al., 2015; Mamun et al., 2020), expensive recourse to the application of diverse kinds of machines learning (ML) models is becoming legitimate. Over the years, several ML models were developed and successfully applied for predicting and forecasting chlorophyll-a concentration in freshwater ecosystems, and it was demonstrated that the nutrient concentration and the dynamics of chlorophyll-a were highly nonlinear (Tian et al., 2019).

Among the proposed machines learning models, Luo et al. (2019) compared the accuracy of three machine learning models for predicting chlorophyll-a concentration (Chl-a: µg/L) using four water quality variables selected as predictors namely, Secchi depth (SD), water turbidity (TU), total nitrogen (TN), and total phosphorus (TP). The proposed models were: (i) adaptive neuro-fuzzy inference systems with fuzzy c-mean-clustering algorithm (ANFIS_F), (ii) ANFIS with grid partition (ANFIS_G), (iii) ANFIS with subtractive clustering (ANFIS_S), and (iv) the multilayer perceptron neural network models (MLPNN). The obtained results were moderate to good, and the best accuracy for the testing dataset was achieved using the ANFIS_S for natural lake, with coefficient of determination (R^2) equals to 0.84. García-Nieto et al. (2015) applied a new hybrid data driven model called multivariate adaptive regression splines optimized using particle swarm optimization (MARS-PSO) for predicting Chl-a concentration using several water quality variables namely, pH, dissolved oxygen concentration (DO), water temperature (TE), irradiance (SR), nitrate (NO_3) and phosphate (Pi) concentration. The authors reported high accuracy of the MARS-PSO model with an R^2 of 0.99. In another study, García-Nieto et al. (2016a) introduced a new hybrid data driven model called MARS optimized using artificial bee colony (ABC) technique (MARS-ABC) for predicting the Chl-a concentration. The authors developed the model using a combination of large amount of water quality and biological variables namely; TE, TU, NO_3, ammonium, DO, specific conductance (SC), pH, total phosphorus, diatoms, cyanobacteria, euglenophytes, dinophlagellata, chrysophytes, chlorophytes, and cryptophytes. The performances of the MARS-ABC were very good with R^2 of 0.84. In addition, the authors have demonstrated that the euglenophytes was

Handbook of HydroInformatics. https://doi.org/10.1016/B978-0-12-821962-1.00015-5

the most significant variable in predicting the Chl-a concentration and ranked in the first order, while the total phosphorus possess the lowest contribution in the estimation of the Chl-a. García-Nieto et al. (2016b) used the hybrid particle swarm optimization in combination with support vector machines (SVM) called SVM-PSO for modeling Chl-a concentration using pH, optical density, DO, NO_3, Pi, salinity, TE and SR. Among three linear, quadratic and radial basis (RBF) kernels, the best accuracy was achieved using SVM-PSO having the RBF kernel with an R^2 of 0.99.

Among the proposed machines learning models, the Gaussian process regression (GPR) was used by Blix et al. (2019) for Chl-a concentration in the high northern latitude optically complex waters, using the ocean and land color instruments (OLCI) remote sensing reflectance (Rrs), and they obtained acceptable results with an R^2 of 0.828. In the same line of investigation, Free et al. (2020) reported that the Chl-a concentration was less well predicted ($R^2 \approx 0.66$) using nonparametric multiplicative regression (NPMR) based on the Sentinel-2 remote sensing imagery. Mamun et al. (2020) compared the MLPNN, SVM, and the multiple linear regression (MLR) for predicting the Chl-a concentration measured at three different zones namely, riverine, transitional and lacustrine zones, in the Imha reservoir South Korea. The proposed models were developed using nine input variables namely, precipitation, DO, BOD, COD, total suspended solid (TSS), SC, TE, TN, TP, and TN/TP ratios. The best accuracy was obtained using the SVM model with R^2 ranging from 0.73 to 0.80. Yajima and Derot (2018) proposed the application of the random forest regression (RF) model for forecasting the Chl-a concentration in the fresh water of the Urayama reservoir and the saline water of lake Shinji, Japan. They reported that the Chl-a concentration was more influenced by the combination of the BOD, COD, pH, and (TN/TP) ratio. In another study, Zhang et al. (2015) collected data from four monitoring sites located at the Yuqiao reservoir, China, and proposed the application of the MLPNN for simultaneous modeling and forecasting Chl-a concentration two weeks in advance. By including the TE, COD, SS, Secchi disk depth (SD), TP, NO_3, pH and DO, the authors demonstrated that the Chl-a was very well predicted with R^2 around 0.98. Yu et al. (2020) used a hybrid model composed of wavelet domain threshold denoising (WDTD), wavelet mean fusion (WMF) and long-short term memory artificial neural network (LSTM) and applied for predicting Chl-a concentration using fifteen water quality variables collected at 10 water quality sites at the Dianchi lake, China, between 2005 and 2012. The authors reported acceptable results with RMSE, MAE, and R^2 of 18.40, 13.56, and 0.63, respectively. Sylaios et al. (2008) applied ANFIS model for predicting Chl-a using DO, TE, dissolved inorganic nitrogen (DIN) and SR, collected at the Vassova Lagoon, located on the northern coast of the Aegean Sea, northern Greece. The ANFIS model was developed using a CHLfuzzy spreadsheet, and excellent results was obtained with R^2 ranging between 0.94 and 0.97, and it was also demonstrated that the sigmoid fuzzy memberships was more suitable than the triangular and trapezoidal functions. The radial basis function artificial neural network (RBFNN) is another kind of machines learning model used for predicting Chl-a concentration. Xiaobo et al. (2014) used the RBFNN model coupled with the principal component analysis (PCA) for predicting the concentration of the Chl-a at the Yuqiao reservoir located in the Haihe River Basin, China. The model was developed using eighteen water quality variables, and they obtained acceptable results with R^2 around 0.607 using only the first five principal component loadings obtained from the PCA analysis. Finally, a hybrid metaheuristic algorithm composed of the SVM and genetic algorithm (SVM-GA) was proposed by Xianquan et al. (2013) for predicting the Chl-a concentration at the Bohai Bay, China. The authors reported that the SVM-GA ($R^2 = 0.850$, RMSE = 6.305) was more accurate compared the standalone SVM ($R^2 = 0.680$, RMSE = 6.960).

In the present study, we use a new nontuned machine learning model called optimally pruned extreme learning machine (OPELM) for predicting the Chl-a concentration at the Charles River and Mystic River Buoys. To the best of our knowledge, this is the first study to apply the OPELM model for predicting the Chl-a using water quality variables as predictor. Results obtained using the OPELM model were compared to those obtained using the MLPNN model. Thus, the main objectives of the present chapter are to: (i) assess the accuracy of the OPELM model compared to the MLPNN model, and (ii) evaluate the sensitivity of the models to changes in input variables. The rest of this chapter is organized as follows: in Section 2, a detailed description of datasets collected at the Charles River and Mystic River Buoys and used in the study is presented. The theoretical description of the developed models is introduced in Section 3. Section 4 presents the results and discussion. The conclusions are summarized in Section 5.

2. Study area and data

Field water quality data were collected at two stations during the full 2019 year at two rivers, USA. The two rivers were respectively: (i) the Charles River Buoy (CR-Buoy) and (ii) the Mystic River Buoy (MR-Buoy). Data were measured at 15 min interval of time, and in total 10,303 data were collected, in addition the incomplete pattern were removed from the dataset, in addition, data that failed in sanity and plausibility checks were also removed. The field data measurements included a set of water quality variables containing the water pH, water temperature (TE), specific conductance (SC), water turbidity (TU), dissolved oxygen (DO), and chlorophyll-a (Chl-a) concentration (see Table 1 for a summary). The Chl-a

TABLE 1 Summary statistics of water quality variables for the two stations.

Variables	Subset	Unit	X_{mean}	X_{max}	X_{min}	S_x	C_v	R
Charles River Buoy (CR-Buoy) station								
TE	Training	°C	22.983	29.080	13.900	3.944	0.172	−0.144
	Validation	°C	22.983	29.080	13.900	3.944	0.172	−0.144
	All data	°C	22.916	29.170	13.860	3.985	0.174	−0.151
SC	Training	mS/cm	0.895	2.020	0.530	0.201	0.224	−0.061
	Validation	mS/cm	0.906	2.020	0.540	0.216	0.238	−0.050
	All data	mS/cm	0.899	2.020	0.530	0.205	0.228	−0.058
pH	Training	/	7.585	9.300	6.980	0.410	0.054	0.105
	Validation	/	7.603	9.290	6.990	0.428	0.056	0.100
	All data	/	7.590	9.300	6.980	0.415	0.055	0.103
TU	Training	FNU	2.339	75.850	1.000	1.715	0.733	−0.092
	Validation	FNU	2.367	26.900	1.040	1.396	0.590	−0.114
	All data	FNU	2.347	75.850	1.000	1.608	0.685	−0.098
DO	Training	mg/L	8.443	13.220	5.410	1.266	0.150	0.354
	Validation	mg/L	8.477	13.270	5.430	1.305	0.154	0.349
	All data	mg/L	8.454	13.270	5.410	1.277	0.151	0.352
Chl-a	Training	*RFU*	2.339	9.170	0.620	1.121	0.479	1.000
	Validation	*RFU*	2.335	8.680	0.610	1.113	0.477	1.000
	All data	*RFU*	2.338	9.170	0.610	1.116	0.478	1.000
Mystic River Buoy (MR-Buoy) station								
TE	Training	°C	24.241	29.450	17.710	2.594	0.107	0.413
	Validation	°C	24.197	29.430	17.790	2.543	0.105	0.388
	All data	°C	24.228	29.450	17.710	2.578	0.106	0.406
SC	Training	mS/cm	1.360	4.010	0.510	0.592	0.435	0.129
	Validation	mS/cm	1.372	3.970	0.520	0.611	0.446	0.112
	All data	mS/cm	1.364	4.010	0.510	0.597	0.438	0.124
pH	Training	/	8.364	9.700	7.060	0.730	0.087	0.071
	Validation	/	8.372	9.710	7.060	0.724	0.086	0.034
	All data	/	8.367	9.710	7.060	0.728	0.087	0.060
TU	Training	FNU	2.880	16.440	0.760	1.314	0.456	0.591
	Validation	FNU	2.891	7.550	0.730	1.305	0.451	0.620
	All data	FNU	2.883	16.440	0.730	1.310	0.454	0.600
DO	Training	mg/L	9.873	16.650	3.940	2.342	0.237	0.114
	Validation	mg/L	9.862	16.730	3.970	2.315	0.235	0.100
	All data	mg/L	9.870	16.730	3.940	2.332	0.236	0.110

Continued

TABLE 1 Summary statistics of water quality variables for the two stations—cont'd

Variables	Subset	Unit	X_{mean}	X_{max}	X_{min}	S_x	C_v	R
Chl-a	Training	*RFU*	3.176	13.710	0.400	2.046	0.644	1.000
	Validation	*RFU*	3.178	13.430	0.430	2.023	0.636	1.000
	All data	*RFU*	3.177	13.710	0.400	2.037	0.641	1.000

Abbreviations: X_{mean}, mean; X_{max}, maximum; X_{min}, minimum; S_x, standard deviation; C_v, coefficient of variation; R, coefficient of correlation with DO; *TE*, water temperature; *DO*, dissolved oxygen; *SC*, specific conductance; *TU*, water turbidity; *mS/cm*, millisiemens per centimeter; *mg/L*, milligram per liter; *RFU*, raw fluorescent units; *FNU*, Formazin Nephelometric Unit.

concentration was used as the predicted variable, and the pH, TE, SC, DO and TU were selected as the independent variables, i.e., the predictors. Chlorophyll was measured in raw fluorescent units (RFU), turbidity was measured in formazin nephelometric units (FNU), DO was measured in milligrams per liter (mg/L), the specific conductance was measured at 25°C, and reported in millisiemens per centimeter (mS/cm), finally, water temperature was measured in degrees Celsius (°C). The concentrations of the main water quality variables (pH, TE, SC, DO, TU, and Chl-a) were assumed to have an average of zero and standard deviation of one, using the z-scores normalization. Briefly, at each sampling station, we provided in Table 1, the mean, maximum, minimum, standard deviation, coefficient of variation values, and the coefficient of correlation of each variable with Chl-a, i.e., X_{mean}, X_{max}, X_{min}, S_x, C_v, and R, respectively. In addition, we split the data into training and validation subset suing a ratio of (70%:30%). Finally, the correlation plot between Chl-a concentration and the water quality variables is shown in Fig. 1. The entire proposed modeling approaches flowchart is shown in Fig. 2.

3. Methodology

3.1 Multilayer perceptron neural networks (MLPNN)

Multilayer perceptron neural networks model (MLPNN) possesses a high ability for approximating high nonlinear and complexes relations between a set of inputs variables and one or more output variable (Hornik et al., 1989; Hornik, 1991). The MLPNN is structured in a series of parallel layers, routes the signal from the input to the output layers and adjusts the parameters before providing the final response. The parameters adjustment is the most and critical task that should be achieved using the MLPNN model to obtain maximum power from the available data, using a training process

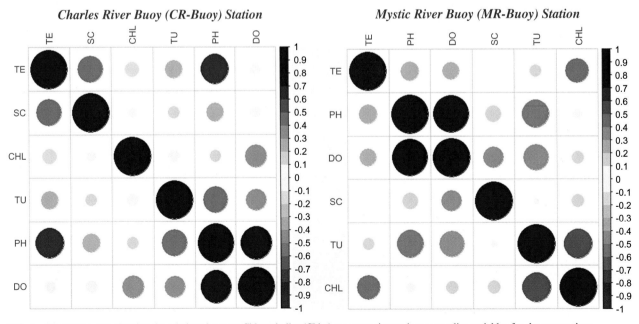

FIG. 1 Correlation plot showing the relations between Chlorophyll-a (Chl-a) concentration and water quality variables for the two stations.

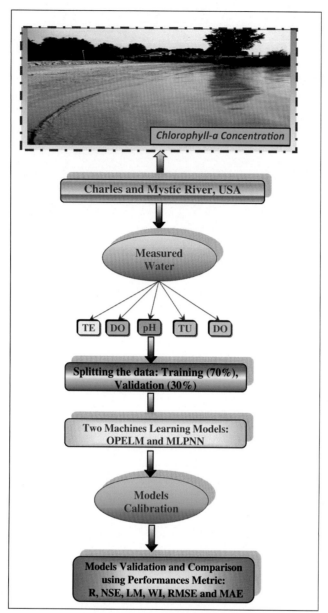

FIG. 2   Flowcharts of the OPELM and MLPNN Frameworks.

leading to a decrease of the total error calculated between the measured (target) and the calculated values of the output variable, based on the backpropagation algorithm. The optimal architecture of the MLPNN model (Fig. 3) is determined by trial and error, and generally the sigmoid and linear transfer functions are used for the hidden and output layers, respectively. From the input to the output layer, a series of neurons play a particular role. The neurons in the input layer were without activation function and mainly used to present the independent variables (predictors) to the model and their number is equal to the number of predictors. The number of neurons in the hidden layer was determined by trial and error and they play a major role, by receiving the signals from the input layer and perform a transformation via the sigmoid activation function. Finally, the unique neuron in the hidden layer corresponds to the final response of the model using a linear transfer function (Hornik et al., 1989; Hornik, 1991).

3.2 Optimally pruned extreme learning machine (OPELM)

Extreme learning machines (ELM) is a training algorithm for the single hidden layer feedforward artificial neural network (SLFN) proposed by Huang et al. (2006a,b). Over the years, he has demonstrated these abilities to guaranties rapid learning speed and high accuracy (Huang et al., 2011; Chen et al., 2019). For a dataset with N patterns, composed of a matrix input

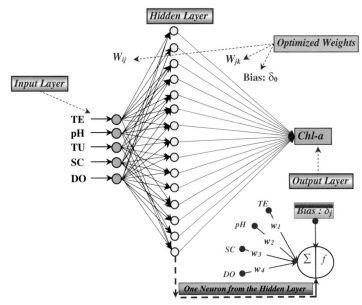

FIG. 3 Architecture of the MLPNN model.

variables X and their corresponding target variable Y and given as $[\{X, Y\} = \{(x_i, y_i) | x_i \in R^d, y_i \in R^c, i = 1, 2, ..., N\}], x_i \in R^d$ is the d-dimensional input data and $y_i \in R^c$ is its corresponding target data (Yu and Webb, 2019; Chen et al., 2019; Sun et al., 2019), the ELM is expressed mathematically as follow:

$$f_L(x_i) = \sum_{j=1}^{L} \beta_j g\left(w_j, b_j, x_j\right) = y_j \quad (i = 1, ..., N) \tag{1}$$

where w_j and b_j respectively denote the weight and bias for the jth hidden layer neuron, β_j is the output weight vector between the jth hidden layer neuron and the unique hidden neuron, and G is the activation function (Yu and Webb, 2019; Sun et al., 2019). Contrary to the standard artificial neural network based on the backpropagation training algorithm, using the ELM model, the weights (w_j) and biases (b) between the input layer and hidden layer were not adjusted, rather they are randomly assigned, while the weight matrix between the hidden and the output layers were analytically determined (Yu and Webb, 2019; Chen et al., 2019; Sun et al., 2019; Huang et al., 2011). Therefore, the ELM determines the output weights matrix (β) by minimizing both the prediction errors and the norm of the output weights simultaneously (Chen et al., 2019). The basic structure of the ELM network is shown in Fig. 4. One of the most and robust version of the ELM is certainly the optimally pruned extreme learning machine (OPELM) introduced by Miche et al. (2008, 2010). For which three kinds of activation functions can be used: Gaussian, sigmoid and linear (Pouzols and Lendasse, 2010). The source code can be found at: http://www.cis.hut.fi/projects/tsp/index.php. OPELM model is built in three major stages: (i) standard ELM with large number of neurons is firstly developed; (ii) the multi-response sparse regression (MRSR) is used for ranking the hidden neurons and (iii) several irrelevant neurons are pruned using leave-one-out cross-validation (LOO-CV) (Miche et al., 2010). During the last few years, the OPELM model has been applied for solving many engineering application problems, e.g., for classification tasks (Eshtay et al., 2020), daily streamflow forecasting (Adnan et al., 2019); monthly streamflow prediction (Adnan et al., 2020), modeling and estimating monthly streamflow (Attar et al., 2020), generating an intelligent velocity module for autonomous underwater vehicle (Lv et al., 2020), predicting daily pan evaporation from dam reservoir (Sebbar et al., 2019), estimating the international roughness index of rigid pavements (Kaloop et al., 2020), and for predicting maximum water temperature in rivers (Zhu and Heddam, 2019).

3.3 Performance assessment of the models

In the present study, a large number of models evaluation metrics have been selected for the evaluation and comparison of models performances. Two kinds of performances metrics were used, namely, the goodness-of-fit measures and the relative and absolute error measures. For the goodness-of-fit measures we selected: (i) the Pearson's correlation coefficient (R), (ii) the Willmott's Index of agreement (WI), (iii) the Nash-Sutcliffe Efficiency (NSE) and (iv) the Legate-McCabes index (LM).

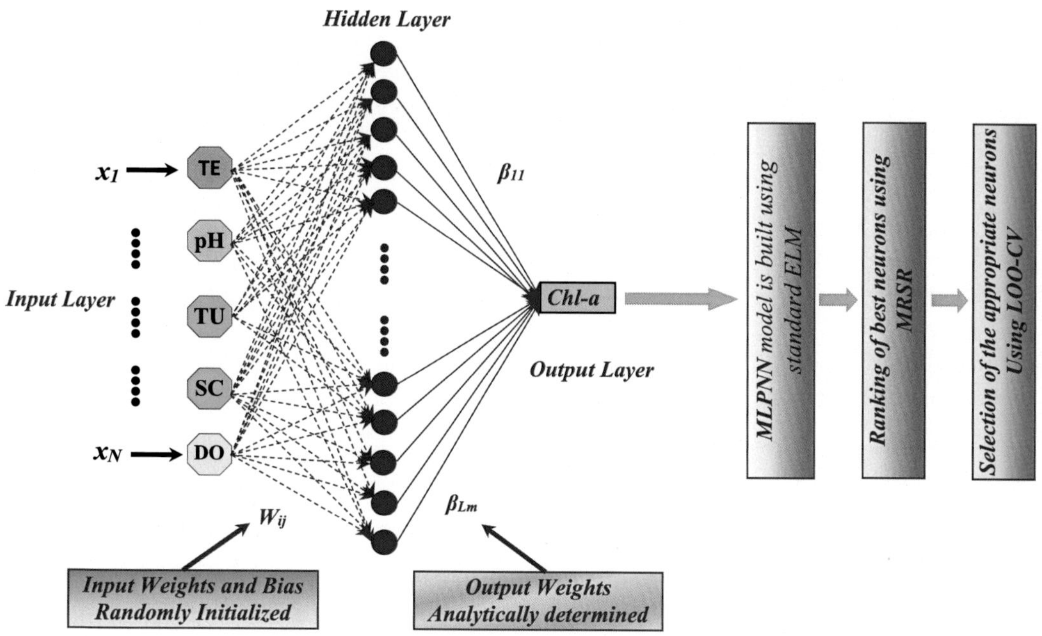

FIG. 4 Flowchart of the OPELM based on the MLPNN model.

For the relative and absolute error measures we selected: (i) the root mean square error (*RMSE*), (ii) the mean absolute error (*MAE*), (iii) the relative root mean square error (*RRMSE*) and (iv) the mean absolute percentage error (*MAPE*). More details about can be found in (Legates and McCabe Jr, 1999; Moriasi et al., 2007).

$$MAE = \frac{1}{N} \sum_{i=1}^{N} |(Chl_0)_i - \left(Chl_p\right)_i|, \; (0 \leq MAE < +\infty) \tag{2}$$

$$RMSE = \sqrt{\frac{1}{N} \sum_{i=1}^{N} \left[(Chl_0)_i - \left(Chl_p\right)_i\right]^2}, \; (0 \leq RMSE < +\infty) \tag{3}$$

$$RRMSE = \frac{\sqrt{\frac{1}{N} \sum_{i=1}^{N} \left[(Chl_0)_i - \left(Chl_p\right)_i\right]^2}}{\frac{1}{N} \sum_{i=1}^{N} (Chl_0)_i} \times 100, \; (0 \leq RRMSE < +\infty) \tag{4}$$

$$MAPE = \frac{1}{N} \sum_{i=1}^{N} \left| \frac{(Chl_0)_i - \left(Chl_p\right)_i}{(Chl_0)_i} \right| \times 100, (0 \leq MAPE < +\infty) \tag{5}$$

$$NSE = 1 - \left[\frac{\sum_{i=1}^{N} \left[(Chl_0)_i - \left(Chl_p\right)_i\right]^2}{\sum_{i=1}^{N} \left[(Chl_0)_i - \overline{Chl_0}\right]^2} \right], \; (-\infty < NSE \leq 1) \tag{6}$$

$$R = \left[\frac{\frac{1}{N} \sum_{i=1}^{N} ((Chl_0)_i - \overline{Chl_0}) \left(\left(Chl_p\right)_i - \overline{Chl_P}\right)}{\sqrt{\frac{1}{N} \sum_{i=1}^{n} ((Chl_0)_i - \overline{Chl_0})^2} \sqrt{\frac{1}{N} \sum_{i=1}^{n} \left(\left(Chl_p\right)_i - \overline{Chl_p}\right)^2}} \right], \; (-1 < R \leq +1) \tag{7}$$

$$WI = 1 - \left[\frac{\sum_{i=1}^{N} \left(\left(Chl_p\right)_i - (Chl_O)_i \right)^2}{\sum_{i=1}^{N} \left(\left| \left(Chl_p\right)_i - \overline{Chl}_0 \right| + \left| (Chl_0)_i - \overline{Chl}_0 \right| \right)^2} \right], \ (0 \le WI \le 1) \tag{8}$$

$$LM = 1 - \left[\frac{\sum_{i=1}^{N} \left(\left| (Chl_0)_i - \left(Chl_p\right)_i \right| \right)}{\sum_{i=1}^{N} \left(\left| (Chl_0)_i - \overline{Chl}_0 \right| \right)} \right], \ (-\infty < LM \le 1) \tag{9}$$

In which, N is the number of data, Chl_O, Chl_p, $\overline{Chl}_0$, and $\overline{Chl}_p$ are the measured, calculated, mean measured, and mean calculated Chlorophyll concentration, respectively.

4. Results and discussion

In the present manuscript, two machines learning were proposed and their suitability to correctly and accurately predict Chlorophyll concentration was investigated. The proposed models were: (i) the optimally pruned extreme learning machine (OPELM) and (ii) the multilayer perceptron neural networks model (MLPNN) models. The proposed models were validated using eight performances metrics namely, RMSE, MAE, MAPE, RRMSE, R, LM WI, and NSE. The models were developed according to several scenarios as seen in Table 2. According to Table 2, the models were trained using two, three, four and five input variables. For the calibration, the models were trained by varying number of neurons in the hidden layer, and by the use of calibrated models, the Chl-a was predicted for the testing dataset. The obtained results will be presented and discussed below for each station separately.

4.1 Results at Charles River buoy (CR-Buoy) station

Table 3 shows the accuracies of the OPELM and MLPNN models according to the nine input combination. According to Table 3, during the validation phase it is clear that, among the nine models the first two input combination (OPELM1, OPELM2) and (MLPNN1, MLPNN2) achieved the best accuracy. Calibrating the models using the first input combination, i.e., OPELM1 and MLPNN1 by including the five water quality variables as input (TE, SC, pH, TU, DO) would likely lead to more accuracy compared to the other input combination, this, despite the difference in the models accuracy is small between the first and the second combination. Therefore we can conclude that the river turbidity (TU) plays a minor role in the prediction of the Chlorophyll concentration. The OPELM1 performed well in overall compared to the MLPNN1 with RMSE = 0.444, MAE = 0.293, RRMSE = 18.99%, and MAPE of 12.815%, and the predicted Chl-a concentration was highly fitted and had high correlation with in situ measured data having a high R, NSE, WI and LM values of 0.917,

TABLE 2 The input combinations of different models.

MLPNN	OPELM	Input combination
MLPNN1	OPELM1	TE, SC, pH, TU, DO
MLPNN2	OPELM2	TE, SC, pH, DO
MLPNN3	OPELM3	TE, pH, TU, DO
MLPNN4	OPELM4	TE, SC, pH, TU
MLPNN5	OPELM5	TE, pH, DO
MLPNN6	OPELM6	TE, SC, DO
MLPNN7	OPELM7	TE, TU, DO
MLPNN8	OPELM8	TE, DO
MLPNN9	OPELM9	TE, pH

TABLE 3 Performances of different models at lower Charles River buoy (CR-Buoy) station.

Models	R	NSE	LM	WI	RMSE	MAE	RRMSE	MAPE
				Training				
OPELM1	0.945	0.893	0.709	0.971	0.366	0.240	15.659	10.335
OPELM2	0.950	0.902	0.719	0.974	0.350	0.231	14.946	10.070
OPELM3	0.939	0.881	0.691	0.968	0.385	0.254	16.457	10.945
OPELM4	0.938	0.881	0.696	0.967	0.386	0.251	16.510	10.911
OPELM5	0.926	0.857	0.656	0.960	0.422	0.284	18.048	12.288
OPELM6	0.939	0.881	0.696	0.967	0.386	0.251	16.488	10.893
OPELM7	0.901	0.812	0.615	0.946	0.484	0.317	20.698	13.697
OPELM8	0.869	0.755	0.557	0.926	0.553	0.365	23.637	15.822
OPELM9	0.836	0.699	0.499	0.904	0.613	0.413	26.203	18.732
MLPNN1	0.927	0.859	0.659	0.961	0.420	0.281	17.935	12.236
MLPNN2	0.916	0.839	0.637	0.954	0.449	0.299	19.184	12.685
MLPNN3	0.916	0.839	0.635	0.955	0.448	0.300	19.142	13.015
MLPNN4	0.907	0.822	0.614	0.949	0.472	0.318	20.157	14.281
MLPNN5	0.894	0.799	0.582	0.941	0.501	0.344	21.438	14.842
MLPNN6	0.926	0.857	0.661	0.960	0.422	0.279	18.059	12.106
MLPNN7	0.907	0.823	0.626	0.949	0.471	0.308	20.120	13.182
MLPNN8	0.876	0.768	0.559	0.931	0.538	0.363	23.013	15.844
MLPNN9	0.838	0.703	0.501	0.906	0.609	0.411	26.041	18.843
				Validation				
OPELM1	0.917	0.839	0.641	0.954	0.444	0.293	18.998	12.815
OPELM2	0.917	0.840	0.630	0.955	0.443	0.302	18.956	13.411
OPELM3	0.904	0.816	0.606	0.948	0.475	0.321	20.332	14.278
OPELM4	0.902	0.810	0.606	0.945	0.483	0.321	20.671	14.281
OPELM5	0.893	0.797	0.589	0.941	0.498	0.336	21.340	14.775
OPELM6	0.889	0.790	0.578	0.940	0.507	0.344	21.717	15.169
OPELM7	0.873	0.761	0.567	0.927	0.541	0.353	23.185	15.464
OPELM8	0.850	0.722	0.522	0.913	0.584	0.390	24.994	16.838
OPELM9	0.812	0.658	0.467	0.889	0.647	0.435	27.703	19.449
MLPNN1	0.903	0.813	0.616	0.944	0.479	0.313	20.512	13.503
MLPNN2	0.901	0.810	0.604	0.943	0.483	0.323	20.669	13.855
MLPNN3	0.897	0.802	0.591	0.943	0.492	0.334	21.090	14.410
MLPNN4	0.889	0.780	0.578	0.933	0.519	0.344	22.220	15.123
MLPNN5	0.880	0.774	0.553	0.933	0.526	0.365	22.523	15.917
MLPNN6	0.896	0.802	0.594	0.941	0.492	0.332	21.090	14.392
MLPNN7	0.873	0.760	0.579	0.928	0.542	0.343	23.223	14.295
MLPNN8	0.844	0.711	0.511	0.909	0.595	0.399	25.486	17.111
MLPNN9	0.813	0.660	0.465	0.888	0.645	0.436	27.626	19.424

0.839, 0.641 and 0.954, respectively. The robust performances of the OPELM1 in the estimation of the Chl-a concentration indicated that OPELM1 has the potential to accurately estimate Chl-a concentration. Then, the performances of the OPELM1 in the estimation of the Chl-a were compared to those obtained using the MLPNN1, showing its superiority. The results suggested that OPELM1 improved the accuracy of the MLPNN1 by decreasing the RMSE, MAE, RRMSE, and MAPE by 7.307%, 3.390%, 7.381%, and 5.095%, respectively. In addition, the OPELM1 ensuring an improvement in terms of R, NSE, WI and LM values close to 1.527%, 3.099%, 3.900%, and 1.048%, respectively.

A comparison between the measured and predicted Chl-a concentration resulting from the models having four input combinations, i.e., combinations 2, 3, and 4 showed a high accuracy and slightly difference in models performances. The R, NSE, WI, and LM indices of the OPELM models were ranged from 0.902 to 0.917, 0.810 to 0.840, 0.606 to 0.630, and 0.945 to 0.955, respectively. Similarly, The R, NSE, WI and LM indices of the MLPNN models were ranged from 0.889 to 0.901, 0.780 to 0.810, 0.578 to 0.604, and 0.933 to 0.943, respectively. Therefore, considering the results reported in Table 3, the capability of the OPELM and MLPNN models can be considered good enough for accurately predicting Chl-a concentration. However, the OPELM show the most robust agreement against in-situ measured *Chl-a* in comparison to the MLPNN model. Furthermore, among the three models having four input variables (i.e., combination 2, 3 and 4), it is clear that form Table 3 that, the models that correspond to the combination 4, i.e., OPELM4 and MLPNN4 show the lowest accuracy in comparison to the combination 2 and combination 3 ones, highlighting the minor importance of the turbidity as input variable.

For the models having only three input variables (i.e., combinations 5, 6, and 7), in terms of errors metrics, the OPELM5 gave the best performances (RMSE = 0.526, MAE = 0.365, RRMSE = 22.52%, and MAPE = 15.917%), slightly less than the MLPNN6 (RMSE = 0.492, MAE = 0.332, RRMSE = 21.09%, and MAPE = 14.392%), while the OPELM7 and MLPNN7 possesses relatively the same accuracy with negligible differences (R = 0.873, NSE = 0.761, LM = 0.567 and WI = 0.927). For numerical comparison, removing the TU and DO concentration from the input variables (i.e., OPELM5), the performances of the OPELM1 model were significantly deteriorated for which the RMSE, MAE, RRMSE, and MAPE were dramatically increased by 10.84%, 12.798%, 10.975%, and 13.266%, respectively. However, the performances of the MLPNN1 were less deteriorated compared to the MLPNN6, and the LM index was remarkably decreased by 3.571% (dropped from 0.616 to 0.594), and the MAE and MAPE were significantly increased by 5.723% and 6.177%, respectively. In summary, OPELM5 and MLPNN6 were the two best performing models having only three input variables. Both performances metrics showed MLPNN6 performed slightly better than OPELM5, but the same performances metrics indicated that OPELM5 had obviously better performances than MLPNN5 and MLPNN7. Further analysis of the obtained results reveals that, reducing the number of input variables from five to two (i.e., combinations 8 and 9) generally decreased the models performances for both MLPNN and OPELM models. However, it is clear from Table 3 that, MLPNN9 and OPELM9 were the worst among all proposed models and worked equally with the same accuracy. Using only TE and pH as input variables decreased the R, NSE, WI and LM indices of the MLPNN1 model by 9.96%, 18.82%, 24.51%, and 5.93%, respectively, and the RMSE, MAE, RRMSE, and MAPE were dramatically increased by 24.74%, 28.21%, 25.75%, and 30.48%, respectively. The performances statistics shows that the OPELM1 model outperforms significantly the OPELM9 (Table 3). For instance, R, NSE, WI and LM indices are improved nearly 11.45%, 21.57%, 27.15%, and 6.18%, respectively. Also, numerical results suggest that OPELM1 guaranteed an improvement in the RMSE, MAE, RRMSE, and MAPE by 31.38%, 32.64%, 31.42% and 34.11%, respectively.

In addition to the numerical results discussed above, graphical representation of the obtained results was provided for further comparison between models accuracy. Fig. 5 shows the scatterplot of measured against calculated Chlorophyll concentration using the validation dataset for the best OPLEM1 and MLPNN1, which clearly highlighting the superiority of the OPELM1 model. The histogram of relative error (RE) for the same models was provided in Fig. 6. In addition Figs. 7 and 8 shows the boxplots and violin plot of the nine MLPNN and OPELM models compared to the measured Chlorophyll concentration using the validation dataset.

4.2 Results at Mystic River buoy (MR-Buoy) station

Results obtained at the Mystic River buoy (MR-Buoy) station are reported in Table 4. First, using only two variables as inputs (i.e., combinations 8 and 9), the performances of the OPELM and MLPNN models were significantly less than the other combination, and the lowest accuracy among all models was obtained using the OPELM8 and the MLPNN8 who have worked equally with the same performances and negligible differences, and the OPELM9 and MLPNN9 were slightly more accurate than the OPELM8 and the MLPNN8, this leads us to conclude that the pH is more suitable as predictor compared to the DO concentration. The estimated and measured Chl-a using the OPELM8 fit poorly with R, NSE, LM and WI equal to 0.691, 0.475, 0.328 and 0.808, respectively, and equally with the MLPNN8. A slight improvement is observed for both the

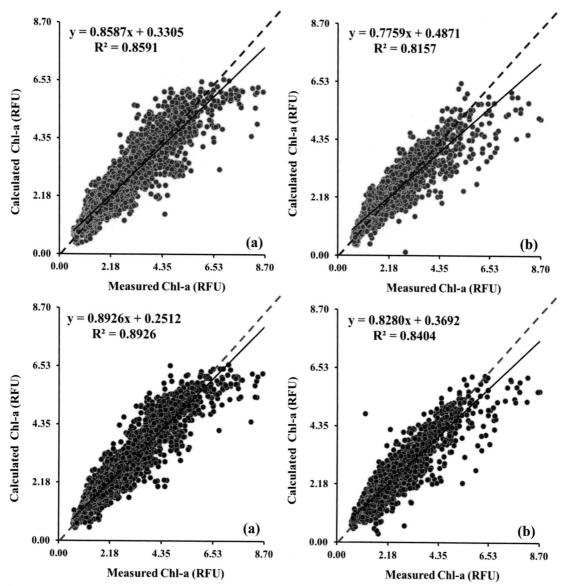

FIG. 5 Scatterplots of measured against calculated Chlorophyll (*Chl-a*) Concentration at the Charles River Buoy (CR-Buoy) station using MLPNN1 *(top panel)* and OPELM1 *(lower panel)*: (A) training and (B) validation.

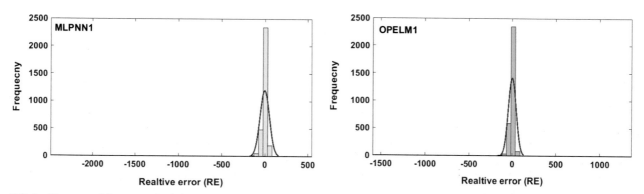

FIG. 6 Histogram of frequency of relative error (RE) for the optimum developed models, at the Charles River Buoy (CR-Buoy) station during the validation phase.

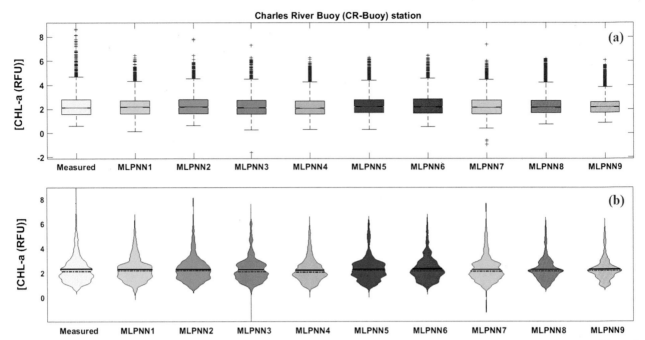

FIG. 7 Boxplot *(top panel)* and Violin plot *(lower panel)* for the MLPNN models at the Charles River Buoy (CR-Buoy) station: (A) training and (B) validation.

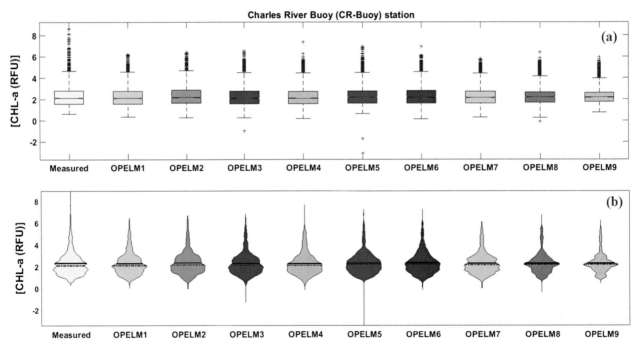

FIG. 8 Boxplot *(top panel)* and Violin plot *(lower panel)* for the OPELM models at the Charles River Buoy (CR-Buoy) station: (A) training and (B) validation.

TABLE 4 Performances of different models at Mystic River buoy (MR-Buoy) station.

Models	R	NSE	LM	WI	RMSE	MAE	RRMSE	MAPE
				Training				
OPELM1	0.959	0.920	0.751	0.979	0.579	0.404	18.229	15.672
OPELM2	0.938	0.880	0.699	0.967	0.706	0.488	22.230	20.260
OPELM3	0.938	0.879	0.697	0.967	0.710	0.493	22.367	18.547
OPELM4	0.959	0.920	0.754	0.979	0.578	0.400	18.187	15.401
OPELM5	0.871	0.758	0.574	0.927	1.004	0.693	31.626	27.979
OPELM6	0.884	0.781	0.598	0.935	0.955	0.653	30.079	26.040
OPELM7	0.899	0.809	0.619	0.945	0.893	0.618	28.120	23.197
OPELM8	0.732	0.536	0.379	0.830	1.391	1.008	43.809	43.885
OPELM9	0.783	0.613	0.447	0.869	1.270	0.898	39.991	37.205
MLPNN1	0.943	0.890	0.703	0.970	0.678	0.483	21.349	19.122
MLPNN2	0.925	0.855	0.670	0.960	0.776	0.537	24.450	21.216
MLPNN3	0.934	0.872	0.685	0.965	0.730	0.511	22.971	19.392
MLPNN4	0.931	0.867	0.674	0.963	0.744	0.530	23.431	20.372
MLPNN5	0.851	0.724	0.533	0.915	1.072	0.759	33.757	31.080
MLPNN6	0.855	0.732	0.538	0.917	1.058	0.750	33.320	30.615
MLPNN7	0.893	0.798	0.602	0.941	0.918	0.646	28.900	24.108
MLPNN8	0.712	0.507	0.355	0.813	1.434	1.047	45.159	45.528
MLPNN9	0.761	0.579	0.418	0.852	1.324	0.945	41.699	39.968
				Validation				
OPELM1	0.952	0.907	0.723	0.975	0.615	0.447	19.354	17.423
OPELM2	0.926	0.857	0.656	0.961	0.763	0.555	24.005	23.490
OPELM3	0.923	0.851	0.668	0.959	0.777	0.536	24.441	19.854
OPELM4	0.946	0.895	0.720	0.972	0.652	0.451	20.522	17.608
OPELM5	0.842	0.707	0.522	0.912	1.091	0.771	34.334	31.619
OPELM6	0.862	0.743	0.554	0.923	1.022	0.720	32.159	29.860
OPELM7	0.892	0.796	0.603	0.941	0.910	0.640	28.643	24.090
OPELM8	0.691	0.475	0.328	0.808	1.460	1.084	45.943	46.642
OPELM9	0.738	0.541	0.390	0.844	1.365	0.985	42.943	40.146
MLPNN1	0.938	0.879	0.686	0.967	0.699	0.507	22.007	19.640
MLPNN2	0.917	0.840	0.642	0.955	0.805	0.577	25.334	23.255
MLPNN3	0.925	0.856	0.665	0.961	0.766	0.541	24.091	20.520
MLPNN4	0.930	0.866	0.668	0.962	0.739	0.535	23.244	20.415
MLPNN5	0.839	0.703	0.508	0.908	1.097	0.794	34.518	31.862

Continued

TABLE 4 Performances of different models at Mystic River buoy (MR-Buoy) station—cont'd

Models	R	NSE	LM	WI	RMSE	MAE	RRMSE	MAPE
MLPNN6	0.843	0.710	0.511	0.909	1.084	0.788	34.121	32.683
MLPNN7	0.886	0.786	0.587	0.938	0.933	0.667	29.352	24.943
MLPNN8	0.684	0.466	0.315	0.799	1.472	1.105	46.306	47.128
MLPNN9	0.737	0.542	0.387	0.837	1.363	0.989	42.890	40.685

OPELM9 (4.70%, 6.60%, 6.20%, and 3.60%) and the MLPNN9 (5.30%, 7.60%, 7.20%, and 3.80%) in terms of R, NSE, LM and WI, respectively (Table 4), but neither of them had the necessary level of accuracy and guaranteed a performances higher than the acceptable average: the NSE values were slightly higher than 0.50 and the LM were below 0.40. Using three input variables, i.e., combinations 5, 6 and 7, a significant improvement is apparent for both MLPNN and OPELM models. In general, the estimated performances metrics values for all three models, e.g., MLPNN5 and OPELM5, MLPNN6 and OPELM6, MLPNN7 and OPELM7, are accurate with comparable accuracies. Nevertheless, the MLPNN7 and OPELM7 seem to be more accurate, leading to conclude that, for modeling CHL-a, combining the TU with DO and TE is more suitable compared to the pH and SC variables. Despite that the performances of different models on estimating CHL-a vary, estimation by the models having three input variables generate comparable results and do not exhibits high variability for which, the R, NSE, LM and WI were ranged from 0.842 to 0.892, 0.707 to 0.796, 0.522 to 0.603 and 0.912 to 0.941, between 86, 0.703 to 0.786, 0.508 to 0.587 and 0.908 to 0.938, between MLPNN5 and MLPNN7. Therefore, using three input variables (i.e., TE, TU, and DO) has had particularly strong positive effects on Chlorophyll estimation for which the OPELM7 improve the accuracy of the OPELM9 by decreasing the RMSE, MAE, RRMSE and MAPE by $\approx 33\%$, $\approx 35\%$, $\approx 33\%$, and $\approx 40\%$, respectively, and the MLPNN7 improve the accuracy of the MLPNN9 by decreasing the RMSE, MAE, RRMSE and MAPE by $\approx 32\%$, $\approx 33\%$, $\approx 32\%$, and $\approx 39\%$, respectively. While the OPELM7 was slightly more accurate than the MLPNN7, the two models demonstrate better performances.

The evaluation of the models having four input variables as predictors (i.e., combinations 2, 3, and 4) revealed that the estimated Chlorophyll was in close agreement with in situ measured data, especially the OPELM4 and MLPNN4 having as predictors four input variables (i.e., TE, SC, pH, TU: Table 4). High R, NSE, LM and WI was achieved and low RMSE, MAE, RRMSE and MAPE was observed for both models, with slightly superiority in favor to the OPELM4. However, when comparing the OPELM4 with OPELM7, it is clear that the performances of the OPELM4 model had markedly improved and that this profit in fact was substantially higher than $\approx 28\%$, $\approx 29\%$, $\approx 28\%$, and $\approx 26\%$, in terms of RMSE, MAE, RRMSE and MAPE reduction. For the MLPNN4 model, the statistics show relatively low improvement, with generally low gain compared to the gain obtained using the OPELM4 for which $\approx 21\%$, $\approx 20\%$, $\approx 21\%$, and $\approx 18\%$ reduction in the RMSE, MAE, RRMSE and MAPE values were achieved, respectively. According to Table 4, the OPELM4 slightly improved the statistical results of the MLPNN4 for which the R, NSE, LM and WI values increased from 0.930 to 0.946 ($\approx 1.6\%$), 0.866 to 0.895 ($\approx 3\%$), 0.668 to 0.720 ($\approx 5\%$), and from 0.962 to 0.972 ($\approx 1\%$), respectively. The improvement was more remarkable taking into account the NSE and LM metric, while the WI index reveals only negligible difference between the OPELM4 and MLPNN4. These findings were further corroborated with the results obtained using the best developed models having the five water quality variables as inputs; the OPELM1 and the MLPNN1. The best correlation between calculated and in situ Chlorophyll concentration was given by the OPELM1 slightly more accurate than the MLPNN1. Although the improvement was less than $\approx 1.4\%$, $\approx 2.8\%$, $\approx 3.7\%$, and $\approx 0.8\%$ (OPELM1 vs. MLPNN1), taking into account the R, NSE, LM and WI, respectively, we assumed that the OPELM1 was considered as the best and promising model for predicting Chlorophyll concentration at the Mystic River buoy (MR-Buoy) station. It is worth noting that, the very negligible difference between the OPELM1 and the OPELM4 models in terms of global performances which is reflected by the negligible improvement rates in terms of R($\approx 0.6\%$), NSE($\approx 1.2\%$), LM($\approx 0.3\%$) and WI($\approx 0.3\%$), respectively lead to consider the OPELM4 as a competitive model for Chlorophyll-a concentration prediction. Graphical representation of the obtained results was provided for further comparison between models accuracy. Fig. 9 shows the scatterplot of measured against calculated Chlorophyll-a concentration using the validation dataset for the best OPLEM1 and MLPNN1, which clearly highlighting the superiority of the OPELM1 model. The histogram of relative error (RE) for the same models was provided in Fig. 10. In addition Figs. 11 and 12 shows the boxplots and violin plot of the

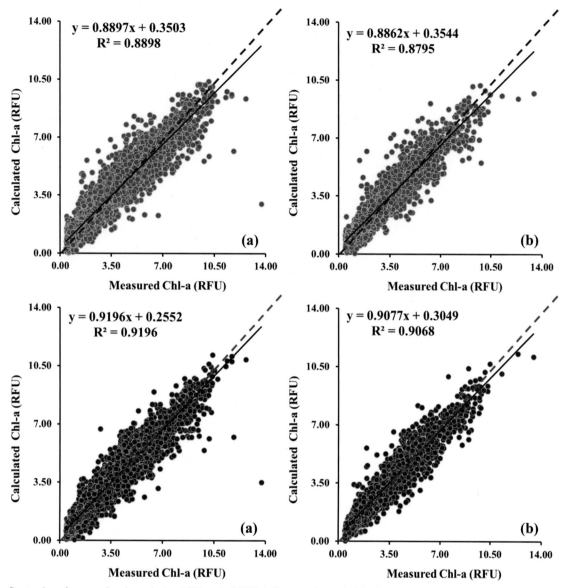

FIG. 9 Scatterplots of measured against calculated Chlorophyll (*Chl-a*) Concentration at the Mystic River Buoy (MR-Buoy) station using MLPNN1 *(top panel)* and OPELM1 *(lower panel)*: (A) training and (B) validation.

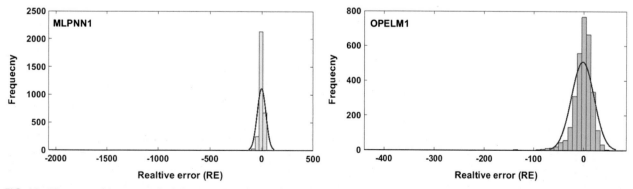

FIG. 10 Histogram of frequency of relative error (RE) for the optimum developed models, at the Mystic River Buoy (MR-Buoy) station during the validation phase.

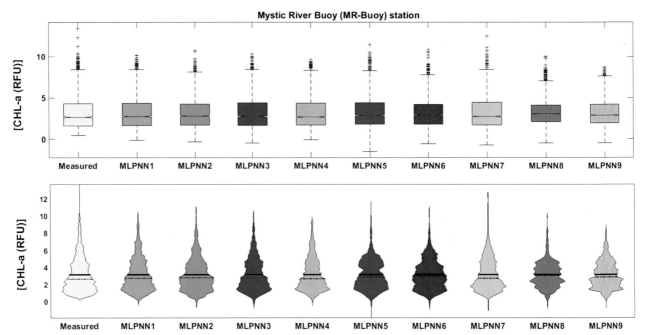

FIG. 11 Boxplot *(top panel)* and Violin plot *(lower panel)* for the MLPNN models at the Mystic River Buoy (MR-Buoy) station: (A) training and (B) validation.

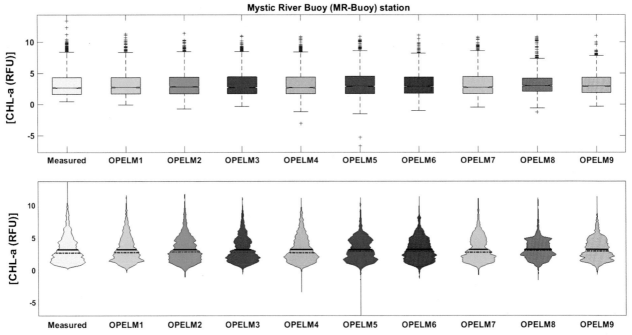

FIG. 12 Boxplot *(top panel)* and Violin plot *(lower panel)* for the OPELM models at the Mystic River Buoy (MR-Buoy) station: (A) training and (B) validation.

nine MLPNN and OPELM models compared to the measured Chlorophyll-a concentration using the validation dataset at the Mystic River buoy (MR-Buoy) station.

5. Conclusions

Results reported in the present paper demonstrate that Chlorophyll concentration can be predicted very well using easily measured water quality variables. Using data measured at fifteen minute intervals of time, the accuracy of the OPELM and the standalone MLPNN models having several inputs combination has been investigated, demonstrating the superiority of the OPELM model. Although the best accuracy was obtained using the OPELM model, good accuracy was also achieved using the MLPNN model, slightly less than the OPELM. Our analysis shows that OPELM model trained using the five water quality variables would result in R and NSE values greater than 0.915 and 0.835 at the Charles River buoy (CR-Buoy) station, and provided an R and NSE values greater than 0.950 and 0.905 at Mystic River buoy (MR-Buoy) station. The present work shows a new method for predicting Chlorophyll concentration that opens new possibilities of application of machines learning paradigms. The accuracy and simplicity of the proposed method makes the OPELM model a valuable and alternative tool for continuous monitoring of Chlorophyll concentration in the absence of the direct in situ measurement. Further development of the proposed model should be performed at different time scale and the application should be extended to other stations.

References

Adnan, R.M., Liang, Z., Heddam, S., Zounemat-Kermani, M., Kisi, O., Li, B., 2020. Least square support vector machine and multivariate adaptive regression splines for streamflow prediction in mountainous basin using hydro-meteorological data as inputs. J. Hydrol. 586, 124371. https://doi.org/10.1016/j.jhydrol.2019.124371.

Adnan, R.M., Liang, Z., Trajkovic, S., Zounemat-Kermani, M., Li, B., Kisi, O., 2019. Daily streamflow prediction using optimally pruned extreme learning machine. J. Hydrol. 577, 123981. https://doi.org/10.1016/j.jhydrol.2019.123981.

ALabdeh, D., Omidvar, B., Karbassi, A., Sarang, A., 2020. Study of speciation and spatial variation of pollutants in Anzali Wetland (Iran) using linear regression, Kriging and multivariate analysis. Environ. Sci. Pollut. Res., 1–14. https://doi.org/10.1007/s11356-020-08126-3.

Attar, N.F., Pham, Q.B., Nowbandegani, S.F., Rezaie-Balf, M., Fai, C.M., Ahmed, A.N., et al., 2020. Enhancing the prediction accuracy of data-driven models for monthly streamflow in Urmia Lake Basin based upon the autoregressive conditionally heteroskedastic time-series model. Appl. Sci. 10 (2), 571. https://doi.org/10.3390/app10020571.

Blix, K., Li, J., Massicotte, P., Matsuoka, A., 2019. Developing a new machine-learning algorithm for estimating chlorophyll-a concentration in optically complex waters: a case study for high Northern Latitude waters by using Sentinel 3 OLCI. Remote Sens. (Basel) 11 (18), 2076. https://doi.org/10.3390/rs11182076.

Chen, C., Jiang, B., Cheng, Z., Jin, X., 2019. Joint domain matching and classification for cross-domain adaptation via ELM. Neurocomputing 349, 314–325. https://doi.org/10.1016/j.neucom.2019.01.056.

Eshtay, M., Faris, H., Obeid, N., 2020. A competitive swarm optimizer with hybrid encoding for simultaneously optimizing the weights and structure of Extreme Learning Machines for classification problems. Int. J. Mach. Learn. Cybern., 1–23. https://doi.org/10.1007/s13042-020-01073-y.

Free, G., Bresciani, M., Trodd, W., Tierney, D., O'Boyle, S., Plant, C., Deakin, J., 2020. Estimation of lake ecological quality from Sentinel-2 remote sensing imagery. Hydrobiologia 847, 1423–1438. https://doi.org/10.1007/s10750-020-04197-y.

García-Nieto, P.J., García-Gonzalo, E., Fernández, J.A., Muñiz, C.D., 2015. Hybrid PSO-MARS-based model for forecasting a successful growth cycle of the Spirulina platensis from experimental data in open raceway ponds. Ecol. Eng. 81, 534–542. https://doi.org/10.1016/j.ecoleng.2015.04.064.

García-Nieto, P.J., García-Gonzalo, E., Fernández, J.A., Muñiz, C.D., 2016a. Using evolutionary multivariate adaptive regression splines approach to evaluate the eutrophication in the Pozón de la Dolores Lake (Northern Spain). Ecol. Eng. 94, 136–151. https://doi.org/10.1016/j.ecoleng.2016.05.047.

García-Nieto, P.J., García-Gonzalo, E., Fernández, J.A., Muñiz, C.D., 2016b. A hybrid PSO optimized SVM-based model for predicting a successful growth cycle of the Spirulina platensis from raceway experiments data. J. Comput. Appl. Math. 291, 293–303. https://doi.org/10.1016/j.cam.2015.01.009.

Hornik, K., 1991. Approximation capabilities of multilayer feedforward networks. Neural Netw. 4 (2), 251–257. https://doi.org/10.1016/0893-6080 (91)90009-T.

Hornik, K., Stinchcombe, M., White, H., 1989. Multilayer feedforward networks are universal approximators. Neural Netw. 2, 359–366. https://doi.org/10.1016/0893-6080 (89)90020-8.

Huang, G.B., Chen, L., Siew, C.K., 2006a. Universal approximation using incremental constructive feedforward networks with random hidden nodes. IEEE Trans. Neural Netw. 17 (4), 879–892. https://doi.org/10.1109/TNN.2006.875977.

Huang, G.B., Zhou, H., Ding, X., Zhang, R., 2011. Extreme learning machine for regression and multiclass classification. IEEE Trans. Syst. Man Cybern. B Cybern. 42 (2), 513–529. https://doi.org/10.1109/TSMCB.2011.2168604.

Huang, G.B., Zhu, Q.Y., Siew, C.K., 2006b. Extreme learning machine: theory and applications. Neurocomputing 70 (1–3), 489–501. https://doi.org/10.1016/j.neucom.2005.12.126.

Kaloop, M.R., El-Badawy, S.M., Ahn, J., Sim, H.B., Hu, J.W., Abd El-Hakim, R.T., 2020. A hybrid wavelet-optimally-pruned extreme learning machine model for the estimation of international roughness index of rigid pavements. Int. J. Pavement Eng. https://doi.org/10.1080/10298436.2020.1776281.

Legates, D.R., McCabe Jr, G.J., 1999. Evaluating the use of "goodness-of-fit" measures in hydrologic and hydroclimatic model validation. Water Resour. Res. 35 (1), 233–241. https://doi.org/10.1029/1998WR900018.

Luo, W., Zhu, S., Wu, S., Dai, J., 2019. Comparing artificial intelligence techniques for chlorophyll-a prediction in US lakes. Environ. Sci. Pollut. Res. 26 (29), 30524–30532. https://doi.org/10.1007/s11356-019-06360-y.

Lv, P.F., He, B., Guo, J., Shen, Y., Yan, T.H., Sha, Q.X., 2020. Underwater navigation methodology based on intelligent velocity model for standard AUV. Ocean Eng. 202, 107073. https://doi.org/10.1016/j.oceaneng.2020.107073.

Mamun, M., Kim, J.J., Alam, M.A., An, K.G., 2020. Prediction of algal chlorophyll-a and water clarity in monsoon-region reservoir using machine learning approaches. Water 12 (1), 30. https://doi.org/10.3390/w12010030.

Miche, Y., Sorjamaa, A., Bas, P., Simula, O., Jutten, C., Lendasse, A., 2010. OP-ELM: optimally pruned extreme learning machine. IEEE Trans. Neural Netw. 21 (1), 158–162. https://doi.org/10.1109/TNN.2009.2036259.

Miche, Y., Sorjamaa, A., Lendasse, A., 2008. OP-ELM: theory, experiments and a toolbox. In: Proceedings of the International Conference on Artificial Neural Networks, Prague, Czech Republic. Lecture Notes in Computer Science, vol. 5163, pp. 145–154, https://doi.org/10.1007/978-3-540-87536-9_16.

Millette, N.C., Kelble, C., Linhoss, A., Ashby, S., Visser, L., 2019. Using spatial variability in the rate of change of chlorophyll a to improve water quality management in a subtropical oligotrophic estuary. Estuar. Coasts 42 (7), 1792–1803. https://doi.org/10.1007/s12237-019-00610-5.

Moriasi, D.N., Arnold, J.G., Van Liew, M.W., Bingner, R.L., Harmel, R.D., Veith, T.L., 2007. Model evaluation guidelines for systematic quantification of accuracy in watershed simulations. Trans. ASABE 50 (3), 885–900. https://doi.org/10.13031/2013.23153.

Munyebvu, F., Mujere, N., Isaac, R.K., Eslamian, S., 2020. Chapter 5: Assessing the microbiological quality of potable groundwater from selected protected and unprotected wells in Murehwa District, Zimbabwe. In: Eslamian, S., Eslamian, F. (Eds.), Advances in Hydrogeochemistry Research. Nova Science Publishers, Inc., USA, pp. 121–138.

Park, Y., Cho, K.H., Park, J., Cha, S.M., Kim, J.H., 2015. Development of early-warning protocol for predicting chlorophyll-a concentration using machine learning models in freshwater and estuarine reservoirs, Korea. Sci. Total Environ. 502, 31–41. https://doi.org/10.1016/j.scitotenv.2014.09.005.

Pouzols, F.M., Lendasse, A., 2010. Evolving fuzzy optimally pruned extreme learning machine for regression problems. Evol. Syst. 1 (1), 43–58. https://doi.org/10.1007/s12530-010-9005-y.

Sebbar, A., Heddam, S., Djemili, L., 2019. Predicting daily Pan evaporation (E pan) from Dam reservoirs in the mediterranean regions of Algeria: OPELM vs OSELM. Environ. Process. 6 (1), 309–319. https://doi.org/10.1007/s40710-019-00353-2.

Sun, Y., Li, B., Yuan, Y., Bi, X., Zhao, X., Wang, G., 2019. Big graph classification frameworks based on extreme learning machine. Neurocomputing 330, 317–327. https://doi.org/10.1016/j.neucom.2018.11.035.

Sylaios, G.K., Gitsakis, N., Koutroumanidis, T., Tsihrintzis, V.A., 2008. CHLfuzzy: a spreadsheet tool for the fuzzy modeling of chlorophyll concentrations in coastal lagoons. Hydrobiologia 610 (1), 99–112. https://doi.org/10.1007/s10750-008-9358-4.

Tian, W., Liao, Z., Wang, X., 2019. Transfer learning for neural network model in chlorophyll-a dynamics prediction. Environ. Sci. Pollut. Res. 26 (29), 29857–29871. https://doi.org/10.1007/s11356-019-06156-0.

Xianquan, X., Xiaofu, X., Jianhua, T., 2013. Modelling chlorophyll-a in Bohai Bay based on hybrid soft computing approach. J. Hydroinf. 15 (4), 1099–1108. https://doi.org/10.2166/hydro.2012.146.

Xiaobo, L., Fei, D., Guojian, H., Jingling, L., 2014. Use of PCA-RBF model for prediction of chlorophyll-a in Yuqiao Reservoir in the Haihe River Basin, China. Water Sci. Technol. Water Supply 14 (1), 73–80. https://doi.org/10.2166/ws.2013.175.

Yajima, H., Derot, J., 2018. Application of the Random Forest model for chlorophyll-a forecasts in fresh and brackish water bodies in Japan, using multivariate long-term databases. J. Hydroinf. 20 (1), 206–220. https://doi.org/10.2166/hydro.2017.010.

Yu, H., Webb, G.I., 2019. Adaptive online extreme learning machine by regulating forgetting factor by concept drift map. Neurocomputing 343, 141–153. https://doi.org/10.1016/j.neucom.2018.11.098.

Yu, Z., Yang, K., Luo, Y., Shang, C., 2020. Spatial-temporal process simulation and prediction of chlorophyll-a concentration in Dianchi Lake based on wavelet analysis and long-short term memory network. J. Hydrol. 582, 124488. https://doi.org/10.1016/j.jhydrol.2019.124488.

Zhang, T., Huang, M., Wang, Z., 2020b. Estimation of chlorophyll-a concentration of lakes based on SVM algorithm and Landsat 8 OLI images. Environ. Sci. Pollut. Res., 1–14. https://doi.org/10.1007/s11356-020-07706-7.

Zhang, Y., Huang, J.J., Chen, L., Qi, L., 2015. Eutrophication forecasting and management by artificial neural network: a case study at Yuqiao Reservoir in North China. J. Hydroinf. 17 (4), 679–695. https://doi.org/10.2166/hydro.2015.115.

Zhang, Y., Wu, L., Ren, H., Liu, Y., Zheng, Y., Liu, Y., Dong, J., 2020a. Mapping water quality parameters in urban rivers from hyperspectral images using a new self-adapting selection of multiple artificial neural networks. Remote Sens. 12 (2), 336. https://doi.org/10.3390/rs12020336.

Zhu, S., Heddam, S., 2019. Modelling of maximum daily water temperature for streams: optimally pruned extreme learning machine (OPELM) versus radial basis function neural networks (RBFNN). Environ. Process. 6 (3), 789–804. https://doi.org/10.1007/s40710-019-00385-8.

Chapter 22

Proposing model for water quality analysis based on hyperspectral remote sensor data

M.V.V. Prasad Kantipudi[a], Sailaja Vemuri[b], N.S. Pradeep Kumar[c], S. Sreenath Kashyap[d], and Saeid Eslamian[e,f]

[a]*Symbiosis Institute of Technology, Symbiosis International (Deemed University), Pune, India,* [b]*Department of Electronics and Communication Engineering, Pragati Engineering College, Surampalem, India,* [c]*Department of Electronics and Communication Engineering, S.E.A.CET, Bangalore, India,* [d]*Department of Electronics and Communication Engineering, Kommuri Pratap Reddy Institute of Technology, Hyderabad, India,* [e]*Department of Water Engineering, College of Agriculture, Isfahan University of Technology, Isfahan, Iran,* [f]*Center of Excellence for Risk Management and Natural Hazards, Isfahan University of Technology, Isfahan, Iran*

1. Introduction

With progressive development in human living standards, rapid population growth and environmental pollution have become a globally severe concern in the last two decades. The massive utilization of natural resources required for human survival and activities has put a tremendous burden on the environment. Among many issues, resolving water pollution is a crucial requirement (Karthe et al., 2015). Water is an essential substance to sustain lives on the earth. It is a key element to ensure food security, ecological fortification, and human health. With the speedy development of society, usage of fertilizer in the agricultural, and rise in the discharge of industrial pollutants into rivers and oceans, causing a gradual increase in the un-wanted concentration level of various biochemicals the water, which brings challenges to water resources protection (Babu and Reddy, 2014). As humans' lifestyle has standardized upward, the requirement for high quantity and quality of water has emerged, posing a severe concern for future suitability (Wang et al., 2013). Therefore, a science-based water quality index (WQI) becomes significant for measuring the level of pollutants concentration in water and recommending whether and to what extent a specific water resource needs to be restored (Lin and Ke, 2019). The main quality parameters comprise d-chlorophyll-a (Chl-a), chemical oxygen demand (COD), DO (dissolved oxygen), nitrate-nitrogen (NO3-N), and dissolved organic carbon (DOC). The estimation of water quality variables is mainly based on three approaches, namely chemical, biological, and physical approach (Liu et al., 2013), as shown in Fig. 1. In chemical technique, electrochemical analysis is carried out to estimate the variable that shows the level of pollutant in the laboratory. In this approach, the tools used are large and expensive and require many reagents, which leads to secondary pollution. Also, the outcomes generated are not real time. In the biological technique, enrichment analysis and biosensor technology are used mainly to quantify water quality parameters. However, the drawback is the low detection accuracy and sensitivity rate compared to other approaches. In physical techniques, remote sensing technologies are used to acquire spectral and hyperspectral water samples. Apart from this, molecular spectroscopy is widely considered for analyzing pollutant contents in the water based on the techniques of emission and spectra of chemical substances (Pu et al., 2019).

Over the years, remote sensing mechanisms have been extensively adopted widely used to quantify water quality standards. However, one of the critical factors is to include the time-series dimension in the prediction analysis task. Different approaches are introduced and explored in the existing literature to perform forecasting of water quality variables that include physical, biological, or chemical factors that significantly influence the water quality. These approaches include statistical analysis, analytical modeling, and predictive models, along the decision-making process is widely introduced by the researchers. To determine the relationship between different water quality parameters, principal component analysis (PCA) is widely adopted in the literature (Cordoba et al., 2014). The work of Taskaya-Temizel and Casey (2005) applied PCA to analyze features water pollutant using spectral matrix. Also, this approach is utilized to matrix dimension reduction, and further reconstruction of new statistical variable is carried out. The chi-square distribution is adopted to assess local outlying degree in the molecular space. Another work toward adoption PCA is seen the study of Hou et al. (2013). In this, authors have performed comparative evaluation outcome based on the PCA dimensionality reduction and outcome without PCA dimensionality reduction. The result analysis exhibited that accuracy of prediction is better with PCA dimensionality

Handbook of HydroInformatics. https://doi.org/10.1016/B978-0-12-821962-1.00007-6

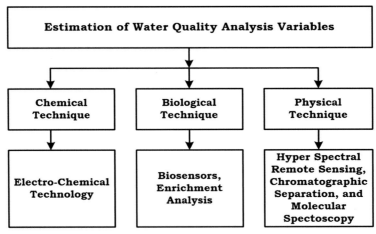

FIG. 1 Different techniques for estimation of water quality.

reduction. Other advanced techniques are also considered in the existing literature based on predictive algorithms like Bayesian Networks (BN), Artificial Neural Networks (ANN) (Ragavan, 2008), Deep learning, Support Vector Regression (SVR) (Environmental Protection Agency, 2001), and Convolution neural network and Long-Short-term Memory (LSTM) (Kantipudi et al., 2021). Prasad and Suresh (2015) suggested hybrid approach based genetic algorithm and SVR. Genetic algorithm is used to searches for the optimized SVR parameters and based on computed parameter SVR modeling is carried out for water quality analysis. The study of Goel et al. (2019) presents the multi-layer prediction model based on CNN for water quality analysis. In this approach transfer learning concept is used to estimate strong correlation between Landsat8 images and in situ water quality levels. The study outcomes exhibit effectiveness of the presented model performance in terms of water quality monitoring. Prasad et al. (2021) used concept of ANN to evaluate chlorine concentration. The modeling is carried based on the integrated approach of Monte-Carlo and ANN. Furthermore, recent literature has shown the effectiveness of hybrid models for water quality prediction but at the cost of computational complexity (Lee and Yoo, 2020). The objective of the proposed study is to introduce an advanced solution toward predicting the water quality with less water quality variables as input. International organizations like WHO EPA have established quality standards, which can be used as benchmarks for estimating water quality. There are a total of 101 quality standards listed by EPA (Banda and Kumarasamy, 2020). However, all quality standards cannot be considered in the prediction task as it is quite a time consuming and tedious task. Also, some quality standards have a significant effect on water quality compared to others. Therefore, the motive of the proposed study is also to address the dependency of predictive models on higher input samples for prediction by implementing LSTM neural network. The advantage of LSTM is that it actually bests suits to time-series data with the ability to learn the context needed to guess in advance for time series prediction problems. The study uses 5 water quality variables in the proposed modeling, i.e., chlorophyll, dissolved oxygen, turbidity, temperature, and specific conductance. The study predicts three water quality parameters (i.e., chlorophyll, dissolved oxygen, and turbidity) based on the input quality variables (specific conductance and temperature).

2. Data collection and study area

In literature, many research works have been done in the context of water monitoring. Existing studies have shown that the accuracy of analysis highly depends on input data's quality and intelligibility (Briciu et al., 2020). Due to the lack of detailed information and observations in most water monitoring organizations (Pashkova and Revenko, 2015), the current research work chooses to obtain data from one of the most reliable water resources organizations. Data set are kept updating over a certain time. The sample data for the case study adopted in the proposed study is obtained from the reposit maintained by the "National Water Information System," an open data repository of the United States Geological Survey (USGS).The study area of this research is located in Island Park Village, in New York State. The study used data from the Hog Island Channel monitoring station, where water samples are acquired and monitored using different mechanisms and technologies (Farrell-Poe et al., 2005; Hou et al., 2015; Tang et al., 2015; Helder, 2015; Zhu and Eslamian, 2020). Satellite telemeters are mainly used for water sample measurements with readings taken from 1.6 Ft. The sample data are obtained in 2014 at the time-interval of every 6min to perform an effective forecasting process. Few statistics about the selected water quality variables collected from USGS are listed in Table 1, including Minimum Value and Maximum Value.

TABLE 1 Statistical summary of the water quality variable from USGS.

Water quality variables	Minimum value	Maximum value
Chlorophyll (µg/L)	0.7	140
Dissolved oxygen (Mg/L)	3.6	18.0
Specific conductance (µs/cm)	38,900	49,100
Water temperature (C)	2.90	29.00
Turbidity (FNU)	< 0.1	120

3. Proposed model

This study focuses on exploring the effectiveness of the advanced leaning model, namely the Long Short-Term Memory (LSTM), in the application of water quality parameters estimation. The estimation is carried based on two water quality variables, namely DO and Chl-a and turbidity concentration. The flow representation of the data model established in this research study is highlighted in Fig. 2.

Fig. 2 depicts the standard approach of developing a computational prediction model. The basic logic behind implementing the long-short terms model is to map prediction and observation parameters into a solution space and conduct a linear relation among these parameters to attains the purpose of prediction modeling. In the first step of modeling, preprocessing operation is carried out over the input samples followed by normalization technique. In this process, the input samples are scaled in the range of [0,1] using the min-max normalization technique. This process prevents the frequent variation in the gradient and facilitates convergence smoothly. The preprocessed data are then dived into testing and training sets. Statistical attributes or features of water quality variables for each set are determined by minimum value, maximum value, mean, standard deviation, skewness, and kurtosis. In the segregation process of data samples, 70% of input samples are considered for training purposes, and 30% are considered for trained model testing purposes. The training and testing datasets are vectorized into frames, followed by a sliding windowing process. The proposed LSTM model is

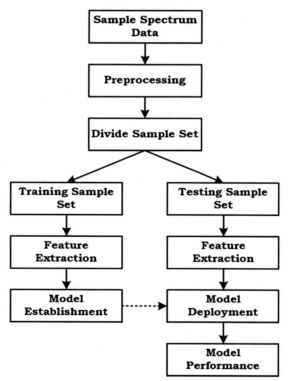

FIG. 2 Flow representation toward establishing a learning model of water quality estimation.

configured with an input layer having input values of Turbidity, Specific Conductance (μS/cm),and temperature. The hidden layer of LSTM consists of 64 and 32 units for predicting Chl-a and DO, respectively. Also, Exponential Linear Unit is used as an activation function in the hidden layer of the LSTM. A dense layer of LSTM is configured using 1 unit with activation function type "linear." Dropout function with a rate of 0.001 is used at each layer of LSTM to avoid overfitting issues. AdaGrad optimizer and MSE loss function is used for model training. Typical LSTM is shown in Fig. 3, and LSTM cell is shown in Fig. 4.

3.1 Long short-term memory (LSTM)

LSTM is a function approximator used for sequence prediction on new data. LSTMs were introduced as an advanced version of the conventional Recurrent Neural Network (RNN). LSTM addresses the issue of vanishing gradients, which impedes the learning of long data series by combining gating functions in its cell. The LSTM model includes layers of cell

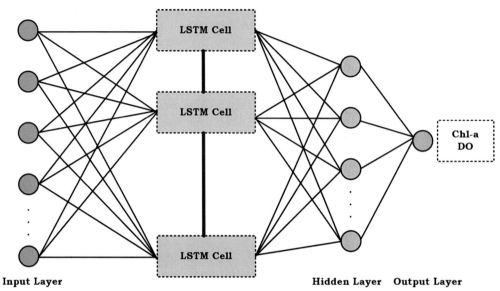

FIG. 3 Schematic of LSTM learning model.

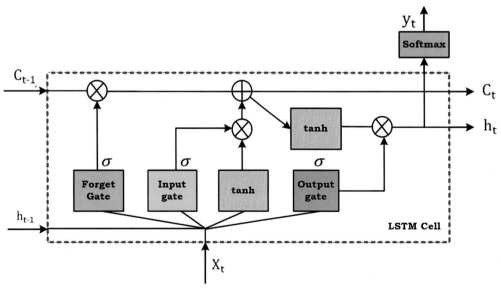

FIG. 4 Schematic of LSTM cell.

blocks having the ability of memory. Each layer consists of three multiplier units (i.e., gates): input gate, forget, and output gate. The core function of the input gate is to convey information to the long-term memory, also called memory state. In this process, a sigmoid function is initially used to adjust the input data fed to the memory state. A 1-dimensional cell array is constructed in the next operation, holding all likely expected information to be fed to the memory state by employing the activation function, i.e., hyperbolic tangent. Finally, the dot product between the adjustment weighting filter and 1-dimensional cell array is carried out, and the information is fed to the memory state. The forget gate then eliminates the information, i.e., not relevant to carry out further processing in the LSTM learning task. Further, the output gate decides what the next information should be selected from the last state to exhibits it as output. The learning architecture of the LSTM model is shown in Fig. 3.

A generic LSTM cell includes three gates integrated with sigmoid function and tangent. The prediction of output variables (y_t) performed by updating gates using summation using bigoplus operator ($\oplus$), and multiplication using bigotimes operator ($\otimes$) on the long-term memory unit (c_t). The fundamental function of LSTM can be expressed as follows:

$$i_t = \sigma(W_i x_i + R_i h_{t-1} + b_i)\ldots \tag{1}$$

$$f_t = \sigma\left(W_f x_t + R_f h_{t-1} + b_f\right)\ldots \tag{2}$$

$$\widetilde{c}_t = \sigma(W_c x_t + R_c h_{t-1} + b_c)\ldots \tag{3}$$

$$y_t = \sigma\left(W_y x_t + R_y h_{t-1} + b_y\right)\ldots \tag{4}$$

$$c_t = f_t c_{t-1} + i_t\widetilde{c}_t)\ldots \tag{5}$$

$$h_t = y_t \times \sigma(c_t)\ldots \tag{6}$$

where, x_t denotes the input data in vector,$W_i, W_f, W_y, R_i, R_f, R_y$ represents a weighting matrix from all three gates to the input. The variables $b_i, b_f,$ and b_y represents a bias of each gate, c_{t-1} represents a previous state, h_{t-1} represents output vector of previous state and h_t is the output vector.

4. Result analysis

The implementation of the proposed model is carried out on a numerical computing tool. The study has designed a prediction model based on the LSTM neural network to predict water quality parameters (i.e., DO turbidity and Chl-a) using a hyperspectral remote sensor water sample. After training the proposed prediction models, the testing data were used to test the model and predict the target water quality parameters. The models were evaluated and compared using statistical indicators, such as correlation coefficient RMSE and MSE. The qualitative observation is shown with MSE for predicted water quality variables. The quantified observation for performance measures for the LSTM model is shown in Table 2.

Fig. 5 exhibits analysis over increasing epochs for water quality variable "Chl-a." The performance analysis exhibits MSE observation for the training phase, testing phase, validation, and best performance. From the analysis, it can be seen that it took about 50 epochs to converge data samples for prediction. From the graph trend, it can be analyzed that the model is not subjected to overfitting issues as it maintains a balance between training, testing, and validation with P-best on 51 epochs. The next figure shows the performance analysis concerning Dissolved Oxygen (DO).

Fig. 6 exhibits analysis over increasing epochs for water quality variable "DO." The performance analysis exhibits the MSE observation for the training phase, testing phase, validation, and best performance. From the analysis, it can be seen that it took about 10 epochs to converge data samples for prediction. From the graph trend, it can be analyzed that the model

TABLE 2 Quantified observation.

Parameter	Units	Model	Training data			Testing data		
			CC	RMSE	MSE	CC	RMSE	MSE
Chlorophyll	µg/L	LSTM	0.87	0.044	0.0019	0.873	0.0470	0.00221
Dissolved oxygen	mg/L	LSTM	0.99	0.037	0.0018	0.970	0.0350	0.00123
Turbidity	FNU	LSTM	0.86	0.019	0.0003	0.782	0.0285	0.00081

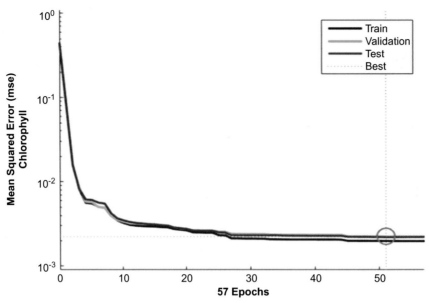

FIG. 5 Performance analysis for Chl-a.

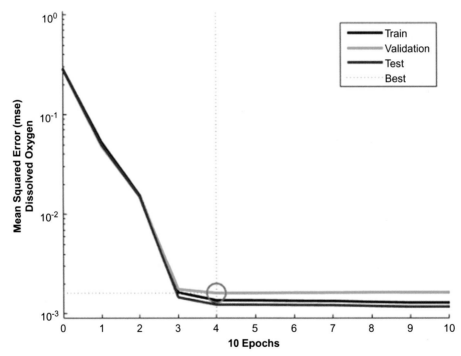

FIG. 6 Performance analysis for DO.

is not subjected to an overfitting issue as it exhibits similar performance among each other. Therefore, it maintains a balance between training, testing, and validation with P-best on 4 epochs. The next figure shows the performance analysis concerning Turbidity.

Fig. 7 exhibits analysis over increasing epochs for water quality variable "Turbidity." The performance analysis exhibits MSE observation for the training phase, testing phase, validation, and best performance (P-best). From the analysis, it can be seen that it took about 22 epochs to converge data samples for prediction. The graph trend shows that the model is subjected to little overfitting issue as it shows lees error in training compared to other metrics testing and validation with P-best on 16 epochs.

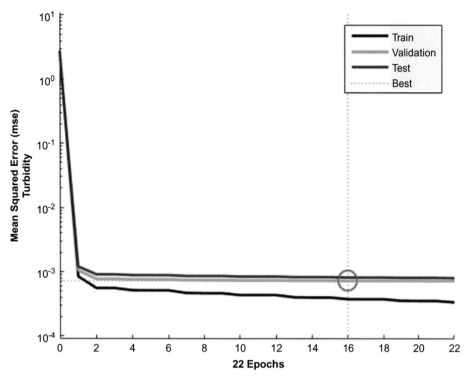

FIG. 7 Performance analysis for turbidity.

5. Conclusions

This book chapter presented a prediction model for the water quality analysis by forecasting water quality variables to determine the concentration level of chlorophyll, dissolved oxygen, and turbidity. The study uses a time-series hyperspectral data sample captured by satellite sensors. The dataset is first normalized using min-max scaling and further segregated into the model implementation testing and training set. The long-short term memory (LSTM) model has been implemented with AdaGrad optimizer and MSE loss function. The CC, MSE, and RMSE are considered the main performance metrics to evaluate the proposed model's effectiveness. The study outcome based on performance analysis provides an insight that the proposed model is efficient for water quality variable prediction in a real-time system.

References

Babu, C.N., Reddy, B.E., 2014. A moving-average filter-based hybrid ARIMA–ANN model for forecasting time series data. Appl. Soft Comput. 23, 27–38.

Banda, D., Kumarasamy, M., 2020. Development of a universal water quality index (UWQI) for South African river catchments. Water 12, 1534.

Briciu, E., Graur, A., Oprea, I., 2020. Water quality index of Suceava River in Suceava City metropolitan area. Water 12, 2111.

Cordoba, G.A.C., Tuhovčák, L., Tauš, M., 2014. Using artificial neural network models to assess water quality in water distribution networks. Proc. Eng. 70, 399–408.

Environmental Protection Agency, 2001. Parameters of water quality. USEDP, USA, p. 133.

Farrell-Poe, K., Payne, W., Emanuel, R., 2005. Arizona Watershed Stewardship Guide: Water Quality & Monitoring. pp. 1–18.

Goel, A.K., Chakraborty, R., Agarwal, M., Ansari, M.D., Gupta, S.K., Garg, D., 2019. Profit or loss: a long short-term memory based model for the prediction of share price of DLF group in India. In: In IEEE 9th International Conference on Advanced Computing (IACC). IEEE, pp. 120–124.

Helder, I., 2015. Water resources meet sustainability: new trends in environmental hydrogeology and groundwater engineering. Environ. Earth Sci. 73, 2513–2520.

Hou, D., Song, X., Zhang, G., Zhang, H., Loaiciga, H., 2013. An early warning and control system for urban, drinking water quality protection: China's experience. Environ. Sci. Pollut. Res. Int. 20 (7), 4496–4508.

Hou, D., Zhang, J., Yang, Z., 2015. Distribution water quality anomaly detection from UV optical sensor monitoring data by integrating principal component analysis with chi-square distribution. Opt. Express 23, 17487–17510.

Kantipudi, M., Kumar, S., Kumar Jha, A., 2021. Scene text recognition based on bidirectional LSTM and deep neural network. Comput. Intel. Neurosci., 1–11. https://doi.org/10.1155/2021/2676780.

Karthe, D., Chalov, S., Borchardt, D., 2015. Water resources and their management in central Asia in the early twenty first century: status, challenges and future prospects. J. Environ. Earth Sci. 73, 487–499.

Lee, C.W., Yoo, D.G., 2020. Decision of water quality measurement locations for the identification of water quality problems under emergency link pipe operation. Appl. Sci. 10, 2707.

Lin, X.U., Ke, L.I.U., 2019. Analysis of Water Quality Monitoring Data Based on LSTM. DEStech Transactions on Computer Science and Engineering, ICCIS, Kingdom of Saudi Arabia.

Liu, S., Tai, H., Ding, Q., Li, D., Xu, L., Wei, Y., 2013. A hybrid approach of support vector regression with genetic algorithm optimization for aquaculture water quality prediction. Math. Comput. Model. 58 (3–4), 458–465.

Pashkova, G.V., Revenko, A.G., 2015. A review of application of total reflection X-ray fluorescence spectrometry to water analysis. J. Appl. Spectrosc. Rev. 50, 443–472.

Prasad, K.M., Pradeep, K.N., Kashyap, S.S., Anusha, V.S., 2021. Time series data analysis using Machine Learning-(ML) approach. Libr. Philos. Pract., 1–7.

Prasad, K.M., Suresh, H.N., 2015. Spectral estimation using improved recursive least square (RLS) algorithm: an investigational study. In: Emerging Research in Computing, Information, Communication and Applications. Springer, New Delhi, India, pp. 363–376.

Pu, F., Ding, C., Chao, Z., Yu, Y., Xu, X., 2019. Water-quality classification of inland lakes using Landsat8 images by convolutional neural networks. Remote Sens. 11, 1674. https://doi.org/10.3390/rs11141674.

Ragavan, A., 2008. SAS Global Forum 2008 Data mining and predictive modeling data mining application of non-linear mixed modeling in water quality analysis SAS Global Forum, Data Mining and Predictive Modeling. Forum Am. Bar Assoc, USA.

Tang, B., Zhao, X., Wei, B., 2015. Optimization method of COD prediction model for detecting water quality by ultraviolet-visible spectroscopy. China Environ. Sci. 35, 478–483.

Taskaya-Temizel, T., Casey, M.C., 2005. A comparative study of autoregressive neural network hybrids. Neural Netw. 18 (5–6), 781–789.

Wang, Y., Wang, Y., Ran, M., Liu, Y., Zhang, Z., Guo, L., Zhao, Y., Wang, P., 2013. Identifying potential pollution sources in river basin via water quality reasoning based expert system. In: 2013 Fourth Int. Conf. Digit. Manuf. Autom, pp. 671–674.

Zhu, B.-Q., Eslamian, S., 2020. Chapter 13: Evaluation of natural water quality in the Jungar Basin in Central Asia and its implications on regional water resource management. In: Eslamian, S., Eslamian, F. (Eds.), Advances in Hydrogeochemistry Research. Nova Science Publishers, Inc., USA, pp. 339–356.

Chapter 23

Real-time flood hydrograph predictions using rating curve and soft computing methods (GA, ANN)

Gokmen Tayfur

Department of Civil Engineering, Izmir Institute of Technology, Izmir, Turkey

1. Introduction

Flood routing is a mean to obtain flow depths, velocities, volumes, and discharges at a river section. When a flood wave enters a river through an upstream section, by the flood routing method(s), one can trace the movement of flood wave along a channel length and thereby he/she can calculate flood hydrograph at any downstream section of the river. This information is needed for designing flood control structures, such as levees and also for channel improvements, navigation, and assessing flood effects.

There are basically two flood routing methods: (1) hydraulic and (2) hydrologic. Hydraulic methods are based on numerical solutions of St. Venant equations of the continuity and momentum. They can handle the lateral flow contributions. Hydrologic methods are, on the other hand, based solely on conservation of mass principle. Both methods require substantial field data, such as cross-sectional surveying, roughness, flow depth and velocity measurements that are costly and time consuming. When the lateral flow comes into picture, the hydrologic model needs to be modified to handle such a case which becomes often problematic.

With the first applications of ANN and GA in hydrology in the late 1990s, researchers have took advantage of these methods for the purpose of flood routing in natural channels, such that these methods have overcome some of major problems of the existing ones.

2. Flood routing methods

2.1 Hydraulic flood routing

It is based on the numerical solutions of the St. Venant equations (Chaudhry, 1993):

$$\frac{\partial A}{\partial t} + \frac{\partial Q}{\partial x} = q_l \tag{1}$$

$$\frac{\partial Q}{\partial t} + \frac{\partial (Qu)}{\partial x} + gA\frac{\partial h}{\partial x} = gA\left(S_o - S_f\right) \tag{2}$$

where A is cross sectional area, Q is the flow rate, q_l is unit lateral flow, g is gravitational acceleration, u is flow velocity, S_o is channel bed slope and S_f is friction slope (energy gradient). There are simplifications of these equations such as the kinematic wave and the diffusion wave. The simplifications are done in the momentum equation (Eq. 2). In the case of the kinematic wave approximation (KWA), all the inertia and the depth gradient terms are neglected, i.e., only the slope terms are preserved. In the diffusion wave approximation (DWA), in addition to the slopes terms, the depth gradient term is also conserved. Friction slope term can be related to flow discharge by using either the Manning or Chezy relations (Henderson, 1966). KWA is employed when flow occurs in steep channels.

St. Venant equations are highly nonlinear and therefore they can be solved numerically by employing either the finite difference, finite volume, or finite elements method. The numerical solutions might have convergence and stability problems (Chaudhry, 1993). Specifically, for flood wave routing, Lax, Lax-Wendroff, and MacCormack schemes were

Handbook of HydroInformatics. https://doi.org/10.1016/B978-0-12-821962-1.00019-2

developed in the literature to handle the numerical solutions of the St. Venant equations (Chaudhry, 1993). The solutions of these equations can give, at any time at any section of a river; the flow depth, flow velocity, and flow rate provided that roughness, slope, and lateral flow information is provided accurately. It can allow the variability in roughness, slope and lateral flow along a channel length. It is quite comprehensive and physics-based and therefore when possible they should be preferred.

2.2 Hydrologic flood routing

It is based on the conservation of mass, i.e., "mass in minus mass out is equal to change in mass per unit time." This, without lateral flow contribution, can be expressed as follows (Henderson, 1966):

$$Q_{in}(t) - Q_{out}(t) = \frac{dS(t)}{dt} \tag{3}$$

where, $Q_{in}(t)$: inflow discharge at time t, $Q_{out}(t)$: outflow discharge at time t, and S is the storage. When Eq. (3) is averaged over a $\triangle t$ time interval, the equation becomes (Henderson, 1966);

$$\frac{Q_{in}(t + \triangle t) + Q_{in}(t)}{2} - \frac{Q_{out}(t + \triangle t) + Q_{out}(t)}{2} = \frac{S(t + \triangle t) - S(t)}{\triangle t} \tag{4}$$

When storage is computed as a function of only outflow using the critical flow assumption, the routing method is called **the storage flood routing method** and it is employed for flood routing in reservoirs. For hydrologic flood routing in rivers, **the Muskingum method** is employed. It is based on Eq. (4) and relates the storage as a function of inflow and outflow rates, as follows (Henderson, 1966):

$$S(t) = K[xQ_{in}(t) + (1 - x)Q_{out}(t)] \tag{5}$$

where, K and x are parameters. K is considered as a wave travel time thus it has a unit of time and x is called the storage coefficient and it is generally taken as 0.2, although there are several methods in the literature to accurately compute x for a given river reach (Henderson, 1966).

Gill (1978) advocated an alternative approach for accounting nonlinearity in the channel routing process by modifying the storage equation of the classical Muskingum method. He replaced the linear storage equation of the classical Muskingum method by the nonlinear storage equation of the following form:

$$S(t) = K[xQ_{in}(t) + (1 - x)Q_{out}(t)]^m \tag{6}$$

where, m is the exponent of the weighted discharge $[\theta I(t) + (1 - \theta)Q(t)]$. In addition to the two parameters (K and x) employed in the classical Muskingum method, Gill's storage equation employs a third parameter in the form of nonlinear exponent m. This method is called **the nonlinear Muskingum (NLM) method**. When $m = 1$, the storage equation of the NLM method reduces to the storage equation of the classical Muskingum method.

Perumal and Price (2013) advanced the NLM by varying the parameters (K, x, and m) at every routing time interval based on the hydrodynamic principles. They called the method as **the Variable Parameter McCarthy-Muskingum (VPMM) method**. Perumal et al. (2013) evaluated and compared the channel routing performance of a physically based variable parameter McCarty-Muskingum (VPMM) method proposed by Perumal and Price (2013), and the NLM method (Mohan, 1997). They concluded that the use of VPMM method, which is capable of accounting the nonlinear behavior of flood wave movement process without involving the model calibration process, is more reliable for field applications than the conceptually based NLM method which performs relatively poorly in calibration as well as in verification modes of the routing process.

2.3 Rating curve method

Moramarco et al. (2005) developed **the Rating Curve Method (RCM)** which can handle the lateral flow contributions. They related discharge at the downstream station to measured flow variables at the upstream station as follows:

$$Q_d(t) = \alpha \frac{A_d(t)}{A_u(t - T_L)} Q_u(t - T_L) + \beta \tag{7}$$

where Q_u is the upstream discharge, Q_d is the downstream discharge, A_d and A_u are the effective downstream and upstream cross sectional flow areas obtained from the observed stages, respectively; T_L is the wave travel time depending on the wave celerity, c; and α β are the model parameters that can be estimated as (Moramarco and Singh, 2001):

$$Q_d(t_b) = \alpha \frac{A_d(t_b)}{A_u(t_b - T_L)} Q_u(t_b - T_L) + \beta \tag{8}$$

$$Q_d\left(t_p\right) = \alpha \frac{A_d\left(t_p\right)}{A_u\left(t_p - T_L\right)} Q_u\left(t_p - T_L\right) + \beta \tag{9}$$

where $Q_d(t_b)$ is the base flow rate at the downstream section; $Q_d(t_p)$ is the peak discharge at the downstream section; t_p and t_b are the times when the peak stage and baseflow occurs at the downstream section, respectively. In particular, t_b is assumed to be the time just before the start of the rising limb of the hydrograph.

Base flow rate $Q_d(t_b)$ can be computed from the velocity measurements during low flows. The peak discharge $Q_d(t_p)$ is surmised as the contribution of two main elements: (i) the upstream discharge delayed for the wave travel time T_L, $Q_u(t_p - T_L)$, with its attenuation, Q^*, due to flood routing along the reach of length L; and (ii) the lateral inflows, q_pL, during the time interval $(t_p - T_L, t_p)$ (Moramarco et al., 2005):

$$Q_d\left(t_p\right) = \left(Q_u\left(t_p - T_L\right) - Q^*\right) + q_pL \tag{10}$$

In Eq. (10), T_L is implicitly assumed as the time to match the rising limb and the peak region of the upstream and downstream dimensionless hydrographs. The flood attenuation (Q^*) is computed from the Price formula (Raudkivi, 1979). The lateral inflow contribution, q_pL, is obtained from the solution of the characteristic form of the continuity equation (Moramarco et al., 2005). q_p is estimated by assuming that along the characteristic corresponding to the downstream peak stage, the following relationship holds (Moramarco and Singh, 2000):

$$\frac{A_d\left(t_p\right) - A_u\left(t_p - T_L\right)}{T_L} = q_p \tag{11}$$

Once $Q_d(t_b)$ and $Q_d(t_p)$ are known, parameters α and β are obtained from the solution of Eqs. (8) and (9).

The hydraulic and hydrologic methods briefly described above require substantial data such as the measurements of flow depth, flow velocity and topographical surveying that are time consuming and costly. When lateral contribution becomes significant, which is the general case in large river basins, the procedures become more complicated and the problem of parameter estimation emerges. For example; Franchini et al. (1999) developed a methodology based on a variable parameter Muskingum–Cunge model with a specific parameterization scheme. The application of this model is however complex and it requires the estimation of nine parameters. The RCM model, for each event that occurs in the same river reach, has to determine the wave travel time and the model parameters. In other words, the model uses different values of model parameters and wave travel time for each event at each river reach. With the advent of the soft computing methods applications in hydrology in the last two decades, the researchers have tried to overcome of the difficulties of the hydrologic and hydraulic flood routing methods, briefly summarized above, by utilizing especially the genetic algorithm (GA) and artificial neural network (ANN).

3. Soft computing methods (GA, ANN) in flood routing

3.1 Genetic algorithm (GA)

The genetic algorithm (GA) was first introduced by Holland (1975) who was inspired by the biological processes of natural selection (i.e., Darwinian evolutionary view), and the genetic operations (i.e., Mendelian genetics). The first engineering application of GA by Goldberg (1983) triggered the use of it in different disciplines, including hydraulic, hydrology, and water resources engineering.

3.1.1 GA basics

GA has four basic units of *bit*, *gene*, *chromosome*, and *gene pool*. *Bit* is the basic building block which is represented by a digit of 1 or 0. The combination of bits forms a *gene* representing a model parameter. The attachment of genes forms a *chromosome* standing for a possible solution. *Gene pool* is formed by many individual chromosomes (Tayfur, 2012).

GA algorithm have four basic operations of *generation of initial gene pool*, *calculation of fitness for each chromosome*, *selection of chromosomes*, *operation of cross-overing* and *mutation* (Tayfur, 2012). Uniform distribution or a normal distribution can be employed to randomly generate initial chromosomes for the gene pool (Sen, 2004). Fitness of each chromosome can be evaluated by first substituting each chromosome into objective function to find their values $f(C_i)$ and then obtaining their fitness by Eq. (1) (Sen, 2004) as follows:

$$F(C_i) = \frac{f(C_i)}{\sum_{i=1}^{N} f(C_i)} \tag{12}$$

where, N is the number of chromosomes in the gene pool, C_i is the chromosome i, $f(C_i)$ is value of the objective function for chromosome i, and $F(C_i)$ is fitness value for chromosome i.

Selection of chromosomes after the evaluation of their fitness values can be performed randomly. There are methods available for the selection process such as roulette wheel and ranking (Sen, 2004). After the selection process, pairs (parent chromosomes) are first formed and they are then subjected to the cross-over operation by interchanging the genes. The last operation in a single iteration is the mutation by which bits are reversed (i.e., 1 to 0 or 0 to 1).

By these operations, it intends to search the solution space thoroughly. Fig. 1 shows an example for crossover and mutation operations. As seen, first two chromosomes (parent chromosome I and parent chromosome II) are subjected to the cross-over by the single cut from the third digit on the left, yielding new chromosomes (off-spring I and off-spring II) at the bottom. The values of 187 and 106 become 123 and 170, respectively after the crossover operation. By the mutation operation, fourth digits from the left of the off-springs are reversed. Thus, the final version of the off-springs becomes 107 and 186, respectively at the bottom. By the basic GA operations, large portion of solution space is searched to reach the global optimum. Fig. 2 shows the flowchart on how the GA algorithm works. One can find more details on GA elsewhere (Goldberg, 1999; Sen, 2004; Tayfur, 2012).

3.1.2 GA-based RCM model for real-time flood hydrograph prediction

The model employs the physics-based formulation (Eq. 7) of the RCM and finds its parameters using basic processes of genetic algorithms. Unlike the RCM, the GA model, for each river reach, employs an average wave travel time and finds single set of average values for the model parameters. Tayfur et al. (2009) employed the rating curve method (RCM) (Eq. 7) as a basis to apply GA for hydrograph predictions. The GA simply finds the optimal values of the coefficients (and) of the RCM by minimizing MAE function.

$$MAE = \frac{1}{N} \sum_{i-1}^{N} |Q_n - Q_p| \tag{13}$$

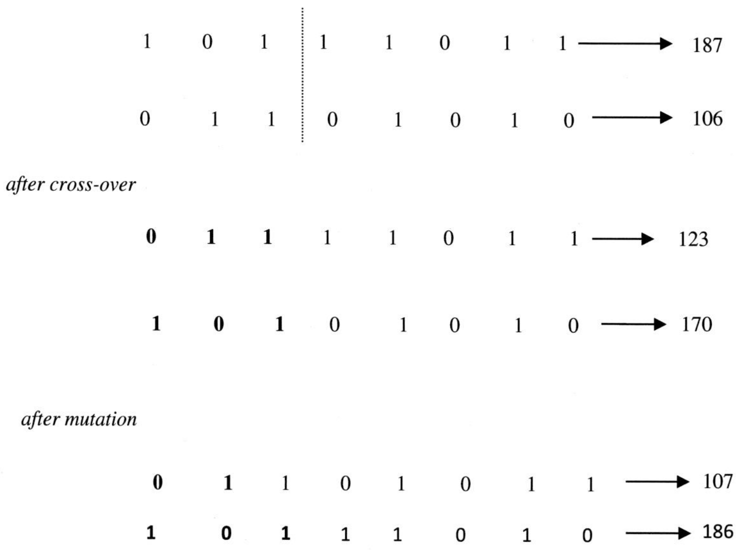

1	0	1	1	1	0	1	1	⟶ 187
0	1	1	0	1	0	1	0	⟶ 106

after cross-over

0	**1**	**1**	1	1	0	1	1	⟶ 123
1	**0**	**1**	0	1	0	1	0	⟶ 170

after mutation

0	**1**	1	0	1	0	1	1	⟶ 107
1	**0**	1	1	1	0	1	0	⟶ 186

FIG. 1 Example for crossover and mutation operations.

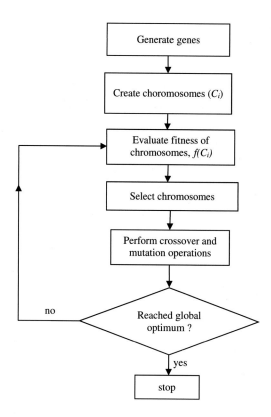

FIG. 2 Flow chart for GA algorithm.

where N is the number of observations, Q_m is measured flow discharge, and Q_p is predicted flow discharge, which is $Q_d(t)$ computed by Eq. (7). They called their model GA_RCM. They tested their model on three equipped river reaches of the Upper Tiber River in Central Italy (Fig. 3).

They considered severe storm events occurred in the three different reaches for GA_RCM model calibration and testing. For the sake of brevity, only events occurred in one of the river reach, namely the reach in between Santa Lucia and Ponte Felcino is presented herein. The main properties of the selected flood events are summarized in Table 1. The wave travel time is about 4 h and α and β parameter values used by the RCM model for each event are also shown in Table 1.

In the GA modeling, Tayfur et al. (2009) employed 100 chromosomes in the initial gene pool, 75% cross-over rate, 5% mutation rate and 10,000 iterations. The range for α was constrained in $[-5$ to $5]$ while β was in $[-50$ to $50]$ for each iteration. Four events marked by * in Table 1 were used for calibrating the model parameters by GA for the reach and their optimal values were obtained as $\alpha = 1.20$ and $\beta = -5.96$.

Fig. 4 presents the simulations measured at Felcino station by the GA_RCM model. As seen, the model shows an excellent performance in capturing the trend, time to peak, and the peak rates of the hydrographs.

The percentage error in peak discharge and the error in time to peak for each event simulated by the GA_RCM and standard RCM models are summarized in Table 2. Note that, in the case of peak rate, a negative error value indicates under-estimation whereas a positive value indicates overestimation. In the case of time to peak, negative error value indicates early rise in reaching the peak rate, while positive value indicates delay. According to Table 2, the GA_RCM and RCM models make about 10% overprediction error of the peak rate of event December 1990. GA_RCM model predicts the peak rates of other two events with less than 4% error, while RCM produces, on average, about 14%. Especially, the peak rate of January 1994 is almost exactly predicted by the GA_RCM model, while RCM over-predicts it with about 24% error. The time to peak for each event is exactly predicted by GA-RCM model, while RCM has, on the average, 40 min delay (Table 2).

GA_RCM model successfully simulates hydrographs at each river reach having different wave travel time and lateral inflows. It can closely capture trends, time to peaks, and peak rates. It outperforms the standard RCM, although using substantially less data. Tayfur et al. (2009) also performed sensitivity analysis by using low peak hydrographs in the calibration stage and predicting the high peak hydrographs. In a similar fashion, they employed shorter wave travel time events in the calibration stage and predicted longer wave travel time events. In all those analysis, they concluded that GA_RCM model has a good extrapolation capability.

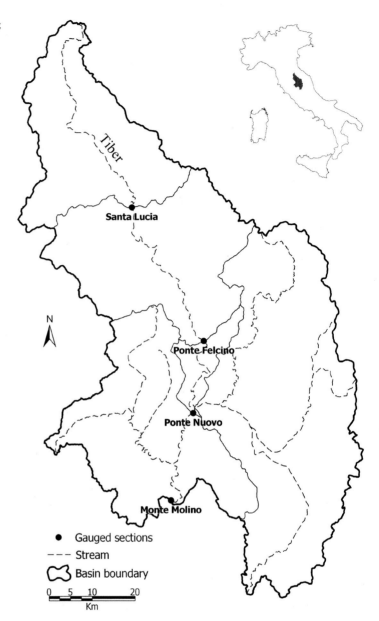

FIG. 3 Upper Tiber river basin with the location of the gauging sites (Tayfur et al., 2009).

Tayfur and Moramarco (2008) took advantage of the GA optimization method by further proposing the following equations:

$$Q_d(t) = \alpha \frac{h_d^\beta(t)}{h_u^\gamma(t - T_L)} + \eta \tag{14}$$

$$Q_d(t) = \alpha_1 h_u^{\beta_1}(t - T_L) + \alpha_2 h_d^{\beta_2}(t) + \eta \tag{15}$$

where h_d is the flow depth at downstream station, h_u is the flow depth at upstream station and α, β, γ, η, α_1, α_2, β_1, β_2 are the model parameters, whose optimal values are found by the GA model. Equation (14) is based on the formulation of Eq. (7) and called as *GA_Stage I* herein. Here, only flow stage information is required. As opposed to the RCM, it does not require velocity, discharge and cross sectional information. From a data reduction point of view, this is a major advantage. Equation (15) is based on the kinematic wave approximation and called as *GA_Stage II* herein.

Tayfur and Moramarco (2008) also proposed a formulation that predicts flow rate at a downstream station using only water surface elevation data. The elevation consists of elevation of a station from a reference datum plus flow depth as

TABLE 1 Main characteristics of flood events observed at Santa Lucia and Ponte Felcino stations.

Date	Santa Lucia station			Ponte Felcino station			α	β
	Q_b (m^3/s)	Q_p (m^3/s)	V (10^6 m^3)	Q_b (m^3/s)	Q_p (m^3/s)	V (10^6 m^3)		
December 1990	9	419	48	5	404	57	1.15	−9
January 1994	36	108	8	51	241	18	1.7	−35
May 1995*	4	71	10	9	139	19	1.24	−2.94
January 1997*	18	120	24	36	225	52	1.28	−3.19
		146			359			
June 1997*	5	345.6	28	11	450	49	1	0.25
January 2003	24	39.7	13	49	113	38	1.68	−24
		57.5			224			
February 2004*	22	91.8	12	55	278	44	1.48	−13
		47.2			153			

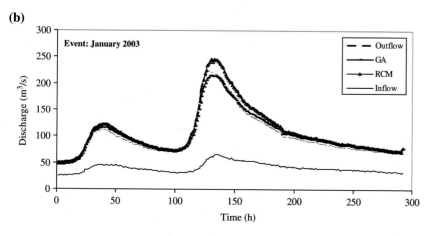

(a)

Event: January 1994

(b)

Event: January 2003

FIG. 4 GA-based RCM and RCM model simulations at Ponte Felcino station on (A) January 1994; and (B) January 2003 (Tayfur et al., 2009).

TABLE 2 Percentage errors in peak discharge (E_{Qp}) and time to peak (E_{Tp}) for the events in Fig. 4 observed at Ponte Felcino station.

Event	E_{Qp} (%)		E_{Tp} (%)	
	GA_RCM	RCM	GA_RCM	RCM
January 1994	−0.55	23.5	0.0	8.3
January 2003	−3.75	9.7	0.0	2.3

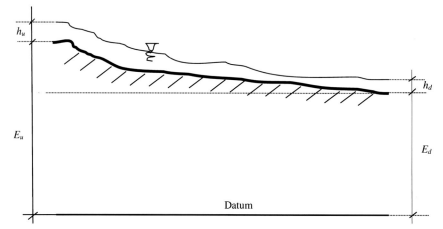

FIG. 5 Schematic representation of water surface elevation from a reference datum [in Eqs. 16 and 17: $E = (E_u - E_d) + h_u$] (Tayfur and Moramarco, 2008).

schematically presented in Fig. 5. Considering the reference datum at the downstream station enables one to need measurement of only flow depth at the downstream station. They proposed the formulations as follows:

$$Q_d(t) = \alpha \frac{h_d^{\beta}(t)}{E^{\gamma}(t - T_L)} + \eta \tag{16}$$

$$Q_d(t) = \alpha_1 E^{\beta_1}(t - T_L) + \alpha_2 h_d^{\beta_2}(t) + \eta \tag{17}$$

where E is the elevation, as shown in Fig. 5. Models, expressed by Eq. (16) and Eq. (7), are called herein as ***GA_Elevation I*** and ***GA_Elevation II*** models, respectively.

These proposed four models were applied to predict real-time flood hydrographs in Tiber River branches, shown in Fig. 3. For each river reach, four events were employed to calibrate the model parameters. The calibrated parameter values were used in the related equations to make predictions of new events. For the sake of brevity, Fig. 6 shows prediction of single events at each river reach.

Above results imply that one can construct a model in the form of Eq. (14) and then find the optimal values of the parameters of the equation using the GA model to perform hydrograph simulations. Such a model would require easily measurable flow stage data at upstream and downstream stations only. In a similar fashion, we can construct a model in the form of Eq. (16) and obtain the optimal values of the parameters of the equation by the GA to do hydrograph simulations. Such a model is even better, since it would require only the elevation data of both the stations. In a way, the elevation data can be easily obtained from satellite maps. This has an important implication for ungauged sites, where flow rates can be easily predicted.

We realize that GAs are strong optimization methods, which can perform quite satisfactorily with substantially less but easily measurable data. This, in turn, has a positive economic implication. That means by this method, we can save time, labor, and money and do, at the same time, a very efficient work.

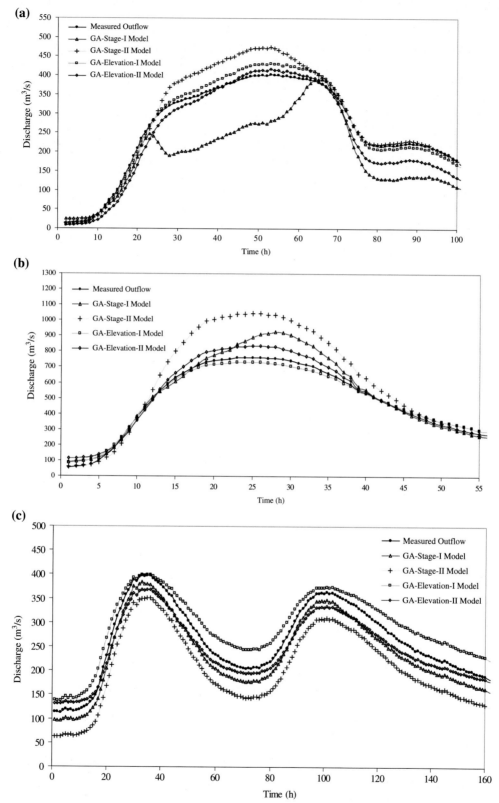

FIG. 6 Hydrograph Predictions (A) at the Lucia–Felcino river reach on Jan 1994, (B) at the Lucia–Nuovo river reach on Feb 1999, and (C) at the Lucia–Molino river reach on January 2001 (Tayfur and Moramarco, 2008).

3.2 Artificial neural network (ANN)

Artificial neural network (ANN) is inspired from the information processing of the brain. There are basically two aspects on how ANN is deduced from the nervous system of a brain. First aspect is the physical deduction where the brain nervous system is overly simplified and resembled in an artificial network consisting of layers of neurons that process information. Second aspect is the learning ability of a brain through experience and experiments. This learning ability of human being and his/her biological nervous system have inspired researchers to form a system made up of artificial layers containing neurons which process information and produce a system output. We call the whole system as the artificial neural network. The developments in the training algorithms have made ANN be employed in many disciplines from finance to engineering to solve classification, prediction, forecasting, and optimization problems. These methods are very powerful in mapping inputs to outputs. They are very powerful to capture trends among several variables. Hence, they are commonly employed in solving highly nonlinear engineering problems, including the ones in water resources engineering field (Matouq et al., 2013).

ANN is attractive for discharge prediction and flood forecasting, because they can accommodate the nonlinearity of the watershed runoff process and uncertainty in parameter estimation, have the capability to extract relationship between input and output of the process without explicitly considering the physics of the process, find relationships between different input samples, and generalize a relationship from small subsets of data while remaining robust in the presence of noisy or missing input.

3.2.1 ANN basics

In hydrologic applications, a three layer-feedforward type of artificial neural network is commonly considered (Fig. 7). In a feedforward network, the input quantities are fed into input layer neurons, which, in turn, pass them on to the hidden layer neurons after multiplication by a weight. A hidden layer neuron adds up the weighted input received from each input neuron, associates it with a bias, and then passes the result on through a nonlinear transfer function. The output neurons do the same operation as does a hidden neuron.

The back-propagation algorithm finds the optimal weights by minimizing a predetermined error function (E) of the following form (Tayfur, 2012):

$$E = \sum_P \sum_p (y_i - t_i)^2 \tag{18}$$

where y_i=component of a network output vector Y; t_i=component of a target output vector T; p=number of output neurons; and P=number of training patterns.

In the back propagation algorithm, optimal weights would generate an output vector $Y = (y_1, y_2, ..., y_p)$ as close as possible to the target values of the output vector $T = (t_1, t_2, ...,t_p)$ with a selected accuracy level. The back propagation algorithm employs the gradient-descent method, along with the chain rule of differentiation, to modify the network weights as (Tayfur, 2012):

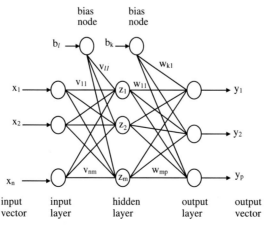

FIG. 7 Representation of three layer feed forward ANN.

$$v_{ij}^{new} = v_{ij}^{old} - \delta \frac{\partial E}{\partial v_{ij}} \tag{19}$$

where v_{ij} = weight from ith neuron in the previous layer to the jth neuron in the current layer and δ = learning rate.

The network learns by adjusting the biases and weights that link its neurons. Before training begins, a network's weights and biases are set equal to small random values. Also, due to the nature of the sigmoid function used in the back-propagation algorithm, all external input and output values before passing them into a neural network are standardized. Without standardization, large values input into an ANN would require extremely small weighting factors to be applied and this could cause a number of problems (Dawson and Wilby, 1998). The details of ANN are available in the literature (Tayfur, 2012).

3.2.2 ANN for real-time flood hydrograph prediction

Tayfur et al. (2007) applied ANN to predict event-based flood hydrographs measured at different gauging stations in Upper Tiber Basin in central Italy (see Fig. 3). Several accurate flow measurements were available which allowed the estimation of the rating curve for each section (Moramarco et al., 2005). Seven severe storm events were available and four events (June 1997, May 1995, January 1997, February 2004) were chosen by Tayfur et al. (2007) for training the ANN. They used the remaining three events for testing the model. The main properties of the selected flood events are summarized in Table 3. It is seen that the lateral inflow contribution was significant in some of the events. The river reach between Santa Lucia and Ponte Felcino gauging stations (Fig. 3) was considered by Tayfur et al. (2007) for testing the models for flood prediction. The ANN model used flow stage data at Santa Lucia station (upstream station) and the flow stage data at Ponte Felcino station (downstream station) to predict the flow discharge at Ponte Felcino station. The travel time between the two stations is about 4 h. As pointed out earlier, since flow stage is an easily measurable variable, engineers usually tend to relate flow rate to flow stage as it is the case in the rating curve method.

Tayfur et al. (2007) trained the ANN with a learning rate of 0.01 and 2000 iterations. The network had two neurons in the input layer, five neurons in the inner layer and one neuron in the output layer. The number of neurons in the hidden layer was decided by the commonly employed trial and error procedure. For this purpose, the mean error (ME) and mean relative error (MRE) were used as error measures. Accordingly, the number of iterations that provided the minimum ME and MRE values was the stopping criterion for terminating the iterations. For example, for this particular problem of flood hydrograph prediction application, the values of the error measures started with $ME = 181.7\,m^3/s$ and $MRE = 63.4$ at the first iteration and rapidly decreased to $31.1\,m^3/s$ and 32.4, respectively, after the 100th iteration. The errors then gradually decreased to and stabilized at $ME = 10.4\,m^3/s$ and $MRE = 16.5$ after 2000 iterations.

Tayfur et al. (2007), then, employed the trained ANN and the linear RCM to predict hydrographs of the three testing events (December 1990, January 1994, January 2003) measured at the Ponte Felcino station (Table 3). Fig. 8 shows the

TABLE 3 Main characteristics of observed flood events at stations on the Tiber river.

Date	Santa Lucia station			Ponte Felcino station			
	Q_b (m³/s)	Q_p (m³/s)	V (10⁶ m³)	Q_b (m³/s)	Q_p (m³/s)	V (10⁶ m³)	T_L (h)
December 1990	8	418	50	10	404	60	2.0
January 1994	36	108	19	51	241	35	3.0
May 1995[a]	4	71	10	9	139	19	4.0
January 1997[a]	18	120	24	36	225	52	3.5
		146			359		
June 1997[a]	5	346	28	11	450	49	5.0
January 2003	24	58	14	50	218	41	3.5
February 2004[a]	22	91	7	55	276	27	3.5

Q_b = base flow; Q_p = peak discharge; V = direct runoff volume; T_L = travel time.
[a]Used for ANN model training.

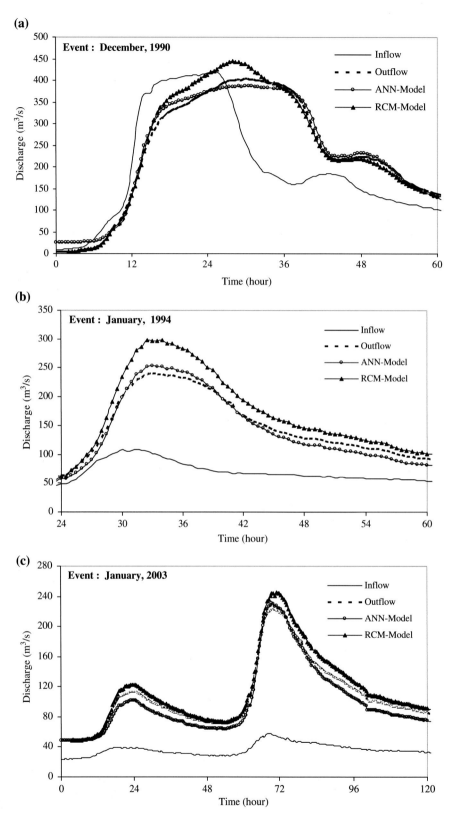

FIG. 8 ANN and RCM model simulations of flood hydrographs measured at Ponte Felcino gauging station on (A) December 1990; (B) January 1994; and (C) January 2003 (Tayfur et al., 2007).

TABLE 4 Percentage errors in peak discharge, E_{Qp}, and time to peak, E_{Tp}.

Event	E_{Qp} (%)		E_{Tp} (%)	
	ANN	RCM	ANN	RCM
Dec 1990	−4	10	0	−6.7
Jan 1994	3	9	0	2.2
Jan 2003	5	24	0	3.0
Average	**4.2**	**14.3**	**0**	**4.0**

predicted hydrographs. It is seen that ANN satisfactorily predicted the hydrographs in terms of the overall trend, time to peak and peak discharges. Overall, it yielded better results than did RCM which, in general, overestimated the discharge.

For the two-peak hydrograph of January 2003, shown in Fig. 8C, ANN underpredicted the lower peak but closely captured the higher peak, whereas RCM better predicted the lower peak but over-predicted the higher peak. The percentage error in peak discharge and time to peak was computed for each event and is given in Table 4. Note that a negative error value indicates underestimation, whereas a positive value indicates overestimation. ANN predicted the peak discharge of each event with less than a 5% error, while RCM had more than 10% error. For the January 2003 event, RCM over-predicted the peak discharge with about 24% error, while ANN had a 5% error. The time to peak was exactly predicted by ANN, while RCM had about a 4% error.

Tayfur et al. (2007) carried out sensitivity analysis by considering flow stage at the upstream station as the only input variable. The results revealed the poor performance of the network, especially for a river reach receiving significant lateral flow. Also, they investigated the extrapolation capability of the network by considering low peak hydrographs in the input vector. The results revealed that ANN is not a good extrapolator. They cannot predict high peak hydrographs when they are trained with low peak ones.

ANN, by using only flow stage data, that is easily measurable, can make good predictions of real flood hydrographs. This is advantages over the RCM model which requires measurements of cross-sections and flow velocities, in addition to flow stage. Also, the RCM employs different travel time and attempts to find values of α and β for each event, even for the same river reach.

4. Conclusions

Flood routing has a long history, more than a century. The original methods are based on the conservation of mass and applied to flood routing in reservoirs and in river reaches. With the advent of the computers, the St. Venant equations were solved in their simplest form. As the computers were developed, so as the numerical solutions of the St. Venant equations, leading to very comprehensive numerical flood routing methods. For these models to be very effective, substantial data and parameter estimation are required, that is often a problematic and costly, apart from the numerical convergence and instability problems. Engineers tend to use simple but problem solver models. For that reason, with the advent of the soft computing methods such as ANN and GA, they were eager to employ them in flood routing.

ANN is a black box model optimization algorithm. Given a pairs of input–output data, it is trained by finding optimal values for the connection weights between neurons located in neighboring layers. It hence cannot reveal any information or yield any mathematical relation between input variables and the output variable. They are strong interpolators but not extrapolators. They need to be trained with the flood hydrograph events for each river reach separately. Despite all these shortcomings, they are powerful to yield answers in a short period of time, which may be crucial for authorities for handling flood situations.

GA is an optimization algorithm, which maximizes or minimizes and objective function under some constraints. It often finds optimal values of some parameters of mathematical equation. Therefore, the relation between input and output variables are given mathematically. Hence, we can say that GA is not a black box model, as ANN. Since there is mathematical relation, it has both interpolation and extrapolation capabilities. One more advantage of GA is that, one can propose a new equation, provided that it physically makes sense, and find its optimal values of the coefficients and exponents.

Soft computing methods (GA, ANN) can make the good simulations of real-time flood hydrographs using substantially less data, such as the easily measurable flow stage. In that sense, they are quite advantages. The use of soft computing methods, together with physics-based models, would be the way forward for solving problems in water resources engineering field, including the flood routing.

References

Chaudhry, M.H., 1993. Open-Channel Flow. Prentice Hall, New Jersey, USA.

Dawson, W.C., Wilby, R., 1998. An artificial neural network approach to rainfall-runoff modeling. Hydrol. Sci. J. 43 (1), 47–66.

Franchini, M., Lamberti, P., Di Giammarco, P., 1999. Rating curve estimation using local stages, upstream discharge data and a simplified hydraulic model. Hydrol. Earth Syst. Sci. 3 (4), 541–548.

Gill, M.A., 1978. Flood Routing by the Muskingum method. J. Hydrol. 36, 353–363.

Goldberg, D.E., 1983. Computer-Aided Gas Pipeline Operation Using Genetic Algorithms and Rule Learning (Ph.D. thesis). University of Michigan, Ann Arbor, MI, USA.

Goldberg, D.E., 1999. Genetic Algorithms. Addison-Wesley, USA.

Henderson, F.M., 1966. Open Channel Flow. MacMillan, USA, New York.

Holland, J.H., 1975. Adaptation in Natural and Artificial Systems. University of Michigan Press, Michigan, USA.

Matouq, M., El-Hasan, T., Al-Bilbisi, H., Abdelhadi, M., Hindiyeh, M., Eslamian, S., Duheisat, S., 2013. The climate change implication on Jordan: A case study using GIS and Artificial Neural Networks for weather forecasting. J. Taibah Univ. Sci. 7 (2), 44–55.

Mohan, S., 1997. Parameter Estimation of Nonlinear Muskingum models using Genetic Algorithm. J. Hydraul. Eng. 123 (2), 137–142.

Moramarco, T., Singh, V.P., 2000. A practical method for analysis of river waves and for kinematic wave routing in natural channel networks. Hydrol. Process. 14, 51–62.

Moramarco, T., Singh, V.P., 2001. Simple method for relating local stage and remote discharge. J. Hydrol. Eng. 6 (1), 78–81.

Moramarco, M., Barbetta, S., Melone, F., Singh, V.P., 2005. Relating local stage and remote discharge with significant lateral inflow. J. Hydrol. Eng. 10 (1).

Perumal, M., Price, R.K., 2013. A fully volume conservative variable parameter McCarthy-Muskingum method: theory and verification. J. Hydrol. 502, 89–102.

Perumal, M., Naren, A., Ch.Madhusudana, R., 2013. Appraisal of two forms of nonlinear Muskingum flood routing methods. In: 6th International Perspective on Water Resources & the Environment. January 7–9, Izmir, Turkey.

Raudkivi, A.J., 1979. Hydrology: An Advanced Introduction to Hydrological Processes and Modeling. Pergamon, New York, USA.

Sen, Z., 2004. Genetic Algorithm and Optimization Methods. Su Vakfı Yayınları, Istanbul, Turkey (in Turkish).

Tayfur, G., 2012. Soft Computing in Water Resources Engineering: Artifical Neural Networks, Fuzzy Logic, and Genetic Algorithm. WIT Press, Southampton, U.K.

Tayfur, G., Moramarco, T., 2008. Predicting hourly-based flow discharge hydrographs from level data using genetic algorithms. J. Hydrol. 352 (1–2), 77–93.

Tayfur, G., Moramarco, T., Singh, V.P., 2007. Predicting and forecasting flow discharge at sites receiving significant lateral inflow. Hydrol. Process. 21, 1848–1859.

Tayfur, G., Barbetta, S., Moramarco, T., 2009. Genetic algorithm-based discharge estimation at sites receiving lateral inflows. J. Hydrol. Eng. 14 (5), 463–474.

Chapter 24

River Bathymetry acquisition techniques and its utility for river hydrodynamic modeling

Azazkhan I. Pathan[a,b], Dhruvesh Patel[a,b], Dipak R. Samal[c], Cristina Prieto[d], and Saeid Eslamian[e,f]

[a]Sardar Vallabhbhai National Institute of Technology, Surat, Gujarat, India, [b]Pandit Deendayal Petroleum University, Gandhinagar, Gujarat, India, [c]CEPT University, Ahmedabad, Gujarat, India, [d]Environmental Hydraulics Institute IH Cantabria- Instituto de Hidraulica Ambiental de la Universidad de Cantabria, Santander, Spain, [e]Department of Water Engineering, College of Agriculture, Isfahan University of Technology, Isfahan, Iran, [f]Center of Excellence for Risk Management and Natural Hazards, Isfahan University of Technology, Isfahan, Iran

1. Introduction

Bathymetry seems to be the only mode to study, discover, and achieve the vast portion of the earth covered under water. For many purposes, large-scale hydrodynamic and hydrologic models operating at local to world-wide scales are used (Bierkens, 2015; Yamazaki et al., 2011). Such aims include climate change predicting effects on water supply, measuring flood hazards (Baldassarre et al., 2020), hydrological prediction (Fan et al., 2014), and assisting hydrological process works. Designing large-scale hydrodynamic models is determining parameter values for the model and is one of the several challenges faced. In situ quantification of physical parameters like bathymetry is time-consuming and erroneous, due to unreachability of large river systems and the size scale. Bathymetry is the science that studies the topography of the seafloor and other bodies of water by measuring their depths. Also, it refers quite accurately to the sea-level depth of the seabed; however, the principles concerned for the bathymetry measurements are distant from this ordinary. Subsequently, it is a measurement of the water level in a body of water such as a sea, a river, or a lake (Kearns and Breman, 2010). Bathymetric maps look very similar to topographic maps which employ contour lines to show the variation in elevation of the landscape. They connect points of equal depth on the bathymetric maps. A smaller circle inside it may show an ocean trench. It may also show an underwater structure or a seamount. In earlier civilizations, scientists would carry out Bathymetric estimates by elevating a heavy rope over a ship's corner and measuring the length of the rope they were taking to touch the seafloor. However, these measurements were incomplete and incorrect. Often the rope traveled in a nonlinear path to the seafloor, as the flows caused it to deviate. Only one point at a time could the rope measure depth. Researchers would also have to make a vast number of rope measurements to get a realistic view of the seafloor (https://www.nationalgeographic.org/encyclo pedia/bathymetry). The mountains, shelves, canyon, and seafloor trenches were mapped to varying degrees of precision from the mid-19th century. Nowadays, seafloor depth can be estimated from kilometer to centimeter using methods as broad as ship multibeam sonar, aircraft and satellite remote sensing, underwater platform, and satellite radar altimetry. Various kinds of bathymetry data were collected and arranged into a grid sequence covering the seabed at 1 arc-minute spatial resolution per pixel (<2 km) of even finer resolution for parts of the globe (e.g., <100 m resolution for U.S. coastal water).

2. History

The Oceans enclose about 75% of the surface of the earth. Various methods such as seafloor photo graphics, remotes acoustic techniques, and geological were established (Lutron). Echo sounding techniques were very popular because of quick depth measurement data acquisition during World War-2. Due to the improvement of designing, technology, and science, high-quality low-cost transducer becomes more widely used during the year of 1950. In 1970, the high-resolution single-beam echo sounder (with high frequency) became available. In the 1980s, GLORIA side-scan sonar, which has an operating frequency of 6 kHz and SEAMARC, which has an operational frequency of 12 kHz were used extensively. Still,

Handbook of HydroInformatics. https://doi.org/10.1016/B978-0-12-821962-1.00025-8

the accuracy of depth information was low. Single-beam echo sounder substituted by lead line approaches since the 1920s onward herewith continuous measurement of depth along the ship track was accessible. Finally, a single-beam echo sounding approach was generally used in the year of 1950, only after improved transducer technology was this possible. The spatial resolution of this kind of system was available at around 30–60° (Gao, 2009; Lane, 2004). For higher resolution, narrow beam width 2–5° methods have been designed much late. Since the 1960s, the system has become completely active at real-time altitudes such as roll pitch correction. The multibeam bathymetry measurement system made available with further facilities similar to data storage and real-time capabilities in the 1970s. Apart from depth information, this sounding system approach is useful for many scientific study purposes include geotechnical properties, and geological surveys to describe the seabed and evaluation research like ridge study, habitant research and gas hydrate. The most recent successor to a single beam approach is the multibeam echo sounder. In the 1970s, seabeam and multibeam transmissions were applicable. Multibeam methodologies for single transmission use multiple narrow beam transmission, giving quality seafloor coverage. The first multibeam Hydro sweep structure was fixed on board Ocean Research Vessel (ORV) Sagar Kanya (Ministry of Earth Science, New Delhi) in India, and functioned by National Oceanography in 1990s. The hydro sweeping system used to form 59 receiver-beams with 2.2° beam breadths cover the seafloor two times the center-beam depth. Once more, the National Oceanography Institute, alteration of the hydro-sweep multibeam system was made in the 1995.

3. Bathymetry measurement techniques used across world

Bathymetric maps made from remote sensing images are becoming progressively popular. Application of remote sensing pixel color to map the depth of water in fluvial surroundings by calibrated depth color relations is sound recognized. Color (Westaway et al., 2003), multispectral (Lyon et al., 1992) and grayscale (Winterbottom and Gilvear, 1997) imagery were used for mapping bathymetric in the river in the UK. Recent attempts to develop hydraulic parameters by remote sensing include estimating river widths utilizing a raster-based arrangement of inundation extent derived from satellite imagery such as LANDSAT (Pathan et al., 2022a). Gao (2009) has used World view-2 multispectral satellite images of coastal nearby California using wave celerity approach and linear dispersion relationship applied to surface gravity wave, together in the surf region and outside the surf region. Remote sensing technique such as LIDAR often estimate the floodplain topography, however, the above mentioned methods are still unable to infiltrate the surface of the water, and thus cannot provide any interpretation of the submerged stream bed.

To minimize this improbability, it is necessary to complement the bathymetric datasets. The river bed is usually studied by assessing the elevation at various points laterally the river through field survey, like the total station in the situation of wadeable rivers and the survey of deep rivers by boats. Logistic constraints on such techniques often limit their application to scale only. Likewise, remote sensing oriented approaches have been developed in New Zealand to estimate bathymetry, but are costly and the uncertainty in their forecasts Grows with rising river depth and turbulence (Gao, 2009). The latest remote sensing methodologies and modern computer-aided program presentation now enable numeric flow modeling across a huge river. (McKean et al., 2014) have differentiated flow model forecasts using LIDAR bathymetry through the ones made using the total station channel field survey the Middle Fork Salmon River, Idaho, USA. The findings indicate that an airborne bathymetric LIDAR could even map the topography of the river with sufficient accuracy to assist a numeric flow model.

Hilldale and Raff (2008) have demonstrated that real time global positioning gives the greatest quality bathymetry datasets for the shallow slow-moving river. LIDAR survey has been done for accuracy and precision on the river Yakima and Trinity River Basins in the United States. Generally, field survey techniques, either land-based or boat-mounted, are employed to obtain elevation point measurements. Land survey methods, like terrestrial laser scanning global positioning systems, total stations, can be used to accomplish high-resolution and accurate bathymetry data. Feurer et al. (2008) have used LIDAR techniques with very high-resolution mapping for depth measurement on the site located on the middle Durance River, France.

4. Bathymetry measurement techniques used in India

Tripathi and Rao (2002) have used Indian remote sensing 1D LISS-III datasets for bathymetry mapping in Kakinada Bay, India, in which a global positioning system has been used to locate precise sea depth sampling points and an echo sounder was used to gather information on the depth of the sea. Image processing methods like remote sensing multispectral data analysis, provide a time-efficient and efficient solution for estimating water depths, are considered quite desirable for bathymetry applications. Pattanaik et al. (2015) have introduced a 4-band image developed by Indian Remote Sensing 1C/1D LISS-III Sensor satellite to provide accurate measurements of depth on Odisha coast, India, which was examined by the radioactive transfer equations that used to obtain remotely sensed water depth data. The devastating effect of the tsunami on the coast of South India on 26 December 2004 has been well recorded. Anandan and Sasidhar (2008) have used the GIS software to extract the data sets from chart to produced three-dimensional bathymetry representation along the near-shore of Kalpakkam, India, of the region offshore before and after the tsunami.

5. Methods of acquiring bathymetry data

The methods used for bathymetry measurements can be divided into two types: field survey methods and remote sensing methods. These shall be addressed in the following parts:

5.1 Field survey methods

Typically point elevation measurements are obtained by field survey techniques. High-resolution and accurate bathymetry data can be accomplished by the land survey methods such as total station and laser scanning and global positioning system (Feurer et al., 2008). Real time global positioning systems and total station approach provide precise bathymetry data for the shallow river (Hilldale and Raff, 2008). Because these methods require manual measurements, the skill of the surveyor has a substantial impact on the measurement accuracy (Bangen et al., 2014). The security concerns of the surveyor, particularly in rapidly-flowing or deeper rivers, impede the application of these survey techniques. The convenience of rivers and time limits also affect precision (McKean et al., 2009). In moderately deeper streams, GPS equipped boat mounted surveying methods that use echo sounders are useful (Hostache et al., 2015). Two common illustrations are Sound Navigation and Range (SONAR) and Acoustic Doppler Current Profiler (ADCP) (Allouis et al., 2010). Echo-sounders were implemented by the Autonomous Underwater Vehicle (AUV) to gather bathymetry in ocean estuaries. These tools provide point measurements of the depth of the river channel immediately under the ship. An echo sounder sends a sound pulse from the bottom of a ship out to the ocean floor. The wave of sound jumps back to the ship. The time it would take to depart and return to the ship for the pulse decides the topography of the seafloor. The longer it takes, the deeper the water. Even so, they need a minimum depth of water that can be reached by a boat, efficient usage and measurements are only taken along the ship's route. This leads to reduced overall measurement accuracy, which may spread the error in measuring the topography. Many limitations include barriers due to vegetative cover, and the inability to assess steep sloping banks etc. (Allouis et al., 2010).

5.2 Remote sensing methods

Logistics and safety challenges have made remote sensing an acceptable prospect in field studies (Legleiter and Overstreet, 2012a, b). These approaches rely on terrestrial or aerial platform-based electromagnetic sensors that measure depth and therefore can be used to survey huge regions in a short period (Casas et al., 2006). Topographic remote sensing has also been employed for the characterization of floodplains. Even so, topographic remote sensing is often not able to penetrate the water surface. So river bathymetry below the surface of the water is described as a level surface (Pathan et al., 2022c).

In modern times, researchers have analyzed the use of green lasers and near-infrared to establish bathymetric LIDAR assessments (Pan et al., 2015). They have a higher potential to penetrate the water body compared to the out-dated LIDARs. Similar to other remote sensing techniques, the efficiency of the LIDAR bathymetry often relies on the ability to penetrate the water and reach the river bed. A good example is the Aquarius LIDAR that can penetrate and identify water column depths of up to 2–3 m, which also decreases to about 1 m for more turbid rivers. (Legleiter et al., 2009). Airborne LiDAR Bathymetry (ALB) has been evolving rapidly in recent years and now facilitates river topography to be mapped with high-resolution (>20 points/m^2) and height accuracy (<10 cm) for both aquatic and riparian areas. (Mandlburger et al., 2015).

The alternative approach is to calculate the surface of the water from remote sensing and then tackle the inverse hydraulic problem by estimating the channel geometry (Roux and Dartus, 2008). Whereas the easiness in data acquisition, particularly in difficult locations, tends to make remote-sensing attractive, the difficulty to obtain precise measurements in deep, turbulent, or murky rivers poses as limitations of these methods. Besides, the price of the execution of these methods is high, making it unfeasible in some cases. For illustration, the LIDAR survey costs about $1200/km^2 (Casas et al., 2006).

6. Approaches for measuring bathymetry

In addition to the collection of further datasets around the world's oceans, new technologies are required to quantify bathymetry, while also defining and processing datasets that were collected previously in the past.

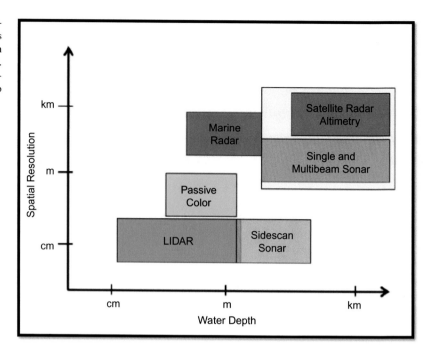

FIG. 1 Schematic demonstration of the applicability of various bathymetry calculation methods is based on the spatial measurement resolution and the range of water depths that can be measured. The *yellow box* shows the data sets which are combined and interpolated at 1 arc minute resolution to form the global girded bathymetry data.

This part highlights the basic modern methods for an estimate of the bathymetry from acoustics to the use of both radio and visible wave parts of the electromagnetic system spectrum. In the case of acoustics, measurements are taken using the water medium, visible light measurements are taken for air medium, using water, and the air medium individually for radar altimetry remote sensing to obtain an estimate of the depth of the sea. Each method has its pros and cons based on the scale of the structures measured, the precision, and the greatest water depth measurable as shown in (Fig. 1). For determining centimeter to kilometer-scale structures. The acoustic method is very effective to validate the radar and optical bathymetry. Active and passive measurements are utilizing for shallow water (<60m depth) because of the limitation of penetrating observable light in water environments. Satellite radar altimetry measurement provides large-scale bathymetry across the deep ocean. They are especially useful in remote locations of the oceans. Marine radar from shore and ship has been used for bathymetry measurement of shallow seawater.

Measures of bathymetry aren't static. Based on the time and location of measurements taken, and also the rise and fall of the tides, the bathymetry measurements tend to deviate by several meters in height. Generally, no tidal corrections are made to ship soundings for bathymetric maps seaward of the continental shelf. Some global navigation maps are widely used in the shallow region like the Great Bahama Bank soundings with the lowest average tide while the lowest mean low water is the U.S. charts datum.

7. Acoustics

Acoustic bathymetry or echo sounding entered the 1920s and was instrumental in assessing the seafloor layout as we know it today. Currently, echo sounders are used to collect bathymetric measurements. An echo sounder from a ship's hull or bottom sends a sound pulse out to the ocean floor, and the sound wave bounces back to the ship. The time taken to transmit the pulse from the ship to the seabed and back to the ship decides the depth of the seafloor. Half the travel time of the round trip multiplied by the sound velocity in seawater is equal to the depth at a given level. Leonardo da Vinci noted in 1490 and later Benjamin Franklin in 1762, the sound travels far and wide in water with little attenuation contrasted with air (Kearns and Breman, 2010). Sound moves often much faster in water than in air, so large ocean depths can be measured without any loss of sound. The speed of sound is around 1500 m/*sec* in seawater, however, the exact speed varies with salinity, pressure, and ocean temperature.

Today, multibeam sonar provides high-resolution measurements of ocean depth and bottom emissivity. Each sound of a multibeam sonar produces a single large swath of sound (equal to 153°) bouncing off the seafloor (Fig. 2). The return echo is transmitted through an array of transducers and the electronic means divided into a series of single beams for each one of which depth is measured. In shallow water, very high resolution is possible but the swath width is reduced. In comparison,

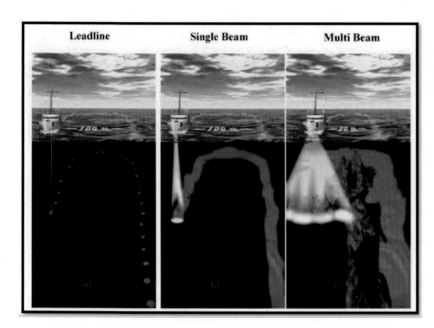

FIG. 2 Illustration of various techniques for assessing bathymetry from (A) line and sinker technology; (B) echo sounding utilizing single-beam acoustics; and (C) multibeam acoustics having the highest possible spatial resolution of sea flood characteristics.

ship operation's efficiency at large depths is enhanced as the swath width increases geometrically while the resolution reduces. Several overlap swaths create a bathymetrical map of the region under survey. Different sound frequencies (i.e., 12–400 kHz) are used for various depth ranges; the lowest the frequency, the deepest the depth measurement is possible, while the higher frequency is related to higher precision but shallow depth. Estimation of bathymetry is obtained through the degree of accuracy and resolution. For example, multibeam shallow water systems can assess bathymetry in 10 m of water at a scale of about 10 cm and have been used to accurately map U.S. coastal waters (Lecours et al., 2016; Poppe and Polloni, 1998). Acoustic sensors can also be mounted on remotely controlled tethered vehicles and autonomous underwater vehicles, such as gliders, which can hold a steady location relative to the seafloor and provide bathymetry of high resolution (Moline et al., 2007). A lot of survey lines with overlapping tracks have to be run to create consistent pictures at high resolution.

The key drawbacks of acoustic measurements are the expenses and time related to making measurements in shallow waters from a small vessel and deep waters from a ship. Since the path width reduces in deeper water, the bays with shallower water and the coastal estuaries need a lot more glider tracks or ship. Acoustic techniques can typically be employed in all sea depths, from shallow estuaries to the deep trenches.

8. Optics

Remote sensing through the visible light spectrum was also used to calculate bathymetry of shallow water, where acoustic techniques are limited. Bathymetry was measured using passive techniques that calculate only the natural light reflecting the seafloor and active techniques that use lasers to measure the distance to the seafloor (Gao, 2009).

9. Radar structure

Satellite marine radars can be employed to capture large-scale bathymetry shifts associated with sea-surface height. Ship and shore imaging radar systems were employed to calculate wave fields and to deduce shallow water bathymetry from the wave theory.

9.1 Satellite altimetry

Involvement of bathymetric aspects, like troughs and ridges, causes variations in the Earth's gravitational field which results in slight variations in sea surface height, like, when water is drawn to the greater mass, sea surface swells somewhat upward in reaction to a seamount (Fig. 3). Satellite-mounted radar altimeters circling the Earth will calculate these minor changes in sea surface elevation by transmitting the high-frequency radio wave bursts, typically within the range of 13 GHz,

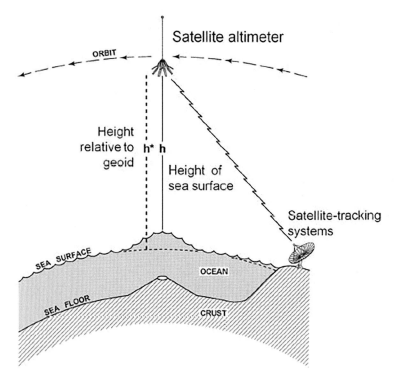

FIG. 3 The illustration that the utilization of radar altimetry to measure bathymetry of the deep ocean floor in broad-scale (Sandwell and Smith, 1997).

to determine variations in sea-surface elevation. The radar pulse scatters off the seafloor, measuring the signal's round-trip travel time. If the location of the satellite in orbit is definite, the pulse's round-trip travel time could be compared to the elevation of the ocean floor for the satellite sensor and is measured in a range of few centimeters (Fig. 3). In the wavelength of 1–200 km, gravity anomaly alterations are strongly associated with the seafloor topography and are used to gravimetrically map world ocean bathymetry using a spatial resolution of approximately 10 km, solving characteristics of about 20 km in scale. The gravimetric anomaly calculations are complied with acoustically resolute soundings to provide accurate estimates of bathymetry to create standardized grids of seafloor topography.

9.2 Imaging radar structure

During the period of 1940s, the wave theory was in practice to derive shallow bathymetry, when the Allied forces used series of satellite images to chart the depth of water along France's Normandy coast. When waves travel through shallow water, their speed reduced and the wavelengths become shorter. Modern methods are based on similar concepts but are using the radar imaging equipment for getting precise wavelength as well as wave speed measurements (Bell, 1999). Marine radar systems installed on coastal stations were adapted to suspect bathymetry of the near shore. Recently, the approach was adapted for radar measurements obtained on moving vessels. More than 64 km^2 of coastal seas inside a bay were mapped with the help of radar data obtained in 2 h from the equipment readily accessible on ship. These were very precise and with a horizontal resolution of 50–100 m pixels, down to 40–50 m water depth (Bell et al., 2006).

10. Methods of river cross-section extraction using DEM with the application of HEC-RAS

10.1 Geometry generation in HEC GeoRAS

A precise representation of ground surface elevations is a prime prerequisite for the creation of any hydrodynamic modeling. A high-resolution digital terrain model accurately demonstrates the elevation point of the river and surrounding floodplain area by adding various physical features including river beds, riverbanks, as well as many features such as embankments wall and roadways by which flow is guided. For the present study availability of data at the preferred interval is a major issue. Additionally, to derive elevation datasets for the entire study area we have used GIS tools in HEC-RAS

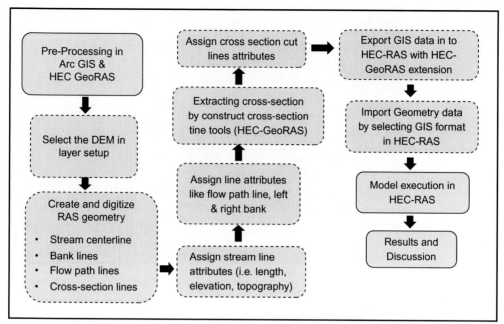

FIG. 4 Flow chart of the methodology adopted in the present study.

(Hydrologic Engineering Center River Analysis System) (to extract the river geometry from Digital Elevation Model (DEM) (Ahmad et al., 2016; Pathan et al., 2022b; Pathan and Agnihotri, 2021).

For the present study, the remote sensing approach has been applied to 20 km reach of the study area is time-consuming and laborious. So, preferred geometries of the present study have been extracted using a remote sensing approach like a DEM and Cartosat-1 satellite with 30 m resolution are used. The step-by-step method for generating geometric data in the HEC-GeoRAS tool is followed (Patel et al., 2020).

Step 1- Preprocessing of Data (Arc-GIS and HEC-GeoRAS).
Step 2- Model Execution (HEC-RAS).

The methodology is represented in the following flowchart (Fig. 4).

10.2 Preprocessing (arc-GIS and HEC-GeoRAS)

Arc-GIS and HEC-GeoRAS interface was utilized to extract river geometries using DEM. HEC-GeoRAS is an extension available in Arc-GIS. It is an interface that permits the geospatial datasets which are utilized in HEC-RAS and permit the formation of import datasets in HEC-RAS (Pathan and Agnihotri, 2019a, b, c).

In the present study, the steps of the creation of geometries are as following.

- Add Cartosat-1 DEM in Arc GIS using Add Data function
- For preprocessing in HEC-RAS, choose the desired DEM and terrain type in Layer setup
- Select Terrain type and desired DEM in Layer setup for HEC-RAS Preprocessing
- Using the RAS layer theme, create and digitize layers including stream centerline, bank lines, flow path lines, and cross-sections line which is shown in (Figs. 5 and 6).
- Based on DEM and base map, first, digitize the stream centerline and assign the river reach code.
- Assign attributes of stream centerline include length, elevation, topography, and station.
- Based on the DEM and base map, digitize bank lines.
- Furthermore, digitize flow path lines, which can be on both sides of the river along the direction of flow.
- Assign attributes of flow path lines includes a channel, left, and right over banks. For this, leftover the bank and right over banks are utilized in the flow direction.
- Create cross-sections by using manually digitizing or XS cut lines tools in HEC-GeoRAS. The cross-section area which is opposite to the flow direction and cut lines must be plot perpendicular to the stream centerline so that the elevation datasets can be obtained from the DEM.

FIG. 5 Representing the RAS theme created in HEC-GeoRAS.

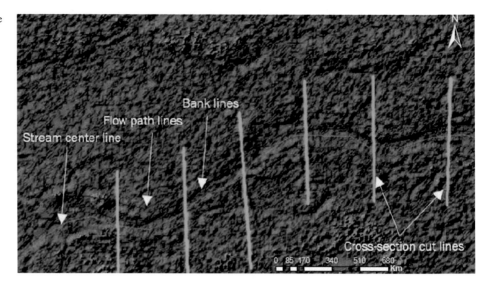

FIG. 6 Cross-sections generated on Purna river in HEC-GeoRAS.

- Assign XS cut line features such as reach name, station name, downstream reach length, and elevation from the RAS theme geometry tool.
- Export geometry datasets that were created in Arc GIS using HEC-GeoRAS extension to HEC-RAS and save the file.
- Now, import geometry datasets by choosing GIS format in the Geometric data editor option tab in HEC-RAS and import file of geometry which is shown in (Fig. 7).
- Import Geometry data by selecting GIS format in Geometric data in HEC-RAS.

10.3 HEC-RAS model execution

It is a program that represents the hydraulics of water moving between the common streams and through the various channels. It is a computer-based program that models the water moving in between frameworks of open channels and it also calculates the water surface profiles using energy equation (Alaghmand et al., 2012). The energy losses are determined using the Manning's equations and contraction or expansion. When the water surface profile is highly fluctuating then the momentum equation is adopted. HEC-RAS identifies the definite possible applications in floodplain mitigation

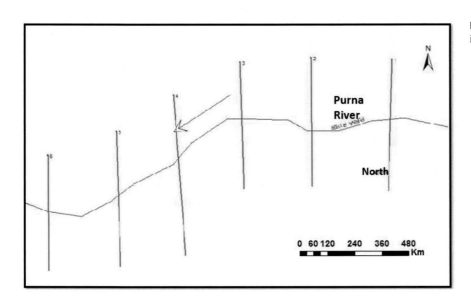

FIG. 7 GIS database import geometry file in HEC-RAS.

solutions. Moreover, it is mainly useful for a steady and unsteady flow analysis of a natural river (Brunner, 1995; Patel and Dholakia, 2010; Memon et al., 2019; Pathan and Agnihotri, 2020).

The main function of HEC-RAS is the present study to simulate the water surface profile at all cross-sections using a steady flow analysis. To simulate the hydrodynamic model in HEC-RAS following data are required:

- Geometric datasets
- Steady flow data
- The slope of the river
- Manning's value
- Peak discharge (steady flow data)

Extracting cross-sections using DEM and geometric data of cross-section which is an import from ArcGIS database file in HEC-RAS are giving station elevation data, left bank and right bank, etc. Geometric details of few cross-sections are presented in Fig. 8. Input boundary condition and run the 1D study flow analysis (Pathan et al., 2021a, b).

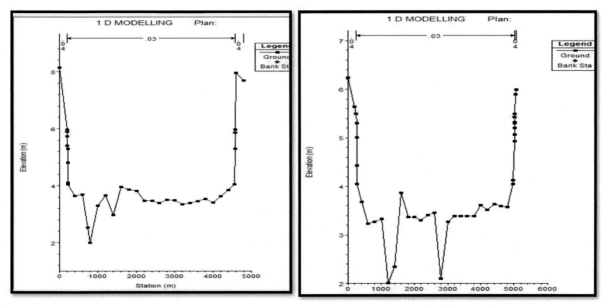

FIG. 8 Purna river cross-sections in HEC-RAS.

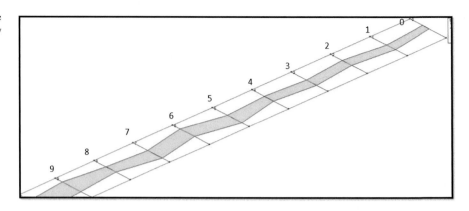

11. Results and discussion

A total 62 cross-sections were extracted from Cartosat-1 DEM for the Purna river's upstream to downstream. However, the cross sections of the river have been simulated that are most affected by flooding during peak discharge. The outcome of the HEC-RAS model revealed the water surface profile of each elevation in context of peak discharge (flood event) of the study area. Moreover, the results showed that the cross-section 19 and 20 which are located the downstream of the Purna river were more affected by flooding event, whereas the cross-section one and two were less affected in context of aforementioned flood event (Fig. 8). Interestingly, the perspective of the whole model simulation (1D HEC-RAS model) with steady flow is depicted in (Fig. 9) and hyperlink is also attached with the Fig. 9 to see the simulation in perspective view, which clearly indicated that the flooded and nonflooded cross-section during peak discharge. Subsequently, to acquire the accuracy of the model, the simulated values were well compared with the observed values illustrated in (Figs. 10 and 11). Results depicts the good correlation and the observed values were closely match with those simulated values (Abdella and Mekuanent, 2021; Pandya et al., 2021; Timbadiya et al., 2014).

The Navsari is the city situated along the coastal side of Arabian Sea. The city experienced flood annually due to heavy rainfall, climate change effect, industrialization, and rapid development of the city around the floodplain of Purna River. The methodology adopted for flood risk assessment using DEM extracted river bathymetry for the study area indeed revealed the novel approach especially in data scares region (i.e., Navsari city). The remote sensing data play a vital role to extract the river bathymetry of the river. Therefore, the approach used for the study is useful for local government and disaster authorities to prepare decision making system, flood management plan, evacuation strategy for the city during peak flood and emergency (Amin et al., 2014; Patel and Dholakia, 2010). The present study is very crucial in current scenarios. However, the applicability of geospatial tool is very effective and solution oriented in the flood management studies, especially it will be very helpful in data scarce region around the world. Many studies have been carried out based on GIS-based

FIG. 10 Comparison of observed and simulated data of CS-19.

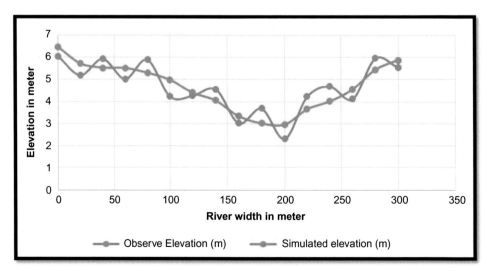

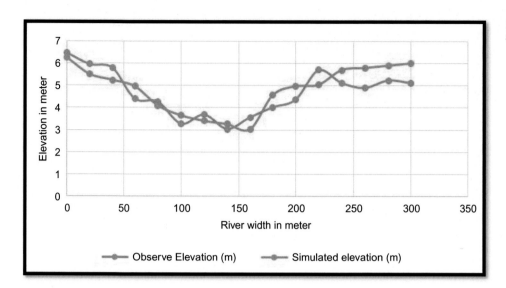

FIG. 11 Comparison of observed and simulated data of CS-20.

flood assessment studies (Ahmad et al., 2016; Patel et al., 2020). The presented study will be helpful for local government and disaster authorities where data availability is very less. As a result, the flood management and resilience approach would be implemented in future research.

12. Conclusions

Even though we enhance research to other areas of science and technology, the global awareness of the situation that we have given to our universe is also expanding to include the ocean. The selection of new datasets and the resources able to make high-resolution sear floor spaces does not give rise to a renewed and increasing appreciation of the underwater topography but also of whatever we can benefit from learning more about it. Modern data collection approach has given military management authorities, a marine scientist, and engineers the capacity to capture high-resolution bathymetry.

The Cartosat-1 DEM was used to extract the river bathymetry for the Purna River, Navsari city in this study. Moreover, 1D hydrodynamic model has been simulated to evaluate the flooding in flood affected area of the study and the outcome of the model showed the water surface profile of each cross-section during peak discharge. Interestingly, to identify the accuracy of the model, the simulated results of the model were compared with the actual observed data which revealed the promising and close results in context of water surface elevation. Certainly, this approach will be useful for the disaster authorities and local government to prepare rescue plan, decision making system, and flood management plan for the similar study areas around the globe especially in data sparse regions.

Awareness of ocean bathymetry has improved significantly in the last century with the development of acoustical, optical, and radar technologies. The seafloor has been mapped to a range of different resolutions from km to cm, but the major challenge remains to obtain the precise measurement in the underwater environment. Many acoustic measurements are needed in remote regions of the world to validate gravimetric bathymetry, and high data on spatial resolution is important for evaluating shallow water bathymetry to delineate shorelines for severe storms and possible sea level rises. The seafloor is regarded as one of the vast unexplored area that requires mapping and visualization to mitigate future challenges.

References

Abdella, K., Mekuanent, F., 2021. Application of hydrodynamic models for designing structural measures for river flood mitigation: the case of Kulfo River in Southern Ethiopia. Model. Earth Syst. Environ. 7 (4), 2779–2791. https://doi.org/10.1007/s40808-020-01057-5.

Ahmad, H.F., Alam, A., Bhat, M.S., Ahmad, S., 2016. One dimensional steady flow analysis using HECRAS–a case of river Jhelum, Jammu and Kashmir. Eur. Sci. J. 12, 340–350. https://doi.org/10.19044/esj.2016.v12n32p340.

Alaghmand, S., Bin Abdullah, R., Abustan, I., Eslamian, S., 2012. Comparison between capabilities of HEC-RAS and MIKE11 hydraulic models in river flood risk modeling (a case study of Sungai Kayu Ara River basin, Malaysia). Int. J. Hydrol. Sci. Technol. 2 (3), 270–291.

Allouis, T., Bailly, J.S., Pastol, Y., Le Roux, C., 2010. Comparison of LiDAR waveform processing methods for very shallow water bathymetry using Raman, near-infrared and green signals. Earth Surf. Process. Landf. 35 (6), 640–650. https://doi.org/10.1002/esp.1959.

Amin, A., Fazal, S., Mujtaba, A., Singh, S.K., 2014. Effects of land transformation on water quality of Dal Lake, Srinagar, India. J. Indian Soc. Remote Sens. 42 (1), 119–128. https://doi.org/10.1007/s12524-013-0297-9.

Anandan, C., Sasidhar, P., 2008. Assessment of the impact of the tsunami of December 26, 2004 on the near-shore bathymetry of the Kalpakkam coast, east coast of India. Sci. Tsunami Haz. 27 (4), 26.

Baldassarre, G.D., Nardi, F., Annis, A., Odongo, V., Rusca, M., Grimaldi, S., 2020. Brief communication: comparing hydrological and hydrogeomorphic paradigms for global flood hazard mapping. Nat. Hazards Earth Syst. Sci. 20 (5), 1415–1419. https://doi.org/10.5194/nhess-20-1415-2020.

Bangen, S.G., Wheaton, J.M., Bouwes, N., Bouwes, B., Jordan, C., 2014. A methodological intercomparison of topographic survey techniques for characterizing wadeable streams and rivers. Geomorphology 206, 343–361. https://doi.org/10.1016/j.geomorph.2013.10.010.

Bell, P.S., 1999. Shallow water bathymetry derived from an analysis of X-band marine radar images of waves. Coast. Eng. 37 (3–4), 513–527. https://doi.org/10.1016/S0378-3839(99)00041-1.

Bell, P.S., Williams, J.J., Clark, S., Morris, B.D., Vila-Concej, A., 2006. Nested radar systems for remote coastal observations. J. Coast. Res. (SI 39), 483–487.

Bierkens, M.F., 2015. Global hydrology 2015: state, trends, and directions. Water Resour. Res. 51 (7), 4923–4947. https://doi.org/10.1002/2015WR017173.

Brunner, G.W., 1995. HEC-RAS River Analysis System. Hydraulic Reference Manual. Version 1.0. Hydrologic Engineering, Center Davis CA, USA.

Casas, A., Benito, G., Thorndycraft, V.R., Rico, M., 2006. The topographic data source of digital terrain models as a key element in the accuracy of hydraulic flood modelling. Earth Surf. Process. Landf. 31 (4), 444–456. https://doi.org/10.1002/esp.1278.

Fan, F.M., Collischonn, W., Meller, A., Botelho, L.C.M., 2014. Ensemble streamflow forecasting experiments in a Tropical Basin: the São Francisco River Case Study. J. Hydrol. 519 (PD), 2906–2919.

Feurer, D., Bailly, J.S., Puech, C., Le Coarer, Y., Viau, A.A., 2008. Very-high-resolution mapping of river-immersed topography by remote sensing. Prog. Phys. Geogr. 32 (4), 403–419. https://doi.org/10.1177/0309133308096030.

Gao, J., 2009. Bathymetric mapping by means of remote sensing: methods, accuracy and limitations. Prog. Phys. Geogr. 33 (1), 103–116. https://doi.org/10.1177/0309133309105657.

Hilldale, R.C., Raff, D., 2008. Assessing the ability of airborne LiDAR to map river bathymetry. Earth Surf. Process. Landf. 33 (5), 773–783. https://doi.org/10.1002/esp.1575.

Hostache, R., Matgen, P., Giustarini, L., Teferle, F.N., Tailliez, C., Iffly, J.F., Corato, G., 2015. A drifting GPS buoy for retrieving effective riverbed bathymetry. J. Hydrol. 520, 397–406. https://doi.org/10.1016/j.jhydrol.2014.11.018.

Kearns, T.A., Breman, J., 2010. Bathymetry-the art and science of seafloor modeling for modern applications. In: Ocean Globe. ESRI Press, pp. 1–36.

Lane, A., 2004. Bathymetric evolution of the Mersey Estuary, UK, 1906–1997: causes and effects. Estuar. Coast. Shelf Sci. 59 (2), 249–263. https://doi.org/10.1016/j.ecss.2003.09.003.

Lecours, V., Dolan, M.F., Micallef, A., Lucieer, V.L., 2016. A review of marine geomorphometry, the quantitative study of the seafloor. Hydrol. Earth Syst. Sci. 20 (8), 3207–3244. https://doi.org/10.5194/hess-20-3207-2016.

Legleiter, C.J., Overstreet, B.T., 2012. Mapping gravel bed river bathymetry from space. J. Geophys. Res. 117 (November), 1–24. https://doi.org/10.1029/2012JF002539.

Legleiter, C.J., Roberts, D.A., Lawrence, R.L., 2009. Spectrally based remote sensing of river bathymetry. Earth Surf. Process. Landf. 34 (8), 1039–1059. https://doi.org/10.1002/esp.1787.

Lyon, J.C., Lunetta, R.S., Williams, D.C., 1992. Airborne Multispectral Scanner Data for Evaluating Bottom Sediment Types and Water Depths of the St. Marys River, Michigan. asprs.org. https://www.asprs.org/wp-content/uploads/pers/1992journal/jul/1992_jul_951-956.pdf.

Mandlburger, G., Hauer, C., Wieser, M., Pfeifer, N., 2015. Topo-bathymetric LiDAR for n monitoring river morphodynamics and instream habitats—A case study at the Pielach River. Remote Sens. 7 (5), 6160–6195. https://doi.org/10.3390/rs70506160.

McKean, J., Nagel, D., Tonina, D., Bailey, P., Wright, C.W., Bohn, C., Nayegandhi, A., 2009. Remote sensing of channels and riparian zones with a narrow-beam aquatic-terrestrial LIDAR. Remote Sens. 1 (4), 1065–1096. https://doi.org/10.3390/rs1041065.

McKean, J., Tonina, D., Bohn, C., Wright, C.W., 2014. Effects of bathymetric lidar errors on flow properties predicted with a multi-dimensional hydraulic model. J. Geophys. Res. Earth Surf. 119 (3), 644–664. https://doi.org/10.1002/2013JF002897.

Memon, N., Patel, D.P., Bhatt, N., Patel, S., 2019. Integrated framework for Flood Relief Package (FRP) allocation in semi-arid region-a case of Rel river flood, Gujarat, India. Nat. Hazards 100, 279–311. Springer https://doi.org/10.1007/s11069-019-03812-z.

Moline, M.A., Woodruff, D.L., Evans, N.R., 2007. Optical delineation of benthic habitat using an autonomous underwater vehicle. J. Field Robot. 24 (6), 461–471. https://doi.org/10.1002/rob.20176.

Pan, Z., Glennie, C., Hartzell, P., Fernandez-Diaz, J.C., Legleiter, C., Overstreet, B., 2015. Performance assessment of high resolution airborne full waveform LiDAR for shallow river bathymetry. Remote Sens. 7 (5), 5133–5159. https://doi.org/10.3390/rs70505133.

Pandya, U., Patel, D.P., Singh, S.K., 2021. A flood assessment in a data-scarce region using an open-source 2D hydrodynamic modeling and Google Earth Image – a case of Sabarmati flood, India. Arab. J. Geosci. 14, 2200. Springer (2021) https://doi.org/10.1007/s12517-021-08504-2.

Patel, D.P., Dholakia, M.B., 2010. Feasible structural and non-structural measures to minimize effect of flood in Lower Tapi Basin. WSEAS Trans. Fluid Mech. 3, 104–121.

Patel, D.P., Srivastava, P.K., Singh, S.K., Prieto, C., Han, D., 2020. One-dimensional hydrodynamic modeling of the river Tapi: the 2006 flood, Surat, India. In: Techniques for Disaster Risk Management and Mitigation. Wiley, pp. 209–235, https://doi.org/10.1002/9781119359203.ch16.

Pathan, A.I., Agnihotri, P.G., 2019a. A combined approach for 1-D hydrodynamic flood modeling by using Arc-Gis, Hec-Georas, Hec-Ras Interface-a case study on Purna River of Navsari City, Gujarat. Int. J. Recent Technol. Eng. 8 (1), 1410–1417.

Pathan, A.I., Agnihotri, P.G., 2019b. One dimensional floodplain modelling using soft computational techniques in HEC-RAS-A case study on Purna basin, Navsari District. In: International Conference on Intelligent Computing & Optimization. Springer, Cham, pp. 541–548, https://doi.org/10.1007/978-3-030-33585-4_53.

Pathan, A.I., Agnihotri, P.G., 2019c. Use of computing techniques for flood management in a Coastal Region of South Gujarat–a case study of Navsari District. In: International Conference on Intelligent Computing & Optimization. Springer, Cham, Switzerland, pp. 108–117, https://doi.org/10.1007/978-3-030-33585-4_11.

Pathan, A.I., Agnihotri, P.G., 2021. Application of new HEC-RAS version 5 for 1D hydrodynamic flood modeling with special reference through geospatial techniques: a case of River Purna at Navsari, Gujarat, India. Model. Earth Syst. Environ. 7 (2), 1133–1144. https://doi.org/10.1007/s40808-020-00961-0.

Pathan, A.I., Agnihotri, P.G., Patel, D., 2021a. River geometry extraction from Cartosat-1 DEM for 1D hydrodynamic flood modeling using HEC-RAS—a case of Navsari City, Gujarat, India. In: Advanced Modelling and Innovations in Water Resources Engineering. Springer, Singapore, pp. 173–185, https://doi.org/10.1007/978-981-16-4629-4_13.

Pathan, A.I., Agnihotri, P.G., Patel, D., 2022a. Integrated approach of AHP and TOPSIS (MCDM) techniques with GIS for dam site suitability mapping: a case study of Navsari City, Gujarat, India. Environ. Earth Sci. 81 (18), 1–19.

Pathan, A.I., Agnihotri, P.G., Patel, D., Prieto, C., 2021b. Identifying the efficacy of tidal waves on flood assessment study—a case of coastal urban flooding. Arab. J. Geosci. 14 (20), 1–21. https://doi.org/10.1007/s12517-021-08538-6.

Pathan, A.I., Agnihotri, P.G., Patel, D., Prieto, C., 2022b. Mesh grid stability and its impact on flood inundation through (2D) hydrodynamic HEC-RAS model with special use of Big Data platform—a study on Purna River of Navsari city. Arab. J. Geosci. 15 (7), 1–23.ces.

Pathan, A.I., Agnihotri, P.G., Said, S., Patel, D., 2022c. AHP and TOPSIS based flood risk assessment-a case study of the Navsari City, Gujarat, India. Environ. Monit. Assess. 194 (7), 1–37.

Pattanaik, A., Sahu, K., Bhutiyani, M.R., 2015. Estimation of shallow water bathymetry using IRS-multispectral imagery of Odisha Coast, India. Aquat. Procedia 4, 173–181. https://doi.org/10.1016/j.aqpro.2015.02.024.

Poppe, L.J., Polloni, C.F., 1998. Long Island Sound Environmental Studies (No. 98-502). US Geological Survey, Coastal and Marine Geology, https://doi.org/10.3133/ofr98502.

Roux, H., Dartus, D., 2008. Sensitivity analysis and predictive uncertainty using inundation observations for parameter estimation in open-channel inverse problem. J. Hydraul. Eng. 134 (5), 541–549. https://doi.org/10.1061/(ASCE)0733-9429, 134:5(541).

Timbadiya, P.V., Patel, P.L., Porey, P.D., 2014. One-dimensional hydrodynamic modelling of flooding and stage hydrographs in the lower Tapi River in India. Curr. Sci. 106, 708–716.

Tripathi, N.K., Rao, A.M., 2002. Bathymetric mapping in Kakinada Bay, India, using IRS-1D LISS-III data. Int. J. Remote Sens. 23 (6), 1013–1025. https://doi.org/10.1080/01431160110075785.

Westaway, R.M., Lane, S.N., Hicks, D.M., 2003. Remote survey of large-scale braided, gravel-bed rivers using digital photogrammetry and image analysis. Int. J. Remote Sens. 24 (4), 795–815. https://doi.org/10.1080/01431160110113070.

Winterbottom, S.J., Gilvear, D.J., 1997. Quantification of channel bed morphology in gravel-bed rivers using airborne multispectral imagery and aerial photography. Regul. Rivers: Res. Manage. 13 (6), 489–499. https://doi.org/10.1002/(SICI)1099-1646(199711/12)13:6<489::AID-RRR471>3.0.CO;2-X.

Yamazaki, D., Kanae, S., Kim, H., Oki, T., 2011. A physically based description of floodplain inundation dynamics in a global river routing model. Water Resour. Res. 47 (4). https://doi.org/10.1029/2010WR009726.

Chapter 25

Runoff modeling using group method of data handling and gene expression programming

Sahar Hadi Pour[a], Shamsuddin Shahid[a], and Saad Sh. Sammen[b]

[a]School of Civil Engineering, Faculty of Engineering, Universiti Teknologi Malaysia, Johor Bahru, Malaysia, [b]Department of Civil Engineering, College of Engineering, University of Diyala, Diyala Governorate, Iraq

1. Introduction

Rainfall–runoff (RR) models provide simulated surface runoff for the area where in-situ data are unavailable or insufficient. A large number of the RR models have been devised in last few decades with varying degrees of complexity and a wide range of applicability (Yaseen et al., 2018a, b). These models have been successfully employed not only for runoff simulations, but also for modeling hydrological changes in response to environmental changes like impacts on land use and climate changes on hydrological regimes and water quality (Yaseen et al., 2019; Nashwan et al., 2018, 2019; Pour et al., 2020b; Ziarh et al., 2021).

RR models developed so far have been classified in different ways (Woolhiser, 1973; Fleming, 1975; Linsley, 1982; El-Kadi, 1989; Beven, 2001) from physically-based to empirical models, deterministic and stochastic models, static and dynamic models, and lumped and distributed models (Clarke, 1973; Beven, 1985, 2001; Wheater et al., 1993; Refsgaard and Storm, 1995). In physical RR models, the physical hydrological processes are characterized in a determinative manner though deplications of various mass balance equations. On the other hand, in empirical RR models, the association between runoff and catchment hydrological properties is represented by simplified mathematical equations. The RR models are also classified as lump and distributed. The first type considers catchment as a single unit and therefore, impacts of any spatial variation in catchment properties like topography, land use, soil, etc. cannot be simulated. The distributed model considers that a catchment is composed of number of small units, and therefore, they can able to simulate the spatial variability of rainfall and catchment properties (Moradkhani and Sorooshian, 2008; Devia et al., 2015; Sa'adi et al., 2017).

Physically based model always simulate same runoff for a particular set of inputs as it estimate the runoff through deterministic process. On the other hand, stochastic model may provide different runoff for same set of inputs. Description of various types of hydrological models can be found in details in Todini (2007), Davision and Kamp (2008), Moradkhani and Sorooshian (2008), and Devia et al. (2015). Stochastic model do not consider the underlying physical processes, and therefore, known as blackbox models (Sammen et al. 2017; Pour et al., 2018; Khan et al., 2019a, b). These types of models are generally considered as the simplest form of hydrological models. The best known blackbox hydrological models are based on statistical methods developed with the support of basic statistical theory (Sharafati et al., 2019; Khan et al., 2020; Ahmed et al., 2020). The regression or transfer functions are usually used in development of such models to determine functional relationships between rainfall and runoff. Examples of such models include different forms of artificial intelligent (AI) such as artificial neural networks (ANNs), support vector machine (SVM) and genetic programming (GP) (Ahmed et al., 2015; Yaseen et al., 2018a, b; Tikhamarine et al. 2020; Mohamadi et al. 2020; Sammen et al. 2021; Malik et al., 2021; Pham et al., 2021a, b). In recent years, few AI models including Group Method of Data Handling (GMDH) and symbolic regression based on Gene Expression Programming (GEP) have been found highly capable in solving intricate nonlinear problems (Amanifard et al., 2008; Sachindra et al., 2018, 2019; Muhammad et al., 2021). However, applications of these methods are still limited in RR modeling.

The GMDH, a inductive statistical learning method was developed by Ivakhnenko (1971) for modeling and characterization of complicated systems (Samsudin et al., 2011). It a self-organized system and has capability to solve complex problems (Amanifard et al., 2008). Several studies reported the potential of GMDH to identify the behavior of nonlinear systems efficiently (Witczak et al., 2006; Srinivasan, 2008; Abdolrahimi et al., 2014). GMDH model has also been found successful in handling uncertainty in a broad range of scientific fields ranging from complicated engineering problem solution to economic modeling (Voss and Feng, 2002). In recent years, GMDH has been used for hydrological modeling (Chang and Hwang, 1999; Onwubolu et al., 2007; Wang et al., 2005; Samsudin et al., 2011; Garg 2015; Ebtehaj et al., 2015a; Pham et al., 2021a, b).

Samsudin et al. (2011) used GMDH for simulating monthly flow of two rivers in peninsular Malaysia and showed that GMDH based hybrid model outperforms other conventional models in forecasting river flow. Garg (2015) investigated the performance of GMDH in modeling sediment yield from annual mean rainfall and geomorphological characteristics such as stream length, catchment area and mean slope of watershed of 20 subcatchments of Arno River Basin in Italy and reported that the model can capture the trend of sediment yield reliably. Ebtehaj et al. (2015a) used GMDH for discharge coefficient prediction and showed that more precisely prediction can be achieved compared to that obtained using other AI models and existing nonlinear regression equations. Ebtehaj et al. (2021) used GMDH for water level prediction and that more precisely prediction can be achieved when the GMDH based model was used compare with the standalone GMDH model.

The symbolic regression is like conventional parametric regression where the basic idea is to find an equation to best describe the relationship between predictand and the predictors (Fernando et al., 2009). It can be used to find the hidden relations in data (Schmidt and Lipson, 2009; Yang et al., 2015, Xu et al., 2016; Muhammad et al., 2021). Therefore, it has been recently employed in solving different kinds of complex scientific and engineering problems in many fields (Keshavarz and Mehramiri, 2015; Wu et al., 2015; Cornforth and Lipson, 2015; Yu et al., 2015; Deng et al., 2016; Faris and Sheta, 2016; Yang et al., 2015). Several approaches based on different philosophies has been used for deriving symbolic regressions including genetic programming (GP), GEP, grammar evolution (GE), analytic programming (AP), etc. (Eslamian et al., 2012). Among all, GEP has been found to perform better compared to others (Chen et al., 2013; Pour et al., 2014).

Studies showed that GEP can be used for modeling nonlinear hydrological phenomena reliably (Yassin et al., 2016; Gholami et al., 2015; Hadipour et al., 2013; Zorn and Shamseldin, 2015; Shoaib et al., 2015; Ebtehaj et al., 2015b). Aytek and Alp (2008) compared ANN and GEP methods to develop RR models and showed that better performance of GEP compared to ANN in RR modeling. Guven and Aytek (2009) and Azamathulla et al. (2011) used GEP in stage-discharge modeling and compared its performance with traditional rating curve and multiple linear regression (MLR). They reported better performance of GEP in comparison to conventional models. Kisi et al. (2012) compared the performances of ANN, adaptive neuro-fuzzy inference system (ANFIS), GEP and autoregressive moving average (ARMA) models for forecasting water level at a lake in Turkey and showed better performance of GEP among all. Kisi et al. (2013) simulated the RR processes using ANN, ANFIS, GEP and MLR models for a small catchment in Turkey and reported GEP as the most suitable one. Hashmi and Shamseldin (2014) employed GEP for modeling flow duration curve (FDC) in ungauged catchments in the Auckland Region of New Zealand. They reported higher efficiency of GEP in deriving relationships between FDCs and catchment properties. Shoaib et al. (2015) investigated the performance of hybrid GEP with simple GEP using discrete wavelet transform (DWT) method and reported the effectiveness of both hybrid GEP and GEP in RR modeling.

The objective of the present study is to develop hydrological model using GMDH and GEP for simulation of runoffs in coastal catchments in the east of peninsular Malaysia. The eastern coastal region is highly susceptible to climate change (Yusuf and Francisco, 2009; Tao et al., 2021). Floods driven by extreme rainfall is a common phenomenon in this region (Noor et al., 2018). It is considered that global warming induced changes in rainfall and temperature will alter hydrological process in the region, which in turn will make the region more prone to hydrological disasters (Muhammad et al., 2019). The model developed in this study can be used for reliable simulation of surface runoff for water resources development and planning of the catchments.

2. Study area

Terengganu, the east coastal state of Malaysia was selected as case study in this research (Fig. 1). The region has an undulating topography which generally increased from the coastline (nearly zero meter elevation) to the inlands (up to 2270.0 m). A considerable part of the interior of Terengganu is mountainous, which influences the climate of the region. The annual average rainfall in the region is about 2900 mm (Mayowa et al., 2015). The area receives significant amount of rainfall even in the driest month. The study area experiences two monsoon seasons namely, the northeast (NE) monsoon from the mid of October to end of February, the south-west (SW) monsoon from April to September (Shahid et al., 2017; Khan et al., 2019a, b). The months of March and October are inter-monsoonal transitional periods. The seasonal variability of rainfall in the area is shown in Fig. 2. Heavy rainfall in the region occurs during NE monsoon. Maximum rainfall in a year

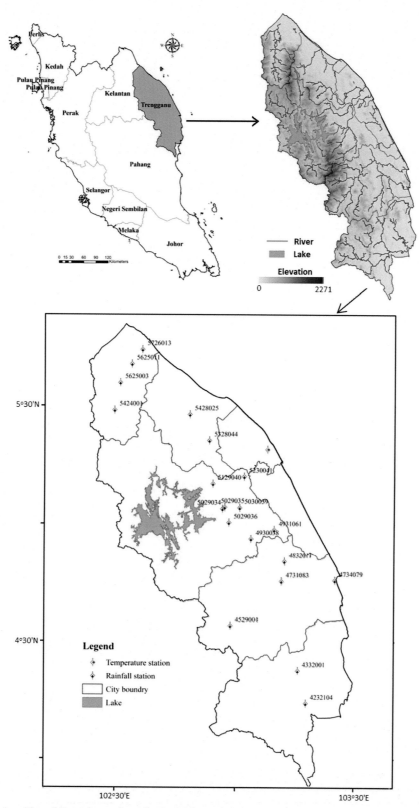

FIG. 1   The geographical position of the study area in peninsular Malaysia.

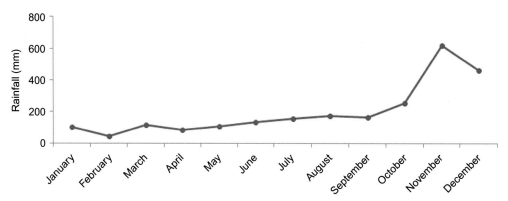

FIG. 2 The seasonal variability of rainfall for the period 1961–2000.

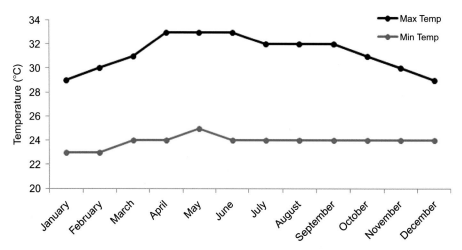

FIG. 3 Seasonal variability of maximum and minimum temperatures for the period 1961–2000.

is usually recorded in the months of November or December. The area often experiences flooding at this time of year (Pour et al., 2020a).

The mean temperature of the area is more or less uniform throughout the year. Seasonal variation of mean temperature (27.0°C) is always less than 2.0°C. The seasonal temperature variability in the area is shown in Fig. 3. The daily maximum temperature varies from 29.0°C in January to 33.0°C in the months of April, while the minimum temperature between 23.0°C in January and 25°C in May (Pour et al., 2020c).

A number of rivers originate in the mountainous interior region of Terengganu. Due to high variation in topography and high amount of rainfall, density of river network in the study area is also very high. Some rivers in the central part of the area are connected to Lake Kenyir, the biggest surface water body in the region. The study area consists of several catchments. In the present study, seven catchments with varying in size between 247 and 3208 km^2 were selected for development of RR models.

3. Data and sources

3.1 Hydrometeorological data

The daily rainfall data for the period 1961–2000 were obtained from the Department of Irrigation and Drainage (DID) and the temperature data were collected from Malaysian Meteorological Department. The study area is covered by more than hundred rainfall gauge stations and one temperature recording station. Considering the length of available records, rainfall data at 20 stations were selected for the present study. Description of rainfall and temperature gauges is provided in Tables 1 and 2, respectively, and their positions are shown in Fig. 1. The streamflow records at seven gauging stations for various time periods were also collected from DID were selected considering their distribution over the study area (Fig. 4). Description of streamflow gauge stations is given in Table 3.

TABLE 1 Description of rainfall stations used in the present study.

NO.	Station ID	Latitude (decimal)	Longitude (decimal)	Annual rainfall (mm)
1	4,232,104	4.24	103.30	2495.0
2	4,332,001	4.38	103.26	3533.4
3	4,529,001	4.57	102.98	2943.1
4	4,731,083	4.76	103.19	3949.6
5	4,734,079	4.76	103.42	2608.3
6	4,832,011	4.84	103.20	3241.5
7	4,930,038	4.94	103.06	3725.8
8	4,931,061	4.98	103.16	3775.6
9	5,029,034	5.07	102.94	3553.2
10	5,029,035	5.07	102.95	3565.9
11	5,029,036	5.01	102.97	3546.7
12	5,030,039	5.07	103.01	3551.0
13	5,129,040	5.17	102.90	3535.4
14	5,230,041	5.20	103.03	3123.4
15	5,328,044	5.36	102.89	3494.8
16	5,424,001	5.48	102.49	2692.8
17	5,428,025	5.47	102.81	3278.7
18	5,625,003	5.60	102.52	3326.2
19	5,625,011	5.68	102.56	2781.1
20	5,726,013	5.74	102.61	2863.1

TABLE 2 Detail of temperature recording station.

Station Name	Type of station	Ht. Above M.S.L.	Latitude	Longitude
Kuala Terengganu Airport	Meteorological	5.2m	5.383N	103.1E

Missing data is a major challenge in working with climatic records, particularly with rainfall data in Malaysia. About 4% to 15% of rainfall data were found missing at different stations in the study area. Distribution of missing data in rainfall time series was found random. Filling of missing data is very important for climate downscaling and hydrological simulation. A state-of-art method known as Expectation Maximization (EM) (McLachlan et al., 2004) was employed for filling missing data using observed data from the nearest stations. The EM employs maximum likelihood approach to develop relationship between model parameters of observed rainfall data and missing rainfall data. The relationship is then used to estimate missing rainfall values. EM uses an iterative process for gradual improvement of prediction (McLachlan et al., 2004). Description of EM method can be found in Krishnan and McLachlan (1997). Several studies reported the effectiveness of EM in estimation of missing values (Ng and McLachlan, 2004; Kalteh and Hjorth, 2009; Firat et al., 2010; Tsidu, 2012).

Quality of rainfall and temperature data is a major requirement for hydrological study. Therefore, a number of methods were used to evaluate the quality of data such as double curve method and various homogeneity tests. Homogeneity assessment using Student's t-test revealed homogeneity of rainfall data at all stations under study at less than 0.05 level of significance. The null hypothesis of homogeneity at the 0.05 significance level could not be rejected at any location. The

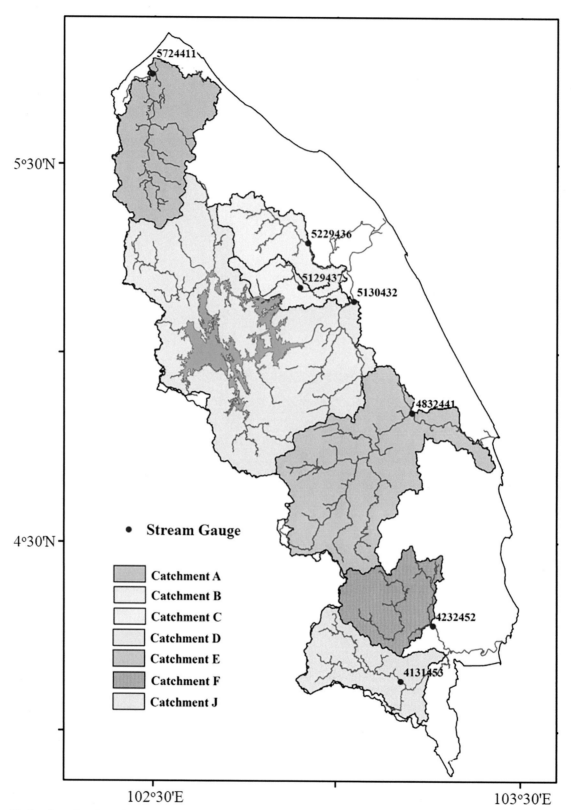

FIG. 4 The locations of study catchments on the map of Terengganu.

TABLE 3 Description of catchments and streamflow gauges.

Catchment ID	Station ID	Name of station	River basin	Lat	Lon	Area (km^2)
A	5,724,411	Sg. Besut di Jambatan Jerteh	Besut	5.74	102.49	990.3
B	5,229,436	Sg. Nerus di Kg. Bukit	Terengganu	5.29	102.92	499.3
C	5,129,437	Sg. Telemong di Paya Rapat	Terengganu	5.17	102.90	274.2
D	5,130,432	Sg. Terengganu di Kg.Tanggol	Terengganu	5.14	103.05	3208.1
E	4,832,441	Sg. Dungun di Jam. Jerangau	Dungun	4.84	103.20	1523.6
F	4,232,452	Sg. Kemaman di Rantau Panjang	Kemaman	4.28	103.26	619.0
J	4,131,453	Sg. Cherul di Ban Ho	Kemaman	4.13	103.18	677.9

double mass curves were found almost straight at all the stations, which indicate sufficient quality of rainfall data for hydrological studies. Therefore, the rainfall and temperature data used in the present study were considered homogeneous.

3.2 Methodology

3.2.1 Group method of data handling (GMDH)

The GMDH model uses a nonlinear function of Volterra series known as Kolmogorov–Gabor polynomial (Ivakhnenko, 1971) to map the relationship between predictors and predictand. The Volterra–Kolmogorov–Gabor (VKG) function is defined as:

$$ y = a_0 + \sum_{i=1}^{m} a_i x_i + \sum_{i=1}^{m} \sum_{j=1}^{m} a_{ij} x_i x_j + \sum_{i=1}^{m} \sum_{j=1}^{m} \sum_{k=1}^{m} a_{ijk} x_i x_j x_k + \dots \tag{1}$$

where $x = (x_1, x_2, \dots, x_m)$ is the input variables vector to the system, m is the number of inputs and $a = (a_0, a_1, a_2, \dots, a_m)$ is the vector of weights or polynomials coefficients. Considering all combinations of input vectors, VKG equation represents the complete polynomial description of a system model. A system of Gaussian normal equations or other adaptive methods are used to compute the coefficients of polynomials for any random sequence of observed data.

In the present study, a quadratic polynomial in terms of combination of two independent variables at a time was used to reconstruct the complete VKG polynomial series through an iterative perceptron type procedure. This approach provides better accuracy because it enables to classify useful information. Therefore, a second order polynomials of following form was utilized,

$$ \widehat{y} = G\left(x_i, x_j\right) = a_0 + a_1 x_i + a_2 x_j + a_3 x_i x_j + a_4 x_i^2 + a_5 x_j^2. \tag{2}$$

The value of the a_i, unknown coefficients can be obtained by regression methods and solving a system of Gauss normal equations for each pair of x_i and x_j variables. On this basis, the a_i coefficients are adjusted by means of the least square error method.

Different learning algorithms are used in GHDH depending on the complexity of model structure or the external criteria. In the present study, polynomial support functions were used as learning algorithm. The number of polynomial terms was determined according to the complexity of the model structure.

3.2.2 Gene expression programming (GEP)

Gene Expression Programming (GEP) is a variant of genetic programming based on evolutionary algorithms (Koza, 1992). GEP consist of two main features namely, the chromosomes and the expression trees for expressing the genetic variations encoded in chromosomes. The genetic variations are introduced by using various genetic operators including mutation, transposition and recombination. The genetic code and the set of rules in GEP are very simple. The chromosomes of the initial population (program) are randomly generated and expressed using expression trees. The fitness of each individual population or program is evaluated using as set of fitness functions. Details about the fitness functions can be found in Ferreira (2006a). The most suitable individuals of the population are selected for the next generation. The chromosomes of selected population

Algebraic Expression

$$\sqrt{(a+b)\cdot(c-d)} + \sqrt{(a-b)\cdot(c+d)}$$

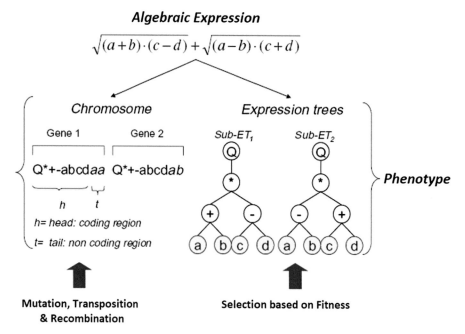

FIG. 5 Example of General GEP model implementation.

is mutated and expressed to generate a new set of population or program. The process is repeated until the expected program is found. It has been reported that genetic diversity and capability to generate valid structure have made GEP a strong search algorithm (Ferreira, 2006b). A typical process of GEP model development is shown in Fig. 5.

The procedure used to develop RR model using GEP is as follows:

1. Selection of fitness function or a set of fitness functions. Different statistical methods used for assessment of errors were used as fitness functions
2. Selection of a set of terminals and a set of functions. The inputs were selected based on their influence on river discharge
3. Creation of chromosomes from the selected terminals and functions
4. Setting the chromosomal architecture
5. Selection of the linking function
6. Selection of genetic operators

The above mentioned steps were repeated until the fittest program is achieved. In this research, the GeneXproTools software package (Gepsoft Inc, 2013) was used to perform symbolic regression operations based on Gene Expression Programming (GEP) for modeling RR process.

3.3 Model development

Rainfall and potential evapotranspiration (PET) data were used to develop the hydrological models for selected catchments in the study area. The PET values were estimated using Hargreaves method using temperature data as mentioned below:

$$ET_P = 0.00023 \times \left(T_{mean} + 17.8\right) \times \left(T_{max} - T_{min}\right)^{0.5} \times R_a \tag{3}$$

where ET_p is daily PET, T_{mean} is daily mean temperature, T_{max} and T_{min} are maximum and minimum daily temperatures, respectively, and R_a is extra-terrestrial radiation (Hargreaves and Samani, 1982; Ouyang et al., 2014).

Separate RR models were developed to simulate monthly and daily streamflow. The input data used for the daily models consists of the daily rainfall $\{R(t)\}$, the antecedent days rainfall $\{R(t-1), R(t-2), ..., R(t-n)\}$, the daily PET $\{ET_O(t)\}$ and the antecedent days PET $\{ET_O(t-1), ET_O(t-2), ..., ET_O(t-n)\}$. Similar inputs were used for development of monthly RR model. However, in that case, antecedent month's rainfall or evapotranspiration data were used. The mathematical representation of the nonlinear models can be expressed as,

$$Q(t) = f\{R(t), R(t-1), R(t-2), ...R(t-n), ET_O(t-1), ET_O(t-2), ..., ET_O(t-n)\} \tag{4}$$

The inputs for Eq. (4) can be presented in matrices form as below:

$$
I = \begin{bmatrix}
R_{t1} & R_{t1} & \cdots & R_{tn} \\
R_{(t-1)1} & R_{(t-1)2} & \cdots & R_{(t-1)n} \\
\vdots & \vdots & \vdots & \vdots \\
R_{(t-m)1} & R_{(t-m)2} & \cdots & R_{(t-m)n} \\
ETo_{(t-1)1} & ETo_{(t-1)2} & \cdots & ETo_{(t-1)n} \\
ETo_{(t-2)1} & ETo_{(t-2)2} & \cdots & ETo_{(t-2)n} \\
\vdots & \vdots & \vdots & \vdots \\
ETo_{(t-m)1} & ETo_{(t-m)2} & \cdots & ETo_{(t-m)n}
\end{bmatrix}
\tag{5}
$$

where I is the input vector matrix; n and m are the number of data points and the number of antecedent data points, respectively. The output is the sequences of runoff, which can be presented as,

$$
Q = [Q_{t1}\ Q_{t2}\ Q_{t3} \cdots Q_{tn}]
\tag{6}
$$

The models were developed using GMDH and GEP techniques. Separate models were developed for each catchment under study. Preliminary assessment using rainfall and evapotranspiration with up to nine antecedent days' data was tried. Performance of the models was assessed by sequentially adding and removing different antecedent days data. Optimum models were obtained with five antecedent days rainfall and evapotranspiration data in modeling daily streamflow, and two antecedent months rainfall and evapotranspiration values in modeling monthly streamflow. Therefore, those inputs were considered in development of RR models using GMDH or GEP.

3.4 Performance evaluation

Standard statistical indices were used for assessment of RR model performance, which include Root Mean Square Error (RMSE), the coefficient of determination (R), and Nash–Sutcliffe Efficiency coefficient (NSE). Methods used to estimate those statistical performance assessment parameters are defined as follows,

$$
RMSE = \left[\frac{1}{n} \sum_{i=1}^{n} (S_i - O_i)^2 \right]^{0.5}
\tag{7}
$$

$$
R = \frac{\sum_{i=1}^{n} (O_i - \overline{O})(S_i - \overline{S})}{\sqrt{\sum_{i=1}^{n} (O_i - \overline{O})^2} \sqrt{\sum_{i=1}^{n} (S_i - \overline{S})^2}}
\tag{8}
$$

$$
NSE = 1 - \frac{\sum_{i=1}^{n} (O_i - S_i)^2}{\sum_{i=1}^{n} (O_i - \overline{O})^2}
\tag{9}
$$

where O_i is the observed values, S_i is the simulated values; n represents the number of observations, $\overline{O}$ and $\overline{S}$ is the corresponding mean value for the observation and simulation, respectively.

4. Results and discussion

Data from different time periods were used for the development of the models at different catchments; however, data of same time periods were used for development of all models at a particular catchment. The models were calibrated with same number of inputs identified using correlation analysis between inputs and output. To find the optimum model of GMDH, different combination of training algorithm, data participation ratio and model structures were investigated for each catchment. The same procedure also used for the development of GEP models. The selected model structures for GMDH and GEP for simulation of daily and monthly streamflow in different catchments are summarized in Table 4. The optimal configuration of GMDH model is summarized in Table 5.

TABLE 4 Structure used for the development of hydrological models at different catchments.

Catchment	Selected daily model structure
A	$Q(t)=f\{R(t-2),R(t-3),R(t-4),ET(t),ET(t-1),ET(t-2),ET(t-3),ET(t-4)\}$
B	$Q(t)=f\{R(t-1),R(t-2),R(t-3),ET(t),ET(t-1),ET(t-2),ET(t-3),ET(t-4)\}$
C	$Q(t)=f\{R(t-1),R(t-2),R(t-3),ET(t),ET(t-1),ET(t-2),ET(t-3),ET(t-4)\}$
D	$Q(t)=f\{R(t-1),R(t-2),R(t-3),ET(t),ET(t-1),ET(t-2),ET(t-3)\}$
E	$Q(t)=f\{R(t),R(t-1),R(t-2),R(t-3),R(t-4),\ R(t-5),ET(t),ET(t-1),ET(t-2),ET(t-3)\}$
F	$Q(t)=f\{R(t-1),R(t-2),R(t-3),ET(t),ET(t-1),ET(t-2),ET(t-3)\}$
J	$Q(t)=f\{R(t-1),R(t-2),R(t-3),R(t-4),R(t-5),ET(t),ET(t-1),ET(t-2),ET(t-3)\}$
Catchment	**Selected monthly model structure**
All catchments	$Q(t)=f\{R(t),R(t-1),\ R(t-2),ET(t)\}$

TABLE 5 Optimum configuration for GMDH models identified for different catchments.

Catchment	Learning algorithms	Neuron function (daily model)	Neuron function (monthly model)	Max. number of layers	Initial layer width	Train/test ratio (daily model)	Train/test ratio (monthly model)
A	GMDH-type neural networks	Polynomial	Polynomial	50	1000	65/35	55/45
B	GMDH-type neural networks	Quadratic polynomial	Polynomial	50	1000	60/40	60/40
C	GMDH-type neural networks	Polynomial	Polynomial	50	1000	55/45	60/40
D	GMDH-type neural networks	Quadratic polynomial	Polynomial	50	1000	65/35	70/30
E	GMDH-type neural networks	Polynomial	Linear	50	1000	60/40	60/40
F	GMDH-type neural networks	Quadratic polynomial	Polynomial	50	1000	50/50	50/50
J	GMDH-type neural networks	Polynomial	Polynomial	50	1000	65/35	60/40

The GMDH and GEP models were used for simulation of both daily and monthly streamflow. Results obtained using the GMDH and GEP models at the same catchment are discussed in the following sections.

Fig. 6 shows the daily observed and GMDH-simulated streamflow at the outlet of catchment B during the model calibration and validation. GMDH was found to simulate the flow efficiently. The low, mean and peak flows were reconstructed by GHMD more accurately during both model calibration and validation. The values for R and NSE were found 0.90 and 0.81 during model calibration; and 0.85 and 0.70 during validation. This also indicates the better performance of GMDH is simulating streamflow.

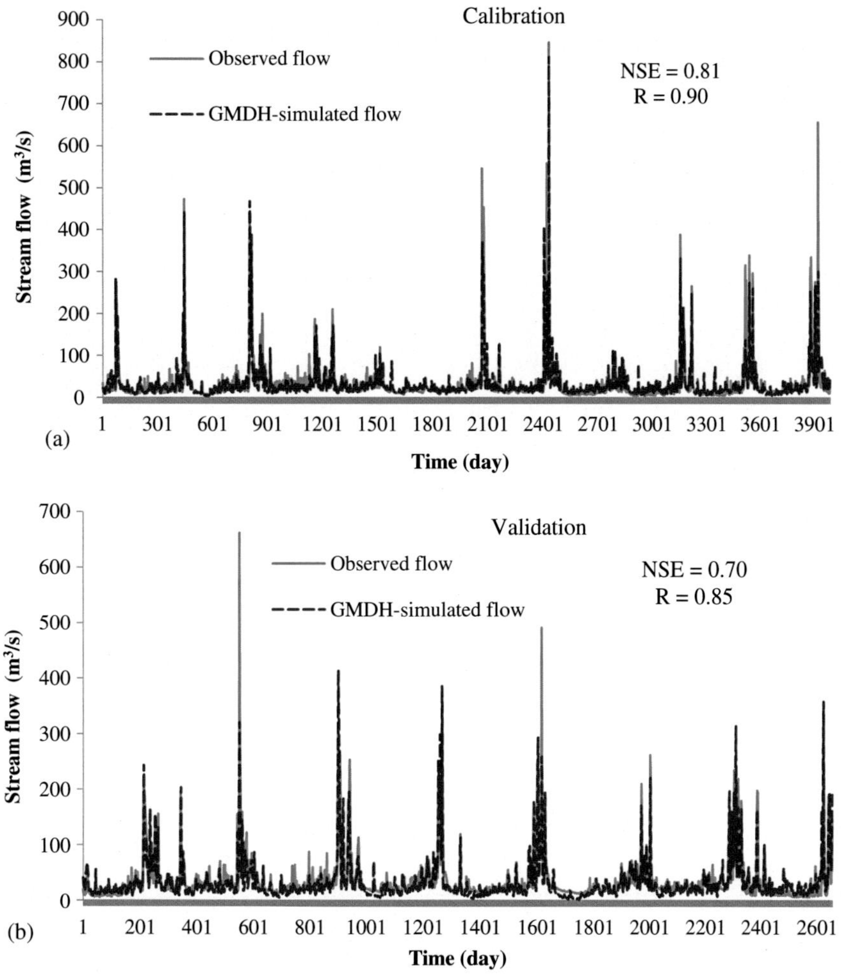

FIG. 6 Daily observed and GMDH simulated streamflow at catchment B during model (A) calibration; and (B) validation.

The daily observed and GEP-simulated streamflow for the same catchment (catchment B) are presented in Fig. 7. The performance of GEP model was found more or less similar to GMDH model. However, the R and NSE values during model calibration and validation were 0.88 and 0.78, and 0.84 and 0.68, respectively, which indicate better performance of GEP than GMDH.

The scatter plots of observed and simulated runoffs by GMDH and GEP during model calibration and validation are presented in Fig. 8. The figures show that all the models were able to replicate the streamflow very well during model calibration. The models were also very good to replicate the low flows during model validation. However, all the models underestimated the peak flow during model validation. Among the two models, GMDH was found to replicate the peak flows more accurately compared to other models during both model calibration and validation.

Both the RR models were used to simulate daily streamflow in all the seven selected catchments. The performance of the models at all the catchments was assessed using standard statistical measures. Obtained results during model calibration and validation are given in Tables 6 and 7, respectively.

The obtained statistics during calibration of GMDH model show that the model was well calibrated for all the catchments. The NSE and R values were more than 0.7 and 0.8, respectively for all the catchments except catchment A and J. The RMSE values were low in most of the catchments during model calibration. The GMDH model was also found satisfactory during validation. The NSE values were found more than 0.5 at all the catchments; the R values were high (>0.78) and RMSE values were reasonably low for all the catchments.

The GEP model was also found to perform satisfactory during both model calibration and validation. The NSE values were found more than 0.5, the R values more than 0.7, and the RMSE values reasonably low for all the catchment. The performance of the model was found satisfactory during model validation at all catchments except for catchment J, where NSE values were found less than 0.5. However, the R values were found more than 0.6 and RMSE value reasonably low for all the catchments. Overall, the performance of GEP was found very near to GMDH model in most of the catchments.

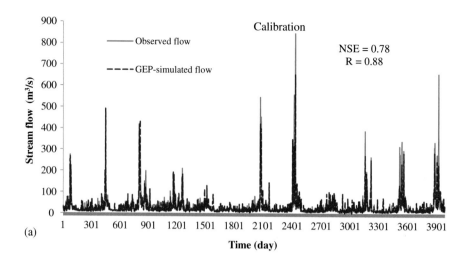

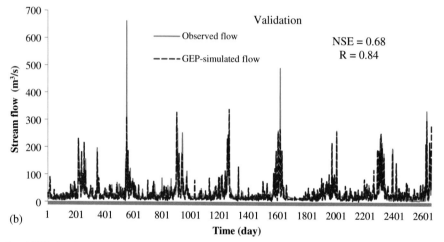

FIG. 7 Daily observed and GEP simulated streamflow at catchment B during model (A) calibration; and (B) validation.

All the models showed their lowest performance at catchment J. The NSE value was 0.53 for GMDH and 0.37 for GEP during validation. The poor performance of the models at this catchment is partly due to low quality of data used for model calibration and validation. The length of streamflow data used for development of hydrological models at this catchment was short compared to other catchments and therefore, the data used during model calibration was unable to capture the necessary information required for the development of RR model. Hence, the models failed to simulate most of the peak flows during validation. Moreover, rain gauges at this catchment are very less. The uncertainty in input rainfall was also responsibility for low performance of the models at this catchment.

Comparison of model performance in different catchments shows that only GMDH was able to show satisfactory results at all the catchment during both model calibration and validation. The performance of GEP was as promising as GMDH at all the catchments except catchment J. Therefore, it can be remarked that GMDH can be used for rainfall–runoff modeling in the study area.

The GMDH model was also used to simulate the monthly river discharge in the study area to show its efficacy. Observed and GMDH simulated monthly flow during model calibration and validation at catchment C is shown in Fig. 9 as an example.

The figure shows that GMDH was able to simulate monthly river flow very well; even the peak flows were reconstructed properly during model calibration and validation. Observed and GEP simulated monthly flow at catchment C is shown in Fig. 10. The figure shows that the performance of GEP is promising like GMDH. GEP was also able to reconstruct the monthly peak flow during model calibration; however, it was less efficient in reconstructing the peaks during model validation.

The performance of GMDH and GEP models measured using standard statistical measures during model calibration and validation at all the catchments is given in Table 8.

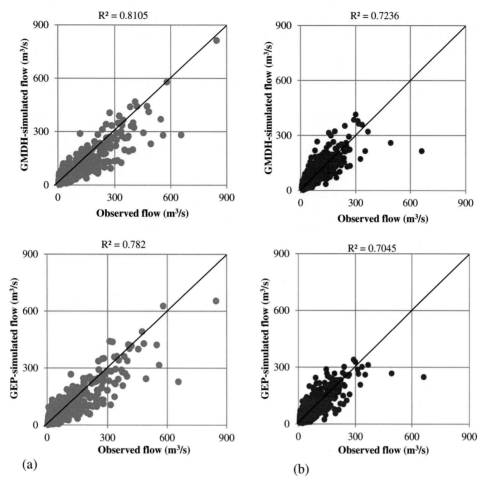

FIG. 8 Scatter plots of daily observed and simulated flow by (A) GMDH; and (B) GEP at catchment B during model calibration (left) and validation (right).

TABLE 6 Performance of rainfall–runoff models during calibration.

Model	GMDH			GEP		
Catchment	R	NSE	RMSE	R	NSE	RMSE
A	0.751	0.561	45.464	0.752	0.566	45.242
B	0.900	0.807	22.007	0.884	0.782	23.436
C	0.918	0.844	10.408	0.896	0.803	11.704
D	0.837	0.701	217.983	0.820	0.672	228.352
E	0.839	0.703	104.479	0.796	0.633	116.279
F	0.842	0.708	22.632	0.823	0.676	23.889
J	0.781	0.609	16.909	0.713	0.507	19.001

The NSE values for GMDH and GEP during calibration of model at different catchments were found in the range of 0.63–0.847 and 0.748–0.897, respectively. On the other hand, those values were found in the range of 0.557–0.814 and 0.496–0.938, respectively during model validation. The R values were found high and RMSE values low for both GMDH and GEP models during both calibration and validation. In term of NSE, GMDH was found to perform satisfactory in all the catchments. The GEP was also found to perform satisfactory in all the catchment except in catchment F, where NSE value was found unsatisfactory (0.496). As GMDH was found to perform well in all the catchments, it can be remarked that

TABLE 7 Performance of rainfall–runoff models during validation.

Model	GMDH			GEP		
Catchment	R	NSE	RMSE	R	NSE	RMSE
A	0.816	0.659	53.289	0.830	0.678	51.815
B	0.850	0.703	22.283	0.839	0.676	23.290
C	0.798	0.604	11.687	0.778	0.517	12.915
D	0.865	0.663	138.053	0.796	0.544	160.570
E	0.812	0.571	99.650	0.775	0.542	103.027
F	0.815	0.654	27.073	0.788	0.610	28.761
J	0.782	0.529	29.266	0.620	0.369	34.130

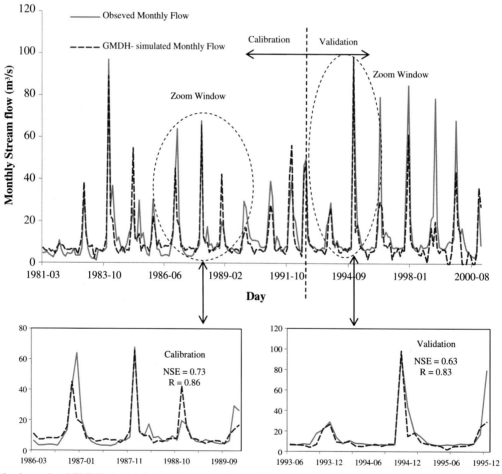

FIG. 9 Monthly observed and GMDH-simulated stream flow at catchment C during model calibration and validation.

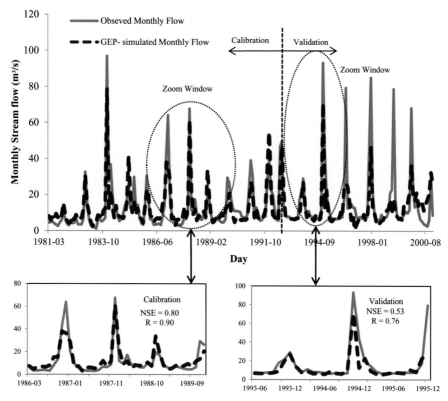

FIG. 10 Monthly observed and GEP-simulated stream flow at catchment C during model calibration and validation period.

TABLE 8 Performance of monthly rainfall–runoff models during calibration and validation.

Catchment	Assessment Criteria	Calibration		Validation	
		GMDH	GEP	GMDH	GEP
A	R	0.826	0.865	0.916	0.903
	NSE	0.679	0.748	0.785	0.795
	RMSE	26.572	23.552	34.270	33.428
B	R	0.931	0.947	0.887	0.882
	NSE	0.890	0.897	0.781	0.776
	RMSE	0.945	10.596	15.169	15.345
C	R	0.858	0.903	0.826	0.762
	NSE	0.729	0.808	0.627	0.532
	RMSE	7.125	6.001	11.232	12.591
D	R	0.906	0.927	0.927	0.918
	NSE	0.819	0.860	0.814	0.805
	RMSE	113.510	100.007	66.257	67.702
E	R	0.806	0.886	0.807	0.831
	NSE	0.630	0.784	0.645	0.559
	RMSE	78.419	59.927	66.657	74.285
	R	0.890	0.918	0.758	0.725

Continued

TABLE 8 Performance of monthly rainfall–runoff models during calibration and validation—cont'd

Catchment	Assessment Criteria	Calibration		Validation	
		GMDH	GEP	GMDH	GEP
F	NSE	0.791	0.842	0.557	0.496
	RMSE	13.503	11.730	20.870	22.392
	R	0.922	0.908	0.964	0.977
J	NSE	0.847	0.825	0.680	0.938
	RMSE	6.442	6.898	40.022	17.644

GMDH is the best model among the two models used in the present study for simulating monthly runoff. However, further analysis based on uncertainty in model simulation was carried out to confirm the performance of GMDH model.

5. Uncertainty assessment of performance of GMDH rainfall-runoff model

The performance of GMDH model in simulating daily and monthly river flow was further investigated by considering uncertainty in the GMDH model simulation through confidence intervals for scatter plot, monthly mean discharge plot, and box plots of percentile flow. Obtained results at three catchments (catchments B, C and E) are presented in following section as examples.

The time series of observed and GMDH predicted streamflow at the outlet of catchment B are presented in previous section. The time series of observed and GMDH simulated streamflow during model calibration and validation at other two catchments (catchments C and E) are shown in Figs. 11 and 12, respectively. Graphs in both the figures show that GMDH was able to simulate the river flow during both model calibration and validation very well. Even most of the flow peaks were simulated by GMDH properly.

The comparison of observed and GMDH simulated monthly mean streamflow at those three catchments are shown in Fig. 13. The graphs in this Figure also show the ability of GMDH in simulating monthly mean streamflow efficiently. Though GHDH slightly under- or over-predicted the mean streamflow in some months, it was able to replicate the seasonal pattern of streamflow properly. The GMDH models were even able to simulate the high streamflow in the monsoon months of November and December.

The scatter plots of observed and GMDH simulated daily streamflow with 95% confidence interval at three catchments (catchment B, C and E) for the whole period (calibration and validation) are shown Fig. 14. The figure shows that most of the data points are within the 95% confidence band. It is very natural that 5% data can be out of 95% confidence interval band. The graphs shows that number of data points outside the confidence band is very less, which indicates that GMDH has simulated the daily river flow well.

The box plots for different flow percentiles of observed and GMDH simulated daily flow at three catchments are shown in Fig. 14. The figure shows that observed and GMDH simulated probability graphs match very well. Even the boxes of observed high flows were very nearly replicated by the GMDH model at all of the three catchments. This again proves the efficiency of GMDH model in replicating flow pattern in the study area.

The performance of GMDH model in simulating monthly average flow was verified similarly by considering the uncertainty in GMDH simulation using confidence interval for scatter plot, monthly mean discharge plot, and box plots of percentile flow. The time series of observed and GMDH simulated monthly streamflow during model calibration and validation at catchments D and B are presented in Figs. 15 and 16, respectively as examples. Graphs in both figures show that GMDH was able to simulate the monthly river flow during both model calibration and validation perfectly. All the peaks were perfectly simulated by GMDH model during validation at catchment D. At catchment B, GMDH was also able to reconstruct most of the peaks during model validation.

The scatter plots of observed and GMDH model simulated monthly average streamflow with 95% confidence interval at three catchments (catchment B, C and D) for the total time period (calibration and validation) are shown in Fig. 17 as

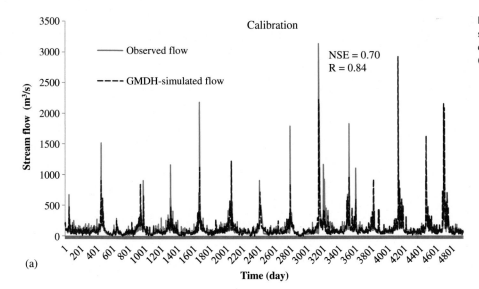

FIG. 11 Daily observed and GMDH-simulated streamflow at catchment E during model (A) calibration; and (B) validation.

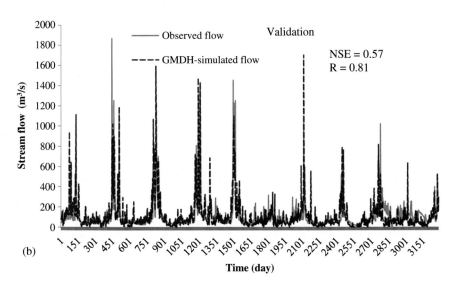

examples. The figure shows that almost all the data points are within the 95% confidence band, particularly. Number of data points outside the confidence band was less than 5% in all the catchments. The box plots of different flow percentiles of observed and GMDH model simulated monthly streamflow (Fig. 17) show good match; even the outliers were very nearly replicated by GMDH model at all the three catchments. The obtained results clearly indicate the capability of GMDH model in simulating monthly river flow in the study area.

6. Conclusions

Two empirical data driven models based on GEP and GMDH were used to simulate streamflow in seven selected catchments in the east coast of peninsular Malaysia. The models were developed to simulate both daily and monthly streamflow at the outlets of the selected catchments. The models were calibrated and validated with historical data. Their performances were evaluated using standard statistical approaches as well as through visual inspection of time series of the simulated and observed streamflow.

Comparison of model performance in different catchments shows that only GMDH was able to perform satisfactory at all the catchments. The performance of the GEP model was found as promising as the GMDH model; however, it failed to

FIG. 12 Daily observed and GMDH-simulated streamflow at catchment C during model (A) calibration; and (B) validation.

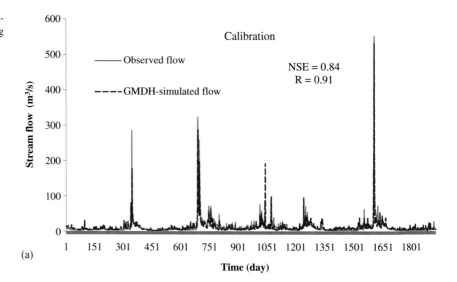

(a)

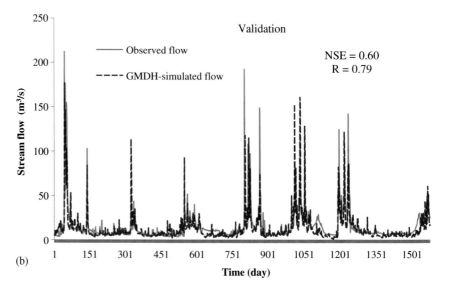

(b)

FIG. 12 Daily observed and GMDH-simulated streamflow at catchment C during model (A) calibration; and (B) validation.

show satisfactory results at all catchments. Similar results were obtained during calibration and validation of daily and monthly models. It means that GMDH perform best in both cases. The scatter plots of observed and GMDH simulated monthly river flow with 95% confidence interval showed that almost all the data points are within the 95% confidence band. All the results together indicate that GMDH performs better in modeling runoff in the study area. Therefore, the GMDH models were used for simulation of future changes in river flow under projected climate in the study area.

Empirical RR models have been widely employed in hydrological modeling in recent years because of their better performance compared with conceptual lumped models and physically based models. The present study also reveals that the empirical models based on GMDH and GEP perform better compared to lumped model. However, GMDH model like other empirical transparent and blackbox models assume that the rainfall is uniformly distributed and the catchment parameters are same over the catchment. As objective of present study was to show the impacts of climate change, more particularly the impacts of the changes in rainfall and temperature on streamflow only, GMDH can be considered as the most suitable model.

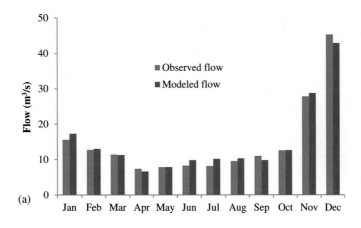

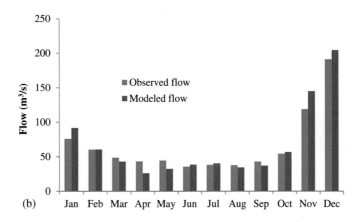

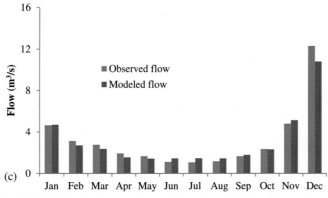

FIG. 13 Monthly mean of observed and GMDH simulated streamflow at (A) catchment B; (B) catchment E; and (C) catchment C.

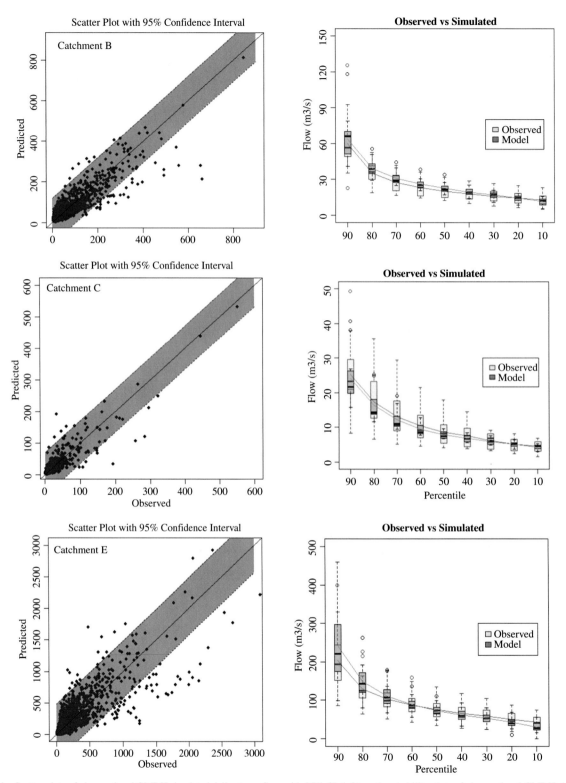

FIG. 14 Scatter plots of observed and GMDH simulated daily streamflow with 95% CI (left); and probability plots of observed and GMDH simulated daily streamflow (right) at Catchments B, C, and E.

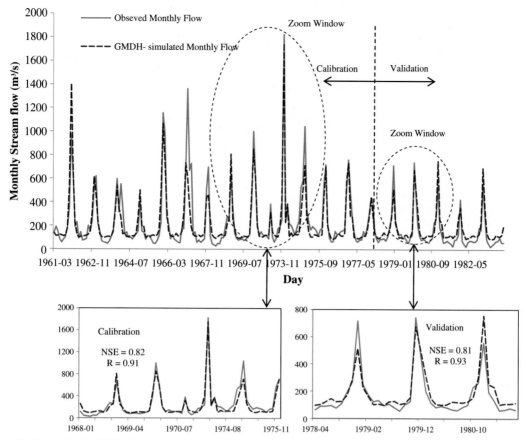

FIG. 15 Monthly observed and GMDH-simulated stream flow at catchment D during model calibration and validation.

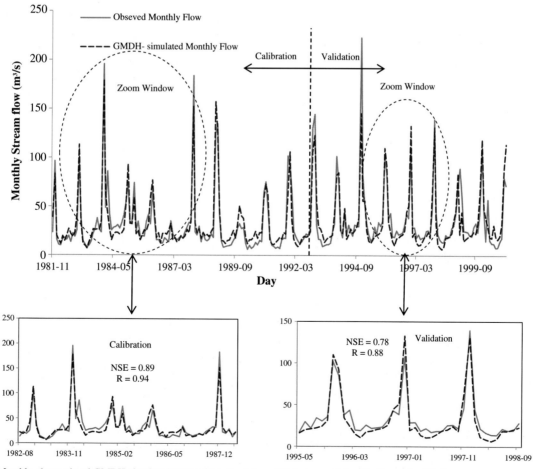

FIG. 16 Monthly observed and GMDH-simulated stream flow at catchment B during model calibration and validation.

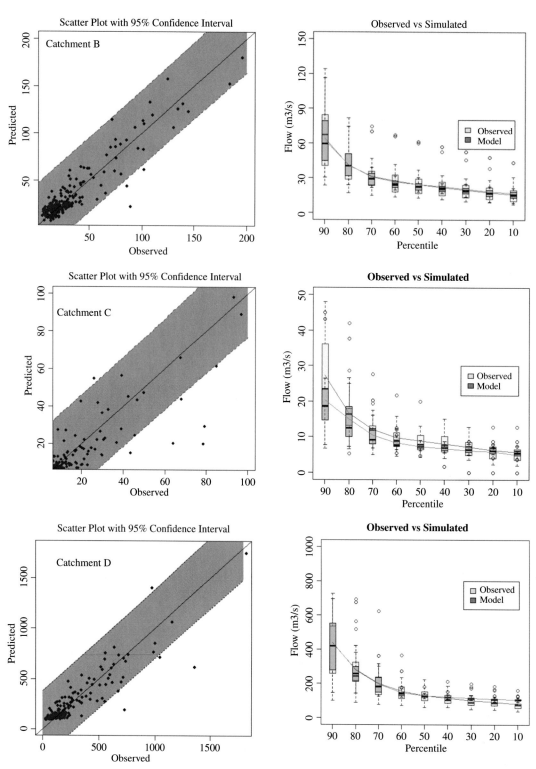

FIG. 17 Scatter plots of observed and GMDH simulated monthly streamflow with 95% CI (left); and probability plots of observed and GMDH simulated monthly streamflow (right) at Catchments B, C, and D.

References

Abdolrahimi, S., Nasernejad, B., Pazuki, G., 2014. Prediction of partition coefficients of alkaloids in ionic liquids based aqueous biphasic systems using hybrid group method of data handling (GMDH) neural network. J. Mol. Liq. 191, 79–84.

Ahmed, K., Shahid, S., Haroon, S.B., Xiao-Jun, W., 2015. Multilayer perceptron neural network for downscaling rainfall in arid region: a case study of Baluchistan, Pakistan. J. Earth Syst. Sci. 124 (6), 1325–1341.

Ahmed, K., Sachindra, D.A., Shahid, S., Iqbal, Z., Nawaz, N., Khan, N., 2020. Multi-model ensemble predictions of precipitation and temperature using machine learning algorithms. Atmos. Res. 236, 104806.

Amanifard, N., Nariman-Zadeh, N., Borji, M., Khalkhali, A., Habibdoust, A., 2008. Modelling and Pareto optimization of heat transfer and flow coefficients in microchannels using GMDH type neural networks and genetic algorithms. Energy Convers. Manag. 49 (2), 311–325.

Aytek, A., Alp, M., 2008. An application of artificial intelligence for rainfall–runoff modeling. J. Earth Syst. Sci. 117 (2), 145–155.

Azamathulla, H.M., Ghani, A.A., Leow, C.S., Chang, C.K., Zakaria, N.A., 2011. Gene-expression programming for the development of a stage-discharge curve of the Pahang River. Water Resour. Manag. 25 (11), 2901–2916.

Beven, K., 1985. Distributed models. Hydrological Forecasting, John Wiley and Sons, New York, p 405–435Beven, K., 1985. Distributed models. In: Hydrological Forecasting. John Wiley and Sons, New York, USA, pp. 405–435.

Beven, K., 2001. Rainfall–runoff Modelling: The Primer. John Wiley & Sons Ltd, p. 360.

Chang, F.J., Hwang, Y.Y., 1999. A self-organization algorithm for real-time flood forecast. Hydrol. Process. 13 (2), 123–138.

Chen, J., Brissette, F.P., Chaumont, D., Braun, M., 2013. Performance and uncertainty evaluation of empirical downscaling methods in quantifying the climate change impacts on hydrology over two North American river basins. J. Hydrol. 479, 200–214.

Clarke, R.T., 1973. A review of some mathematical models used in hydrology, with observations on their calibration and use. J. Hydrol. 19 (1), 1–20.

Cornforth, T.W., Lipson, H., 2015. A hybrid evolutionary algorithm for the symbolic modeling of multiple-time-scale dynamical systems. Evol. Intel. 8 (4), 149–164.

Davison, B., van der Kamp, G., 2008. Low-flows in deterministic modelling: a brief review. Can. Water Resour. J. 33 (2), 181–194.

Deng, S., Yue, D., Yang, L.C., Fu, X., Feng, Y.Z., 2016. Distributed function mining for gene expression programming based on fast reduction. PLoS One 11 (1), e0146698.

Devia, G.K., Ganasri, B.P., Dwarakish, G.S., 2015. A review on hydrological models. Aquat. Procedia 4, 1001–1007.

Ebtehaj, I., Bonakdari, H., Khoshbin, F., Azimi, H., 2015a. Pareto genetic design of GMDH-type neural network for predict discharge coefficient in rectangular side orifices. Flow Meas. Instrum. 41, 67–74.

Ebtehaj, I., Bonakdari, H., Zaji, A.H., Azimi, H., Sharifi, A., 2015b. Gene expression programming to predict the discharge coefficient in rectangular side weirs. Appl. Soft Comput. 35, 618–628.

Ebtehaj, I., Sammen, S.S., Lariyah, M.S., Malik, A., Parveen, S., Al-Janabi, A.M.S., Kwok-Wing, C., Bonakdari, H., 2021. Prediction of daily water level using new hybridized GS-GMDH and ANFIS-FCM models. Eng. Appl. Comput. Fluid Mech. 15 (1), 1343–1361. https://doi.org/10.1080/19942060.2021.1966837.

El-Kadi, A.I., 1989. Watershed models and their applicability to conjunctive use management. Water Resour. Bull. 25, 125–137.

Eslamian, S., Abedi-Koupai, J., Zareian, M.J., 2012. Measurement and modelling of the water requirement of some greenhouse crops with artificial neural networks and genetic algorithm. Int. J. Hydrol. Sci. Technol. 2 (3), 237–251.

Faris, H., Sheta, A., 2016. A comparison between parametric and non-parametric soft computing approaches to model the temperature of a metal cutting tool. Int. J. Comput. Integr. Manuf. 29 (1), 64–75.

Fernando, T.M.K.G., Maier, H.R., Dandy, G.C., 2009. Selection of input variables for data driven models: an average shifted histogram partial mutual information estimator approach. J. Hydrol. 367, 165–176.

Ferreira, C., 2006a. Gene Expression Programming: Mathematical Modeling by An Artificial Intelligence. vol. 21 Springer.

Ferreira, C., 2006b. Designing neural networks using gene expression programming. In: Applied Soft Computing Technologies: The Challenge of Complexity. Springer, Berlin Heidelberg, Germany, pp. 517–535.

Firat, M., Dikbas, F., Koç, A.C., Gungor, M., 2010. Missing data analysis and homogeneity test for Turkish precipitation series. Sadhana 35 (6), 707–720.

Fleming, G., 1975. Computer simulation techniques in hydrology. In: Environmental Science Series. Elsevier, p. 333.

Garg, V., 2015. Inductive group method of data handling neural network approach to model basin sediment yield. J. Hydrol. Eng. 20 (6). https://doi.org/10.1061/(ASCE)HE.1943-5584.0001085.

Gepsoft Inc, 2013. GeneXproTools Release 5.0 (Computer Program). Gepsoft, Inc, Bristol, United Kingdom.

Gholami, A., Bonakdari, H., Zaji, A.H., Akhtari, A.A., Khodashenas, S.R., 2015. Predicting the velocity field in a 90° open channel bend using a gene expression programming model. Flow Meas. Instrum. 46, 189–192.

Guven, A., Aytek, A., 2009. New approach for stage–discharge relationship: gene-expression programming. J. Hydrol. Eng. 14 (8), 812–820.

Hadipour, S., Shahid, S., Harun, S.B., Wang, X.J., 2013. Genetic programming for downscaling extreme rainfall events. In: 2013 1st International Conference on Artificial Intelligence, Modelling and Simulation. IEEE, pp. 331–334.

Hargreaves, G.H., Samani, Z.A., 1982. Estimating potential evapotranspiration. J. Irrig. Drain. Div. 108 (3), 225–230.

Hashmi, M.Z., Shamseldin, A.Y., 2014. Use of gene expression programming in regionalization of flow duration curve. Adv. Water Resour. 68, 1–12.

Ivakhnenko, A.G., 1971. Polynomial theory of complex systems. IEEE Trans. Syst. Man Cybern. 4, 364–378.

Kalteh, A.M., Hjorth, P., 2009. Imputation of missing values in a precipitation-runoff process database. Hydrol. Res. 40 (4), 420–432.

Keshavarz, A., Mehramiri, M., 2015. New gene expression programming models for normalized shear modulus and damping ratio of sands. Eng. Appl. Artif. Intell. 45, 464–472.

Khan, N., Pour, S.H., Shahid, S., Ismail, T., Ahmed, K., Chung, E.S., Wang, X., 2019a. Spatial distribution of secular trends in rainfall indices of Peninsular Malaysia in the presence of long-term persistence. Meteorol. Appl. 26 (4), 655–670.

Khan, N., Shahid, S., Juneng, L., Ahmed, K., Ismail, T., Nawaz, N., 2019b. Prediction of heat waves in Pakistan using quantile regression forests. Atmos. Res. 221, 1–11.

Khan, N., Sachindra, D.A., Shahid, S., Ahmed, K., Shiru, M.S., Nawaz, N., 2020. Prediction of droughts over Pakistan using machine learning algorithms. Adv. Water Resour. 139, 103562.

Kisi, O., Shiri, J., Nikoofar, B., 2012. Forecasting daily lake levels using artificial intelligence approaches. Comput. Geosci. 41, 169–180.

Kisi, O., Shiri, J., Tombul, M., 2013. Modeling rainfall–runoff process using soft computing techniques. Comput. Geosci. 51, 108–117.

Koza, J.R., 1992. Genetic Programming: On the Programming of Computers by Means of Natural Selection. vol. 1. MIT Press, UK.

Krishnan, T., McLachlan, G., 1997. The EM Algorithm and Extensions. Wiley, pp. 58–60.

Linsley, R.K., 1982. Rainfall–runoff models—an overview. In: Singh, V.P. (Ed.), Proceedings of International Symposium on Rainfall–Runoff Modelling. Water Resources Publications, Littletown, Colorado, USA, pp. 3–22.

Malik, A., Tikhamarine, Y., Sammen, S.S., Abba, S.I., Shahid, S., 2021. Prediction of meteorological drought by using hybrid support vector regression optimized with HHO versus PSO algorithms. Environ. Sci. Pollut. Res. 28, 39139–39158.

Mayowa, O.O., Pour, S.H., Shahid, S., Mohsenipour, M., Harun, S.B., Heryansyah, A., Ismail, T., 2015. Trends in rainfall and rainfall-related extremes in the east coast of peninsular Malaysia. J. Earth Syst. Sci. 124 (8), 1609–1622.

McLachlan, G.J., Krishnan, T., Ng, S.K., 2004. The EM Algorithm, No. 24, Papers/Humboldt-Universität Berlin. Center for Applied Statistics and Economics (CASE), Germany.

Mohamadi, S., Sammen, S.S., Panahi, F., et al., 2020. Zoning map for drought prediction using integrated machine learning models with a nomadic people optimization algorithm. Nat. Hazards 104, 537–579. https://doi.org/10.1007/s11069-020-04180-9.

Moradkhani, H., Sorooshian, S., 2008. General review of rainfall–runoff modeling: model calibration, data assimilation, and uncertainty analysis. In: Hydrological Modelling and the Water Cycle. Springer, Berlin Heidelberg, Germany, pp. 1–24.

Muhammad, M.K.I., Nashwan, M.S., Shahid, S., Ismail, T.B., Song, Y.H., Chung, E.S., 2019. Evaluation of empirical reference evapotranspiration models using compromise programming: a case study of Peninsular Malaysia. Sustainability 11 (16), 4267.

Muhammad, M.K.I., Shahid, S., Ismail, T., Harun, S., Kisi, O., Yaseen, Z.M., 2021. The development of evolutionary computing model for simulating reference evapotranspiration over Peninsular Malaysia. Theor. Appl. Climatol. 144 (3), 1419–1434.

Nashwan, M.S., Ismail, T., Ahmed, K., 2018. Flood susceptibility assessment in Kelantan river basin using copula. Int. J. Eng. Technol. 7 (2), 584–590.

Nashwan, M.S., Ismail, T., Ahmed, K., 2019. Non-stationary analysis of extreme rainfall in peninsular Malaysia. J. Sustain. Sci. Manag. 14 (3), 17–34.

Ng, S.K., McLachlan, G.J., 2004. Speeding up the EM algorithm for mixture model-based segmentation of magnetic resonance images. Pattern Recogn. 37 (8), 1573–1589.

Noor, M., Ismail, T., Chung, E.S., Shahid, S., Sung, J.H., 2018. Uncertainty in rainfall intensity duration frequency curves of peninsular Malaysia under changing climate scenarios. Water 10 (12), 1750.

Onwubolu, G.C., Buryan, P., Garimella, S., Ramachandran, V., Buadromo, V., Abraham, A., 2007. Self-organizing data mining for weather forecasting. In: IADIS European Conference Data Mining, pp. 81–88.

Ouyang, F., Lü, H., Zhu, Y., Zhang, J., Yu, Z., Chen, X., Li, M., 2014. Uncertainty analysis of downscaling methods in assessing the influence of climate change on hydrology. Stoch. Env. Res. Risk A. 28 (4), 991–1010.

Pham, Q.B., Sammen, S.S., Abba, S.I., et al., 2021a. A new hybrid model based on relevance vector machine with flower pollination algorithm for phycocyanin pigment concentration estimation. Environ. Sci. Pollut. Res. 28, 32564–32579. https://doi.org/10.1007/s11356-021-12792-2.

Pham, Q.B., Mohammadpour, R., Linh, N.T.T., et al., 2021b. Application of soft computing to predict water quality in wetland. Environ. Sci. Pollut. Res. 28, 185–200. https://doi.org/10.1007/s11356-020-10344-8.

Pour, S.H., Harun, S.B., Shahid, S., 2014. Genetic programming for the downscaling of extreme rainfall events on the East Coast of Peninsular Malaysia. Atmosphere 5 (4), 914–936.

Pour, S.H., Shahid, S., Chung, E.S., Wang, X.J., 2018. Model output statistics downscaling using support vector machine for the projection of spatial and temporal changes in rainfall of Bangladesh. Atmos. Res. 213, 149–162.

Pour, S.H., Abd Wahab, A.K., Shahid, S., 2020a. Physical-empirical models for prediction of seasonal rainfall extremes of Peninsular Malaysia. Atmos. Res. 233, 104720.

Pour, S.H., Abd Wahab, A.K., Shahid, S., Asaduzzaman, M., Dewan, A., 2020b. Low impact development techniques to mitigate the impacts of climate-change-induced urban floods: current trends, issues and challenges. Sustain. Cities Soc. 62, 102373.

Pour, S.H., Abd Wahab, A.K., Shahid, S., Ismail, Z.B., 2020c. Changes in reference evapotranspiration and its driving factors in peninsular Malaysia. Atmos. Res. 246, 105096.

Refsgaard, J.C., Storm, B., 1995. MIKE SHE. In: Singh, V.P. (Ed.), Computer Models of Watershed Hydrology. Water Resources Publications, Littletown, Colorado, USA, pp. 806–846.

Sa'adi, Z., Shahid, S., Ismail, T., Chung, E.S., Wang, X.J., 2017. Distributional changes in rainfall and river flow in Sarawak, Malaysia. Asia-Pac. J. Atmos. Sci. 53 (4), 489–500.

Sachindra, D.A., Ahmed, K., Shahid, S., Perera, B.J.C., 2018. Cautionary note on the use of genetic programming in statistical downscaling. Int. J. Climatol. 38 (8), 3449–3465.

Sachindra, D.A., Ahmed, K., Rashid, M.M., Sehgal, V., Shahid, S., Perera, B.J.C., 2019. Pros and cons of using wavelets in conjunction with genetic programming and generalised linear models in statistical downscaling of precipitation. Theor. Appl. Climatol. 138 (1), 617–638.

Sammen, S.S., Mohamed, T.A., Ghazali, A.H., et al., 2017. Generalized regression neural network for prediction of peak outflow from dam breach. Water Resour. Manag. 31, 549–562. https://doi.org/10.1007/s11269-016-1547-8.

Sammen, S.S., Ehteram, M., Abba, S.I., et al., 2021. A new soft computing model for daily streamflow forecasting. Stoch. Environ. Res. Risk Assess. 35, 2479–2491. https://doi.org/10.1007/s00477-021-02012-1.

Samsudin, R., Saad, P., Shabri, A., 2011. River flow time series using least squares support vector machines. Hydrol. Earth Syst. Sci. 15 (6), 1835–1852.

Schmidt, M., Lipson, H., 2009. Distilling free-form natural laws from experimental data. Science 324 (5923), 81–85.

Shahid, S., Pour, S.H., Wang, X., Shourav, S.A., Minhans, A., Bin Ismail, T., 2017. Impacts and adaptation to climate change in Malaysian real estate. Int. J. Clim. Change Strategies Manage. 9 (1), 87–103. https://doi.org/10.1108/IJCCSM-01-2016-0001.

Sharafati, A., Khosravi, K., Khosravinia, P., Ahmed, K., Salman, S.A., Yaseen, Z.M., Shahid, S., 2019. The potential of novel data mining models for global solar radiation prediction. Int. J. Environ. Sci. Technol. 16 (11), 7147–7164.

Shoaib, M., Shamseldin, A.Y., Melville, B.W., Khan, M.M., 2015. Runoff forecasting using hybrid Wavelet Gene Expression Programming (WGEP) approach. J. Hydrol. 527, 326–344.

Srinivasan, D., 2008. Energy demand prediction using GMDH networks. Neurocomputing 72 (1), 625–629.

Tao, H., Al-Bedyry, N.K., Khedher, K.M., Shahid, S., Yaseen, Z.M., 2021. River water level prediction in coastal catchment using hybridized relevance vector machine model with improved grasshopper optimization. J. Hydrol. 598, 126477.

Tikhamarine, Y., Souag-Gamane, D., Ahmed, A.N., Sammen, S.S., Kisi, O., Huang, Y.F., El-Shafie, A., 2020. Rainfall–runoff modelling using improved machine learning methods: Harris hawks optimizer vs. particle swarm optimization. J. Hydrol. 589, 125133.

Todini, E., 2007. Hydrological catchment modelling: past, present and future. Hydrol. Earth Syst. Sci. 11 (1), 468–482.

Tsidu, G.M., 2012. High-resolution monthly rainfall database for Ethiopia: homogenization, reconstruction, and gridding. J. Clim. 25 (24), 8422–8443.

Voss, M.S., Feng, X., 2002. A new methodology for emergent system identification using particle swarm optimization (PSO) and the group method data handling (GMDH). In: GECCO, pp. 1227–1232.

Wang, X., Li, L., Lockington, D., Pullar, D., Jeng, D.S., 2005. Self-Organizing Polynomial Neural Network for Modelling Complex Hydrological Processes. The University of Sydney, Sydney, Australia, p. 30.

Wheater, H.S., Jakeman, A.J., Beven, K.J., 1993. Progress and Directions in Rainfall–runoff Modelling. FAO, pp. 101–132.

Witczak, M., Korbicz, J., Mrugalski, M., Patton, R.J., 2006. A GMDH neural network-based approach to robust fault diagnosis: application to the DAMADICS benchmark problem. Control. Eng. Pract. 14 (6), 671–683.

Woolhiser, D.A., 1973. Hydrologic and watershed modeling-state of the art. Trans. ASAE 16 (3), 553–0559.

Wu, P., Walker, B.A., Broyl, A., Kaiser, M., Johnson, D.C., Kuiper, R., van Duin, M., Gregory, W.M., Davies, F.E., Brewer, D., Hose, D., 2015. A gene expression based predictor for high risk myeloma treated with intensive therapy and autologous stem cell rescue. Leuk. Lymphoma 56 (3), 594–601.

Xu, J., Wang, J., Wei, Q., Wang, Y., 2016. Symbolic regression equations for calculating daily reference evapotranspiration with the same input to Hargreaves-Samani in Arid China. Water Resour. Manag. 30, 2055–2073.

Yang, G., Li, X., Wang, J., Lian, L., Ma, T., 2015. Modeling oil production based on symbolic regression. Energy Policy 82, 48–61.

Yaseen, Z.M., Awadh, S.M., Sharafati, A., Shahid, S., 2018a. Complementary data-intelligence model for river flow simulation. J. Hydrol. 567, 180–190.

Yaseen, Z.M., Ehteram, M., Sharafati, A., Shahid, S., Al-Ansari, N., El-Shafie, A., 2018b. The integration of nature-inspired algorithms with least square support vector regression models: application to modeling river dissolved oxygen concentration. Water 10 (9), 1124.

Yaseen, Z.M., Mohtar, W.H.M.W., Ameen, A.M.S., Ebtehaj, I., Razali, S.F.M., Bonakdari, H., Shahid, S., 2019. Implementation of univariate paradigm for streamflow simulation using hybrid data-driven model: case study in tropical region. IEEE Access 7, 74471–74481.

Yassin, M.A., Alazba, A.A., Mattar, M.A., 2016. A new predictive model for furrow irrigation infiltration using gene expression programming. Comput. Electron. Agric. 122, 168–175.

Yu, Z., Lu, H., Si, H., Liu, S., Li, X., Gao, C., Cui, L., Li, C., Yang, X., Yao, X., 2015. A highly efficient gene expression programming (GEP) model for auxiliary diagnosis of small cell lung cancer. PLoS One 10 (5), e0125517.

Yusuf, A.A., Francisco, H., 2009. Climate Change Vulnerability Mapping for Southeast Asia. Economy and Environment Program for Southeast Asia (EEPSEA), Singapore, pp. 10–15.

Ziarh, G.F., Asaduzzaman, M., Dewan, A., Nashwan, M.S., Shahid, S., 2021. Integration of catastrophe and entropy theories for flood risk mapping in peninsular Malaysia. J. Flood Risk Manage. 14 (1), e12686.

Zorn, C.R., Shamseldin, A.Y., 2015. Peak flood estimation using gene expression programming. J. Hydrol. 531, 1122–1128.

Chapter 26

Sediment transport with soft computing application for tropical rivers

Mohd Afiq Harun[a], Aminuddin Ab. Ghani[a], Saeid Eslamian[b,c], and Chun Kiat Chang[a]

[a]*River Engineering and Urban Drainage Research Centre, Universiti Sains Malaysia, George Town, Malaysia,* [b]*Department of Water Engineering, College of Agriculture, Isfahan University of Technology, Isfahan, Iran,* [c]*Center of Excellence for Risk Management and Natural Hazards, Isfahan University of Technology, Isfahan, Iran*

1. Introduction

Information regarding sediment transport is essential to maintain a fluvial environment to the river and creates equilibrium between erosion and sedimentation. Prediction of sediment transport is vital to design water-related structures (Constantine et al., 2014; Ghani et al., 2010; Harun and Ab. Ghani, 2020; Meshkova and Carling, 2012) and assess the environmental impact (Constantine et al., 2014; Rajaee and Jafari, 2020). Accurate sediment transport prediction helps the river authorities to manage and plan necessary water body intervention (Fakhri et al., 2014). The commonly used sediment transport equation does not apply to the river in the tropical regions because many of the previous equations were developed by using river and flume data from Europe and the United States (Ahmad Abdul Ghani et al., 2019; Ariffin, 2004; Harun and Ab. Ghani, 2020; Syvitski et al., 2014). For instance, the equation developed by Ackers and White (1973), Engelund and Hansen (1967), and Yang (1976) used data from a flume experiment where water depth was less than 0.5 m. Table 1 shows the commonly used sediment transport equations by research.

Ariffin (2004), Sinnakaudan et al. (2006), and Ahmad Abdul Ghani (2019) explained that the rivers in the tropical regions have different hydraulic characteristics and sediment properties, thus limiting the sediment transport equations' effectiveness. This is supported by the research done by Gunawan et al. (2019), whereby the best-reported discrepancy ratio (DR) in the tropical neighborhood country (Indonesia) was found to be below 28%. To overcome this limitation, Ariffin (2004) and Sinnakaudan et al.(2006) developed a sediment transport equation based on earlier research of predicting sediment transport in sewers done by Ab Ghani (1993). The multiple linear regression (MLR) technique was adopted to generate the sediment transport equation representing the tropical rivers' hydraulic characteristics in Malaysia. Table 2 lists the developed equation for the rivers in Malaysia.

Recent studies conducted by Saleh et al. (2017) and Teo et al. (2017) suggested that the commonly used total bed material load equations cannot predict sediment transport with high accuracy results. According to Teo et al. (2017), the average DR for the commonly used sediment equation is below 42%, whereas the local researcher's equation returns with much lower DR of 30% (Harun et al., 2020). Tables 3 and 4 list down the results of the performance of the commonly used sediment transport equation for Malaysian rivers. The findings from Teo et al. (2017), Saleh et al. (2017), Ahmad Abdul Ghani et al. (2019), and Ahmad Abdul Ghani et al. (2020) suggested that there is a lack of accuracy in predicting the total material load by applying the conventional method (MLR), particularly for higher river data range which resulted in low accuracy with low R^2 (coefficient of determination) and MAE (mean absolute error). Given the low accuracy of the current sediment transport equation, the revised equation of sediment transport, particularly for a tropical river like Malaysia, needs further study to increase the accuracy of sediment transport equation. Apart from that, the incorporation of machine learning into the sediment transport prediction model should be explored to enhance the existing total bed material load equation. Research done by Khan et al. (2018) and Rajaee and Jafari (2020) showed a good accuracy prediction results by utilizing machine learning program.

Handbook of HydroInformatics. https://doi.org/10.1016/B978-0-12-821962-1.00017-9

TABLE 1 Commonly used total sediment load equations (Garcia, 2008).

Researcher	Equation
Ackers and White (1973)	$C_w = \frac{\left(G_{gr}d\frac{\gamma_s}{\gamma}\right)}{\gamma}\left(\frac{U^*}{V}\right)^{-n}$
Brownlie (1981)	$C_{ppm} = 9022(F_g - F_{go})^{1.978}S^{0.6601}\left(\frac{R_b}{d_{50}}\right)^{-0.3301}$
Engelund and Hansen (1967)	$C_w = 0.05\left(\frac{G}{G-1}\right)\left(\frac{VS_f}{[(G-1)gd]^{1/2}}\right)\left(\frac{RS_f}{(G-1)d}\right)$
Yang (1979)	$\log C_{ppm} = 5.435 - 0.286\log\frac{\omega_s d_{50}}{\upsilon} - 0.457\log\frac{U^*}{\omega} + \{(1.799 - 0.409\log\frac{\omega d_{50}}{\upsilon} - 0.314\log\frac{U^*}{\omega}) \times \log\left(\frac{VS}{\omega} - \frac{V_c S}{\omega}\right)\}$

Note: C_w is the concentration by weight, C_{ppm} is the concentration by weight in ppm, G_{gr} is the general dimensionless sediment transport function, Υ is the specific weight of the water, d is the particle size distribution of bed materials, U^* is the shear velocity, V is the average flow velocity, n is the roughness coefficients, F_g is the grain related Froude number, F_{go} is the critical grain related Froude number, R is the hydraulic radius, S is the bed slope, R_b is the hydraulic radius for bed, d_{50} is the median size of bed material, G is the specific gravity of the sediment, S_f is the friction slope, g is the gravitational acceleration, ω is the fall velocity of the bed material, VS is the unit stream power and V_cS is the critical unit stream power.

TABLE 2 Total bed material load equation for Malaysian rivers.

Reference	Equation
Ariffin (2004)	$C_v = 1.156\times10^{-5}(\frac{R}{d_{50}})^{0.716}(\frac{U^*}{\omega_s})^{-0.975}(\frac{U^*}{V})^{0.507}(\frac{V^2}{gy})^{0.524}$
Sinnakaudan et al.(2006)	$C_v = 1.811\times10^{-4}(\frac{VS_o}{\omega_s})^{0.293}(\frac{R}{d_{50}})^{1.390}(\frac{\sqrt{g(Ss-1)d_{50}3}}{VR})$

Note: C_v is the sediment concentration by volume, Ss is the specific gravity of the sediment, R is the hydraulic radius, d_{50} is the median size of bed material, U^* is the shear velocity, S_o is the bed slope, ω_s is the fall velocity of the bed material, V is the average flow velocity, g is the standard gravity and y is the average depth of the water.

TABLE 3 Performance of the commonly used sediment transport equation in past research.

Researcher	No. of data	Yang (1979)	Engelund and Hansen (1967)	Ackers and White (1973)	Graf (1971)
Abu Hassan (1998)	11	54.5	27.3	45.5	45.5
Yahaya (1999)	60	65.0		3.3	61.7
Ibrahim (2002)	108	28	41.5	20.0	27.0
Ariffin (2004)	165	32.7	19.4	11.5	3.6
Chang (2006)	14	42.86	50	7.14	28.57
Sinnakaudan et al. (2006)	181	35.91		12.71	17.13
Saleh et al. (2017)	35	25.71	17.14	40.0	11.43
Average		40.31	32.86	20.02	27.85

(Modified after Saleh, A., Abustan, I., Mohd Remy Rozainy, M.A.Z., Sabtu, N., 2017. Assessment of total bed material equations on selected Malaysia rivers. In: AIP Conf. Proc. pp. 070002 (1–7). https://doi.org/10.1063/1.5005720.)

2. Application of machine learning in sediment transport

The use of machine learning to predict a model in water resources engineering has been the method of choice for many researchers in recent times (Azamathulla et al., 2013a; Basri et al., 2018; Chang et al., 2012; Kargar et al., 2019; Khan et al., 2018; Mohammad-Azari et al., 2020; Muhammad et al., 2018). This method is successfully applied in predicting sediment transport in pipes (Ebtehaj et al., 2016; Montes et al., 2020; Safari et al., 2018) and rivers (Ab. Ghani et al., 2010; Safari et al., 2019; Zangeneh Sirdari et al., 2014). Several studies were also conducted by using gene expression programming (GEP), artificial neural network (ANN), adaptive neuro fuzzy inference system (ANFIS) and evolutionary polynomial

TABLE 4 Comparison of the performance of the total bed material load equations (note: No. = number) (Teo et al., 2017).

Equation	Location	No. of data	DR between 0.5 and 2.0 (no. of data)	Percentage of accuracy (%)	Average accuracy (%)
Engelund and Hansen (1967)	Muda River	76	19	25.00	41.80
	Langat River	60	31	51.67	
	Kurau River	78	38	48.72	
Yang (1979)	Muda River	76	16	21.05	37.79
	Langat River	60	30	50.00	
	Kurau River	78	33	42.31	

regression (EPR) to increase the accuracy of sediment transport in Malaysia. Table 5 summarizes the sediment transport equation developed by the previous researchers for Malaysian rivers.

Many of the equations were developed by using a lot of variables (up to 8 variables) and many complicated functions. According to Bonakdari et al. (2020) and Ab. Ghani and Azamatulla (2014), the equation should be more straightforward and efficient by having less variables. Complicated equations are often very sensitive and challenging to apply in different data sets (Bonakdari et al., 2020). A significant model should be carefully identified to avoid the low accuracy problems outside of the range of model prediction.

TABLE 5 Development of machine learning incorporation in past studies in sediment transport.

Research	Findings	Limitation
Zakaria et al. (2010)	Development of new equation for Malaysian river by using GEP	• Many variables involved (Q, V, B, Y_o, R, S_o, W_s, d_{50}) • Use complicated function (sqrt, exp., /, *, +, sin, Atan, ln)
Ab. Ghani et al. (2010)	Prediction of total bed material load by using ANN	• Many variables involved (Q, V, B, Y_o, R, S_o) • ANN does not provide a functional equation
Chang et al. (2012)	The equation produced by GEP has the highest accuracy, followed by FFNN and ANFIS	• Many variables involved (Q, V, B, Y_o, R, S_o, W_s, d_{50}) • Use complicated function (sqrt, power, /, *, +, sin, Atan, log
Ab.Ghani and Azamathulla (2012)	Simplified the equation produced by Zakaria et al. (2010)	• Fewer variables involved ($\frac{V}{\sqrt{gd_{50}\,(S_s-1)}}$, $\frac{R}{d_{50}}$, $\frac{R}{Y_o}$) • Use complicated function (sqrt, power, /, *, +, sin, exp., ln)
Ahmad Abdul Ghani et al. (2019)	A new equation developed for Malaysian rivers by using EPR	• The data range is smaller compared to Ab. Ghani and Azamathulla (2014) • Not applicable to a large river

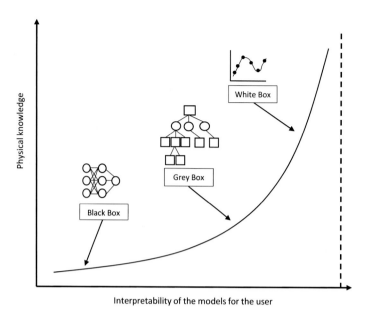

FIG. 1 Classification of machine learning program (Ahmad Abdul Ghani et al., 2020).

3. A hybrid method by using soft computing technique

Relevant literature (Ebtehaj et al., 2019; Shaghaghi et al., 2018b) reveals that machine learning techniques could produce better model prediction because they are more complex and can evolve to better suit the model, unlike the traditional regression method. Pintelas et al. (2020) classified machine learning into three categories: white box, gray box, and the black box (Fig. 1). The workings and programming steps used in the models determine the type of the machine learning program. A more transparent step which leans toward a more comfortable means of interpreting the model, such as the linear regression model, belongs to the white box. An inner workings model that is not known and hard to understand is referred to as the black box model (Pintelas et al., 2020). The gray box model is the combination of both black and white boxes where the model is semisupervised.

Often, researchers used single and hybrid methods as an approach to improve the accuracy of the predictions (Yahaya, 2019). A single method such as MLR, artificial neural networks (ANN) or gene expression programming (GEP) utilizes one method to optimize the model parameters (Ab. Ghani and Azamathulla, 2014; Ara Rahman and Chakrabarty, 2020; Chang et al., 2012), whereas hybrid models combine the methods to get the most appropriate model for the model predictions (Ab. Ghani et al., 2010; Ab. Ghani and Azamathulla, 2014). According to Yahaya (2019), the hybrid methods' performance is better than the single method in most cases. In water resources engineering, the application of the hybrid method is widely used to predict stable channel dimension, flow discharge, sediment transport modeling, scour depth, and rainfall forecasting (Balouchi et al., 2015; Danandeh Mehr et al., 2019; Khosravi et al., 2020; Nourani et al., 2019, 2016, 2012; Safari and Danandeh Mehr, 2018; Shaghaghi et al., 2018a; Sharghi et al., 2019; Shiri et al., 2020; Tayfur et al., 2013, 2003; Tayfur and Guldal, 2006; Ulke et al., 2009). This study combines MLR with the soft computing technique. The parameters that are developed earlier in the MLR were used as the input parameters in the soft computing program. Researchers have widely applied this method to increase the model prediction accuracy (Azamathulla et al., 2013b; Ebtehaj et al., 2019; Safari et al., 2019; Yahaya, 2019). There are many methods available to increase model prediction accuracy, including evolutionary polynomial regression (EPR), multigene genetic programming (MGGP), and M5 tree model (M5P).

4. Evolutionary polynomial regression (EPR)

EPR can be considered as a data processing tool driven by the hybrid regression technique (Giustolisi and Savic, 2009, 2006). This method uses a single genetic algorithm to concentrate on the formula symbol space to provide a few alternatives models for prediction purposes (Giustolisi and Savic, 2009, 2006). It is a nonlinear stepwise regression that involved non-linear function among variables but, linear to the regression parameters (Zahiri and Najafzadeh, 2018). EPR has a unique general structure that combines additive terms multiplied by many coefficients that can be described as follows:

$$\widehat{Y} = a_o + \sum_{j=1}^{m} a_j (X_1)^{ES(j,1)} \ldots (X_k)^{ES(j,k)} \cdot f\left((X_1)^{ES(j,k+1)} \ldots (X_k)^{ES(j,2k)}\right) \tag{1}$$

where m can be defined as the maximum number of additive terms, X_1 and $\widehat{Y}$ are model input and output variables. Function f is the exponents of the variables, and ES can be chosen by the user beforehand (Giustolisi and Savic, 2009, 2006). Ultimate regression expressions are linear to the coefficient a_j, and often estimated using classical numeral regression (Giustolisi and Savic, 2009, 2006).

5. Multi-gene genetic programming (MGGP)

Originating from the GP, the MGGP enhances the fitness of solutions by combining low depth GP to the monolithic GP (Danandeh Mehr et al., 2019; Danandeh Mehr and Safari, 2020; Safari and Danandeh Mehr, 2018). Danandeh Mehr et al. (2018) explained that the smaller tree application in MGGP is more straightforward compared to the monolithic GP. Summation of weighted outputs of two or more GP trees in a multigene program produces the output variable, meanwhile for the bias; it depends on the stochastic term. Pseudo linear MGGP model is represented by the output variable $\widehat{Y}$ which combines three genes. Each gene comprised the function of a given input variable x_1 and x_2. Fig. 2 shows an example of how MGGP operated. In this example, each multigene consists of three genes. Eq. (2) describe the MGGP mathematical expression where d_o is bias term, and d_1 and d_2 represent the gene weight.

$$\widehat{Y} = d_o + d_1(x_1 \times \cos x_2 + x_2 \times \sin x_1) + d_2(x_1 \times x_2 \sin x_{12}) + d_3(C_1 \times x_{12} + x_1 + x_2) \tag{2}$$

Linear regression was applied in the MGGP to suit the physical system's nonlinear condition (Danandeh Mehr et al., 2018). Danandeh Mehr et al. (2018) also explained that any data preprocessing technique that can enhance the accuracy of the results could optimize the gene weight.

6. M5 tree model (M5P)

M5P is a linear tree based model introduced by Quinlan (1992). M5P decision tree is convenient because multivariate linear models can be operated within the model, and indeed, it is very flexible to be managed (Balouchi et al., 2015; Khosravi et al., 2020). The main steps involved in developing M5P are constructing the tree, pruning the tree, and smoothing the tree. The best model was achieved in growing the trees by maximizing the standard deviation reduction (SDR). SDR is explained in Eq. (3), where E is defined as the set of cases, E_i is the ith subset of cases splitting the tree, SDE is the standard deviation of E and $SD(E_i)$ is the standard deviation of E_i.

$$SDR = SDE - \sum_i \frac{|E_i|}{|E|} x SD(E_i) \tag{3}$$

The overfitting problem in which the model is excellent in the dataset but does not perform well in the testing dataset can be solved through the pruning step. In this step, subtrees were eliminated to maximize the results, and the attribute will be

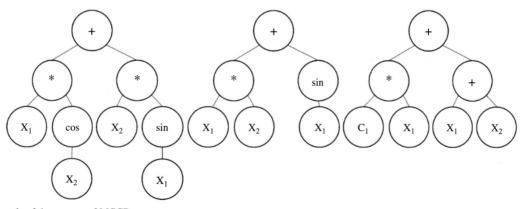

FIG. 2 Example of three genes of MGGP.

TABLE 6 Range of river data for the current study compared to that of Sinnakaudan et al. (2006) and Ariffin (2004).

Variable/parameter	Sinnakaudan et al. (2006)	Ariffin (2004)	The current study (Muda River, Langat River, and Kurau River)
Number of data sets	346	165	234
Discharge, Q (m³/s)	0.74–87.79	0.74–87.79	0.478–343.710
Average velocity, V (m/s)	0.19–1.42	0.19–1.18	0.140–1.450
River width, B	13.50–30.00	13.80–33.00	6.30–90.00
Flow depth, y	0.22–3.23	0.228–3.25	0.244–3.093
Area (m²)	3.42–96.83	3.4–96.80	1.430–278.34
Hydraulic radius, R (m)	0.22–2.66	0.22–2.66	0.174–3.900
Bed slope, S_o	0.0004–0.0167	0.0004–0.0167	0.00008–0.0009
Median sediment bed material, d_{50} (mm)	0.37–4.00	0.542–2.288	0.29–3.00
Total load Q_t (kg/s)	0.10–118.95	0.06–118.95	0.0178–99.40

reduced to minimize the error. Next, the smoothing step will continue to occur by adjusting the discontinuity at the pruned tree leaves (Khosravi et al., 2020). More details can be found by referring to the research done by Shaghaghi et al. (2018a) and Kargar et al. (2020).

7. Results and discussion

The current study has a broader range of data than the previous research by Sinnakaudan et al. (2006) and Ariffin (2004). The previous researches covered Pari River, Kinta River, Raia River, Kampar River, Kerayong River, Kulim River, Langat River, and Lui River. The maximum and minimum value of the table parameters such as discharge, average velocity, river width, flow depth, area, hydraulic radius, bed slope, median sediment bed material, and total load is different compared to the previous study. Table 6 summarizes the comparison between the data range of the earlier studies and the current study.

The equations from both Ariffin (2004) and Sinnakaudan et al. (2006) were revised using MLR technique. The dependent variable for equation revised Ariffin (2004) is $\ln C_v$, and the independent variables are $\ln U^*/W_s$, $\ln U^*/V$, $\ln R/d_{50}$, and $\ln V^2/gy$. Based on the analysis done by using ANOVA for the original equation developed by Ariffin (2004), it is observed that the parameters R/d_{50} and U_x/W_s have p-value greater than $\alpha = 0.05$, thus reflecting the insignificant effect on the overall regression results. Therefore the parameters are revised, and only U^*/V and V^2/gy are considered for the revised equation.

The revised Ariffin (2004) equation yields the following new equation

$$C_v = 4.032 \times 10^{-2} \left(\frac{U^*}{V}\right) 2.178 \left(\frac{V^2}{gy}\right) 0.795 \tag{4}$$

As for revised equation of Sinnakaudan et al. (2006), the study used log phi (ϕ) as the dependent variable and log R/d_{50}, log VS_o/w_s as the independent variables.

The Sinnakaudan et al. (2006) equation can be rewritten as:

$$C_v = 6.237 \times 10^{-3} \left(\frac{VS_o}{\omega_s}\right) 0.712 \left(\frac{R}{d_{50}}\right) 1.068 \left(\frac{\sqrt{g(S_s-1)d_{50}^3}}{VR}\right) \tag{5}$$

The revised equations are compared with the existing commonly used equation (Table 1) to examine the predicted equations' effectiveness and accuracy. Table 7 summarizes the performance of the revised equation.

TABLE 7 Summary of performance of the revised equations and the current commonly used equations.

Equation	Mean (μ)	Standard deviation (σ)	Mean absolute error (MAE)	R^2	DR average accuracy (%)
Revised Ariffin (2004)	2.062	9.189	2.526	0.616	66.34
Revised Sinnakaudan et al. (2006)	1.915	3.070	2.784	0.465	64.49
Ariffin (2004)	13.338	35.222	11.955	0.021	25.70
Sinnakaudan et al. (2006)	8.599	12.637	6.577	0.260	30.37
Engelund and Hansen (1967)	6.195	15.813	4.996	0.295	41.80
Yang (1979)	4.630	8.890	3.650	0.355	37.79

In terms of DR, the revised Ariffin (2004) equation has the highest value of 66.34%, followed by the revised Sinnakaudan et al. (2006) equation of 64.49%, Engelund and Hansen (1967) 41.80%, Yang (1979) 37.79%, Sinnakaudan et al. (2006) 30.37% and Ariffin (2004) 25.70%. Among all the equations, the revised Sinnakaudan et al. (2006) equation has the lowest mean and standard deviation value. Conversely, Ariffin (2004) has the highest mean and standard deviation value. As for the mean and standard deviation, the revised Sinnakaudan et al. (2006) equation is the lowest among all ($\mu = 1.915$, $\sigma = 3.070$), but in terms of R^2 and MAE, the revised Ariffin equation yielded a better R^2 results with a value of 0.616 and 2.526. The results show that the best equation is revised Ariffin (2004), followed by Sinnakaudan et al. (2006), Engelund and Hansen (1967), Yang (1979), Sinnakaudan et al. (2006), and Ariffin (2004). Both equations were analyzed further by using a soft computing technique. The parameters used in both equations were kept permanent during the execution. The function of total bed material load is in the form

$Q_t = f(Q, \frac{U^*}{V}, \frac{V^2}{gy})$ for revised Ariffin (2004) and $Q_t = f(Q, \frac{VS_o}{\omega_s}, \frac{R}{d_{50}}, \frac{\sqrt{g(Ss-1)d_{50}3}}{VR})$ for revised Sinnakaudan et al. (2006). Both model applied the same output model Q_t. In general, there are 6 models altogether that has been developed in this study. The developed models are as follows:

(1) Revised Ariffin (2004) EPR: $Q_t = f(Q, \frac{U^*}{V}, \frac{V^2}{gy})$
(2) Revised Sinnakaudan et al. (2006) EPR: $Q_t = f(Q, \frac{VS_o}{\omega_s}, \frac{R}{d_{50}}, \frac{\sqrt{g(Ss-1)d_{50}3}}{VR})$
(3) Revised Ariffin (2004) MGGP: $Q_t = f(Q, \frac{U^*}{V}, \frac{V^2}{gy})$
(4) Revised Sinnakaudan et al. (2006) MGGP: $Q_t = f(Q, \frac{VS_o}{\omega_s}, \frac{R}{d_{50}}, \frac{\sqrt{g(Ss-1)d_{50}3}}{VR})$
(5) Revised Ariffin (2004) M5P: $Q_t = f(Q, \frac{U^*}{V}, \frac{V^2}{gy})$
(6) Revised Sinnakaudan et al. (2006) M5P: $Q_t = f(Q, \frac{VS_o}{\omega_s}, \frac{R}{d_{50}}, \frac{\sqrt{g(Ss-1)d_{50}3}}{VR})$

The data set is split into two separate parts, which are training and testing. The data for training and testing was chosen by adopting the Kennard-Stone algorithm. The training process employs 70% of the data, and the testing process uses the remaining 30% of the data.

Equations (6) and (7) yielded results using the EPR program for revised Ariffin (2004) and revised Sinnakaudan et al. (2006) equations. The values of $\beta i, x_1, x_2, x_3, and x_4$ are shown in Tables 8 and 9. The equation is further analyzed in the training and testing dataset. P1–P13 are referring to the number of model's output. The results for both revised Ariffin (2004) and Sinnakaudan et al. (2006) using EPR are shown in Figs. 3 and 4.

$$C_v = \sum_{i=1}^{13} P_i P_i = \beta_i \times Q^{x_1} \times \left(\frac{u^*}{V}\right)^{x_2} \times \left(\frac{V^2}{gy}\right)^{x_3} \tag{6}$$

(Revised Ariffin, 2004.)

$$C_v = \sum_{i=1}^{13} P_i P_i = \beta_i \times Q^{x_1} \times \left(\frac{R}{d_{50}}\right)^{x_2} \times \left(\frac{VS_0}{w_s}\right)^{x_3} \times \left(\frac{\sqrt{g(S_s-1)d_{50}^3}}{VR}\right)^{x_4} \tag{7}$$

(Revised Sinnakaudan et al., 2006.)

TABLE 8 Value of βi, x_1, x_2, x_3, and x_4 for revised Ariffin (2004).

	(β_i)	(x_1)	(x_2)	(x_3)
P1	1.03E−06	4	1	0
P2	2.95E+00	3	2	2
P3	4.49E−04	4	0	3
P4	−1.39E−03	4	1	2
P5	−8.01E+03	0	0	7
P6	4.56E+03	2	3	2
P7	5.06E+01	2	0	5
P8	−3.82E−08	5	2	0
P9	−5.79E−05	3	1	0
P10	−4.10E+02	2	2	2
P11	1.39E−11	6	0	1
P12	−2.30E−01	2	1	1
P13	7.54E+03	0	5	1

TABLE 9 Value of βi, x_1, x_2, x_3, and x_4 for revised Sinnakaudan et al. (2006).

	(β_i)	(x_1)	(x_2)	(x_3)	(x_4)
P1	−0.00454	4	0	2	0
P2	0.00140	0	3	4	0
P3	1.80E−14	7	0	0	0
P4	5.33E+15	0	0	5	2
P5	807.276	4	0	2	1
P6	−5.33E−10	6	0	1	0
P7	17,428,117,874	1	1	0	5
P8	−436,934,172.4	1	0	6	0
P9	−3.32E+15	1	0	0	6
P10	2.02E−13	5	1	0	0
P11	−7.34E−12	6	0	0	0
P12	−51,610.829	3	0	1	2
P13	−1.38064E+11	1	0	5	1

For the revised Ariffin (2004), the R^2 for training and testing are 0.949 and 0.892, respectively. Meanwhile, for RMSE, the training and testing are 2.564 and 4.596, respectively. As for the revised Sinnakaudan et al. (2006), the R^2 for the training stage is 0.946, and for the testing, the value is 0.806. In terms of RMSE, the value is 2.912 (training) and 6.646 (testing). Overall, based on the R^2 and RMSE results, the EPR models can produce good prediction accuracy, especially for the modeling of the total bed material load in the tropical regions like Malaysia.

MGGP, on the hand, has depicted the following equation for revised Ariffin (2004) and revised Sinnakaudan et al. (2006):

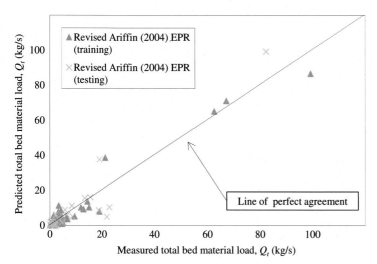

FIG. 3 Training and testing results of the revised Ariffin (2004) by using EPR.

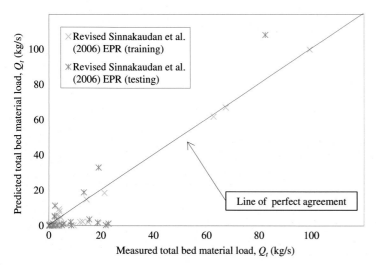

FIG. 4 Training and testing results of the revised Sinnakaudan et al. (2006) by using EPR.

$$C_v = 25.8 \frac{u^*}{V} \left(e^{e^{\frac{u^*}{V}}} - 0.869 \frac{V^2}{gy} log \left(\frac{V^2}{gy} \right) \right) - 200 \frac{u^*}{V} - 4.2 log(Q) - 6.15 log \left(\frac{V^2}{gy} \right) - 0.787Q \frac{V^2}{gy} - 0.135Q$$
$$+ 1311Q \left(\frac{u^*}{V} \right)^2 \frac{V^2}{gy} - 1.96 \tag{8}$$

(Revised Ariffin, 2004.)

$$C_v = 0.953 \frac{\sqrt{g(S_s - 1)d_{50}^3}}{VR} \left(Q \left(Q + \frac{R}{d_{50}} \right) + Q^{\frac{\sqrt{g(S_s-1)d_{50}^3}}{VR}} \right) - 10.7Q \frac{VS_0}{w_s}$$

$$-0.0724Q \, log \left(\frac{\sqrt{g(S_s - 1)d_{50}^3}}{VR} \right) - 0.00157Q^2 + 1.16Q^2 \, log \left(\left(\frac{\sqrt{g(S_s - 1)d_{50}^3}}{VR} \right)^{\frac{\sqrt{g(S_s-1)d_{50}^3}}{VR}} \right)$$

$$2000Q^2 \frac{VS_0}{w_s} \frac{\sqrt{g(S_s - 1)d_{50}^3}}{VR} \, log \left(\frac{\sqrt{g(S_s - 1)d_{50}^3}}{VR} \right) \tag{9}$$

(Revised Sinnakaudan et al., 2006.)

FIG. 5 Training and testing results of the revised Ariffin (2004) by using MGGP.

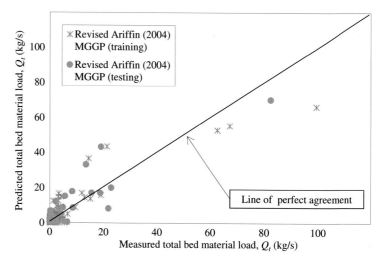

Results from the modeling by using MGGP show the moderate R^2 value for both revised equations. Revised Ariffin (2004) R^2 value for training is 0.796, and for the testing stage, the value is 0.781. RMSE for both training and testing stages are found as 10.578 and 12.727, respectively. The R^2 value for revised Sinnakaudan et al. (2006) is slightly higher than the revised Ariffin (2004) equation, which is 0.815 for training and 0.740 for testing stages. However, the RMSE is observed to be slightly more in the revised Sinnakaudan et al. (2006), whereas the values for testing and training are found as 10.689 and 12.383, and more details can be found in Figs. 5 and 6.

M5P generate mixed predictions for both the revised equations. Figs. 7 and 8 give an outlook for the predicted and observed of both revised equations. The R^2 for training were observed to be higher compared to the training stage. The R^2 value is 0.939 (training) and 0.553 (testing) for revised Ariffin (2004) equation. RMSE values for training and testing stages were 11.388 and 14.108, respectively. The revised Sinnakaudan et al. (2006) equation, in turn, produced R^2 values of 0.718 (training) and 0.443 (testing). In comparison to the revised Ariffin (2004), RMSE for the revised Sinnakaudan et al. (2006) are 12.383 for training and 11.405 for the testing.

The two revised equations using EPR were compared with the existing revised equation and the revised models by implementing MGGP and M5P machine learning algorithms. The overall machine learning performance is summarized in Table 10.

Clearly, all machine learning models can increase prediction accuracy with low errors compared to the existing revised equations. The revised model using EPR was found to produce better prediction results compared to the MGGP and M5P models. Interestingly, the revised Sinnakaudan et al. (2006) M5P has the highest errors RMSE (6.961), and the revised Ariffin (2004) return the highest of MAE (3.054). All machine learning appears to be able to increase the accuracy of the model. However, in terms of DR, only revised Ariffin (2004) M5P and revised Sinnakaudan et al. (2006) M5P give better DR prediction results than the revised MLR results. Figs. 9–14 explain the results in terms of DR for the respective machine learning program.

FIG. 6 Training and testing results of the revised Sinnakaudan et al. (2006) by using MGGP.

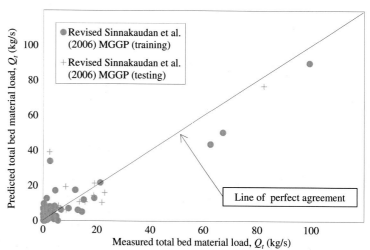

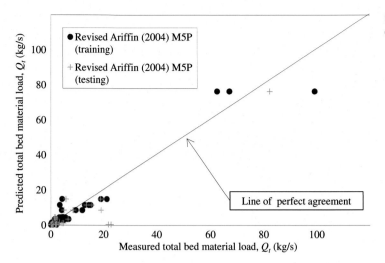

FIG. 7 Training and testing results of the revised Ariffin (2004) by using M5P.

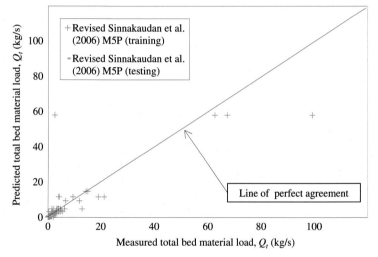

FIG. 8 Training and testing results of the revised Sinnakaudan et al. (2006) by using M5P.

TABLE 10 Summary of performance of the models.

Model	R^2	NSE	RMSE	MAE	DR (0.5%–2.0%)
Revised Ariffin (2004)	0.616	0.228	9.462	2.526	66.36
Revised Sinnakaudan et al. (2006)	0.482	0.221	9.902	2.784	64.49
Revised Ariffin (2004) EPR	0.922	0.913	3.305	1.552	34.58
Revised Sinnakaudan et al. (2006) EPR	0.884	0.848	4.377	2.137	14.49
Revised Ariffin (2004) MGGP	0.787	0.784	5.217	3.054	21.03
Revised Sinnakaudan et al. (2006) MGGP	0.787	0.784	5.207	3.011	31.31
Revised Ariffin (2004) M5P	0.786	0.762	5.467	1.561	72.43
Revised Sinnakaudan et al. (2006) M5P	0.622	0.615	6.961	1.994	73.36

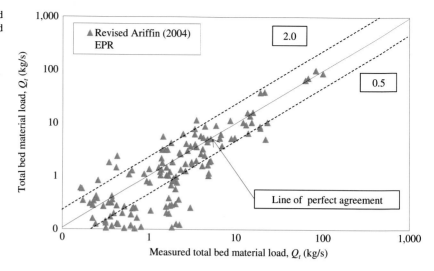

FIG. 9 Comparison results between measured and predicted total bed material load for the revised Ariffin (2004) by using EPR.

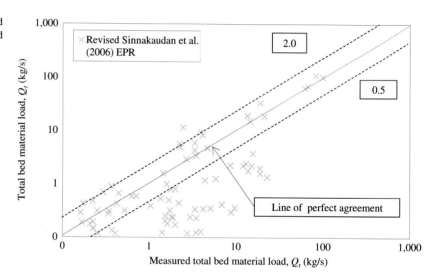

FIG. 10 Comparison results between measured and predicted total bed material load for the revised Sinnakaudan et al. (2006) by using EPR.

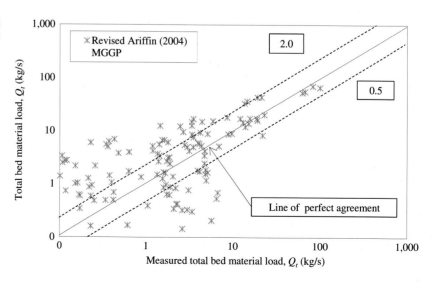

FIG. 11 Comparison results between measured and predicted total bed material load for the revised Ariffin (2004) by using MGGP.

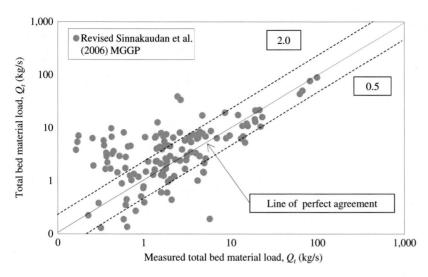

FIG. 12 Comparison results between measured and predicted total bed material load for the revised Ariffin (2004) and Sinnakaudan et al. (2006) by using MGGP.

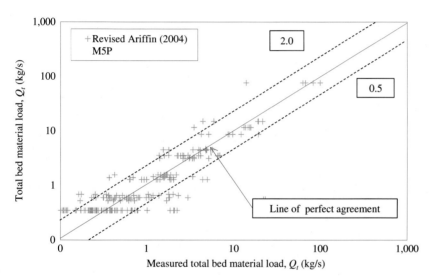

FIG. 13 Comparison results between measured and predicted total bed material load for the revised Ariffin (2004) by using M5P.

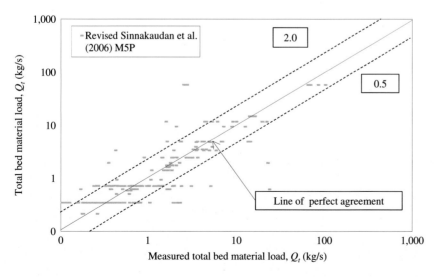

FIG. 14 Comparison results between measured and predicted total bed material load for the revised Sinnakaudan et al. (2006) by using M5P.

From Table 10, revised Sinnakaudan et al. (2006) M5P turn to be the highest DR of 73.36%, followed by revised Ariffin (2004) M5P (72.43%), revised Ariffin (2004) 66.36% revised Sinnakaudan et al. (2006) 64.49%, revised Ariffin (2004) EPR 34.58%, revised Sinnakaudan et al. (2006) MGGP 31.31%, revised Ariffin (2004) MGGP 21.03% and revised Sinnakaudan et al. (2006) EPR 14.49%. It is also important to note that even though the DR for M5P is considerably good, which exceeds 73%, but the data did not distribute well and is rather flattening at the lower total bed material load rate. The results showed the inconsistency in the prediction accuracy model. The lack of accuracy as explained by Ng et al. (2020), Cui and Gong (2018), and Moghaddam et al. (2020) may mainly due to the sample size used in this study which only consist of 234 data. To get a result with good model prediction accuracy, the data has to be more than 2000 in total.

8. Conclusions

From the summary, EPR was found to improve the prediction distribution value with higher R^2 and NSE and lower RMSE and MAE, followed by MGGP and M5P. EPR can predict better results from the evidence of both revised Ariffin (2004) and Sinnakaudan et al. (2006) equations that gave a better prediction model compared to the MGGP and M5P counterparts. More importantly, despite the lack of accuracy in the model prediction accuracy in terms of R^2 and the NSE values by using the M5P program, but in terms of the DR, M5P shows better prediction accuracy and gives better prediction results compared to the revised equations. The results also suggested that machine learning is very sensitive and better at predicting total bed material load at a high value compared to the lower value.

Acknowledgments

The authors would like to acknowledge the financial assistance from the Ministry of Higher Education Malaysia under the TRGS 2015 Flood Management Grant No. 203/PREDAC/6767003 entitled "Model-Based Morphological Prediction for Large Scale River Basin in Raising Flood Protection Levels for Sungai Pahang" and the support by the Higher Institution Centre of Excellence Program (HICoE) at REDAC (Grant no. 311/PREDAC/4403901). Appreciation also goes to the Department of Irrigation and Drainage Malaysia for providing data and reports, as well as the staff of REDAC, Universiti Sains Malaysia, for their involvement in the research work. Acknowledgement also goes to the Public Service Department of Malaysia for the first author's scholarship under the Hadiah Latihan Persekutuan (HLP) program.

References

Ab. Ghani, A., Azamathulla, H.M., 2014. Development of GEP-based functional relationship for sediment transport in tropical rivers. Neural. Comput. Appl. 24 (2), 271–276. https://doi.org/10.1007/s00521-012-1222-9.

Ab.Ghani, A., 1993. Sediment Transport in Sewers (Ph.D. thesis). University of Newcastle upon Tyne, UK.

Ab.Ghani, A., Azamathulla, H.M., 2012. Development of GEP-based functional relationship for sediment transport in tropical rivers. Neural Comput. Appl. 24, 271–276. https://doi.org/10.1007/s00521-012-1222-9.

Ab.Ghani, A., Azamathulla, H.M., Chang, C.K., Zakaria, N.A., Hasan, Z.A., 2010. Prediction of total bed material load for rivers in Malaysia: A case study of Langat, Muda and Kurau Rivers. Environ. Fluid Mech. 11, 307–318. https://doi.org/10.1007/s10652-010-9177-9.

Abu Hassan, Z., 1998. Evaluation of Scour and Deposition in Malaysian Rivers Undergoing Training Works: Case Studies of Pari and Kerayong Rivers (M. Sc. thesis). Universiti Sains Malaysia, Penang, Malaysia.

Ackers, P., White, W.R., 1973. Sediment transport: new approach and analysis. J. Hydraul. Eng. 99, 2041–2060.

Ahmad Abdul Ghani, N.A., Tholibon, D.A., Ariffin, J., 2019. Robustness analysis of model parameters for sediment transport equation development. ASM Sci. J. 12. https://doi.org/10.32802/asmscj.2019.268.

Ahmad Abdul Ghani, N.A., Kamal, N.A., Ariffin, J., 2020. Improving total sediment load prediction using genetic programming technique (case study: Malaysia). In: IOP Conference Series: Materials Science and Engineering. IOP Publishing, https://doi.org/10.1088/1757-899X/736/2/022108.

Ara Rahman, S., Chakrabarty, D., 2020. Sediment transport modelling in an alluvial river with artificial neural network. J. Hydrol. 588, 125056. https://doi.org/10.1016/j.jhydrol.2020.125056.

Ariffin, J., 2004. Development of Sediment Transport Models for Rivers in Malaysia Using Regression Analysis and Artificial Neural Network (Ph.D. thesis). Universiti Sains Malaysia, Penang, Malaysia.

Azamathulla, H.M., Ahmad, Z., Ab.Ghani, A., 2013a. An expert system for predicting Manning's roughness coefficient in open channels by using gene expression programming. Neural Comput. Appl. 23, 1343–1349. https://doi.org/10.1007/s00521-012-1078-z.

Azamathulla, H.M., Cuan, Y.C., Ghani, A.A., Chang, C.K., 2013b. Suspended sediment load prediction of river systems: GEP approach. Arab. J. Geosci. 6, 3469–3480. https://doi.org/10.1007/s12517-012-0608-4.

Balouchi, B., Nikoo, M.R., Adamowski, J., 2015. Development of expert systems for the prediction of scour depth under live-bed conditions at river confluences: application of different types of ANNs and the M5P model tree. Appl. Soft Comput. J. 34, 51–59. https://doi.org/10.1016/j.asoc.2015.04.040.

Basri, H., Sidek, L.M., Shih, D.S., Lloyd, H.C., Azad, W.H., Abdul Razad, A.Z., 2018. One dimensional shallow water equation streamflow modeling using WASH123D model. Int. J. Eng. Technol. 7, 880. https://doi.org/10.14419/ijet.v7i4.35.26274.

Bonakdari, H., Gholami, A., Sattar, A.M.A., Gharabaghi, B., 2020. Development of robust evolutionary polynomial regression network in the estimation of stable alluvial channel dimensions. Geomorphology 350, 106895. https://doi.org/10.1016/j.geomorph.2019.106895.

Brownlie, W.R., 1981. Prediction of Flow Depth and Sediment Discharge in Open Channels. Report No. KH-R-43A. California Institute of Technology, Pasadena, California, USA.

Chang, C.K., 2006. Sediment Transport in Kulim River, Kedah (M.sc. thesis). Universiti Sains Malaysia, Penang, Malaysia.

Chang, C.K., Azamathulla, H.M., Zakaria, N.A., Ghani, A.A., 2012. Appraisal of soft computing techniques in prediction of total bed material load in tropical rivers. J. Earth Syst. Sci. 121, 125–133. https://doi.org/10.1007/s12040-012-0138-1.

Constantine, J.A., Dunne, T., Ahmed, J., Legleiter, C., Lazarus, E.D., 2014. Sediment supply as a driver of river meandering and floodplain evolution in the Amazon Basin. Nat. Geosci. 7, 899–903. https://doi.org/10.1038/ngeo2282.

Cui, Z., Gong, G., 2018. The effect of machine learning regression algorithms and sample size on individualized behavioral prediction with functional connectivity features. NeuroImage 178, 622–637. https://doi.org/10.1016/j.neuroimage.2018.06.001.

Danandeh Mehr, A., Safari, M.J.S., 2020. Application of soft computing techniques for particle Froude number estimation in sewer pipes. J. Pipeline Syst. Eng. Pract. 11, 1–8. https://doi.org/10.1061/(ASCE)PS.1949-1204.0000449.

Danandeh Mehr, A., Nourani, V., Kahya, E., Hrnjica, B., Sattar, A.M.A., 2018. Genetic programming in water resources engineering: a state-of-the-art review. J. Hydrol. 566, 643–667. https://doi.org/10.1016/j.jhydrol.2018.09.043.

Danandeh Mehr, A., Jabarnejad, M., Nourani, V., 2019. Pareto-optimal MPSA-MGGP: a new gene-annealing model for monthly rainfall forecasting. J. Hydrol. 571, 406–415. https://doi.org/10.1016/j.jhydrol.2019.02.003.

Ebtehaj, I., Bonakdari, H., Zaji, A.H., 2016. An expert system with radial basis function neural network based on decision trees for predicting sediment transport In sewers. Water Sci. Technol. 74, 176–183. https://doi.org/10.2166/wst.2016.174.

Ebtehaj, I., Bonakdari, H., Safari, M.J.S., Gharabaghi, B., Zaji, A.H., Riahi Madavar, H., Sheikh Khozani, Z., Es-Haghi, M.S., Shishegaran, A., Mehr, A.D., 2019. Combination of sensitivity and uncertainty analyses for sediment transport modeling in sewer pipes. Int. J. Sediment Res. 35, 157–170. https://doi.org/10.1016/j.ijsrc.2019.08.005.

Engelund, F., Hansen, E., 1967. A Monograph on Sediment Transport in Alluvial Streams. Teknisk Forlag, https://doi.org/10.1007/s13398-014-0173-7.2.

Fakhri, M., Dokohaki, H., Eslamian, S., Fazeli Farsani, I., Farzaneh, M.R., 2014. Flow and sediment transport modeling in rivers. In: Eslamian, S. (Ed.), Handbook of Engineering Hydrology. Modeling, Climate Changes and Variability, 2. Taylor and Francis, CRC Group, USA, pp. 233–275. Chapter 13.

Garcia, M.H., 2008. Sedimentation Engineering: Processes, Measurements, Modeling, and Practice. ASCE, Reston, Virginia, https://doi.org/10.1061/9780784408230.

Giustolisi, O., Savic, D., 2006. Symbolic data-driven technique based on Evolutionary Polynomial Regression. J. Hydroinf. 8, 207–222. https://doi.org/10.2166/hydro.2006.020b.

Giustolisi, O., Savic, D., 2009. Advances in data-driven analyses and modelling using EPR-MOGA. J. Hydroinf. 11, 225–236. https://doi.org/10.2166/hydro.2009.017.

Graf, W.H., 1971. Hydraulics of sediment transport. McGraw-Hill, New York, USA.

Gunawan, T.A., Daud, A., Haki, H., Sarino, 2019. The estimation of total sediments load in river tributary for sustainable resources management. IOP Conf. Ser. Earth Environ. Sci. 248, 11. https://doi.org/10.1088/1755-1315/248/1/012079.

Harun, M.A., Ab.Ghani, A., 2020. Revised equations of total bed material load for rivers in Malaysia. In: Mohd Sidek, L. (Ed.), WRDM. Springer Singapore, pp. 332–340, https://doi.org/10.1007/978-981-15-1971-0.

Harun, M.A., Ab.Ghani, A., Mohammadpour, R., Chan, N.W., 2020. Stable channel analysis with sediment transport for rivers in Malaysia: a case study of the Muda, Kurau, and Langat rivers. Int. J. Sediment Res. 35, 455–466. https://doi.org/10.1016/j.ijsrc.2020.03.008.

Ibrahim, N., 2002. Penilaian dan pembangunan persamaan pengangkutan enapan sungai-sungai di Malaysia (M.Sc. thesis). Universiti Sains Malaysia, Penang, Malaysia.

Kargar, K., Safari, M.J.S., Mohammadi, M., Samadianfard, S., 2019. Sediment transport modeling in open channels using neuro-fuzzy and gene expression programming techniques. Water Sci. Technol. 79, 2318–2327. https://doi.org/10.2166/wst.2019.229.

Kargar, K., Samadianfard, S., Parsa, J., Nabipour, N., Shamshirband, S., Mosavi, A., Chau, K., 2020. Estimating longitudinal dispersion coefficient in natural streams using empirical models and machine learning algorithms. Eng. Appl. Comput. Fluid Mech. 14, 311–322. https://doi.org/10.1080/19942060.2020.1712260.

Khan, M.Y.A., Tian, F., Hasan, F., Chakrapani, G.J., 2018. Artificial neural network simulation for prediction of suspended sediment concentration in the River Ramganga, Ganges Basin, India. Int. J. Sediment Res. https://doi.org/10.1016/j.ijsrc.2018.09.001.

Khosravi, K., Cooper, J.R., Daggupati, P., Pham, B.T., Bui, D.T., 2020. Bedload transport rate prediction: application of novel hybrid data mining techniques. J. Hydrol., 124774. https://doi.org/10.1016/j.jhydrol.2020.124774.

Meshkova, L.V., Carling, P.A., 2012. The geomorphological characteristics of the Mekong River in northern Cambodia: a mixed bedrock-alluvial multi-channel network. Geomorphology 147–148, 2–17. https://doi.org/10.1016/j.geomorph.2011.06.041.

Moghaddam, D.D., Rahmati, O., Panahi, M., Tiefenbacher, J., Darabi, H., Haghizadeh, A., Tien Bui, D., 2020. The effect of sample size on different machine learning models for groundwater potential mapping in mountain bedrock aquifers. Catena 187 (2020), 104421. https://doi.org/10.1016/j.catena.2019.104421.

Mohammad-Azari, S., Bozorg-Haddad, O., Loáiciga, H.A., 2020. State of art of genetic programming applications in water resources systems analysis. Environ. Monit. Assess. 192. https://doi.org/10.1007/s10661-019-8040-9.

Montes, C., Berardi, L., Kapelan, Z., Saldarriaga, J., 2020. Predicting bedload sediment transport of non-cohesive material in sewer pipes using evolutionary polynomial regression-multi-objective genetic algorithm strategy. Urban Water J. 17, 154–162. https://doi.org/10.1080/1573062X.2020.1748210.

Muhammad, M.M., Yusof, K.W., Ul Mustafa, M.R., Zakaria, N.A., Ghani, A.A., 2018. Artificial neural network applications for predicting drag coefficient in flexible vegetated channels. J. Telecommun. Electron. Comput. Eng. 10 (1 – 12), 99–102.

Ng, W., Minasny, B., de Sousa Mendes, W., Melo Dematté, J.A., 2020. The influence of training sample size on the accuracy of deep learning models for the prediction of soil properties with near-infrared spectroscopy data. Soil 6 (2), 565–578. https://doi.org/10.5194/soil-6-565-2020.

Nourani, V., Kalantari, O., Baghanam, A.H., 2012. Two semidistributed ANN-based models for estimation of suspended sediment load. J. Hydrol. Eng. 17 (12). https://doi.org/10.1061/(ASCE)HE.1943-5584.0000587.

Nourani, V., Alizadeh, F., Roushangar, K., 2016. Evaluation of a two-stage SVM and Spatial Statistics methods for modeling monthly river suspended sediment load. Water Resour. Manag. 30, 393–407.

Nourani, V., Molajou, A., Najafi, A.D.T.H., 2019. A wavelet based data mining technique for suspended sediment load modeling. Water Resour. Manag. 33, 1769–1784.

Pintelas, E., Livieris, I.E., Pintelas, P., 2020. A Grey-Box ensemble model exploiting Black-Box accuracy and White-Box intrinsic interpretability. Algorithms 13. https://doi.org/10.3390/a13010017.

Quinlan, J.R., 1992. Learning with continuous classes. In: 5th Australian Joint Conference on Artificial Intelligence, pp. 343–348.

Rajaee, T., Jafari, H., 2020. Two decades on the artificial intelligence models advancement for modeling river sediment concentration: state-of-the-art. J. Hydrol. 588, 125011. https://doi.org/10.1016/j.jhydrol.2020.125011.

Safari, M.J.S., Danandeh Mehr, A., 2018. Multigene genetic programming for sediment transport modeling in sewers for conditions of non-deposition with a bed deposit. Int. J. Sediment Res. 33, 262–270. https://doi.org/10.1016/j.ijsrc.2018.04.007.

Safari, M.J.S., Mohammadi, M., Ab Ghani, A., 2018. Experimental studies of self-cleansing drainage system design: a review. J. Pipeline Syst. Eng. Pract. https://doi.org/10.1061/(asce)ps.1949-1204.0000335.

Safari, M.J.S., Ebtehaj, I., Bonakdari, H., Es-Haghi, M.S., 2019. Sediment transport modeling in rigid boundary open channels using generalize structure of group method of data handling. J. Hydrol. 577, 123951. https://doi.org/10.1016/j.jhydrol.2019.123951.

Saleh, A., Abustan, I., Mohd Remy Rozainy, M.A.Z., Sabtu, N., 2017. Assessment of total bed material equations on selected Malaysia rivers. AIP Conf. Proc., 070002 (1–7) https://doi.org/10.1063/1.5005720.

Shaghaghi, S., Bonakdari, H., Gholami, A., Kisi, O., Binns, A., Gharabaghi, B., 2018a. Predicting the geometry of regime rivers using M5 model tree, multivariate adaptive regression splines and least square support vector regression methods. Int. J. River Basin Manag. 17, 333–352. https://doi.org/10.1080/15715124.2018.1546731.

Shaghaghi, S., Bonakdari, H., Kisi, O., Shiri, J., Binns, A., Gharabaghi, B., 2018b. Stable alluvial channel design using evolutionary neural networks. J. Hydrol. 566, 770–782. https://doi.org/10.1016/J.JHYDROL.2018.09.057.

Sharghi, E., Nourani, V., Najafi, H., Gokcekus, H., 2019. Conjunction of a newly proposed emotional ANN (EANN) and wavelet transform for suspended sediment load modeling. Water Supply 19, 1726–1734. https://doi.org/10.2166/ws.2019.044.

Shiri, N., Shiri, J., Nourani, V., Karimi, S., 2020. Coupling wavelet transform with multivariate adaptive regression spline for simulating suspended sediment load: independent testing approach. ISH J. Hydraul. Eng. 00, 1–10. https://doi.org/10.1080/09715010.2020.1801528.

Sinnakaudan, S.K., Ghani, A.A., Ahmad, M.S.S., Zakaria, N.A., 2006. Multiple linear regression model for total bed material load prediction. J. Hydraul. Eng. 132, 521–528. https://doi.org/10.1061/(ASCE)0733-9429(2006)132:5(521).

Syvitski, J.P.M., Cohen, S., Kettner, A.J., Brakenridge, G.R., 2014. How important and different are tropical rivers? – an overview. Geomorphology 227, 5–17. https://doi.org/10.1016/j.geomorph.2014.02.029.

Tayfur, G., Guldal, V., 2006. Artificial neural networks for estimating daily total suspended sediment in natural streams. Hydrol. Res. 37, 69–79.

Tayfur, G., Ozdemir, S., Singh, V.P., 2003. Fuzzy logic algorithm for runoff-induced sediment transport from bare soil surfaces. Adv. Water Resour. 26, 1249–1256. https://doi.org/10.1016/j.advwatres.2003.08.005.

Tayfur, G., Karimi, Y., Singh, V.P., 2013. Principle component analysis in conjuction with data driven methods for sediment load prediction. Water Resour. Manag. 27, 2541–2554.

Teo, F.Y., Noh, N., Ab.Ghani, A., Zakaria, N.A., 2017. River sand mining capacity in Malaysia. In: Proceedings of the 37th IAHR World Congress, pp. 538–546.

Ulke, A., Tayfur, G., Ozkul, S., 2009. Predicting suspended sediment loads and missing data for Gediz River, Turkey. J. Hydrol. Eng. 14 (9), 954–965.

Yahaya, N.K., 1999. Development of Sediment Rating Curve for Rivers in Malaysia: Case Studies of Pari, Kerayong and Kulim Rivers (M.Sc. thesis). Universiti Sains Malaysia, Penang, Malaysia.

Yahaya, A.S., 2019. Application of statistical techniques in environmental modelling. AIP Conf. Proc. 2129, 020074-1–020074-17. https://doi.org/10.1063/1.5118082.

Yang, C.T., 1976. Minimum unit stream power and fluvial hydraulics. J. Hydraul. Div. 102, 919–934.

Yang, C.T., 1979. Unit stream power equations for total load. J. Hydrol. 40, 123–138.

Zahiri, A., Najafzadeh, M., 2018. Optimized expressions to evaluate the flow discharge in main channels and floodplains using evolutionary computing and model classification. Int. J. River Basin Manag. 16, 123–132. https://doi.org/10.1080/15715124.2017.1372448.

Zakaria, N.A., Azamathulla, H.M., Chang, C.K., Ghani, A.A., 2010. Gene expression programming for total bed material load estimation-a case study. Sci. Total Environ. 408, 5078–5085. https://doi.org/10.1016/j.scitotenv.2010.07.048.

Zangeneh Sirdari, Z., Ab Ghani, A., Hassan, Z.A., 2014. Bedload transport of small rivers in Malaysia. Int. J. Sediment Res. 29, 481–490. https://doi.org/10.1016/S1001-6279(14)60061-5.

Index

Note: Page numbers followed by *f* indicate figures, *t* indicate tables, and *b* indicate boxes.

Printed in the United States
by Baker & Taylor Publisher Services